W9-BEU-000

Wide Bandgap Semiconductors

Kiyoshi Takahashi, Akihiko Yoshikawa
and Adarsh Sandhu (Eds.)

Wide Bandgap Semiconductors

Fundamental Properties and Modern Photonic and Electronic Devices

With 394 Figures and 36 Tables

Editors

Kiyoshi Takahashi
Professor Emeritus
Tokyo Institute of Technology

Chairman (1996–2005)
The 162nd Committee on Wide Bandgap Semiconductor Photonic and Electronic Devices
Japan Society for the Promotion of Science
e-mail: k-taka@apost.plala.or.jp

Director
R & D Center
Nippon EMC Ltd
e-mail: takahashi@n-emc.co.jp

Akihiko Yoshikawa
Professor
Chiba University

Chairman (2006–)
The 162nd Committee on Wide Bandgap Semiconductor Photonic and Electronic Devices
Japan Society for the Promotion of Science
e-mail: yoshi@faculty.chiba-u.jp

Adarsh Sandhu
Associate Professor
Tokyo Institute of Technology
e-mail: sandhu.a.a.@m.titech.ac.jp

Associate Editors

Yoshihiro Ishitani
Associate Professor
Chiba University

Yoichi Kawakami
Associate Professor
Kyoto University

Library of Congress Control Number: 2006935324

ISBN-10 3-540-47234-7 Springer Berlin Heidelberg New York
ISBN-13 978-3-540-47234-6 Springer Berlin Heidelberg New York

This work is subject to copyright. All rights are reserved, whether the whole or part of the material is concerned, specifically the rights of translation, reprinting, reuse of illustrations, recitation, broadcasting, reproduction on microfilm or in any other way, and storage in data banks. Duplication of this publication or parts thereof is permitted only under the provisions of the German Copyright Law of September 9, 1965, in its current version, and permission for use must always be obtained from Springer. Violations are liable to prosecution under the German Copyright Law.

Springer is a part of Springer Science+Business Media.

springer.com

© Springer-Verlag Berlin Heidelberg 2007

The use of general descriptive names, registered names, trademarks, etc. in this publication does not imply, even in the absence of a specific statement, that such names are exempt from the relevant protective laws and regulations and therefore free for general use.

Typesetting by SPi using a Springer LaTeX macro package
Cover design: Mr. Kirchner, Heidelberg

Printed on acid-free paper SPIN 11688655 62/3100/SPi 5 4 3 2 1 0

Editorial Committee Members

Fumio Hasegawa
Professor Emeritus
Tsukuba University
Professor, Kogakuin University

Hideo Kawanishi
Professor, Kogakuin University

Katsumi Kishino
Professor, Sophia University

Hiroyuki Matsunami
Professor Emeritus, Kyoto University
Director, Innovation Plaza Kyoto
Japan Science and Technology
Agency (JST)

Kentaro Onabe
Professor, University of Tokyo

Sadafumi Yoshida
Professor, Saitama University

Preface

The p–n junction was invented in the first half of the twentieth century and the latter half saw the birth of light emitting diodes: red and yellow/green in the 1960s and yellow in the 1970s. However, theoretical predictions of the improbability of synthesizing p-type wide bandgap semiconductors cast a long shadow over hopes for devices emitting in the elusive blue part of the electromagnetic spectrum, which would complete, with red and green, the quest for the primary colors making up white light. At a time when many researchers abandoned their efforts on nitrides, Professor Isamu Akasaki of Nagoya University at this time remained committed to his belief that "synthesis of high quality GaN crystals would eventually enable p-type doping" and in 1989 he succeeded in fabricating the world's first GaN p–n junction light emitting diode.

Professor Isamu Akasaki kindly accepted our invitation to contribute to this book and describes his journey 'from the nitride wilderness' to the first experimental results of blue emission from GaN p–n junctions: Japan's major contribution to the development of wide bandgap semiconductor devices.

The discovery of blue emission from GaN p–n junctions in 1989 was the major technological turning point during the development of wide bandgap emission devices with wide reaching scientific, industrial and social implications. In Japan, I assembled a group of academics and industrialists to discuss the feasibility of setting up a research committee under the Japanese Society for the Promotion of Science (JSPS) to study nitride semiconductor optical devices. Together, we submitted a proposal to the JSPS on establishing an 'academic/industrial/government' committee to assess nitride optical devices. The JSPS accepted our proposal, and a JSPS committee on "Wide Band Gap Semiconductors and Optoelectronic Devices", the *162 Committee*, was formally set up in 1996. The Committee consists of internationally recognized leaders who were instrumental in triggering the 'blue-tsunami' that led to the renaissance of blue-light semiconductor research and the birth of a viable technology as exemplified by the ubiquitous GaN LEDs.

In 2001, the Committee's areas of interest were widened to include wide bandgap electronic devices in addition to optical structures. In April 2006, Professor Akihiko Yoshikawa of Chiba University was appointed Chairman of the 162 Committee and under his guidance new topics such as nanostructures, biotechnology and high temperature devices, will also be covered.

The invention of the blue LED and laser in the 1990s led to the establishment of new industries based on blue technology including solid state traffic signals, UV nitride lasers in high density DVD recorders, solid state white light illumination, biotechnology, nano-structures and high temperature devices.

The six chapters of this unique book were written by innovators from Japanese academia and industry recognized as having laid the foundations of the modern wide bandgap semiconductor industry. Apart from the in-depth description of recent developments in the growth and applications of nitride semiconductors, this book also contains chapters on the properties and device applications of SiC, diamond thin films, doping of ZnO, II–IVs and the novel BeZnSeTe /BAlGaAs material systems. Practical issues and problems such as the effect of defects on device performance are highlighted and potential solutions given based on recent research.

I expect that this book will be a valuable source of up to date information on the physical properties, current trends and future prospects of wide bandgap semiconductors for professional engineers, graduate students, industrial planners responsible for charting projects and think-tank managers monitoring developments in the semiconductor electronics industry.

The contents of this book are based on the Japanese language, *Wide Gap Semiconductors, Optical and Electron Devices*, published in March 2006 by Morikita Shuppan Co. Ltd.

I would like to express my sincere thanks to the Industry Club of Japan for their financial contribution to a part of the cost of preparation of this book by "Special Fund for the Promotion of Sciences" which was made available by the Japan Society for the Promotion of Sciences (JSPS).

Finally, I would like to thank the members of the 162 Committee, JSPS, Morikita Shuppan Co. Ltd and Springer Publishing, for their support, time and energy spent in the production and publication of this book.

Kiyoshi Takahashi
Chairman of the 162 Committee, JSPS (1996–2005)
Tokyo Institute of Technology
Nippon EMC Ltd

Lifetime of Research on Nitrides – Alone in the Wilderness

I. Akasaki

Blue light emitting devices fabricated using nitride semiconductors are ubiquitous. They are used in mobile phones, traffic signals and outdoor screens in soccer stadiums. The situation 30 years ago was a lot different when many groups withdrew from this subject of research due to unfavorable results. In spite of the lack of progress in this field at the time, I made a conscious decision to make research on nitride semiconductors my life work. This is a recollection of my thoughts and memories about my contribution to the blue renaissance [1–3]. I hope that the mistakes and moments of jubilation will inspire young scientists and engineers to have confidence in your own beliefs, not be swayed by trends and fashions, and move forward with passion and commitment.

First Meeting with Light Emission and Compound Semiconductors

In 1952, I graduated from the Faculty of Science of Kyoto University and joined Kobe Kogyo Corp. (now Fujitsu Ltd). At the time, the company employed many talented researchers and was committed to a wide range of research activities. Two years after my arrival, the company was one of the first to begin manufacturing cathode ray tubes for televisions; I was put in charge of the fluorescent screens. Polycrystalline, powder Zn(Cd)S-based phosphors were applied to the interior of the CRT face plate and electron irradiation enabled production of images. Now, Zn(Cd)S is a group II–VI compound semiconductor and this was my first 'meeting' with these materials and luminescence. Looking back, this work had a great effect on my research career implicitly and explicitly, publicly and privately, in every possible way.

In the early 1950s, industrial researchers were pre-occupied with the implications of the invention of the transistor and Kobe Kogyo already had a team working on the development of transistors using single crystal Ge. The term 'single crystal' resonated with me because of the problems I was having with the poor reproducibility of experimental data due to polycrystalline Zn(Cd)S powder for the CRT screens. I seriously considered, and dreamt, about the possibility of using single crystal Zn(Cd)S and transparent Zn(Cd)S thin films for the CRT screens. Perhaps this unconscious interest in single crystal semiconductors was part of the reason I worked on single crystal GaAs, GaP and GaN, which emit light, during my days at Matsushita Research Institute Tokyo, Inc. (MRIT).

In 1959, I accepted an invitation to move to the Department of Electronics at Nagoya University where I worked with Professor Tetsuya Arizumi on setting up a series of technologies ranging from production of Ge ingots by reduction of germanium oxide and subsequent purification by zone refining, growth of single crystals, impurity doping, and device fabrication. At the time, p–n junctions were produced by diffusing arsenic (or antimony) into slices of p-type Ge. However, I was anxious about this method: the n-region in p–n junctions formed by 'diffusion' was compensated with p-type impurities which were introduced beforehand, and the impurity distribution is limited to the complementary error function. In order to overcome these drawbacks, I thought a new method about which was later known as 'epitaxial growth' of thin film semiconductors by which single crystalline Ge films could be grown through gas phase reactions of Ge compounds on Ge substrates. In collaboration with Professor Nishinaga (a prospective graduate student at that time and current president of Toyohashi University of Technology), we studied vapor phase epitaxial growth (VPE) of Ge from its very inception. I think that this research laid the foundations for my work at MRIT on epitaxial compound semiconductors including metalorganic vapor phase epitaxial growth (MOVPE) of nitride semiconductors. In fact my work on Ge epitaxy was the reason for an invitation by MRIT to join their new institute in 1964.

The Nitride Challenge

At Matsushita, I was put in charge of a laboratory with freedom to decide my research activities. My basic policy for semiconductor research was "make our own crystals, characterize their physical properties, use the results to optimize growth conditions, produce extremely high quality crystals and develop novel device applications". At the time, I first focused on the growth of so-called 'magic crystal', namely GaAs and related III–V compounds such as GaP, GaAsP, GaInP and their development for red and green/yellow light emitting diodes. In spite of the fact that I managed to produce the best results for such materials at the time, for a variety of reasons, these semiconductors were not used for fabrication of commercial devices. However, I was determined to develop blue light emitting devices using p–n junctions; something that no one had yet succeeded in producing.

Realization of blue light emitting devices requires growth of high quality single crystals of semiconductors with bandgaps (E_g) larger than 2.6 eV and control of their electrical conductivity, in particular realization of p-type material. At the time, however, theoretical studies indicated the improbability of achieving p-type ZnSe and other such wide bandgap semiconductors due to self compensation; almost all the major groups were pessimistic about realization of ZnSe and in particular GaN p–n junctions. The only example of success was SiC p–n junctions used to produce green/yellow emitting devices. However, I had absolutely no interest in this material for photonic device applications because of its indirect bandstructure did not enable the possibility of achieving high emission efficiency and hence laser oscillation would be impossible. Soon after my move to Matsushita, my belief in the potential of nitrides led me to start research on the VPE growth of AlN (1966) [4] in parallel to the GaAs-based work. I confirmed blue emission from AlN using CL and PL and was enjoying working with light emitting materials, but EL was not possible due to the excessively large bandgap.

Thus, in 1973 I decided to focus first on GaN growth.

In fact, in 1969, Maruska et al. [5] had reported on the growth of single crystals of GaN by hydride vapor phase epitaxy (HVPE) and used optical absorption to determine its bandstructure to be direct with Eg of $\sim$3.39 eV. Then in 1971, Pankove et al. fabricated MIS-type blue LEDs [6]. These reports triggered a sudden increase in research on blue light emitting devices [Fig. 1, period (A)]. However, the surfaces of GaN crystals were very rough with cracks and pits, and p-type GaN was impossible to produce. Thus many groups retired from GaN research and some moved on to other materials such as ZnSe and activities on nitrides declined [Fig. 1, period (B)]. From my experience of VPE growth of high quality GaAs [7], I was confident that drastic improvements in the crystal quality of GaN would eventually enable p-type conduction of this material as well. My view was, "It is too early to discuss self-compensation in poor quality GaN. High quality GaN crystals (residual donor density of at least less than $10^{15}\,\mathrm{cm}^{-3}$) must first be produced before discussing such physical properties."

Then in 1973, I decided to make "realization of blue light emitting devices by GaN p–n junctions", an idea abandoned by many, my life's work.

A slight digression, but at the time, apart from groups working on SiC, almost all researchers were studying ZnSe which has a direct bandgap of $\sim$2.7 eV and shows bright CL and PL emission. In addition, its lattice constant is similar to GaAs thus enabling epitaxial growth on GaAs substrates. Thus many researchers were reasonable in thinking of ZnSe as being the first choice as a material for blue or green light emitting devices. (Later, in 1986, very few people showed interest in GaN even after we reported on the successful growth of high quality GaN crystals.) However, apart from ZnSe being 'softer' than GaN, I was also anxious about the crystallinity and stability of ZnSe because of its low growth temperature. On the other hand, the melting point and vapor pressure of GaN are both much higher than ZnSe making its crystal growth extremely difficult. In addition, the E_g of GaN is very large compared with ZnSe, which makes it much more difficult to produce p-type GaN compared

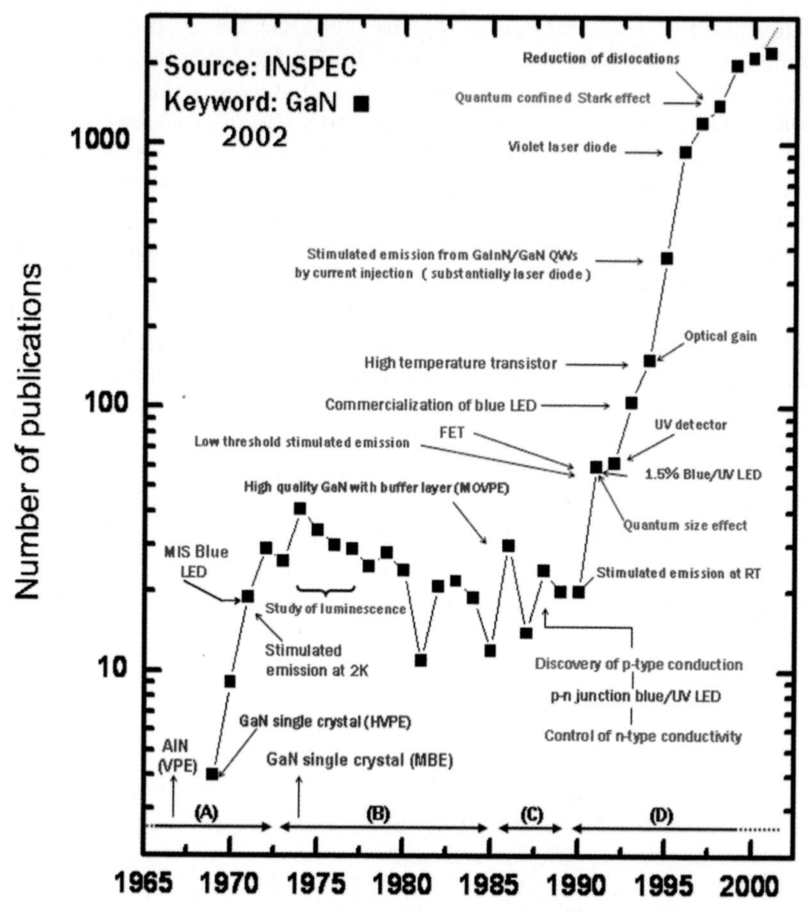

Fig. 1. Number of Publications (INSPEC) and activities related to nitrides between 1965 and 2000. All events are marked in the years, when they were first achieved. Most of important results were achieved by MOVPE using LT-buffer layer after 1986. It is clear that the start of the steep increase of numbers of publications and accomplishments is due to the key inventions (high quality GaN, conductivity control and p–n junction blue LED etc.) in the late 1980s. Green: Crystal Growth, Blue: Devices, Red: Conductivity Control and Physics

with ZnSe crystals. However, I was not discouraged by the problems with GaN because I still believed in the potential of GaN and once the problems were resolved, then compared with ZnSe, nitrides would enable production of more robust and shorter wavelength p–n junction light emitting devices.

Choice of MOVPE as the Preferred Growth Method

Until 1973, GaN was mostly grown by HVPE and no one used molecular beam epitaxy (MBE). However, I decided to use MBE for the first time. Using Ga and NH_3 as sources, I was able to produce single crystal GaN, albeit material with non-uniformities. I sent a proposal for a project entitled, 'Development of blue light emitting devices by ion implantation into single crystal GaN' to the Ministry of International Trade and Industry (MITI) at the time and in 1975, was awarded a three-year research grant to conduct the project [8].

Following many difficulties, and with the assistance of my team of researchers at MRIT, we were able to produce blue GaN LEDs with much better emission characteristics than ever reported, and succeeded in achieving the aims of the project. Approximately 10,000 GaN blue light emitting test devices with the newly developed as-grown cathode electrodes were manufactured (an example is shown in Fig. 2). But due to poor surface uniformity and low yield of MIS structures, the devices were not sold commercially. My goal was p–n junction devices, so I was not discouraged by the lack of success of these MIS LEDs. Two years later, we reported our results at an international conference [9] (where we were the only group to report on GaN related work) but without much of a reaction since most researchers had lost interest in GaN work.

Fig. 2. Fountain display using tricolor LEDs, which was later in 1981 demonstrated in Chicago by Matsushita Electric Industrial Co. Ltd

However, as a result of my increasing experience with handling GaN crystals, fluorescence microscopy showed the existence of high quality microcrystals in parts of larger crystals containing cracks and pits. Also, we found that clusters of needle-like crystals ('GaN-fungus!') left inside the growth reactors exhibited highly efficient light emission. From these experiences, I reconfirmed the great potential of GaN and believed that p-type GaN could indeed be produced if a whole wafer could be made of the same quality as the GaN microcrystals left in the growth reactor.

I then decided to go back to basics and reconfirm the fundamentals of crystal growth: I had experience of growing at least ten kinds of semiconductors and intuitively I knew that crystal quality depended greatly on growth conditions and hence the choice of growth method would be the critical factor in determining the future of the research.

Epitaxial GaN can be grown by MBE, HVPE and MOVPE. MBE was prone to nitrogen desorption and also the growth rate was slow at the time. In HVPE, the growth rate was too fast and crystal quality was affected by reversible reactions. So I thought that this method was not suitable for producing high quality crystals.

On the other hand, the MOVPE method, which at the time was hardly used for GaN growth, uses thermal dissociation reactions at a single temperature and has negligible reversible reactions. Also, the growth rate was intermediate between the other two methods thus enabling growth of nitrides on substrates with a large lattice mismatch. Thus in 1979, I decided to adopt MOVPE as the most suitable method for the growth of GaN. Another advantage of this method was that alloy composition and impurity doping could be readily controlled by varying the source flow rate. Looking back, this decision had a tremendous effect on the development of nitride semiconductors (Fig. 1, caption).

The next problem was the choice of substrate. I experimented with growth on Si, GaAs and sapphire substrates. Eventually, in 1979 I decided on using sapphire because it was stable at temperatures above 1,000^{o}C as well as being able to withstand NH_3 during growth. This choice, at the time, was also appropriate and afterwards led to the first demonstration of p–n junction blue LEDs.

Two Important Breakthroughs

In 1981, I returned to my old nest at Nagoya University. One of the most important research themes of my group was 'research and development of blue light emitting devices using GaN based p–n junctions'. Prospective graduate students who showed interest in this theme included Hiroshi Amano (now professor at Meijo University) and Yasuo Koide (now group leader (Optical Sensor Group) of National Institute for Materials Science)). With the support of my colleague, associate professor Nobuhiko Sawaki (now professor and dean of Graduate School of Engineering, Nagoya University) and Koide and Amano, I set up a clean room and MOVPE growth facilities. In spite of our efforts, however, MOVPE growth did not yield favorable results.

I thought that the main reason was because of the extremely large interfacial energy between GaN and sapphire due to the large mismatches between these materials. Then I had an idea: the insertion of a soft, 'low temperature (LT) buffer layer' to relax the strain. For the buffer layers, I thought about using AlN, GaN, ZnO and SiC because these materials had similar physical properties to those of GaN and sapphire. In a discussion with Amano, I said that the buffer layers should be less than 50 nm and deposited at less than 500°C. Amano, in spite of the poor results up until that time, worked intensely on the use of MOVPE for growth of GaN. Quite accidentally, when the growth reactor was not functioning properly, he deposited a thin AlN layer at low temperature (according to my suggestion) followed by the growth of GaN on top. He showed me the sample he had just grown: the GaN surface was flat and optically transparent [Fig. 3 (right)] [10]. I can still remember the excitement at seeing the reality of the dreams of my Matsushita days: mirror-like, crack and pit free, transparent GaN crystals.

The residual donor density of the GaN crystals grown with LT-buffer layer was also drastically reduced [10,11] but in spite of repeated efforts on acceptor-doping, it was not possible to produce conducting p-type materials using Zn as a dopant. In 1987, when Amano was carrying out CL measurements on high quality Zn-doped GaN crystals grown with the LT-buffer layer, he said that "the emission intensity continues to increase the longer we irradiate the Zn-doped GaN sample with the electron beam during the CL measurements". I thought that this was an important observation, which might be related to the p-type conduction, and asked for more detailed measurements. The results did not show p-type conduction, but the discovery was that: Zn-related

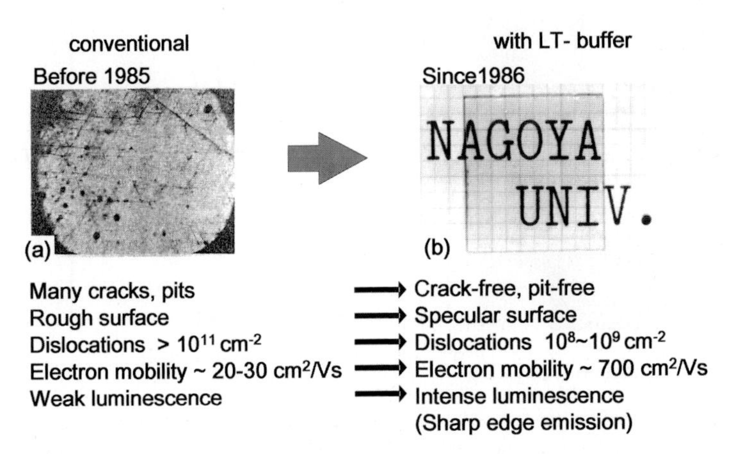

Fig. 3. Optical images of GaN grown on sapphire by MOVPE without (**a**) and with (**b**) LT-AlN buffer layer. The latter is transparent and crack and pit free

blue luminescence from Zn-doped high quality GaN crystals grown with the LT-buffer layers increased as a result of electron beam irradiation. I called this phenomenon the LEEBI effect. This discovery was reported in 1988 [12]. Meanwhile at the beginning of 1988, we noticed that Mg would have smaller ionization energy than Zn and started doping experiments using Mg instead of Zn. Even though at the time the delivery time was several months, we used Cp_2Mg and MCP_2Mg, as Mg sources and grew Mg-doped GaN layers on sapphire covered with the LT-buffer layers [13]; these experiments were mainly carried out by a prospective graduate student, Masahiro Kito (now manager, Intellectual Property Center of Semiconductor Company, Matsushita Electric Industrial Co., Ltd). The Mg-doped GaN samples were irradiated with electron beams in the same way as the Zn-doped samples. We found greatly increased Mg-related blue emission from such Mg-doped GaN samples as well as the samples being low resistivity p-type GaN from Hall Effect measurements: the discovery of p-type conduction in nitride semiconductors. Soon afterwards, we fabricated a GaN p–n junction blue/UV LED (Fig. 4) for the first time, and reported its encouraging I–V characteristics [14].

In 1989, we also succeeded in controlling the conductivity of n-type nitrides using high quality nitride crystals grown with the LT-buffer layer in combination with SiH_4 doping [15]. This control of the conductivity of n-type nitrides is very important as well as the realization of p-type conduction. In 1990, we demonstrated room temperature stimulated emission in the UV region for the first time, which is indispensable for laser operation, from the high quality GaN grown with the LT-buffer layers [16].

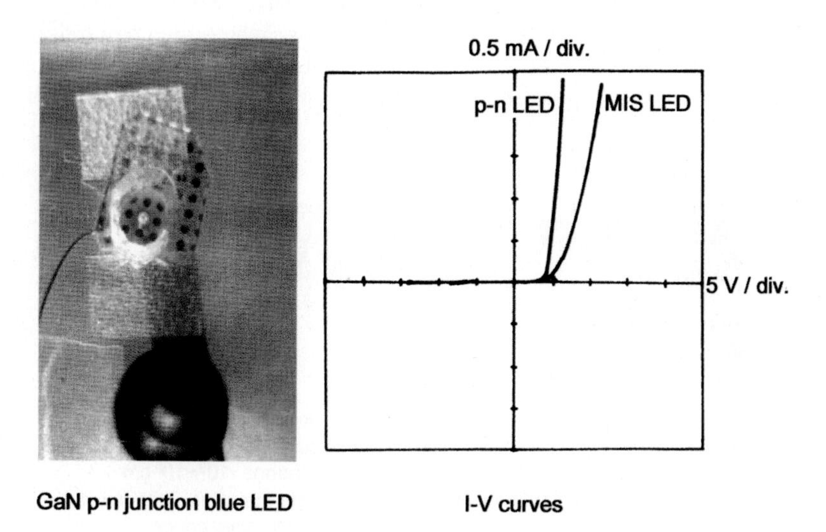

Fig. 4. The world's first GaN p–n junction blue LED on a sapphire substrate developed in 1989. Only one LED is emitting light. The I–V characteristics of a GaN p–n junction LED and MIS-LED

These breakthroughs (achieved in period (C) in Fig. 1) inspired nitride researchers around the world to greater efforts and eventually led to the commercialization of high performance blue LEDs, violet laser diodes and related devices as shown in period (D) in Fig. 1.

Conclusion

Thirty years ago, the fabrication GaN p–n junction blue light emitting devices was a dream which was finally realized ten years ago. This technology will have profound implications in the future. Of course, this technology is the result of contributions from not only myself, but many people including my colleagues at Matsushita Research Institute Tokyo, collaborators and talented students at Nagoya University and Meijo University and industrialists whose favorable evaluation of our findings led to commercial devices.

I am truly grateful to the many friends and research collaborators who I had the good fortune to work with over the last 40 years.

Finally, to young researchers I would like to say, "mistakes accompany new challenges but are overtures to discovery."

References

1. I. Akasaki: Mater. Res. Soc. Symp. Proc. 639, (2001) G8.1.1
2. I. Akasaki: Oyo Buturi 71 (2002) 814 (in Japanese)
3. I. Akasaki: Oyo Buturi 73 (2004) 1060 (in Japanese)
4. I. Akasaki and M. Hashimoto: Solid State Commun. 5 (1967) 851
5. M.P. Maruska and J.J. Tietjen: Appl. Phys. Lett. 15 (1969) 327
6. J.I. Pankove, E.A. Miller, and J.E. Berkeyheiser: RCA Rev. 32 (1971) 383
7. I. Akasaki and T. Hara: Proc. 9th Int. Conf. Phys. Semicond. IX (1968) 787
8. For example I. Akasaki and I. Hayashi: Kogyo Gijutsu 17 (1976) 48 (in Japanese)
9. Y. Ohki, Y. Toyoda, H. Kobayashi, and I. Akasaki: Inst. Conf. Ser. 63 (1981) 479
10. H. Amano, N. Sawaki, I. Akasaki, and Y. Toyoda: Appl. Phys. Lett. 48 (1986) 353
11. I. Akasaki, H. Amano, Y. Koide, K. Hiramatsu, and N. Sawaki: J. Cryst. Growth 98 (1989) 209
12. H. Amano, I. Akasaki, T. Kozawa, K. Hiramatsu, N. Sawaki, K. Ikeda, and Y. Ishii: J. Lumin. 40 & 41 (1988) 121
13. H. Amano, M. Kito, K. Hiramatsu, and I. Akasaki: J. Electrochem. Soc. 137 (1990) 1639
14. H. Amano, M. Kito, K. Hiramatsu, and I. Akasaki: Jpn J. Appl. Phys. 28 (1989) L2112
15. H. Amano and I. Akasaki: Ext. Abstr. Mater. Res. Soc. EA-21 (1990) P165
16. H. Amano, T. Asahi, and I. Akasaki: Jpn J. Appl. Phys. 29 (1990) L205

Contents

List of Contributors

Akasaki, Isamu
Professor Emeritus
Nagoya University
Meijo University

Akimoto, Katsuhiro
University of Tsukuba

Amano, Hiroshi
Meijo University

Arai, Manabu
New Japan Radio Co., Ltd

Egawa, Takashi
Nagoya Institute of Technology

Fujimori, Naoharu
National Institute of Advanced Industrial Science and Technology (AIST)

Fujita, Shizuo
Kyoto University

Fujiwara, Shinsuke
Sumitomo Electric Industries, Ltd

Hanada, Takashi
Tohoku University

Hasegawa, Fumio
Professor Emeritus
University of Tsukuba
Kogakuin University

Hatakoshi, Gen-Ichi
Toshiba Corporation

Hiraki, Akio
Professor Emeritus
Osaka University
Kochi University of Technology

Hiraki, Hirohisa
Kochi University of Technology

Hiramatsu, Kazumasa
Mie University

Hirano, Akira
Osaka Gas Co., Ltd

Hirayama, Hideki
RIKEN

Ishitani, Yoshihiro
Chiba University

Ito, Toshimichi
Osaka University

Kamiyama, Satoshi
Meijo University

Kanie, Hisashi
Tokyo University of Science

Kashiwaba, Yasube
Iwate University

Katayama, Masao
Tokyo Institute of Technology

Kawakami, Yoichi
Kyoto University

Kawanishi, Hideo
Kogakuin University

Kawarada, Hiroshi
Waseda University

Kawasaki, Masashi
Tohoku University

Kimoto, Tsunenobu
Kyoto University

Kishino, Katsumi
Sophia University

Kohmoto, Kohtaro
Teknologue Co., Ltd

Koinuma, Hideomi
Professor Emeritus
Tokyo Institute of Technology
Japan Science and Technology Agency (JST)

Kumano, Hidekazu
Hokkaido University

Kuramoto, Masaru
Sony Corporation

Matsunami, Hiroyuki
Professor Emeritus
Kyoto University
Director, Innovation Plaza Kyoto
Japan Science and Technology Agency (JST)

Miyajima, Takao
Sony Corporation

Miyamoto, Hironobu
Sumitomo Electric Industries, Ltd

Mizutani, Takashi
Nagoya University

Mukai, Takashi
Nichia Corporation

Nakamura, Takao
Sumitomo Electric Industries, Ltd

Nanishi, Yasushi
Ritsumeikan University

Narukawa, Yukio
Nichia Corporation

Niikura, Ikuo
Iwate University

Nomura, Ichirou
Sophia University

Ohkawa, Kazuhiro
Tokyo University of Science

Ohtani, Noboru
Nippon Steel Corporation

Ohtomo, Akira
Tohoku University

Okumura, Hajime
National Institute of Advanced Industrial Science and Technology (AIST)

Okushi, Hideyo
National Institute of Advanced Industrial Science and Technology (AIST)

Onabe, Kentaro
University of Tokyo

Saito, Hiroshi
Okayama University of Science

Sakai, Shiro
University of Tokushima

Sandhu, Adarsh
Tokyo Institute of Technology

Sarayama, Seiji
Ricoh Co., Ltd

Shibata, Naoki
Toyoda Gosei Co., Ltd

Shinohe, Takashi
Toshiba Corporation

Suemune, Ikuo
Hokkaido University

Sumiya, Hitoshi
Sumitomo Electric Industries, Ltd

Suzuki, Nobuo
Toshiba Corporation

Tadatomo, Kazuyuki
Yamaguchi University

Takahashi, Kiyoshi
Professor Emeritus
Tokyo Institute of Technology
Nippon EMC Ltd

Taniguchi, Takashi
National Institute for Materials Science (NIMS)

Taniyasu, Yoshitaka
NTT Corporation

Watanabe, Kenji
National Institute for Materials Science (NIMS)

Yamada, Yoichi
Yamaguchi University

Yanamoto, Tomoya
Nichia Corporation

Yao, Takafumi
Tohoku University

Yoshida, Sadafumi
Saitama University

Yoshida, Seikoh
Furukawa Electric Co., Ltd

Yoshikawa, Akihiko
Chiba University

1

Development and Applications of Wide Bandgap Semiconductors

1.1 Optical Devices (A. Yoshikawa)

1.1.1 Wide Bandgap Semiconductors Indispensable for Short Wavelength Optical Devices

As the name implies, a "wide bandgap" semiconductor is one having a large bandgap energy (forbidden bandgap), which is directly related to the emission/absorption wavelength of optical devices. Typical wide bandgap semiconductors exhibit emission/absorption wavelengths in the green/blue part of the visible spectrum and on into the shorter wavelengths of violet/ultraviolet light. For example, the "blue light emitting diode" is a well known application of wide bandgap semiconductors. Thus in general, wide bandgap semiconductors can be defined as having fundamental optical absorption edges that are of shorter wavelengths than the color red.

Examples of optical devices include light emitting diodes, laser diodes, photodiodes, photoconductive sensors, electro-modulation devices, and optical–optical modulation devices. Semiconductor light emitting devices are ultrasmall, light weight, high efficiency, and have much longer lifetimes than other light sources. In particular, wide bandgap semiconductors have become increasingly important in the electronics industry as optical sources for full color displays, white light illumination, UV/deep UV light sources, and blue–violet laser diodes for high density DVDs. But in spite of the intense world-wide interest in the use of wide bandgap semiconductors for blue LEDs, a long time passed before technological breakthroughs led to the fabrication of pn junctions in GaN. After 1990, the number of researchers involved in wide bandgap semiconductors increased dramatically following the first successful operation of the blue LED.

Wide Bandgap Semiconductor Materials

Figure 1.1 is a comparison of the bandgap energies and lattice constants (a) of wide bandgap semiconductors and other well known semiconductors including silicon (Si), GaAs and ZnSe.

The three main types of wide bandgap semiconductors are:

- Group III nitrides such as GaN
- Group II oxides such as ZnO
- Group II chalcogenides such as ZnSe

As described in Sect. 1.2, other important wide bandgap semiconductors such as silicon carbide and diamond are being studied for electronic device applications.

In Fig. 1.1, the hexagon and square symbols represent hexagonal and cubic crystal structures, respectively. Materials shown in plain and italicized text refer to direct and indirect bandgap semiconductors, respectively. Further, by multiplying the lattice constants of hexagonal crystal structures by $\sqrt{2}$, it is possible to directly compare those values to the lattice constants of cubic crystal structures in terms of the atomic bond lengths. A bandgap energy of 2.25 eV corresponds to a wavelength of 550 nm (green light emission) and semiconductors with bandgap energies larger than 2 eV are said to be "wide bandgap." Thus it is apparent that wide bandgap semiconductors are essential for fabricating optical devices emitting visible-green, blue, and UV/deep UV wavelengths.

As described in Chap. 2, direct bandgap semiconductors, such as GaN, ZnO, and ZnSe are necessary for fabricating high efficiency light emitting

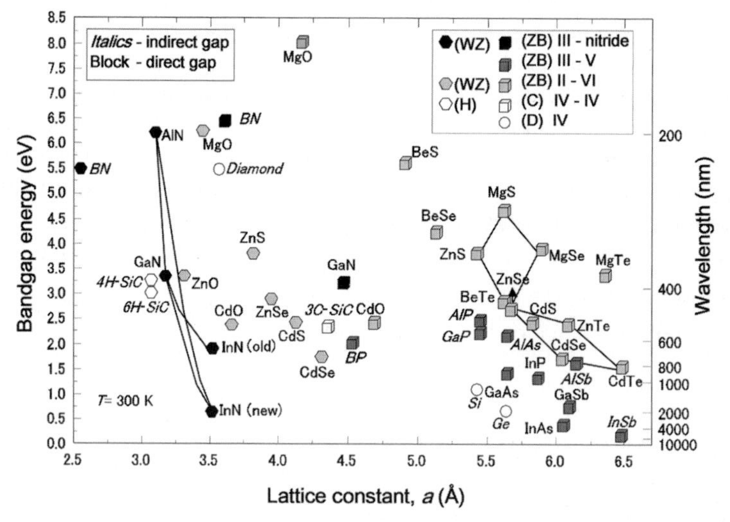

Fig. 1.1. Relationship between forbidden energy gap and lattice constant of wide bandgap semiconductors (Courtesy of Dr. Wang, Chiba University)

devices. SiC and diamond are important indirect wide bandgap semiconductors, and prior to recent developments in nitrides, SiC pn junctions were used for fabricating blue LEDs operating in 460–470 nm wavelength range. But efforts to commercialize such SiC LEDs were thwarted due to the poor emission efficiency which is more than two orders of magnitude less than that of nitride semiconductors.

The points in Fig. 1.1 are seen to increase toward the left hand side of the graph which indicates that semiconductors with small lattice constants exhibit large bandgap energy. The small lattice constant shows that the material exhibits strong interatomic forces, with the outermost shell electrons, that are responsible for chemical bonding, being strongly bound to the lattice thus leading to a large bandgap energy.

As described in Chap. 2, compared with covalently bonded group IV elemental semiconductors, group III–V compound semiconductors exhibit slightly ionic bonding while group II–VI semiconductors exhibit greater ionic bonding.

In the case of ionic bonding, the bonding electrons are localized to the composite elements, resulting in weak bonding forces and large bandgap energies. ZnSe and ZnS are typical examples of II–VI compound semiconductors exhibiting such bonding and they have a "soft" crystalline structure that can be scratched during handling with tweezers.

Thus the bandgap energy tends to increase for compound semiconductors composed of light elements and also having high ionicity in bonding. That is, the bandgap energy increases for elements toward the top right hand region of the periodic table.

As shown in Fig. 1.1, the crystal structures of wide bandgap semiconductors are cubic (diamond and zincblende), hexagonal (wurtzite), and rock salt (NaCl). Generally, as the ionic bonding component of elements constituting compound semiconductors increases, so the interelemental attractive forces increase resulting in shorter distances between the elements with the result that crystal structures change from zincblende to wurtzite. This is the main reason that many wide bandgap semiconductors exhibit a hexagonal crystal structure. Further increases in the ionic bonding component eventually lead to rock salt crystal structures, as seen in MgO.

Overview of Optical Devices with Emission Wavelengths in the Visible Short and UV Regions

Historically, phosphide compound semiconductors were used for fabricating short wavelength light emitting devices. For example, high efficiency LEDs emitting up to wavelengths of 580 nm can be fabricated using quaternary AlInGaP epilayers grown on GaAs or GaP substrates. For shorter wavelengths, there have been reports on nitrogen-doped GaP that has an indirect bandgap, to produce isoelectronic traps, via which bound exciton recombination paths yield green emission in GaP diode structures. However, the emission efficiency of GaN diodes at 630 nm is reduced to 1/50. Thus it is not possible to produce emission at shorter wavelengths using conventional III–V compound semiconductors.

The AlN–GaN–InN system of compound semiconductors has been extensively studied for fabricating light emitting devices for wavelengths shorter than green. As described later, it is still not possible to reproducibly produce p-type ZnO although there have been reports of emission from ZnO diode structures [1]. ZnSe is widely recognized as being an excellent material system for fabricating blue-green LEDs and laser diodes. In spite of the tremendous effort expended in the development of ZnSe-based devices it was found that the lifetime of ZnSe optical devices was too short for commercialization and this field of research has largely been abandoned with developments in nitride semiconductors showing greater promise. There are still groups using ZnSe for fabricating laser diodes emitting at longer wavelengths. A combination of homoepitaxial technology and new device structures are being used for the development of ZnSe-based LEDs emitting white light [2].

Figure 1.2 shows the wavelength dependence (ultraviolet to red) of the external quantum efficiency of nitride-based light emitting devices. A detailed analysis of each wavelength region will be given in a later chapter but this graph shows the wide range of emission wavelengths being studied, ranging from blue and green LEDs to devices emitting in the amber region. Since blue, green, and other such semiconductor light emitting diodes are solid, robust, and highly efficient they are increasingly being used as light sources in displays and traffic signals. Further, it is also possible to fabricate short wavelength, high efficiency light sources emitting at 365 nm, corresponding to *i*-line wavelength of high pressure mercury lamps, that are widely used in industry, so the prospect of a long lifetime, solid state light source is of great industrial significance. Optical devices emitting at less than 350 nm are of interest for biological applications with increasing activity in their development.

For reference, Fig. 1.2 also includes the wavelength dependence of the efficiency of AlInGaP-based LEDs. It can be seen that nitride semiconductors

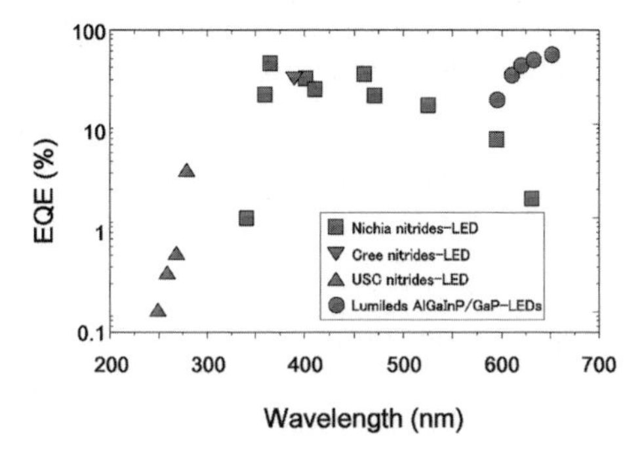

Fig. 1.2. Comparison of the wavelength dependence of the emission efficiency (external quantum efficiency) of nitride and phosphorus-based semiconductors (as of September, 2005)

can be used for fabricating high efficiency light emitting devices in the short wavelength spectral range. ZnO-based light emitting devices are still in the development stage and do not compare favorably with nitrides at the present time. However, although ZnO-based devices cover almost the same wavelength range as nitrides, they are expected to enable the fabrication of lower threshold laser diode devices using excitons to operate at higher efficiencies and higher temperatures. Further advantages over nitrides include the plentiful supply of raw materials, the ease with which substrates are produced, and the wide range of device applications that could use the transparency of ZnO [3].

Nitride-based light emitting devices are being used as white light sources in the illumination industry. The aim is to manufacture nitride devices having efficiencies higher than fluorescent phosphor light sources. The efficiencies of nitride light sources are already much higher than incandescent tungsten lamps and they are being increasingly used for localized, spot illumination and flash lights. Uniformly superimposing red, green, and blue (RGB) light sources would produce white light, but in practice it is more common to use a single light source to excite fluorescent phosphors. This is because it is easier to control the intensity of white light using a single light source than three independent RGB sources. There are two methods for producing white light using a single light source:

- Use high efficiency blue LEDs to excite fluorescent phosphors with complementary emission wavelengths with respect to blue.
- Use a UV light source to excite a fluorescent phosphor of the three primary colors.

As described in a later chapter, the main issues to resolve are high efficiency and high color rendering. Nitride and ZnSe-based white diodes both have color rendering figures of merit that are similar to incandescent light sources.

As described in a following chapter, nitride-based wide bandgap semiconductors are being used in high density DVD players as short wavelength laser diodes (LDs) having wavelengths in the range 405–410 nm. A range of nitride-based laser diodes have been fabricated extending continuous wave operation to wavelengths between 364 and 482 nm.

Short Wavelength Photodetectors, Modulators, and Novel Optical Devices

Ultraviolet (UV) sensors exploiting transparency in the visible region are examples of wide bandgap semiconductor photodetectors. Such solar blind UV sensors that are insensitive to visible light and respond only to UV irradiation are used as flame sensors and to detect harmful UV radiation in sunlight. High efficiency, photovoltaic type sensors have been fabricated and their simple structure has led to interest in photoconductive types of sensors as well.

Recently, the bandgap of InN has been clarified as being 0.64 eV, which corresponds to the near infra-red region and enables AlN–GaN–InN-based nitride materials to be used for fabricating devices operating from the UV to near infra-red range. These properties of nitride semiconductors are being used to develop high efficiency tandem-type solar cells.

The band offsets of wide and small bandgap semiconductor heterostructures can be several electron volts. For example, the conduction band discontinuity of AlN/GaN based heterostructures is approximately 2 eV, which is much larger than the bandgap itself of conventional semiconductors. If the thickness of the quantum well layer in such semiconductor heterostructures is grown to be only a few atomic layers, then the energy difference between the subband electron levels in the conduction band well layer is larger than the photon energy of optical communication wavelengths. The relaxation time of electrons from the excited level is extremely short at approximately 10 fs, thus enabling the possibility of fabricating inter-subband transition (ISBT) optical–optical switches and modulators.

The use of wide bandgap semiconductors in optical communication requires development of technology for the growth of superlattice structures with atomically sharp interfaces. Further, wide bandgap semiconductors could also be used for fabricating quantum cascade lasers. However, homo and heteroepitaxial growth methods for producing bulk crystals of GaN and ZnO and wide bandgap semiconductor heterostructures still need to be developed to the same level of precision and quality as for conventional group III–V compound semiconductors.

It is possible to grow wide bandgap semiconductors that are transparent in the visible spectrum, in layers of semi-insulating, semiconducting and conducting thin films. Such properties of ZnO are being studied for use as transparent displays.

1.1.2 Control of the Physical Properties of III–V Nitrides and II–VI Semiconductors

Nitride materials are environmentally friendly as well as being resistant to harsh environments. These materials contain nitrogen, a light element often participating in strong chemical bonding. Oxide materials are also composed of a light element, oxygen, and exhibit similar physical properties.

Irrespective of the choice of semiconductors, in order to fabricate high performance optical devices, it is essential to produce pn junctions. However, it is often difficult to control both p- and n-type conduction in materials with strong chemical bonding or large bandgaps. Further, it is necessary to use multiple quantum well structures in order to improve the emission efficiency and other device characteristics, and the material properties are not always conducive to the fabrication of such structures.

Wide bandgap semiconductors have the following common physical properties and problems:

Difficulties in Controlling both p-Type and n-Type Conductivity

The development of the blue LED using wide bandgap semiconductors took a long time primarily due to difficulties encountered in making them amphoteric, i.e., controlling both the p- and n-type conduction. Wide bandgap semiconductors generally show these trends. For example, it is relatively easy to produce n-type GaN and ZnSe but p-type films were difficult to produce for a long time.

Figure 1.3 shows the theoretical models of the energy band structure of the ternary AlN–GaN–InN alloy where the Fermi level stabilizing energy, E_{fs}, is used as a reference in calculating the positions of the conduction and valence band edges [1]. The Fermi level stabilizing energy is a measure of the average dangling bond energy and shows the position where the Fermi level is pinned when defects are introduced [2]. These calculations show that controlled, n-type doping of GaN is easy to achieve but p-type is difficult. In the case of AlN, it is difficult to control both types of doping, with p-type being particularly challenging. Further, InN, which has recently been shown to have a bandgap of 0.64 eV, a value much smaller than the 1.9 eV previously reported, can be readily made as the n-type by introducing crystal defects. However, difficulties in producing the p-type conductivity are expected.

The results of Fig. 1.3 is also extremely important for the design of band-lineups of nitride materials. That is, the formation of heterojunctions does not depend only on bandgap differences but also on the resulting relative proportions of the conduction and valence band discontinuities. The band edge discontinuities are important parameters for forming superlattice structures.

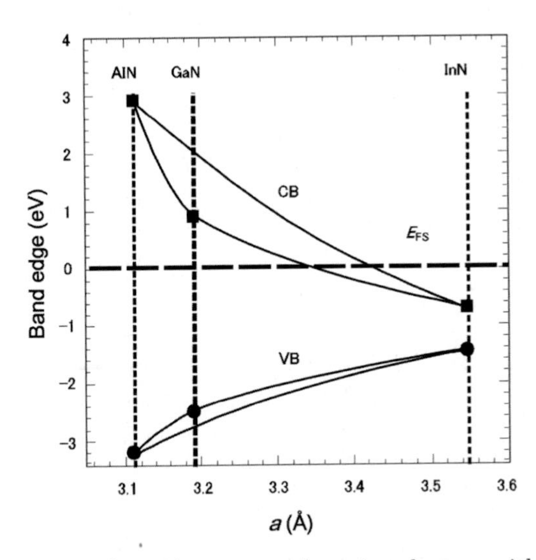

Fig. 1.3. Band line up of nitride compound semiconductors with reference to the stabilized Fermi energy of defects [1]

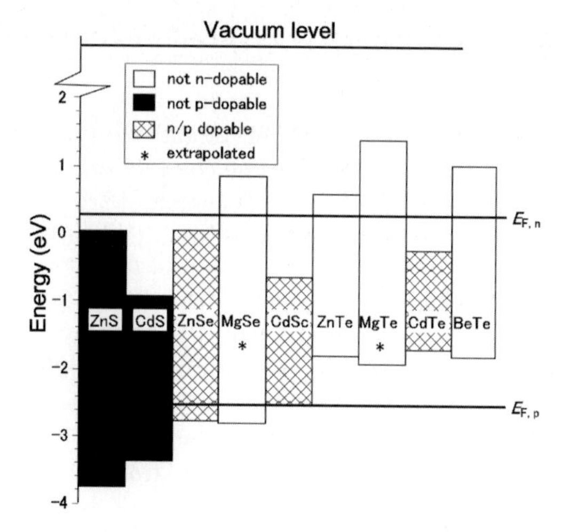

Fig. 1.4. Band line up and ease of doping of II–VI based compound semiconductors [3]

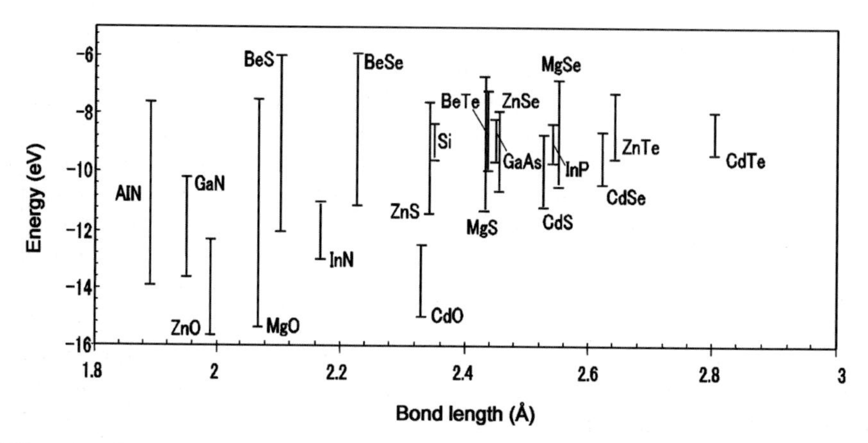

Fig. 1.5. Band line ups of a several wide bandgap semiconductors based on the Harrison model [4]

Figure 1.4 shows the band alignment and ease of doping of II–VI chalcogenide semiconductors [3], where it can be seen that it is possible to produce both n- and p-type ZnSe. The limited data for ZnO oxides are shown in Fig. 1.5 together with band alignments for a selection of other semiconductors calculated according to the Harrison model [4]. There have been many reports on the difficulties encountered in p-type doping of ZnO. There are also considerable hurdles to overcome in producing both p- and n-type conduction in other wide bandgap semiconductors.

Control of Material Properties and Effect of Epitaxial Crystal Structure

The hexagonal wurtzite crystal structure is the most stable form of GaN and ZnO type compound semiconductors. This crystal structure does not have symmetry in the *c*-axis direction and is polarized in this direction. Further, nonpolar sapphire substrates are used for growing these semiconductors because of the difficulties involved in obtaining appropriate homoepitaxial substrates. Almost all device structures are epitaxially grown on *c*-plane substrates. Both GaN and ZnO grow perpendicular to the *c*-plane of the substrate where the direction of the polarity of the *c*-axis affects the processing and properties of devices. Further, as seen from Fig. 1.1, a large lattice mismatch during growth results in the generation of rotation domains (regions with rotated crystal directions) and the introduction of strain.

Effect of Crystal Polarity on Epitaxy

For the growth of GaN and ZnO on nonpolar *c*-plane sapphire substrates, the epitaxy relationship for crystalline lattice between the epilayer and substrate, in particular the polarity direction of epilayers, are uncertain. Thus a comprehensive understanding of polarity is necessary for preparing the surfaces of sapphire substrates prior to epitaxial growth. For MOCVD growth of GaN-based nitrides, sapphire substrates are annealed at 1,100°C in a hydrogen atmosphere producing an excess of Al on the surface and resulting in the growth of GaN with Ga-polarity (+c polarity). Further, nitridation of sapphire surface results in the formation of a thin layer of AlN with a nitrogen polarity (-c polarity) and nitrides grown on these surfaces have N-polarity.

There are extremely large differences in the vapor pressures of elements constituting GaN-based nitride crystals. In general, +c polarity leads to enhanced surface diffusion of metal elements. This is used in epitaxial lateral overgrowth (ELO) technology for drastically improving crystalline quality and surface flatness. Further, polarity is also known to affect impurity incorporation mechanisms.

The epitaxial growth of ZnO-based oxides on sapphire substrates is more complicated than the growth of nitrides in spite of the presence of oxygen in both the substrate and epitaxial layer. There are several reasons for this. First, during the initial bonding of Zn or O in the ZnO layer with Al or O in the substrate, the complexity of the sapphire crystal structure and the large differences between the lattice constants of sapphire and ZnO, results in only small differences in bonding energies among several permissible epitaxy relationships. This affects not only the polarity but also the generation of rotation domains in the ZnO. Another reason for difficulties during the growth of ZnO on sapphire is that oxide compounds are prone to grow with an amorphous structure, a trend that is more pronounced at low growth temperatures when accurate information about crystal polarity during growth becomes difficult to ascertain.

Effect of Crystal Natures and Large Lattice Mismatch on the Characteristics of Optical Devices

Several problems arise due to the large lattice mismatch between epitaxial layers and the substrates used for fabricating wide bandgap semiconductor optical devices. As a result of this extremely large lattice mismatch, GaN-based blue LEDs grown on c-surfaces of sapphire consist of columnar crystal structures with approximately $10^{10}\,\mathrm{cm}^{-2}$ lattice mismatch dislocations. The dislocation density in LEDs made of conventional III–V compound semiconductors is approximately five to six orders of magnitude less than that in wide bandgap semiconductors. Thus in spite of the extremely high density of dislocations, the optical emission efficiency of nitride semiconductor devices is miraculously high.

These physical properties can be explained by the short diffusion length of minority carriers. The effective mass of carriers in wide bandgap semiconductors is generally large. That is, since the diffusion length of injected minority carriers is short, it is thought that they undergo high efficiency emission transitions before reaching grain boundaries. Further, it is easy to control the electrical conduction of hexagonal wide bandgap semiconductors having even polycrystalline structures, leading to the conclusion that crystalline defects do not trap carriers in such semiconductors. Further, due to the low crystal symmetry (anisotropy), the gliding direction of the dislocation is also controllable, which enables alignment of the dislocations generated during operation in a direction parallel to the active layer. This dramatically reduces degradation of laser diodes operated at high carrier injection. The following properties enable significant improvements in device performance:

- Relatively large contribution of ionic bonding in wide bandgap semiconductors
- Low crystal symmetry
- Heavy effective mass of charge carriers
- Short diffusion length

However, crystal polarity and the large internal strain are sources of new problems in wide bandgap semiconductors. That is, an extremely large electric field arises in the quantum well of the active layer due to polarization resulting from the crystal strain. Thus the large band bending in the active layer causes spatial separation of electrons and holes, which subsequently results in a reduction of efficiency for emission, a peak shift to longer wavelengths comparing to the calculated values, and dependence of the emission wavelength on the magnitude of the current. Homoepitaxy and nonpolar growth are being studied to reduce such problems.

1.1.3 Other Characteristics and Trends of Wide Bandgap Semiconductor Optical Devices

The environment and healthcare are becoming central themes for future research and development. Conventional optical devices emitting at red and infrared wavelengths are fabricated using poisonous elements such as P and As. At the present time, there are no alternative materials for producing light sources for optical communications and information networks. On the other hand, devices made using nitride and oxide compound semiconductors are not harmful to humans and the environment, and show stable operation at elevated temperatures and under exposure to radiation. Then a number of investigations are underway for developing red LEDs and green LDs by III-nitrides. Further, research and development for solid-state UV and deep UV light source using wide bandgap semiconductors, in particular III-nitrides, are intensively undertaken all over the world now.

The large effective mass of electrons and holes in wide bandgap semiconductors could also enable realization of excitonic devices operating at room temperature if the binding energy of free excitons in the material is larger than 26 meV, the thermal energy of room temperature, 300 K. The free exciton binding energies of GaN and ZnO at room temperature are 28 meV and 60 meV, respectively, physical properties that raise expectations for room temperature exciton devices.

The physical properties of wide bandgap semiconductors are still not completely understood, thus raising the possibility for development of novel functionality devices. However, process development for fabricating nitride devices is also important. Many researchers concur that the development of growth methods for high quality and cheap GaN and ZnO bulk crystals and/or substrates will become imperative for major advances in this field.

1.2 Electronic Devices

1.2.1 Silicon Carbide Electronic Devices (H. Matsunami)

Expectation for Wide Bandgap Semiconductors [5–8]

Electronics based on semiconductors has directly contributed to information, energy and global environment. Electronics is the core of information technology. Computers and their networks have spread all over the world, which will promote further development of our society. The semiconductors which have been used in the progress of electronics are silicon (Si), gallium arsenide (GaAs) and related III–V materials. The energy bandgaps of these semiconductors are 1.12 eV for Si and 1.43 eV for GaAs, which are mid-range values. According to the progress of electronics, strong demand to use the technology has long been continued. However, those devices from semiconductors with mid-range bandgaps cannot be used for a wide range of applications owing to the performance limit imposed by the size of the bandgap. Nowadays, wide bandgap semiconductors take strong attention for electronic devices such as high-power, high-frequency/high-power, robust devices under harsh environment such as high temperature and heavy radiation influence, optoelectronic devices for ultraviolet detection, and light-emitting devices in the short wavelength region.

Power Devices and Limits of Si Technology

The consumption of electric power has steadily been increasing by the rapid shift to the age of information. From the viewpoint of environmental protection, we should not solve the problem simply by increasing electric power supply. Cutting down the consumption of electric energy and its efficient use are urgently required, and especially highly-efficient use of electric power is very important, because it enables us to decrease the consumption of electric energy remarkably.

Many power electronic devices are used at important places such as voltage and frequency control for transmitting and converting electric power. Accordingly, reduction in loss and performance improvement of these devices will directly lead to considerable reduction in electric power consumption. Present power semiconductor devices, for example inverters, are all made of Si. The performance of Si power devices has been improved using microfabrication technology that has aided the progress of Si VLSI (very large scale integration). But now, the performance of Si power devices is approaching its limit due to physical properties, energy bandgap, and we cannot expect dramatic development in Si power devices.

Expectations of SiC

Table 1.1 shows the physical properties and figures of merit of Si, GaAs, and SiC, together with other important requirements in device processes. The large energy bandgap, breakdown field strength, saturation electron velocity and thermal conductivity (similar to Cu) of SiC enables this material to be used for fabrication of devices that can be operated at high voltages and large drive currents, in high frequency, and at high temperatures.

Conductive and semi-insulating SiC substrates can be obtained. The conductivity of SiC layers can be readily controlled over a wide range,

Table 1.1. Properties of Si, GaAs, and SiC (4H)

	Si	GaAs	SiC (4H)
energy gap (eV)	1.12	1.43	3.26
electron mobility ($cm^2 V^{-1} s^{-1}$)	1,350	8,000	1,000
breakdown field ($MV cm^{-1}$)	0.3	0.4	3.0
saturation electron velocity ($cm s^{-1}$)	1×10^7	1×10^7	2×10^7
thermal conductivity ($W cm^{-1} K^{-1}$)	1.5	0.5	4.9
Johnson figure of merit	1	1.8	400
Baliga figure of merit	1	15	603
conductivity control	easy	easy	easy
oxidization	easy	difficult	easy
conducting wafer	yes	yes	yes
semi-insulating wafer	no	yes	yes

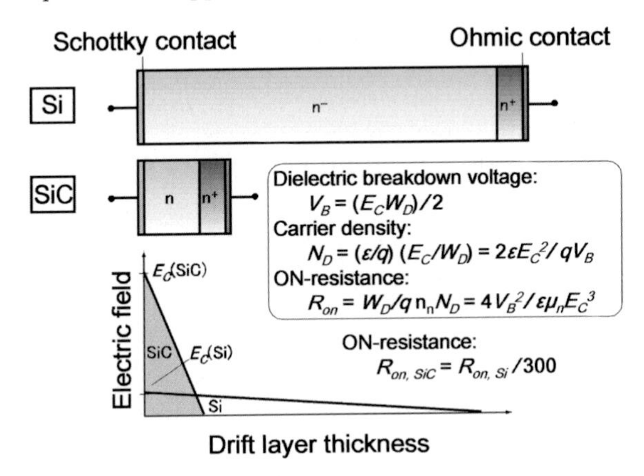

Fig. 1.6. Comparison of performance of Si and SiC devices in the case of Schottky diodes

10^{14}–10^{19} cm^{-3} both for p- and n-types by impurity doping. Further, the surface of SiC can be covered with high-quality oxide layers by thermal oxidation, an essential factor for device fabrication.

Figure 1.6 shows a comparison of the characteristics of Si and SiC Schottky diodes, majority-carrier power devices. The upper figure is a comparison of the device length, which shows that the one order of magnitude larger breakdown field strength of SiC enables a reduction of its device length to 1/10 of that required for Si. The lower figure is a comparison of the relation between drift-layer thickness and electric field strength, where the gradient of the straight line shows the impurity concentration; it is possible to introduce two orders of magnitude more impurities into SiC than Si.

Because SiC has electric breakdown field strength of about ten times larger than that of Si, SiC power electronic devices can be made one tenth thinner than Si. And because handling current density is larger, the SiC devices can be smaller over all. In the case of popularly used MOSFETs (metal-oxide-semiconductor field effect transistors) for power electronic devices, the specific on-resistance that causes thermal loss during power conversion is inversely proportional to the third power of the electric breakdown field strength; consequently the on-resistance of SiC MOSFET becomes smaller than that of Si by two or more orders of magnitude. Also, because the thermal conductivity of SiC is about three times that of Si, the equipment using SiC power electronic devices can use simple cooling systems. Thus, the future of SiC is very promising because of its smaller size, lower loss, higher efficiency, and easier heat-dissipation.

The physical properties put SiC power devices to practical use with power losses of less than 1/300 that of Si power devices. High-frequency operation of SiC devices can solve the unpleasant noise in low-frequency operation of present-day Si power devices. The upper limit of operation temperature for SiC powerdevices can be increased up to 300°C rather than at 150°C for Si power devices. That is, SiC offers potential for miniaturization, low power losses, high efficiency and simple cooling equipment.

Historical Events in the Development of SiC Technology

SiC was first synthesized in the nineteenth century, and the first light emission was observed upon electrical excitation of a point contact structure at the beginning of the twentieth century. Following the invention of the Ge transistor and increase in related research on semiconducting materials in early 1950s, light emission from SiC was shown to be due to minority carrier injection and recombination in a semiconductor – a finding that triggered the race for the development of visible light-emitting diodes. In the middle 1950s, the possibility of using wide bandgap semiconductors at high temperatures led to tremendous investment in the USA for SiC as an alternative to Ge for electronic applications. It is noteworthy, that during this period of time, SiC was being compared with Ge and not Si. However, it was difficult to produce high-quality single crystal SiC due to its stability at high temperatures. In the 1970s there was hardly any interest in SiC research due to the well documented advances in Si-based electronics.

Then toward the end of the 1970s, a Russian group reported a breakthrough in the growth of bulk single crystal SiC using the "seeded sublimation method"; in the latter half of the 1980s, a group in Japan proposed "step-controlled epitaxy" to produce high-quality layers on lower-quality SiC substrates. At the beginning of the 1990s, these two events, together with the realization of the physical limits of Si power electronic devices, led to renewed research on SiC by groups in the USA, Germany, Sweden, France, and Japan.

During this period, the major contributions from Japan have been the clarification of growth mechanism in step-controlled epitaxy; control of impurity doping during growth; formation of oxide layers and control of the interface properties of MOS devices; and the establishment of ion implantation technology. Further, a report in 2004 on the production of ultrahigh quality SiC substrates by the "Repeated A-face Growth Method" has enabled the elimination of micropipe defects and reduction of dislocation density to less than $10^2\,\mathrm{cm}^{-2}$ – a major advance in this field.

Applications of SiC Power Devices

SiC Schottky diodes reported by Japanese researchers in 1993 and 1995 are commercialized in 2001, and now available as commercial products. They are slowly replacing Si pn diodes in switching power supplies as fast switching diodes. Nowadays switching transistors are being widely investigated in the world. There are also signs of progress in improving electron mobility in the inversion channel of SiC MOSFETs. In a few years, normally-off MOSFETs will probably become commercially available. Devices for small power capacity with small area capable of handling several amperes are at the stage of being fabricated on 3-in. substrates for commercial production. The basic technological infrastructure for fabricating devices with larger power capacity handling several tens and above hundred amperes for power conversion (converters and inverters) is being prepared.

Power electronics encompasses a very wide range of applications including electrical and electronic equipment, household appliances, industrial production equipment, uninterrupted power supplies, tractions for trains, and high-voltage DC transmission equipment. Use of SiC power electronic systems with high efficiency and simple cooling for these applications will strongly reduce electric energy consumption.

Further, there will be many major changes in the automobile industry with the introduction of hybrid electric cars based on conventional gasoline engine with electric motors and, probably in future, supported by fuel cells, which can reduce exhausted gases and reliance on fossil fuels. Considering a limited space of trunk, existence of high temperature around engine, and long-term use of batteries in automobiles, a big revolution in power electronic system is strongly expected. A new power electronic system, in which SiC power devices is the central core, is the best suited technology.

Figure 1.7 shows the main potential applications of power semiconductor devices. As shown in the figure, in the present Si power electronics, various types of power devices are used in individual applications, caused by historical step in the development of each device. SiC MOSFETs will replace Si IGBT (insulated gate bipolar transistor) technology, because the power loss in SiC MOSFETs is two orders of magnitude less compared with Si MOSFETs, which covers Si IGBT power ranges. In addition, high-frequency switching is possible with MOSFETs than IGBTs owing to the majoritycarrier operation. It is not

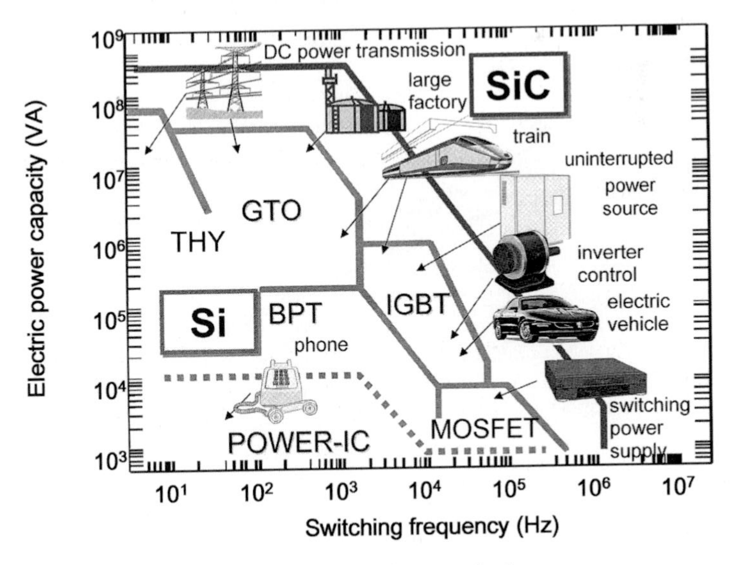

Fig. 1.7. Types of power devices

a simple replacement, but it gives advantages such as miniaturization, low loss, and simple cooling system.

In communication, due to the generational change of cellular phones, new base stations operating at high capacity and high speed are being constructed continuously. There is also strong interest in the development of SiC MESFETs (metal-semiconductor field effect transistors) as compact, high power output microwave oscillators operating at less than 10 GHz for wireless base stations. Power outputs of several ten watts for these devices have been reported and field tested.

1.2.2 Nitride Compound Semiconductor Electron Devices (Y. Nanishi)

Nitride semiconductors are of interest for more effective device solutions with respect to energy and the environment, as well as in the development of the information technology industry. This section reviews applications of nitride semiconductors and highlights the potential problems and trends.

Applications of Nitride Semiconductors

Figure 1.8 shows the device frequencies expected to be used for portable phones, satellite communication, wireless systems, and intelligent transportation systems (ITS). With increasing frequency, f, wave propagation losses

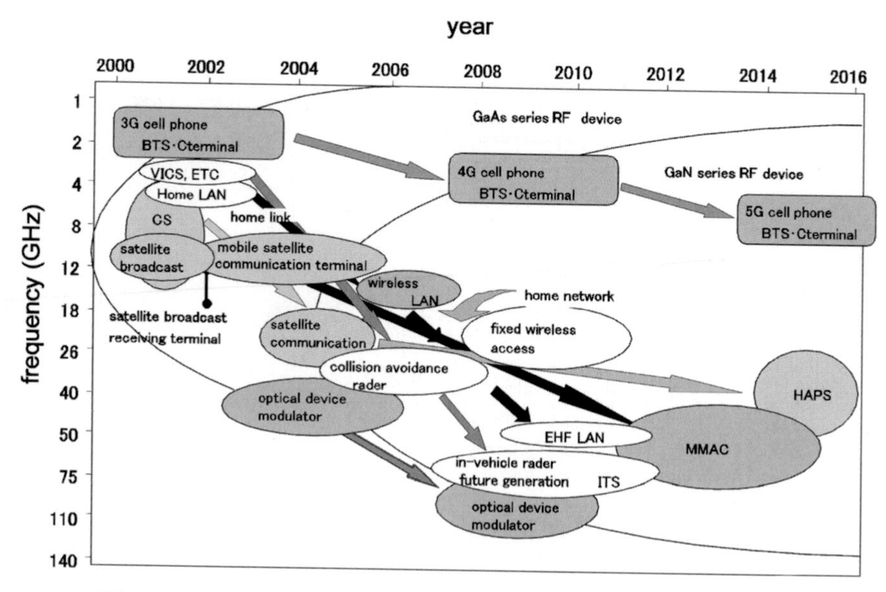

Fig. 1.8. Wireless systems and related operating frequency ranges

generally show an approximate $\sim f^{2.6}$ dependence. Further, the high frequency power increases linearly with increases in the volume of information. On the other hand, an extremely high degree of linearity is required in order to stand up to intense modulation and demodulation processes. These types of systems can only be realized with high frequency devices having high power output, high efficiency, good linearity, and low power consumption. To date, high frequency and low noise devices, with relatively low power output have been fabricated using conventional compound semiconductors such as GaAs and InP, but wide bandgap nitride semiconductors are required for high power output devices operating in the GHz range.

Characteristics of Nitride Semiconductors and Electron Devices

The Ga–N atomic bonding is strong because GaN is composed of N, a light element. As a result the lattice constant of nitrides is smaller than conventional III–V semiconductors. Large, bulk crystals of GaN cannot be supplied as lattice matched substrates for epitaxial growth. This is a major difference compared with SiC device technology. However, the strong atomic bonding results in nitrides being thermally, chemically, and mechanically robust. Nitrides also show greater ionicity than conventional III–V semiconductors such as GaAs and exhibit spontaneous polarization due to lack of reverse symmetry along the c-axis of the hexagonal wurtzite crystal structure. Additional piezoelectric polarization arises when strain is introduced into the crystal structure.

Table 1.2 shows the main physical properties of commonly used wide bandgap and conventional III–V compound semiconductors. At room temperature, GaN has a large bandgap of 3.4 eV and a breakdown electric field of $3.3 \times 10^6\,\mathrm{V\,cm^{-1}}$, which is 8–10 times larger than GaAs. Further, the saturation drift velocity of GaN is $2.5 \times 10^7\,\mathrm{cm\,s^{-1}}$, approximately 1.2 times larger than that of GaAs. The thermal conductivity is a little less than that of SiC, but 4 times larger than that of GaAs.

Table 1.2. Physical properties of widely used semiconductors

material	bandgap (eV)	relative dielectric constant	effective electron mass	electron mobility ($\mathrm{cm^2\,V^{-1}s^{-1}}$)	break-down field ($10^6\,\mathrm{V\,cm^{-1}}$)	saturation electron velocity	thermal conductivity ($\mathrm{W\,cm^{-1}K^{-1}}$)	Transition n type ($10^7\,\mathrm{V\,cm^{-1}}$)
InN	0.6 ∼ 0.7	15.0	0.04	4,000	2.0	4.2	0.8	direct
GaN	3.4	9.5	0.22	1,200	3.0	2.5	2.1	direct
AIN	6.2	8.5	0.29	–	–	2.0	2.9	direct
Si	1.1	11.8	0.19	1,500	0.3	1.0	1.5	indirect
GaAs	1.4	12.8	0.067	8,500	0.4	2.0	0.5	direct
4H–SiC	3.3	10.0	0.3	1,000	3.0	2.0	4.9	indirect
diamond	5.5	5.5	0.2	1,800	4.0	2.5	20.9	indirect

One of the main features of nitrides when compared with SiC is the possibility of fabricating heterostructures having large band discontinuities using AlGaN and AlInGaN. A two-dimensional electron gas (2DEG) is formed at this heterointerface due to spontaneous and piezoelectric polarization. The presence of the 2-DEG leads to a large drift velocity and a relatively high electron mobility. These structures are used to fabricate heterostructural FETs (HFET) which can be driven at large current densities. The large breakdown electric field implies that a large electric field can be applied without damaging the devices. Thus the capacity to pass a large drive current and to apply large voltage enables fabrication of HFETs capable of large output power. Further, assuming a constant applied voltage, devices fabricated with shorter gate lengths would further enhance the electric field enabling ultrahigh frequency operation.

In addition to the high power output and high frequency advantage over GaAs, nitride semiconductors and those dopants contain elements that are less harmful to the environment.

Due to the large band offset and the effects of spontaneous and piezoelectric polarization at the AlGaN/GaN heterostructure interface, it is relatively easy to produce structures with sheet electron densities of $(1 \sim 3) \times 10^{13}\,\mathrm{cm}^{-1}$ for Al compositions of 30%. This electron density is 4–5 times larger than in AlGaAs/GaAs heterostructures. Also, the saturation drift velocity of GaN is 1.2 times that of GaAs. The maximum current density of HFETs is proportional to the product of the carrier density and the saturation drift velocity. A first-order calculation shows that the maximum current density that is possible per unit gate width for AlGaN/GaN HFETs is 5–6 times larger than for AlGaAs/GaAs structures. Also, it is possible to apply a voltage between the source and drain that is approximately one order of magnitude larger than in GaAs-based structures. The high frequency output power density per unit gate width of Class A amplifiers is given by

$$\mathrm{P_{OUT}} = 1/8 \times \mathrm{I_{DS}} \times \mathrm{V_{DS}}$$

where, $\mathrm{V_{DS}}$ is the voltage applied between the source and drain and $\mathrm{I_{DS}}$ is the maximum current density. From this relationship, it can be seen that the $\mathrm{P_{OUT}}$ of an AlGaN/GaN HFET is expected to be 50–60 times larger than for AlGaAs/GaAs devices.

The figure of merit defined by $\mathrm{FM} = \mathrm{P_A f_{10}}^2\mathrm{R_L}$ is also used in high frequency, high output power amplifiers. Here, $\mathrm{P_A}$ is the output for class A amplifiers; $\mathrm{f_{10}}$ is the frequency at 10 dB gain; $\mathrm{R_L}$ is the load resistance. For GaN, FM is expected to be $1.2 \times 10^{23}\,\mathrm{WHz^2\Omega}$, a value 56 times that of GaAs.

The performance of high output power, high frequency, low loss devices can also be defined by the Johnson and Baliga figures of merit. As shown in Table 1.2, the values of the breakdown electric field, saturation drift velocity and electron mobility of nitride semiconductors are greater than SiC, again

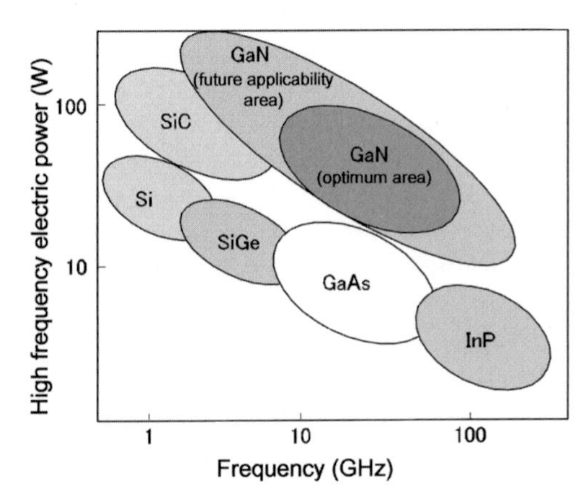

Fig. 1.9. Operating frequencies and output power of electron devices made using a range of materials

highlighting the potential of nitrides for power electronics [9, 10]. Figure 1.9 shows the frequency and output power of electron devices fabricated using a wide selection of semiconductors.

History and Present Status of Nitride Semiconductor Devices

Worldwide efforts to develop nitride semiconductor heterostructure electron devices started in 1991. In 1992, the formation of a 2DEG at the AlGaN/GaN interface was confirmed. Then, in 1994, the successful operation of HFETs with 11 GHz f_T and 35 GHz f_{max} was announced. This was immediately after the report of high brightness blue LEDs. After this, researchers in the USA led the world in the development of X-band (10 GHz) nitride semiconductor electron devices for high output radar applications. Initially, using sapphire substrates, the output power density of nitride devices at 2–3 W mm^{-1}, was only slightly greater than GaAs-based devices. Then in 1998, HFETs grown on SiC substrates showed output powers of 9–10 W mm^{-1}; performances that surpassed SiC electron devices. However, it was found that the power density of devices with larger gate widths was drastically reduced with a limit in total power of only ~10 W being attainable from a single chip.

In Japan, during this time, research was focused on the development of optical devices and groups in the USA had a 2–3 year lead on power devices. The initial areas of research on AlGaN/GaN heterostructures were: improvement of the interface, production of high quality GaN layers with high resistivity, the effect of an AlN spacer on electron mobility, surface passivation, formation of low resistance ohmic contacts, dry etching and formation of Schottky gates, ion implantation for device isolation, and electrodeposition of low resistivity wiring.

The main issues investigated for improving the power density and increasing the total power were the collapse of the drain current and the lower than expected breakdown field of devices. A trade-off was found where an improvement in the current collapse characteristics resulted in lower breakdown voltages and vice-versa.

The current collapse effect was found to be due to a charge trapped in the surface layer. Silicon nitride passivation has been shown to be effective in reducing this problem.

Improvements in the breakdown voltage were achieved using field plate and recessed gate structures to alleviate the concentration of electric field at the drain terminal of the gate. Suppression of gate leakage currents, by control of the Al composition and the thickness of AlGaN layer and improvement of surface morphology, is also important for improving the breakdown voltage.

A recent report described the fabrication of high power devices with a breakdown voltage of 120 V, and a high power density of 32.2 W mm^{-1}, at 4 GHz, on high resistance buffer layers grown on high quality SiC substrates [11].

Recently there have been tremendous improvements in the total power output in devices employing field plate/recess structures and an optimized Al composition of the AlGaN layer. These devices achieve 230 W (4.8 W mm^{-1}) output power in CW operation at 2 GHz with a 48 mm gate width [12]. This level of output power is only possible from combinations of 2–4 chips in the case of GaAs-based devices.

Further, Fujitsu has shown that a device structure with an n-GaN layer located below the gate is effective for improving breakdown voltages and reducing the current collapse . Devices with such structures combined onto two chips were reported to produce an output of 250 W [13]. Trial samples of these devices are available for WCDM amplifiers.

Figure 1.10 shows the recent dramatic increases in the high frequency power output of L-band (2 GHz) devices.

A power output of 5.8 W mm^{-1} has already been reported for 30 GHz, quasimillimeter band power amplifiers [14]. Figure 1.11 shows the variation of output power with frequency.

In a recent report, a device with a gate length of 0.06 μm, grown by MBE and with an SiN passivation layer deposited by CAT–CVD, resulted in 163 GHz f_T and 184 GHz f_{max} [15]. In another device structure, the piezoelectric effect of an InGaN strain layer buried underneath the GaN active layer was used to increase confinement of the 2DEG layer, and for a gate length of 0.16 μm, the device demonstrated 153 GHz f_T, 230 GHz f_{max} [16].

The problem of large gate leakage currents must be solved yet for fabrication of high frequency power devices. A MIS structure HFET was reported for reducing the gate leakage in a device with an Al_2O_3/SiN two-layered insulating gate structure with a length of 0.1 μm; the device showed 70 GHz f_T and 90 GHz f_{max} [17].

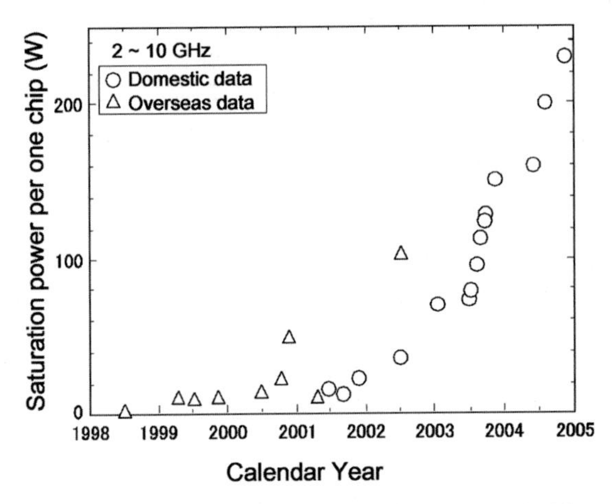

Fig. 1.10. Status of the development of high frequency nitride power devices

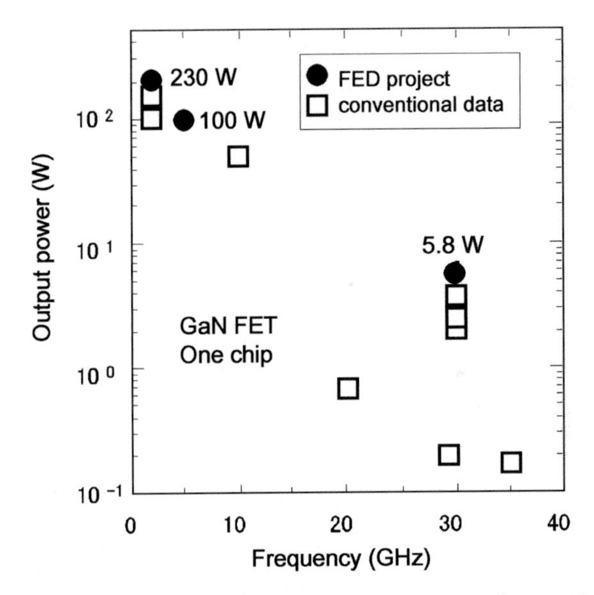

Fig. 1.11. Relationship between the RF power output and operating frequency

There have also been important advances in exploiting the low on-resistance, high-speed switching capabilities of nitride semiconductor power devices for use in inverters and other such power transformers. Nitride semiconductor devices fabricated on sapphire substrates have an on-resistance of $4\,\mathrm{m}\Omega\,\mathrm{cm}^2$ at 370 V, a quarter of the value for Si MOSFETS, and have been confirmed to operate at 20 A [18]. Devices grown on Si substrates have been reported to have an on-resistance of 1.9 mΩ cm^2 at 350 V, operating at 150 A [19].

The use of nitride semiconductors for fabricating heterobipolar transistors (HBT) is also being studied because of their large dynamic range, good linearity, and good uniformity in threshold voltages. Typically, npn-type HBT structures employ a low resistance p-type InGaN base layer and re-growth is used for fabricating an external base. The best figures of merit reported to-date for npn devices are: a maximum current gain of 3,000, 120 V breakdown voltage, 6.7 kA cm^{-2} maximum current density and 270 kW cm^{-2}maximum power output density [20]. For pnp, the corresponding figures are: a maximum current gain of 85, 7.7 kA cm^{-2} maximum current density, 232 kW cm^{-2} maximum power output density and 194 V breakdown voltage [21].

Remaining Issues and Trends

The performance of high power nitride devices operating in the 2 GHz band is considered as being satisfactory and the main areas of concern are reliability, production yield, and cost. Improvements in yield and reductions in cost will only be possible with a stable supply of high quality 4-in. substrates. Manufacturing of semi-insulating SiC substrates is not easy and there are concerns about the costs as well. Use of n-type SiC substrates is also being investigated because they are easier for growth and less expensive. Large area Si wafers and AlN, which has a large thermal conductivity, are also being considered as substrates.

Reliability issues will require improvements in the crystalline quality, surface passivation, ohmic and gate electrodes, and efficient heat dissipation. However, compared with conventional devices fabricated using Si and GaAs, there is no doubt about the superior performance of nitride semiconductor devices, which will eventually be used for third and fourth generation mobile phone base stations.

Nitride-based devices have already surpassed the performance of GaAs for devices operating in the millimeter band range, and are expected to be used in fixed wireless access systems and for high speed wireless communication between base stations.

Figure 1.12 shows other potential applications of nitride semiconductor devices such as satellite communication, the high speed intelligent traffic systems, home networks, and sensing.

The wide bandgap of nitrides also has potential applications for devices operating at high temperature. Si is limited to operation below about 150°C. There have been reports of nitride devices operating at 300°C, with the possibility of operation even at 400–500°C.

Normally-off devices are important for applications in power electronics, such as power sources.

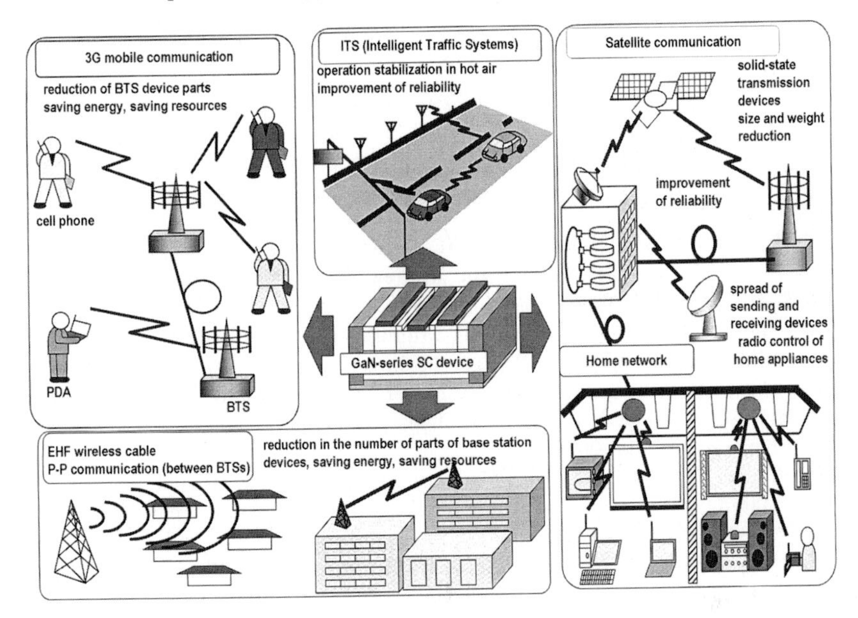

Fig. 1.12. Applications of electron devices fabricated using nitride compound semiconductors

References

1. W. Walukiewicz, S.X. Li, J. Wu, K.M. Yu, J.W. Ager, E.E. Haller, H. Lu, and W.J. Schaff: J. Cryst. Growth 269 (2004) 119
2. W. Walukiewicz: Physica B Condens. Matter 302 (2001) 123
3. W. Faschinger: in Wide Bandgap Semiconductors – Growth, Processing and Applications (S.J. Pearton, ed.), William Andrew Publishing/Noyes Publications, Norwich, NY, 2000, pp. 1–41
4. W.A. Harrison and J. Tersoff: J. Vac. Sci. Technol. B 4 (1986) 1068
5. H. Matsunami (ed.): Technology of Semiconductor SiC and Its Application, The Nikkan Kogyo Shimbun, Tokyo, 2003 (in Japanese)
6. K. Arai and S. Yoshida: Fundamentals and Applications of SiC Devices, Ohmsha, Tokyo, 2003 (in Japanese)
7. W.J. Choyke, H. Matsunami, and G. Pensl (eds.): Silicon Carbide – A Review of Fundamental Questions and Applications to Current Device Technology, Wiley, New York, 1997
8. W.J. Choyke, H. Matsunami, and G. Pensl (eds.): Silicon Carbide – Recent Major Advances, Springer, Berlin Heidelberg New York, 2003
9. S. Yoshida: OYO BUTURI 68 (1999) 787
10. S. Yoshida: IEICE Trans. Electron. (Japanese Edition) J86-C (2003) 412
11. Y.F. Wu, A. Saxler, M. Moore, R.P. Smith, S. Sheppard, P.M. Chavarkar, T. Wisleder, U.K. Mishra, and P. Parikh: IEEE Electron Device Lett. 25 (2004) 117

12. Y. Okamoto, Y. Ando, K. Hataya, T. Nakayama, H. Miyamoto, T. Inoue, M. Senda, K. Hirata, M. Kosaki, N. Shibata, and M. Kuzuhara: IEEE Trans. Microw. Theory Tech. 52 (2004) 2536
13. T. Kikkawa, T. Maniwa, H. Hayashi, M. Kanamura, S. Yokokawa, M. Nishi, N. Adachi, M. Yokoyama, Y. Tateno, and K. Joshin: in IEEE MTT-S Int. Microw. Sym. Digest (D.B. Denniston, ed.), Vol. 3, IEEE, Fort Worth, Texas, 2004, p. 1347
14. T. Inoue, Y. Ando, H. Miyamoto, T. Nakayama, Y. Okamoto, K. Hataya, and M. Kuzuhara: in IEEE MTT-S Int. Microw. Sym. Digest (D.B. Denniston, ed.), Vol. 3, IEEE, Fort Worth, Texas, 2004, p. 1649
15. M. Higashiwaki, T. Matsui and T. Mimura: The 6th International Conference on Nitride Semiconductors, V.A-9, 2005
16. T. Palacios, A. Chakraborty, S. Keller, S.P. DenBaars, and U.K. Mishra: 63rd Device Research Conference Digest, 2005. DRC '05, Vol. 1, IEEE, Santa Barbara, California, 2005, p. 181
17. N. Maeda, T. Makimura, T. Maruyama, C.X. Wang, M. Hiroki, H. Yokoyama, T. Makimoto, T. Kobayashi, and T. Enoki: Jpn J. Appl. Phys. Part 2: Lett. Express Lett. 44 (2005) L646
18. S. Yoshida, D.L. Wang, and M. Ichikawa: Jpn J. Appl. Phys. Part 2: Lett. 41 (2002) L820
19. M. Hikita, M. Yanagihara, K. Nakazawa, H. Ueno, Y. Hirose, T. Ueda, Y. Uemoto, T. Tanaka, D. Ueda, and T. Egawa: Electron Devices Meeting, 2004 – IEDM Technical Digest, IEEE International, San Francisco, California, 2004, p. 803
20. T. Makimoto, Y. Yamauchi, and K. Kumakura: Appl. Phys. Lett. 84 (2004) 1964
21. K. Kumakura, Y. Yamauchi, and T. Makimoto: Phys. Status Solidi C 2 (2005) 2589

2

Fundamental Properties of Wide Bandgap Semiconductors

2.1 Crystals and Band Structure

2.1.1 IV–IV Group Semiconductors

SiC (S. Yoshida)

Crystal Structure

Silicon carbide is a binary $A^N B^{8-N}$ compound with eight valence electrons per atom and as shown in Fig. 2.1a, the four nearest neighbor atoms form a regular tetrahedral crystal structure. Since Si and C are both group IV atoms, they are covalently bonded. However, according to Pauling [1], the differences in the electronegativity of Si and C results in the compound having ionicity of 12%. On the other hand, the ionicity according to Phillips [2] is $f_{\mathrm{i}} = C^2/{E_{\mathrm{g}}}^2 = 0.177$, where C is the ionic bonding energy gap (3.85 eV) and from ${E_{\mathrm{g}}}^2 = {E_{\mathrm{h}}}^2 + C^2$, E_{h} is the covalent energy gap (9.12 eV). By contrast, SiGe, another IV–IV group compound, is completely covalent and it is possible to produce alloys of all ratios of Si:Ge composition.

Many different polytypes of SiC exist, and have different physical properties from each other. Therefore, SiC is not a single semiconductor. The existence of the different crystal structures for the material with same composition is called *polymorphism*, and the term *polytypism* refers to one-dimensional polymorphism. When the tetrahedron structure shown in Fig. 2.1a is extended in the c-direction (corresponding to the [111] direction for cubic crystal structures and the c-axis direction for hexagonal crystal structures), the chemical bonding alternates as three bonds and then one bond. That is, a Si–C bilayer is formed by bonding in the c-direction with atomic positions as shown in Fig. 2.1(b) (cubic sites, k-sites) or (c) (hexagonal sites, h-sites). Crystals composed of only k-sites have cubic zinc blende (ZB) structures, while, those composed of only h-sites have hexagonal wurtzite (WZ) structures. When k- and h-sites are mixed, the crystal structure is either

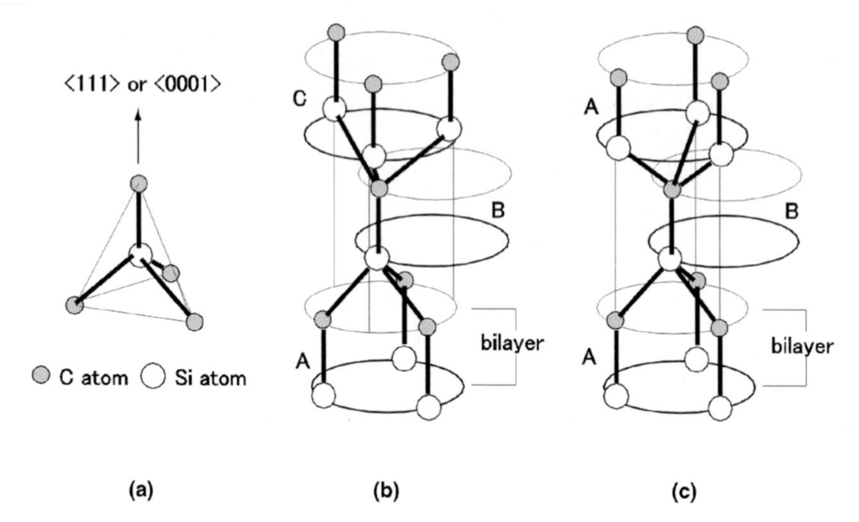

Fig. 2.1. (**a**) Tetrahedron structure; (**b**) cubic sites; (**c**) hexagonal sites.

hexagonal or rhombohedral. More than 200 different polytypes of SiC exist depending on the repetition period of the Si–C bilayer. Figure 2.2 shows the atomic arrangements on the $(11\bar{2}0)$ planes of 3C-, 2H-, 4H-, and 6H-SiC. Here C refers to cubic crystals and H to hexagonal. (R is used for rhombohedral structures.) The numerals indicate the number of bilayers including in one period. The structure in Fig. 2.2(a) is denoted as 3C, being composed of k-sites only, where one period consists of three bilayers, ABC. The hexagonal structures shown in (b), (c), and (d) have periods of AB, ABCB, and AB-CACB, respectively, and are each referred to as 2H, 4H, and 6H. The cubic 3C structure is sometimes denoted as β-SiC, while all other polytypes are designated α-SiC. The surface denoted by the arrow is called as *Si face*, and the other side as *C face*.

The reasons for the existence of large numbers of SiC polytypes has been theoretically studied by many researchers and it has been thought to be due to the small differences in the energy of hexagonal and cubic crystal layers. Thus even small energy fluctuations, resulting, for example, from growth temperature variations, affect the crystal structure. The energy for hexagonal stacking is slightly smaller than that for cubic stacking, though the later is smaller on the growth surface. Then high temperature annealing of cubic crystals results in the formation of hexagonal structures [3]. The hexagonality parameter, h, defined as the ratio of h-sites to all the sites composed, is used to understand the dependence of the properties on polytypes. For example, Choyke et al. [4] found the bandgap of SiC to have a linear relationship with the hexagonality. The h-values of 3C(kkk), 6H (kkhkkh), 4H(khkh), and 2H-SiC(hhh) are 0, 30, 50, and 100%, respectively.

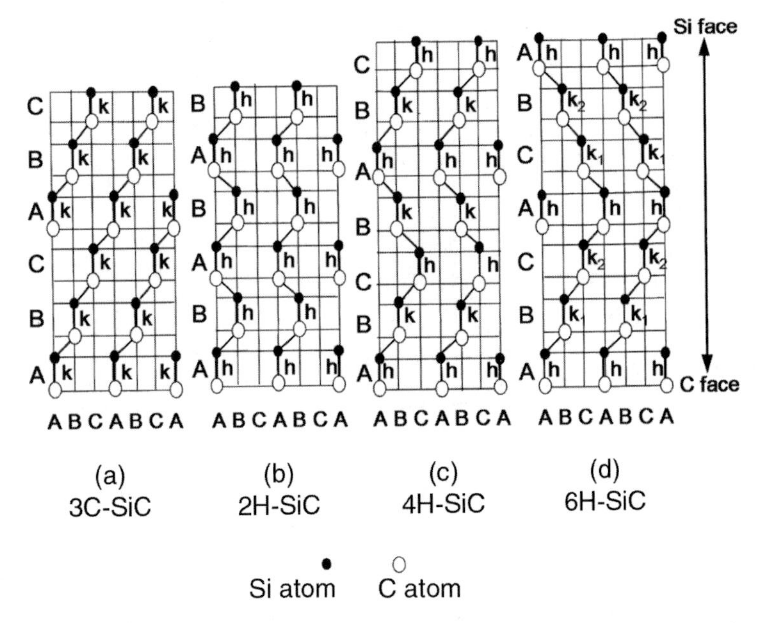

Fig. 2.2. The main SiC-polytype structures (atomic ordering viewed from the $(11\bar{2}0)$ surface)

Band Structure

The band structure of SiC depends on the particular polytype. The band structures of 3C, 2H, and other main polytypes, such as 6H, 4H, and 15R, have been calculated using first-principle methods [5]. The characteristics of the band structures of the polytypes can be understood by considering the zone folding effects to the band structure of 2H, because the polytypes can be regarded as the superstructures having long repetition period. The valence band maximum for all polytypes is at the zone center (Γ-point). However, the conduction band minimum is not the Γ-point for any polytype, so that all versions of SiC are indirect bandgap type semiconductors. The conduction band minimum is at the X-point for 3C (corresponding to 2/3 of the M–L line), near the M-point for 4H, 6H (in the M–L line), and the K-point for 2H. Figure 2.3 shows the variation of the bandgap of SiC polytypes with hexagonality. For polytypes with h less than 50%, the bandgap is seen to be almost proportional to h. The bandgap value for 2H-SiC is not on this linear line, because the energy at the K-point is lower than that at the M-point for large h. An increase in doping concentration results in SiC crystals with green/blue coloration. This is due to transitions within the conduction band, known as Biedermann absorption [6]. Based on these band structure calculations, the values for effective mass, impurity levels, and phonon energies have been derived, which agree well with experimental observations [7]. The impurity energy levels depend on whether the impurity atoms occupy at

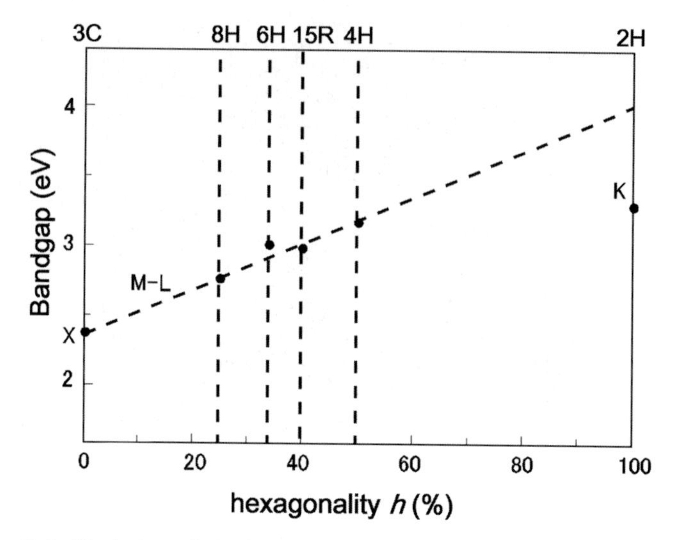

Fig. 2.3. Variation of the bandgap of SiC polytypes with hexagonality

h- or k-sites. There are two types of k-site, denoted as k_1 and k_2, as shown in Fig. 2.2(d).

Since the band structure of SiC depends on the polytype, the formation of heterostructure junctions using different polytypes results in band discontinuities at the interfaces. As the discontinuities of the conduction (ΔE_c) and valence band (ΔE_v) are different sign, the junction is type II [8]. The origin of the degradation of the forward voltage characteristics of α-SiC bipolar devices is thought to be due to the generation and migration of stacking faults in the SiC crystal. As the stacking faults can be regarded as 3C polytypes (that have smaller bandgaps) sandwiched into the α-SiC crystal, these 3C polytype regions form quantum wells that may act as electron traps, resulting in poor forward characteristics of devices [9].

Diamond (T. Ito, A. Hiraki)

Crystal Structure

The crystalline structure of diamond is shown in Fig. 2.4. The same structure is exhibited by Si and Ge. The lattice constant of diamond is smaller than those of the other materials with diamond crystalline structures. The sp^3 hybrid orbital bonding of C–C atoms results in an interatomic separation of 1.544 Å (approximately 66% of the Si–Si separation). According to Harrison [10], the covalent bonding energy (V_2) is 10.35 eV (∼2.3 times larger than that of silicon). Diamond crystals are extremely hard and have large coefficients of thermal conductivity. Table 2.1 lists the main physical properties of diamond [11].

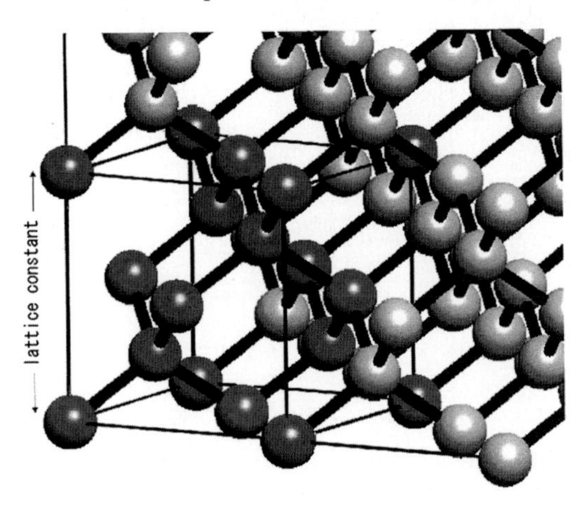

Fig. 2.4. Crystal structure of diamond

Table 2.1. Main physical property values of diamond (at room temperature)

bandgap (indirect)		5.47 eV
bond length/lattice constant		0.1544/0.357 nm
specific thermal conductivity		$20.9\,\mathrm{W\,cm^{-1}\,K^{-1}}$
sound velocity (longitudinal)		$1.82 \times 10^6\,\mathrm{cm\,s^{-1}}$
dielectric constant (static)		5.7
drift mobility[a]	electron	$2{,}400\ \mathrm{cm^2\,V^{-1}\,s^{-1}}$
	hole	$2{,}100\ \mathrm{cm^2\,V^{-1}\,s^{-1}}$
saturation drift velocity	electron	$1.50 \times 10^7\,\mathrm{cm\,s^{-1}}$
	hole	$1.05 \times 10^7\,\mathrm{cm\,s^{-1}}$
breakdown electric field		$1 \times 10^7\,\mathrm{V\,cm^{-1}}$
free-exciton binding energy		80 meV
optical phonon energy (at the conduction band edge)	TO	141 meV
	LO	163 meV

[a]The following large values of carrier mobilities were recently reported for undoped high-quality CVD diamond (electron: 4,500 $\mathrm{cm^2\,V^{-1}\,s^{-1}}$, hole: 3,800 $\mathrm{cm^2\,V^{-1}\,s^{-1}}$ [?])

Band Structure

As shown in Fig. 2.5, diamond has an indirect bandgap, E_g, of 5.470 ± 0.005 eV at room temperature (thermal coefficient $\mathrm{d}E_g/\mathrm{d}T = -5.4 \times 10^{-5}\,\mathrm{eV\,K^{-1}}$). The energy band structure near the conduction band minimum is similar to that of Si and can be approximated by rotational ellipsoids using longitudinal and transverse effective masses [12]. The direct bandgap is approximately 7.3 eV and associated with the Γ_{25}–Γ_{15} transition at the Γ-point. The electron mobility of high-quality diamond has been reported to be 2,400 $\mathrm{cm^2\,V^{-1}\,s^{-1}}$ at room temperature. The effective mass (ratio with respect to the free-electron

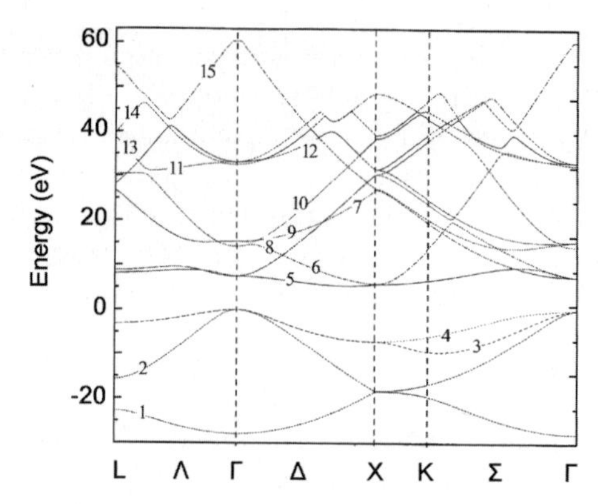

Fig. 2.5. Band structure of diamond determined by the pseudopotential method

rest mass) near the Γ-point is 0.3 for light holes while 1.1 for heavy holes. Further, the spin–orbit coupling induces band splitting, the energy difference of which is 0.006 eV at the Γ-point. The hole mobility at room temperature is 2,100 $cm^2 V^{-1} s^{-1}$. The binding energy of excitons is large due to the large E_g, with free excitons having a value of 80 meV. Clear excitonic emissions have been observed at room temperature from high-quality diamond films homoepitaxially grown on synthetic diamond (100) substrates by microwave-plasma chemical-vapor-deposition (CVD) methods [13].

Crystal Growth and Impurity Doping

Boron is widely used as a p-type dopant in diamond. At low concentrations, the acceptor level is located at 0.37 eV from the top of the valence band (E_v) [11]. Nitrogen forms a deep level in diamond (1.7 eV from the conduction band minimum, E_c) and does not behave as a donor at ambient temperatures [11]. In spite of recent reports that phosphorus forms a relatively shallow donor level at approximately 0.6 eV from the conduction band minimum for (111) surfaces [14] and that a limited P doping to (100) surfaces can be realized [15], it is still difficult to sufficiently control the n-type conductivity. The main reason should be related to the fact that E_c exists at nearly the same energy level as the vacuum.

Unique Characteristics

Figures of Merit

As listed in Table 2.1, diamond has a large breakdown electric field, high saturation drift velocities, high thermal conductivity, and low dielectric constant. These physical properties make diamond promising for high-power-density and high-frequency power devices. Further, the excellent chemical stability of diamond is being exploited for fabrication of devices applicable in extreme environments (e.g., high temperature and harsh radiation), sensors, optical windows, and electrodes in corrosive environments. Diamond layers are also being studied for applications including heat sinks for electronic devices or solid state lasers using diamond's high thermal conductivities, and optoelectronic devices or UV lithography masks exploiting diamond's optical transparency between the ultraviolet (UV) to infrared wavelengths. Surface-acoustic-wave (SAW) devices can also be fabricated using diamond.

Negative Electron Affinity

Since the conduction band edge, E_c, of diamond is located at almost the same level as the vacuum, the sign of the electron affinity depends on the structure of the surface layer. For example, the electron affinity is negative for hydrogen-terminated surfaces but positive for oxygen-terminated surfaces [16]. Such materials with negative electron affinities are essential as high-brightness emitters (cold cathodes) for displays and vacuum microelectronics. More details are given in Sect. 4.3.1 of this book.

Surface Conducting Layers

Terminating the diamond surface with hydrogen creates degenerate, p-type conducting layers. Explanations of the formation mechanism include transfer doping due to the adsorption of positive ions from the atmosphere and the presence of acceptor-like levels in the layer containing hydrogen [17]. Such conducting surface layers have been used for fabricating FET structures [18], details of which are described in Sect. 4.1.3 of this book.

2.1.2 II–VI Semiconductors

ZnSe Semiconductors (H. Saito)

Crystal Structure of Zinc Blende II–VI Compounds

In II–VI compound semiconductors, the ns^2 electrons of the outer orbitals of the group II atoms and the ns^2p^4 electrons (n is the principal quantum number) of the group VI atoms rearrange to form $(2\times)sp^3$ hybrid orbitals. Two adjacent atoms (A and B) share the two electrons resulting in each atom forming tetrahedral coordination with eight electrons in the outermost orbital. This configuration leads to highly stable atoms with strong covalent bonding. In semiconductors that are composed of only one atom, such as silicon, the maximum value of the electron distribution is located at the center of each atom. In the case of compound semiconductor crystals the electron distribution is localized at the atom with the higher electronegativity, the group VI atom in this case. That is, the group II atoms have positive ionicity and the group VI atoms negative ionicity.

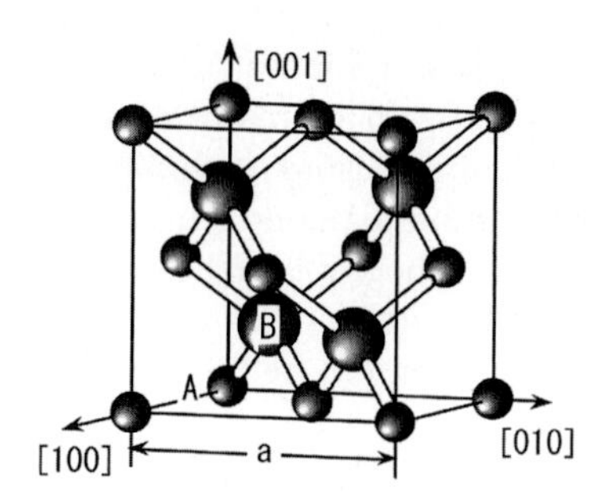

Fig. 2.6. Structure of zinc blende crystals

The zinc blende structure of ZnSe is shown in Fig. 2.6 and consists of two interpenetrating face-centered cubic sublattices with one displaced by a/4 in each direction from the other sublattice. Each atom is at the center of a regular tetrahedron formed by four atoms of the opposite type. Crystal structures with high ionicity are thought to be wurtzite in nature.

Energy Band Structure

The electronic structure of zinc blende crystals is composed of the p-like bonding states of the valence band and s-like antibonding states. Figure 2.7 shows a schematic representation of the energy band near the Γ-point of the Brillouin zone, which plays an important role in the design of device applications. Due to the free electron-like properties near the Γ-point, the electron energy is proportional to k^2 (square of the wavevector, $\boldsymbol{k}$). Figure 2.7b is for a zinc blende structure and Fig. 2.7a, 2.7c for structures under the action of external forces.

The bottom of the conduction band, Γ_6, has s-like properties and is doubly degenerate. The top of the valence band, Γ_8, has p-like properties and is fourfold degenerate. A doubly degenerate split state, Γ_7, resulting from the spin–orbit coupling, exists at an energy Δ(SO) beneath Γ_8. Away from the Γ-point ($\boldsymbol{k} = 0$), $\boldsymbol{k}\cdot\boldsymbol{p}$ perturbation theory splits the Γ_8 hole band into two – giving the heavy (HH, with heavy effective mass) and the light (LH, light effective mass) hole bands.

In Fig. 2.7b, the energy difference between the Γ_8- and Γ_6-points becomes the direct bandgap (E_g). Zinc Selenide and most II–VI compound semiconductors have direct bandgaps. However, dipole transitions are forbidden depending on the symmetry conditions between the conduction and valence bands. In semiconductors such as BeSe and BeTe, the bottom of the conduction band is not at the Γ-point but rather at the X-points (the band edge in the [100] direction of the Brillouin zone) while the top of the valence band is at the Γ-point. Such semiconductors have an indirect bandgap. For example, BeSe has a minimum bandgap of 2.4 eV, but a very large transition of 5.1 eV at the Γ-point.

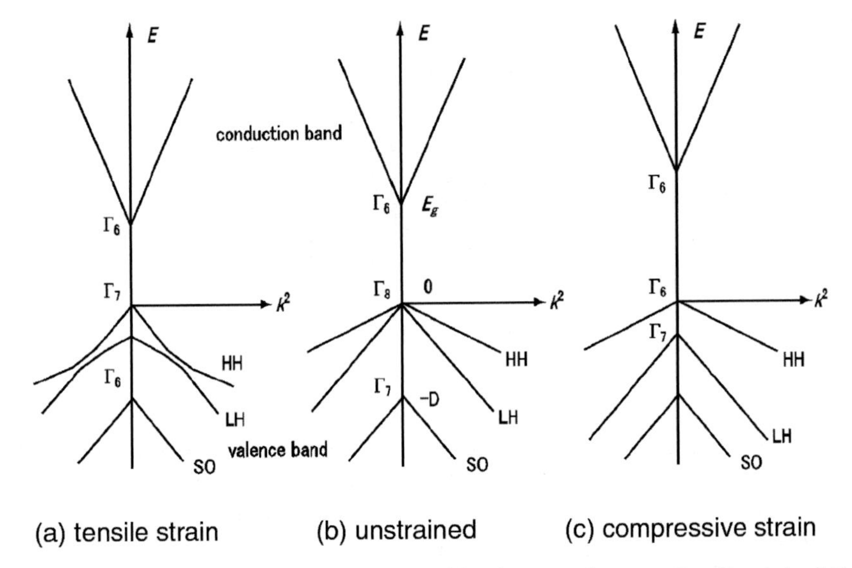

Fig. 2.7. Energy band structure of zinc blende crystals near the Γ-point of the Brillouin zone. (**a**) and (**c**) are for zinc blende structures under two-dimensional external tensile and compressive strain, respectively

Energy Band Structure of ZnSe Alloys

ZnSe has a room temperature bandgap of 2.7 eV ($\lambda \approx 460$ nm; blue-emission region) and high-quality films can be grown on GaAs substrates due to the small lattice mismatch of only 0.26%. Figure 2.8 shows the relationship between the lattice constants and bandgaps (E_g) of II–VI group semiconductors and III–V substrates used for their growth. ZnSSe and ZnMgSSe alloys are used to minimize the lattice mismatch between the epitaxial layers and substrates.

Since the radius of Be ions is extremely small compared with group VI ions (except for oxygen), localization of the electron distribution in the bond branches is low, resulting in even stronger covalent bonding. That is, these crystals are mechanically very robust. II–VI alloys containing Be have direct bandgaps only when Be densities are low, where the bottom of the conduction band is at the Γ-point as shown in Fig. 2.7b.

Differences in the thermal expansion coefficients of epitaxial layers and substrates create stress in such materials. For growth on (100) substrates, if the average lattice constant of the epitaxial film is smaller than that of the substrate, tensile stress acts in the (100) plane. Where the lattice constant of the epitaxial layer is larger than that of the substrate, the compressive stress results. In both cases, the crystal symmetry is reduced from T_d to D_{2d}, and Γ_8 splits into Γ_6 and Γ_7. The top of the valence band is Γ_7 in the case of tensile stress (Fig. 2.7a) and Γ_6 for compressive stress (Fig. 2.7c). For the Γ_6–Γ_6 transition, only the $\boldsymbol{E} \perp [100]$ properties ($\boldsymbol{E}$ is the electric vector of

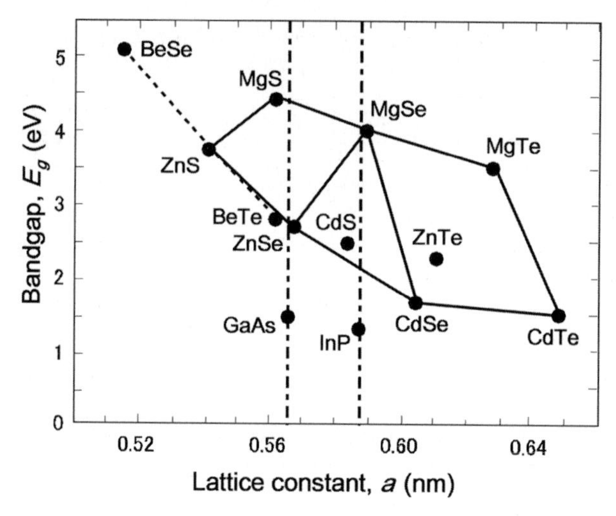

Fig. 2.8. Relationship between the lattice constants and bandgap energies of a selection of II–V-based and III–V-based semiconductors

light) are observable. For the Γ_7–Γ_6 transition, the mutual $\boldsymbol{E} \perp [100]$ and $\boldsymbol{E} \parallel [100]$ transitions are allowed. Optical polarization is observed in absorption and emission transitions.

ZnO Semiconductors (S. Fujita)

Zinc oxide semiconductors are expected to find applications for fabrication of excitonic optical devices, gas sensors, magnetic devices, and transistors. Further, combination of this material with CdO and MgO creates opportunities for bandgap engineering and multilayer structures. The basic parameters of the crystal structures and physical properties of these materials are shown in Table 2.2. The main problem with these alloys is that whereas ZnO forms a hexagonal crystal structure, CdO and MgO have stable rock salt (RS) structures, resulting in crystalline defects and phase separation. Thus the use of such materials for devices requires control of these properties.

ZnO is a direct bandgap semiconductor. The band edge at the Γ-point is triply degenerate, and for the case of GaN, is split due to the effect of the crystalline electric field and the spin–orbit coupling. Table 2.3 shows the atmospheric pressure bandgap, the pressure dependence of the bandgap, the exciton binding energy, and exciton energy levels at 6 K [20]. Further, the splitting energies due to the crystalline electric field, Δ_{cf}, and the spin–orbit coupling, Δ_{sc}, are 39.4 meV and –3.5 meV, respectively [20]. The negative value of Δ_{sc} is thought to be due to coupling with the Zn 3d-orbital [21–23].

The ZnCdO alloy also functions as a blue-emission layer because it has a bandgap smaller than ZnO. Phase separation has been observed for Cd contents greater than 10% due to the different crystal structures and bond

Table 2.2. Crystal structures and fundamental parameters of electronic properties of ZnO, CdO, and MgO

Material	ZnO	CdO	MgO
bandgap energy (RT) (eV)	3.37	2.3	7.8
exciton binding energy (meV)	59–63		
biexciton binding energy (meV)	15		
effective electron mass	0.24–0.28m_0		
effective hole energy mass (HH)	1.8m_0		
crystal structure	hexagonal	rock salt	rock salt
lattice constant (Å)	a = 3.250, c = 5.207		
bond length (Å)	1.99	2.35	2.1075

Table 2.3. Band parameters of ZnO at 6 K

valence band	atmospheric pressure bandgap (eV)	pressure dependence of bandgap (dEg/dP) (meV GPa^{-1})	exciton binding energy (meV)	exciton energy level (eV)
A	3.4410	24.7	63.1	3.37785 (1s), 3.4252 (2p)
B	3.4434	25.3	50.4	3.39296 (1s), 3.4308 (2p)
C	3.4817	26.8	48.9	3.4327 (1s), 3.4694 (2p)

lengths of ZnO and CdO [24, 25]. However, plasma-assisted metal–organic chemical-vapor-deposition (MOCVD) recently enabled the growth of large hexagonal crystals of ZnCdO with a Cd composition of 69.7% and a bandgap of 1.85 eV [26].

On the other hand, the larger bandgap of ZnMgO alloys compared with ZnO will enable its use for the growth of multiple quantum-well optical devices (lasers, photodetectors) operating in the ultraviolet region and heterojunction transistors. MgO has a stable rock salt crystal structure but if it were possible to produce hexagonal crystal structures, then one would expect that the Mg–O bond length (1.96 Å) would be closer to Zn–O bond lengths (1.99 Å) [21].Thus the use of high-quality ZnO buffer layers is expected to enable the growth of hexagonal crystal structures. Growth of large hexagonal crystals with Mg content of 51% (bandgap of 4.44 eV) has been reported and used for fabrication of solar-blind optical detectors [27], the two-dimensional electron gas (2DEG) ZnMgO/ZnO heterostructures [28] and transistors [29]. Further, hexagonal ultra-thin films of MgO have been grown on hexagonal ZnO substrates and used to fabricate MgO/ZnO superlattice structures [30]. Heterostructures consisting of ZnMgO/ZnO form type I band structures [21].

The lattice constant of ZnO is nearly the same as GaN (a = 3.180 Å, c = 5.128 Å) and thus studies are underway to grow GaN semiconductors on bulk ZnO substrates, and also to grow ZnO on bulk layers of GaN for the upper transparent layer of GaN optical devices.

2.1.3 III–V Nitride Semiconductors (Y. Ishitani)

Binary Semiconductors: GaN, AlN, and InN

Group III nitrides are compounds composed of elements from group III and nitrogen of group V of the periodic table. BN, AlN, GaN, and InN are examples of binary nitrides for B, Al, Ga and In. These crystals exhibit wurtzite, zinc blende, and rock salt structures.

The crystal structure depends on the degree of ionicity [31]. The potential energy related to the bandgap is divided into covalent and ionic bonding components. By considering the 90° difference in the phase parameter between these components, we see that the average energy bandgap, E_{ga}, is the sum of the covalent (E_{h}) and ionic bonding components

$$E_{\mathrm{ga}}{}^2 = E_{\mathrm{h}}^2 + C^2, \tag{2.1}$$

where C is the ionic bonding energy gap, and the ionicity is defined as:

$$f_{\mathrm{i}} = C^2/E_{\mathrm{ga}}{}^2. \tag{2.2}$$

Figure 2.9 shows the relationship between various crystal structures, E_{h} and C. RS structures are distributed in the upper area of the boundary shown by the line $f_{\mathrm{i}} = 0.785$. Although a straight line cannot represent the boundary of the area for ZB and WZ structures, crystals with high ionicity tend to show WZ structures. At room temperature and pressure, the fourfold coordination of WZ structures in AlN, GaN, and InN is stable and the fourfold coordination of ZB structures is quasistable. At high pressure, these compounds undergo a phase transition into the sixfold coordination of the RS structure.

As shown in Fig. 2.10, the symmetries of the WZ and ZB crystal structures can be divided into the $\mathrm{C_{6v}}^4$ and $\mathrm{T_d}^2$ space groups, respectively. The large spheres are the group III elements, and the smaller spheres are atoms of the group V element, nitrogen. Extensive literature discusses the differences between the cubic and hexagonal crystal structures [32–45]. Here, the effect of polarity on the crystal structures and planes will be discussed.

The WZ [0001] (c-axis) and ZB [111] axes are the equivalent axes, and polarization occurs along these axes because of differences in the electronegativities of the group III and group V elements. In the illustration showing the WZ structure, a thick hexagonal line surrounds the group III (+c) polarized (0001) surface. In cubic crystals, the {100} planes are nonpolarized, and are

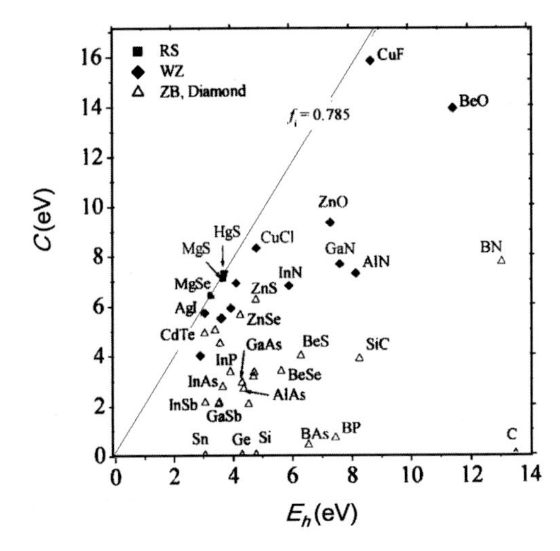

Fig. 2.9. Relationship between the covalent and ionic bonding components of various crystal structures [31]

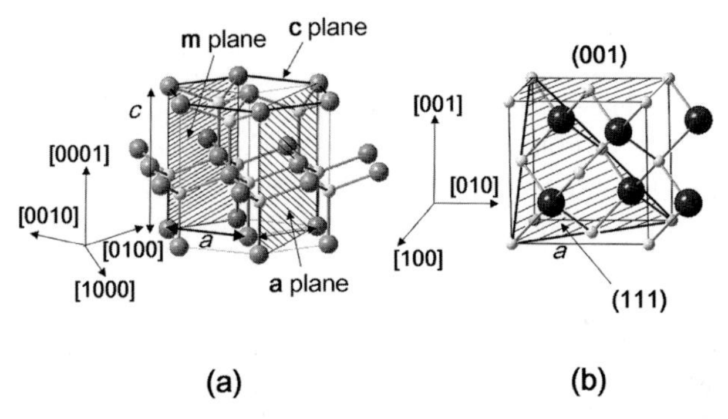

Fig. 2.10. Schematic diagrams of wurtzite (**a**) and zinc blende (**b**) crystal structures. These can be divided into the C_{6v}^4 and T_d^2 space groups, respectively. The large spheres are the group III elements and the smaller spheres are atoms of the group V element, nitrogen

usually used as crystal growth surfaces. Growth conditions for hexagonal crystals depend on the polarization (+c, −c) of the c surfaces. In heterostructures, interface charges are generated as a result of differences in polarization, leading to polarization fields within the crystal. As shown in Fig. 2.10, the m and a surfaces are not polarized. It is thought that internal electric fields are not generated when these surfaces are used for the growth of heterostructures. Now there are many of studies on the crystal growth and optical characterization of these surfaces for high-efficiency light-emitting device applications.

Table 2.4. Band parameters of binary group III nitride semiconductors

		Eg (eV) (WZ)	Δ_{cr} (meV)	Δ_{SO} (meV)	Eg (eV) (ZB)	Δ_{SO} (meV)
GaN	calculation		72.9[36]	15.6		20[36]
	experiment	3.39 (300 K)[35]	21[37]	16	3.302 (300 K)[39]	17[39]
		3.35 (1.6 K)[35]	22[38]	15		
AlN	calculation		−58.5[36]	20.4	5.11[41]	20[36]
	experiment	6.2 (300 K)[40]			5.34 (indirect)[42]	19[43]
		6.28 (15 K)[40]				
InN	calculation	0.81[44], 0.85[48]			0.59[44], 0.65[45]	
	experiment (300 K)	0.64[46], 0.63[47]				
BN	experiment				6.4 eV (indirect)[33]	

Band Structure

The hexagonal crystals of AlN, GaN, and InN all have direct-transition bandgap structures, whereas the cubic structures of AlN and BN exhibit indirect bandgaps. The bandgap values are shown in Table 2.4. InN was initially reported to have an energy gap of 2 eV, but since 2001, values from 0.67 (low-temperature) to 0.63 eV (room temperature) have been reported based on the optical properties of high-quality hexagonal InN epitaxial layers with low background electron densities grown by molecular beam epitaxy (MBE) [46, 47]. This bandgap is smaller than InP (1.35 eV), although AlN and GaN have very wide energy bandgaps. The small energy bandgap is a unique property of InN. The first-principle calculation of the band structure by the full-potential linearized augmented plane wave (FLAPW) method [32] was used by Suzuki et al. [33] to determine the structure of GaN assuming that the valence electrons in GaN are strongly localized at the N atoms, and also to point out that the quasipotential method is inappropriate.

The difference between AlN, GaN, and InN is the existence of d-orbital electrons in the latter two compounds that interact with the N2s electrons. Wei et al. [48] used the local density approximation (LDA) method for an analysis of InN and reported that the origin of its small bandgap resulted from the repulsion between the N2p and In4d levels, the energy difference between the In5s and N2s levels, and the large bond length compared with GaN and AlN. These factors were also reported by the authors to affect the bandgaps of ZnO and ZnSe. Further, InN was also reported to exhibit superconducting properties [49], results that illustrate the present incomplete understanding of InN's band structure.

Figure 2.11 shows the bandgap of WZ and ZB near the Γ-point. For WZ, the energy at the Γ-point, Γ_{15}, is split into three levels through the action

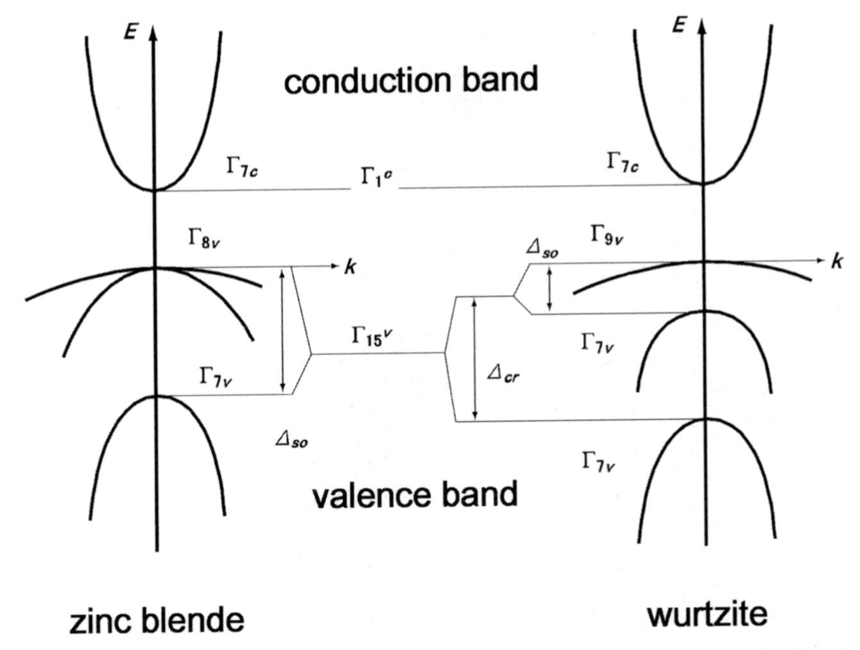

Fig. 2.11. Band structures of wurtzite and zinc blende structures near the Γ-point

of the crystal field and spin–orbit coupling. The total angular momentum J is 3/2 for $\Gamma_{9\mathrm{v}}$ and $\Gamma_{7\mathrm{v}}$ and 1/2 for $\Gamma_{7\mathrm{v}}$. The three resulting bands are called heavy hole (HH, A band), light hole (LH, B band), and crystal field hole (CH, C band), respectively. On the other hand, ZB is divided into the doubly degenerate $\Gamma_{8\mathrm{v}}$ ($J = 3/2$, HH, LH) and $\Gamma_{7\mathrm{v}}$ ($J = 1/2$, CH) because of spin–orbit splitting.

Spectroscopic ellipsometry has been used to determine the band structure of WZ-type InN [50]. Polarized photoluminescence (PL) and optical absorption measurements show the energy separation of these states to have the values for GaN at most [51]. Oscillator-strength and momentum matrix elements have also been studied [52]. Improvements in crystal quality and reductions in electron density will lead to more information about the valence band structure as well.

The temperature dependence of the bandgap energy is given by the semi-empirical Varshni formula (2.3) resulting in the following relationship [53, 54]

$$E_{\mathrm{g}}^{\Gamma}(T) = E_0 - E_{\mathrm{v}}(T) - \sum_{i=1}^{m} E_i \left(\frac{1}{e^{\theta_{\mathrm{p}i}/T} - 1} + \frac{1}{2} \right), \tag{2.3}$$

where $E_{\mathrm{v}}(T)$ is a compensation term for the effect of volume expansion and the third term is included as a secondary compensating term for the electron coupling to lattice vibrations (phonons). E_{i} shows the magnitude of the electron–phonon coupling. Generally, the third term is known to be the main factor affecting temperature dependence. For P-based semiconductors there is a change of 70–100 meV between ambient and cryogenic temperatures [54], compared with a value of 80 meV for the A band of GaN. The electron–phonon coupling is well represented by the sum of the self-energy and Debye–Waller terms. These two factors tend to partially cancel each other out. InN has been reported to have values ranging from 40 to 50 meV [46, 47].

AlGaInN Alloys

The difference between the lattice constants of AlN and GaN is 2.5%, and relatively high-quality $Al_xGa_{1-x}N$ alloys can be grown. The difference between the lattice constants of InN–GaN is 11.0% and that of InN–AlN is 13.5%. It is difficult to grow high-quality crystals of these alloys because of their strong immiscibility and large differences in optimal growth temperatures.

The lattice constants of the alloys obey Vergard's law in that they are proportional to the mole fractions of binary crystals. Apart, however, from some alloys such as AlInP [55], their bandgap is not proportional to the mole fraction. The existence of a so-called *bowing effect* is also known to be present in nitride semiconductors. The energy bandgap, $E_g(x)$, of a semiconductor composed of two semiconductor compounds, each with bandgaps E_{gA} and E_{gB}, can be approximated as

$$E_g(x) = E_{gA}x + E_{gB}(1-x) - b \cdot x(1-x), \qquad (2.4)$$

where x is the molar fraction and b the bowing constant.

The bowing parameter of $Al_xGa_{1-x}N$ has been reported to be 1.0 [56], 0.30 [57], and 0.82 eV [58].

The immiscibility of this alloy is less than for the other two, and the bowing constant is also small. For $Al_xIn_{1-x}N$, the bowing constants have been reported as b = 3.0–3.1 eV for a bandgap of 0.7 eV [59, 60] and as b = 4.78 (±0.3) eV for a bandgap of 0.65 eV [61]. For $Ga_xIn_{1-x}N$, reports show b = 1.4 eV at $E_g = 0.77$ eV [59], b = 2.5 eV at $E_g = 0.7$ eV [62], and b = 0.97 eV at $E_g = 0.85$ eV [63]. The crystalline quality of these alloys is poor, with electron densities greater than $10^{18}\,\mathrm{cm}^{-3}$ for $x \leq 0.5$. There is still no consensus about accurate values for the band edge, and there are still likely to be fluctuations in the reported values for the bowing constant.

Other Nitride Semiconductors

There is also interest in GaNAs and GaNP nitride semiconductors. Sharp optical emission at energies less than the bandgap has been observed for nitrogen-doped GaP. The emission is thought to be due to localized isoelectronic carrier traps, potential wells that form because of the large differences in the electronegativities of P (1.64) and N (3.00).

GaNP is known to have a large bowing constant. This is because of large differences in the atomic bond radii of the binary crystals and because of the stability of the WZ- and ZB-type structures. The bowing parameters at the X- and Γ-points of ZB-type crystals grown on GaP have been reported as 14 eV [64], 10 eV, and 3.9 eV [65]. Tight-binding calculations have yielded results that coincide with values of 3.9 eV [66]. PL and oscillator-strength studies [67] show hybridization of the electron wave functions at X–Γ resulting in an anticrossing.

The two binary crystals constituting GaNAs are direct bandgap semiconductors. Crystals can be produced for GaN mole fractions, x, of up to 0.15. Studies show that the transition from the isoelectronic trap band structure to alloy structures occurs at $x = 0.02$ [68]. According to analysis of electron states using first-principle calculations, the wave function of the band edge is localized at the As sublattice for the conduction band and at the N sublattice for the valence band. Further, it has been shown that for x = 0.125–0.5 the bowing parameter is in the range 7–16 eV [69, 70]. Experimentally, for $x \leq 0.15$ the bowing parameter has been found to range from 5 to 20 eV and to be dependent on the alloy composition. The nonlinearity factor has been reported to be $20.4x^2$–$100x^3$ eV [65,71].

2.1.4 Heterostructures and Band Structure (S. Sakai)

Band Discontinuities

In binary semiconductors, the bandgap is determined as shown in Fig. 2.12, but the magnitude of the band discontinuity is not. The band discontinuities can be deduced using the Harrison method [72], LMTO, LAPW, and PWP methods [73]. The Harrison calculation gives

$$E_v = \frac{\varepsilon_p^c + \varepsilon_p^a}{2} \sqrt{\left(\frac{\varepsilon_p^c - \varepsilon_p^a}{2}\right)^2 + V_{xx}^2} \quad V_{xx} = \frac{2.16\hbar^2}{md^2}, \tag{2.5}$$

where ε_p is the energy of the outer p-orbital, the c and a superscripts refer to cations and anions, respectively, m is the electron mass, d the interatomic separation. Values for ε_p have been reported by Herman and Skillman [74].

Figure 2.12 shows the variation of band discontinuities with interatomic separation for III–V semiconductors [75,76]. The valence band energy is seen to increase with increasing interatomic separation (the opposite behavior can also be found, as in the case of BAs and InN). These semiconductors exhibit a type I heterojunction and carrier confinement, while types II and III heterojunctions do not occur [72]. At the AlN/GaN interface, 27% (0.756 eV) of the bandgap difference occurs in the valence band and 73% in the conduction band. For GaN/InN interfaces, 43% of the bandgap difference is formed in the valence band and 53% in the conduction band. There is extremely good agreement between the calculated and experimental results for the discontinuities

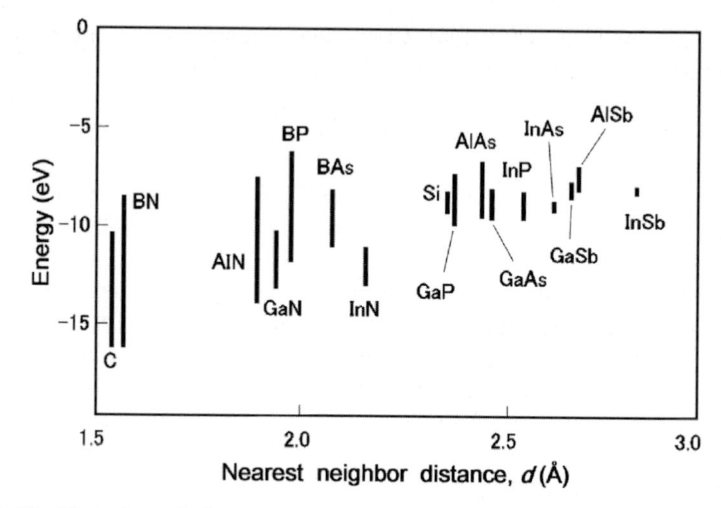

Fig. 2.12. Variation of the energy and interatomic separation for a selection of semiconductors

Table 2.5. Deduced interface band edge discontinuities (InN is assumed to have a bandgap of 1.9 eV, although recent studies have shown it to be about 0.7–0.8 eV)

interface	discontinuity (eV)
InN/GaN	0.93 ± 0.25
GaN/InN	0.59 ± 0.24
GaN/AlN	0.60 ± 0.24
AlN/GaN	0.57 ± 0.22
InN/AlN	1.71 ± 0.20
AlN/InN	1.32 ± 0.14

at these interfaces. But experimental values obtained by Martin et al. [77] show large differences for InN/GaN, GaN/InN, AlN/InN, and InN/AlN interfaces (Table 2.5). The differences are thought to be due to piezoelectric effects that arise with their growth on c-sapphire substrates.

Further, results published by King et al. [73], based on UPS, show that the band discontinuities of AlN on GaN and GaN on AlN are 0.5 ± 0.2 and 0.8 ± 0.2 eV, respectively. These values are for MBE growth on SiC. Hang et al. [78] reported a conduction-to-valence band ratio for $Al_{0.14}Ga_{0.86}N$/GaN quantum wells as 65%:35%. Data obtained by LMTO, LAPW, and PWP methods [73] also agree with these results. Results published by Morkoç [79] are in between the other studies.

Electrical Conduction at Heterojunction Interfaces

In optical devices, emission occurs over a surface area in the range $300\,\mu m^2 - 1\,mm^2$, with current flowing in a direction parallel to the substrate surface. The resistance of n- and p-type semiconducting optical devices can be calculated from Ohm's law, expressed as

$$V = IR = I\frac{1}{ne\mu}\frac{L}{S_1 S_2},$$

where I, n, e, μ, L, S_1, S_2 are the current, carrier density, electron charge mobility, and size of current path, respectively. For n-type semiconductors, $\mathrm{I} = 20\,\mathrm{mA}$ giving a voltage drop over the resistance in the range 0.01–0.1 V, and for p-type materials, about an order of magnitude smaller.

Figure 2.13 shows the variation of the operating voltage with wavelength of a UV LED. The bandgap, E_g/e, was deduced from the emission wavelength of an LED operating at 20 mA [80]. Although there is insufficient data for commercial LEDs, at the emission wavelength of 470 nm ($E_g/e = 2.64\,\mathrm{V}$, $I = 20\,\mathrm{mA}$), a first approximation for the voltage is in the range 2.8–2.9 eV. That is, for almost all nitride optical devices, the voltage is about 0.1–0.5 V larger than the value of E_g/e. If the ohmic contact resistance of p-GaN is assumed to be about $10^{-2}\,\Omega\,\mathrm{cm}^2$, then for a current of 20 mA, the voltage drop for a $300 \times 300\,\mu\mathrm{m}$ device will be 0.2 V. Thus the heterojunction does not have a significant effect on the electrical properties of LEDs. Further, for the 265 nm wavelength range, the voltage drop is a few volts larger than E_g/e which shows that there are still problems with the structure of electrodes, high n- and p-type doping and the formation of heterojunctions. Also, when insulating substrates such as sapphire are used, the n- and p-type electrodes are formed on Ga surfaces. In vertical device structures where the sapphire substrate is removed and current flows from the p- to n-type electrodes, such as lasers, it is necessary to optimize the electrodes to match the polarity of the top layer [72].

The band structure of field effect devices is shown in Fig. 2.14. A carrier density of $\sim 10^{13}\,\mathrm{cm}^{-2}$ can be obtained in the GaN layer even without doping the AlGaN barrier. Figure 2.14c shows theoretical results reported by

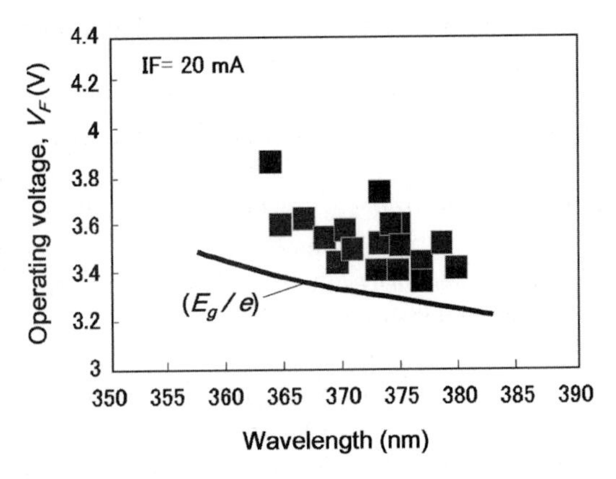

Fig. 2.13. Variation of the operating voltage with wavelength of a UV LED

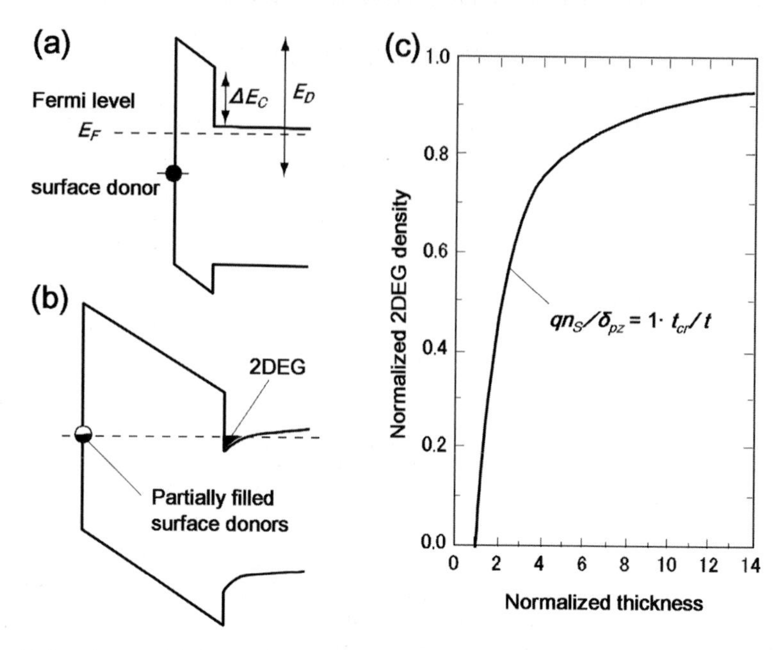

Fig. 2.14. (**a**) and (**b**) band structure of field effect devices. A carrier density of $\sim 10^{13}\,\mathrm{cm}^{-2}$ can be obtained in the GaN layer even without doping the AlGaN barrier. (**c**) shows theoretical results reported by Ibbetson et al. [81] on the effect of the thickness of the AlGaN layer on the density of the two-dimensional electron gas (2DEG) formed at an $Al_{0.34}Ga_{0.66}N$/GaN interface

Ibbetson et al. [81] on the effect of the thickness of the AlGaN layer on the density of the 2DEG formed at an $Al_{0.34}Ga_{0.66}N$/GaN interface. The density is seen to increase up to a normalized AlGaN layer thickness of 0.8, and saturate for greater values. These calculations are in good agreement with experimental results, showing that in FETs, electrons originate from surface states. FETs require an ohmic contact resistance of $10^{-7}\,\Omega\,\mathrm{cm}^2$ for optimal operation. Contact to n-GaN is usually made using a Ti/Al/Ni/Au alloy [82].

Formation of Electrodes and Work Functions of Metals

The contacts for n-type materials are readily formed without significant post-metal deposition processing. Mg is widely used for p-type doping. Generally, metals with large work functions result in a Schottky barrier contact, and ones with low work functions yield ohmic contacts. Figure 2.15 shows the variation of the barrier height with work function for n-GaN [82]. Here, rectifying electrodes are offered by Pt, Ni, Pd, Au, Co, Ru, Ag, and ohmic contacts come from Hf, Zn, Al, V. The use of Nb, Ti, Cr, W, Mo yield slightly rectifying contacts. The majority of optical devices employ Al/Ti [84] for forming n-type contacts. For FETs, Ti/Al/Ni/Au [82] and Ti/Al/Mo/Au [85] alloys are used.

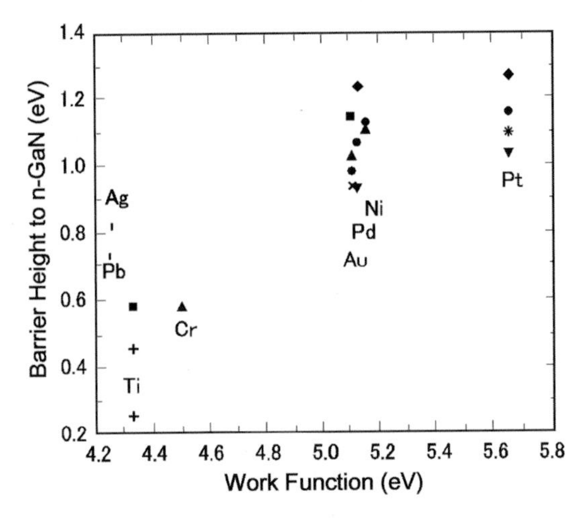

Fig. 2.15. Relationship between the work function and barrier height with n-GaN

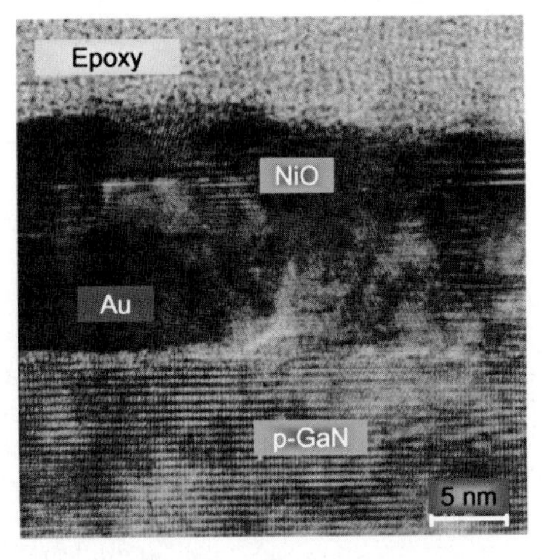

Fig. 2.16. Transmission electron microscope image of p-type GaN after annealing

Electrodes for p-GaN are formed by annealing Au/Ni in air [86], where 5 nm/5 nm, transparent electrodes have a typical contact resistance of $10^{-6}\,\Omega\,\mathrm{cm}^2$. Figure 2.16 is a transmission electron microscopy (TEM) image of an Au/Ni ohmic contact. The formation of the NiO and metal network is thought to be the reason for the low resistance. Further, contacts formed using Pt have low resistances of $\sim 10^{-5}\,\Omega\,\mathrm{cm}^2$ [87], but have limited lifetimes. Indium tin oxide (ITO) has low contact resistance when used with Ni/Au [88]. Contacts formed with Ni/Ag/Ru/Ni/Au have contact resistance of $5.2 \times 10^{-5}\,\Omega\,\mathrm{cm}^2$ and 91% transparency [89, 90]. Ru forms a barrier layer with Ag and Ni/Au layers.

Strained Band Structures

Layers of AlGaInN and GaInN grown on thick substrates are strained due to thermal and lattice strain. The band structure of ternary compounds can be approximated as

$$E_{AB} = (1 - x) E_A + xE_B - cx(1 - x),$$

where E_A and E_B are the bandgaps of the binary compounds and E_{AB} is the bandgap energy of the ternary compound.

GaN

Even a single layer of GaN is strained when grown on thick sapphire and SiC substrates. Figure 2.17 shows the relationships between the excitonic levels of the A, B, and C bands [91]. Compressive and tensile stress increases in the order, sapphire, GaN, 6H-SiC, and Si. Here, E_C, E_B, E_A are three bands related to A, B, and C. The strain is seen to depend on the type of substrate used.

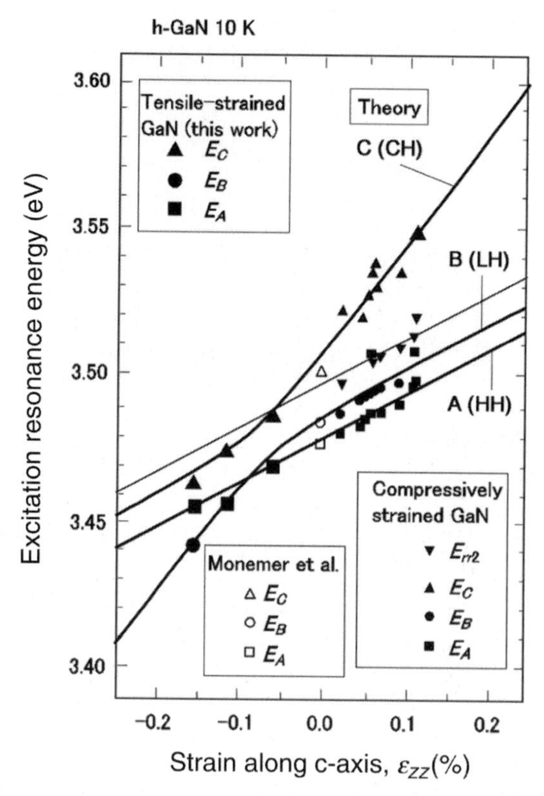

Fig. 2.17. Variation of excitonic resonant energy with strain in n-GaN at 10 K

AlGaN

The bandgap energy of AlGaN depends on the Al content, and at room temperature, ranges from 3.5 to 6.2 eV. The bandgap is direct for all compositions. Yoshida et al. [92] used EPMA (for composition) and optical transmission (the bandgap) techniques to characterize the bandgap bowing of AlGaN grown by MBE. Koide et al. [93] measured the bandgap of AlGaN by optical absorption and deduced the bandgap bowing to be given by $c = 1.0 \pm 0.3$ eV.

InGaN

InGaN is a ternary compound composed of InN and GaN. Recent studies have shown the bandgap of InN to be 0.7–0.8 eV, but no satisfactory value of c covers all compositions. Takeuchi et al. [94] have clarified the compositional relationship of the bandgap energy of a 40 nm layer of InGaN (In = 0–0.2) grown on GaN with compressive strain acting in the plane of the film. They determined the bowing parameter to be $c = 3.2$ eV. McCluskey et al. [95] measured the composition (X-ray diffraction) and bandgap (optical absorption) and obtained $dE_g/dx = 4.1$ (x < 0.12). Parker et al. [96] calculated the bowing for strained and unstrained layers to be 3.42 and 4.11 eV (x < 0.25), respectively. Their results are in close agreement with those reported by Takeuchi et al. [94].

2.1.5 Lattice Defects in Wide Bandgap Semiconductors (T. Miyajima)

Control and reduction of lattice defects is an essential aspect of the development of wide bandgap semiconductors.

Line Defects: Edge Dislocations, Screw, and Mixed Dislocations in GaN

High-quality GaN layers can be grown by MOCVD or MBE on c-sapphire and SiC substrates using AlN or GaN buffer layers [97,98]. Figure 2.18 shows a typical cross-sectional TEM image of a GaN layer grown on a sapphire substrate. Many types of lattice defects (stacking faults, c-surface defects) can be seen within a region of about 100 nm of the sapphire/GaN heterointerface. These defects result in complicated lattice relaxation effects at the substrate/epilayer interface. However, above 100 nm, only about 10^8–10^{10} cm^{-2} line defects (dislocations) remain [99–101]. The line defects can be divided into the following three types: edge dislocations, screw dislocations, and mixed dislocations [102].

Figure 2.19 shows the structure of edge and screw dislocations. The dislocation lines and Burgers vector cross in the case of edge dislocations and are parallel for screw dislocations. Edge dislocations result from introduction of

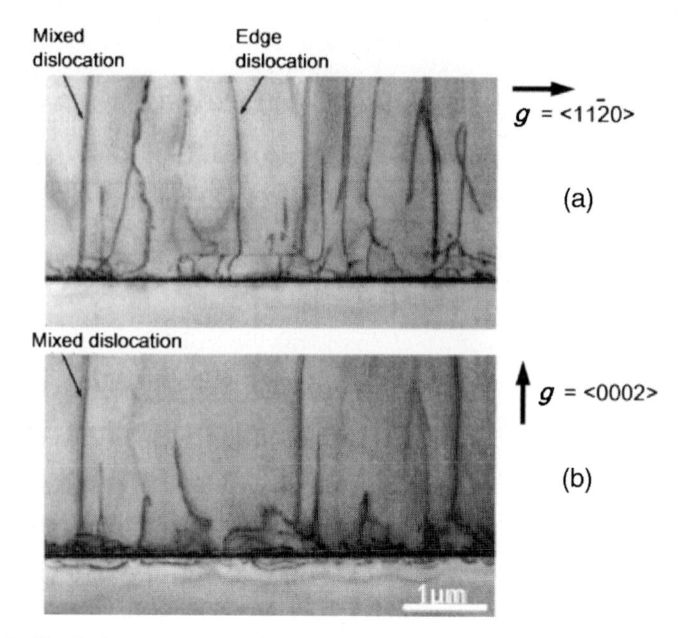

Fig. 2.18. Typical cross-sectional transmission electron microscopy (TEM) image of a GaN layer

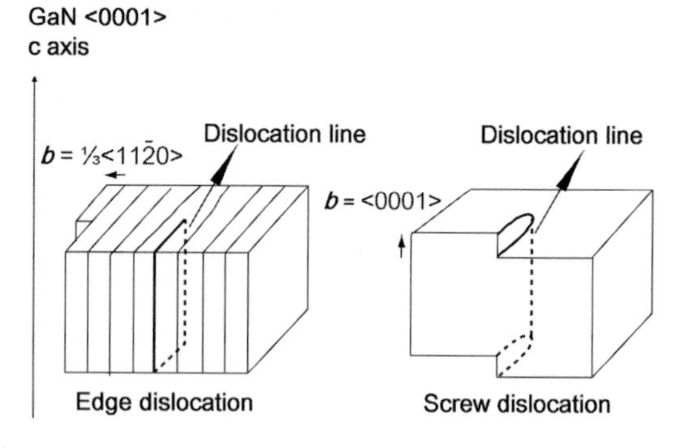

Fig. 2.19. Structures of edge and screw dislocations

an excess $(11\bar{2}0)$ plane and have a Burgers vector of $\boldsymbol{b} = 1/3 < 11\bar{2}0 >$. On the other hand, screw dislocations are formed due to the localized atomic steps and have a Burgers vector of $\boldsymbol{b} =< 0001 >$. Mixed dislocations are a combination of edge and screw dislocations and have a Burgers vector of $\boldsymbol{b} = 1/3 < 11\bar{2}3 >$.

The line dislocations shown here are formed when hexagonal crystal structures are grown on c-surfaces. GaN can exist as either the zinc blende (cubic) or wurtzite (hexagonal) crystal structures but stable layers have wurtzite structures. GaN is probably the first semiconductor with a wurtzite structure used for optical devices.

X-ray rocking curves are used for characterizing the crystalline structure of GaN. The half widths of the X-ray diffraction peaks are strongly dependent on the density of line defects. The FWHM of symmetry diffraction such as (0 0 0 2) or (0 0 0 4) broadens with increasing densities of screw and mixed dislocations which can increase the tilts of the c-axis of GaN. On the other hand, the FWHM of asymmetry diffraction increases with increasing densities of edge and mixed dislocations which can increase the twists of the c-axis.

Observation of Line Defects (Edge, Screw, and Mixed Dislocations)

TEM is extensively used for evaluating line defects. The need for the specialized equipment and long time necessary for preparing the samples are the main drawbacks for this method. The alternative and simpler method employing HCl dry etching enables the classification and observation of line defects in GaN.

Transmission Electron Microscopy

It is relatively easy to define the various types of line defects using the extinction condition that $\boldsymbol{g} \cdot \boldsymbol{b} = 0$, where $\boldsymbol{g}$ is the reciprocal lattice vector of the diffraction spot used for TEM observation and $\boldsymbol{b}$ the line defect Burgers vector. That is, as shown in the cross-sectional TEM image of Fig. 2.18b, the screw dislocation, with the Burgers vector $\boldsymbol{b} = <0001>$ crossing $\boldsymbol{g} = < 11\bar{2}0 >$, is not observed, but edge and mixed dislocations are observed. Further, as shown in Fig. 2.18a, the edge dislocation, with the Burgers vector $\boldsymbol{b} = 1/3 < 11\bar{2}0 >$ crossing $\boldsymbol{g} = < 0002 >$, is not observed, but screw and mixed dislocations could have been seen. However, the screw dislocations cannot be seen here because they have a much smaller density than edge dislocations.

Plan-view TEM observation (c-surface (0001)) – and not cross-sectional TEM (a-surface $(11\bar{2}0)$ or m-surface $(1\bar{1}00)$) – is used for the accurate measurement of the density of line defects. This is because the estimation of the line defect density depends on the thickness of the sample in the cross-sectional TEM observation.

Observation of Etch Pits by Etching with HCl Vapor [102–104]

The three distinctive types of etch pits shown in Fig. 2.20 were obtained by heating GaN to 600°C and etching its surface using HCl gas diluted with nitrogen. The dilution ratio, $[HCl]/[N_2]$, was 0.2. The etching time and rate were 30 min and 2 nm/min, respectively. The three types of etch pits were confirmed to correspond to edge, screw, and mixed line defects by TEM observations.

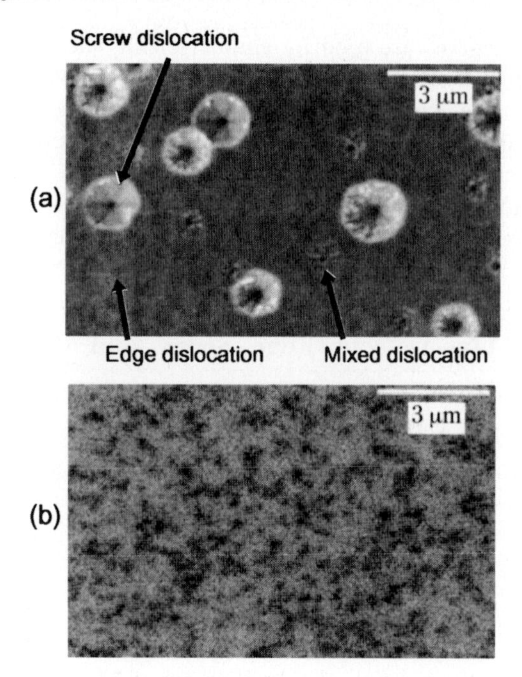

Fig. 2.20. (**a**) Three distinctive types of etch pits obtained by heating GaN:Si to 600°C and etching its surface using HCl gas diluted with nitrogen. (**b**) is the surface topography of the same GaN:Si sample observed by room temperature cathode luminescence

The following line defects are found in GaN layers grown under optimized conditions on sapphire substrates by MOCVD: edge dislocations $3.0 \times 10^8\,\mathrm{cm}^{-2}$ (96.0%), mixed $1.1 \times 10^7\,\mathrm{cm}^{-2}$ (3.5%), and screw $1.4 \times 10^6\,\mathrm{cm}^{-2}$ (0.5%). The numbers in the brackets show the density of the given type as a percentage of the total defect density. The defect densities depend on the tilt and twist of crystalline grains formed at low-temperature during the initial growth, and on the size of grains expanded upon subsequent annealing. The density ratios of these three types of line defects also strongly depend on the growth method [103, 105–107]. Raised-pressure MOCVD [108] is useful for growing GaN-based semiconductors with high crystalline quality because this growth method enables expansion of the size of single grains during temperature rising processes after growing low-temperature buffer layers and to reduce the density of the line defects especially screw and mixed dislocations.

Effect of Line Defects on the Optical Properties of GaN

Etch pits have been used to analyze the correlation between the type/density of defects (10^8–$10^{10}\,\mathrm{cm}^{-2}$) and PL integrated emission intensity from GaN films [102–104]. Because PL emission intensity in undoped-GaN is known to

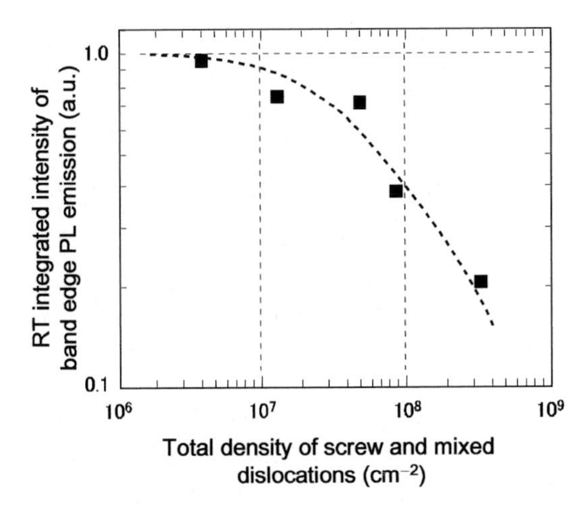

Fig. 2.21. Variation of integrated PL intensity and density of screw and mixed dislocations for GaN:Si sample

be determined by the concentration of residual impurities, GaN is intentionally doped with Si at a density of 2×10^{18} cm^{-3}. Figure 2.21 shows the variation of integrated PL intensity and density of screw and mixed dislocations. These results show that in spite of screw and mixed dislocations only being approximately 10% of the total density of line defects, they act as strong nonradiative centers. The effect of edge dislocations on emission from GaN could not be clarified since all the samples studied had the same density ($2–3 \times 10^8$ cm^{-2}) of edge dislocations. At this time, it was believed that, in GaN the mixed and screw dislocations (the minority in terms of dislocation density) are the only type of defects to act as nonradiative centers and edge dislocations (the majority in terms of dislocation density) may not act as nonradiative centers. This model was supported by the fact that in spite of the presence of high densities of crystalline defects in GaN, there have been many reports of high-brightness LEDs and room temperature emission from GaN-based lasers. Further, this model was in agreement with a theoretical prediction of Elsner et al. [109]. However, based on the high spatial resolution CL measurements shown in Fig. 2.20b, edge dislocations have also been found to act as nonradiative centers. Figure 2.20 shows room temperature CL images of an HCl-etched sample (a) and as-grown sample (b), cut from the same GaN:Si wafer. In Fig. 2.20a, only about ten screw and mixed dislocations each are seen, in contrast with the extremely high density of edge dislocations covering the surface. The topographic CL image of Fig. 2.20b shows nonradiative points (black points) with a density almost equal to that of the edge dislocations shown in Fig. 2.20a. These results lead to the conclusion that edge dislocations also act as nonradiative centers.

Recently, Fall et al. [110] theoretically (using a model that assumed a much larger crystal structure than that used by Elsner) showed that edge dislocations form mid-bandgap energy levels and act as nonradiative centers. Further, Northrup [111] showed, also theoretically, that screw dislocations act as nonradiative centers.

2.2 Optical, Mechanical, and Thermal Properties of Wide Bandgap Semiconductors

2.2.1 Optical Properties (Y. Yamada)

Optical Absorption

The main optical absorption processes responsible for electron transitions in semiconductors are inter- and intraband transitions, and transitions via impurities or lattice defects. This section describes the important aspects of interband transitions (fundamental absorption) and excitonic absorption.

Fundamental Absorption Edge

When a semiconductor is illuminated with light, electrons absorb photons and undergo transitions to higher energy levels. If the energy of the incident light ($\hbar\omega$) is greater than the bandgap of the semiconductor ($\hbar\omega \geq E_g$), then valence band electrons are excited into the conduction band. Optical absorption resulting from such band-to-band transitions is known as *fundamental absorption*, and the minimum energy at which such absorption takes place is called the *fundamental absorption edge*. The fundamental absorption edge energy is determined by the semiconductor bandgap.

The percentage of incident light absorbed by a particular semiconductor depends on the transition probability of electrons between the valence and conduction bands. The probability strongly depends on semiconductor band structures. The optical absorption intensity is experimentally measured in terms of the absorption coefficient (α). The photon energy dependence of the absorption coefficient is given by $\alpha \propto (\hbar\omega - E_g)^{1/2}$ for direct transitions (direct allowed transitions) and $\alpha \propto (\hbar\omega - E_g)^2$ for indirect transitions [112]. The rise of the absorption coefficient at the fundamental absorption edge is steeper due to the higher probability of interband electron transitions in direct bandgap semiconductors than in indirect band structures.

Excitonic Absorption

Attractive Coulomb forces act between optically excited electrons in the conduction band and the corresponding holes created in the valence band by optical absorption. Electron–hole pairs bound by Coulomb attraction are known as *excitons*, which can be thought of as excited states, where the Bohr radii for Frenkel excitons are similar to the crystal lattice constants, while Wannier excitons have Bohr radii of approximately ten times larger. Electron–hole pairs in wide bandgap semiconductors are Wannier excitons because the exciton Bohr radius in wide bandgap semiconductors has the value of a few nanometers.

The binding energy of an exciton is quantized in the same way as for an isolated hydrogen atom. Excitonic absorption results in a sharp line at the ground state energy (E_{n1}) given by, $E_{n1} = E_g - G_{ex}$, where E_g is the bandgap energy and G_{ex} is the exciton binding energy. Further, excitonic absorption lines corresponding to higher excited states ($n > 1$) are observed in high purity crystals. Table 2.6 shows the exciton binding energies of some representative wide bandgap semiconductors (G_{ex}). The large effective masses of electrons and holes, and small dielectric constant of wide bandgap semiconductors result in large exciton binding energies. It is possible to observe clear exciton absorption at room temperature in semiconductors where the exciton binding energy is larger than the thermal energy (kT = 26 meV).

Photoluminescence

Carriers excited in semiconductors return to their ground state after a given time (lifetime) via a process known as the *carrier recombination.* The carriers recombine via the process of radiative recombination (emitting light with energy equal to the difference between the ground and excited states), phonon emission, and nonradiative recombination through dissipation of thermal energy. Emission of light due to radiative recombination is known as *luminescence* and in particular, the optical emission due to optical excitation, is referred to as *photoluminescence.* This section describes excitonic radiative recombination processes observed in wide bandgap semiconductors.

Free Exciton Emission

Free exciton emission refers to the light emitted due to the recombination of free electron–hole pairs (free excitons). The process of optical emission due to free excitons can be understood by considering the interaction of excitons and transverse electromagnetic waves (polaritons). The polariton dispersion

Table 2.6. Bandgap energy (E_g) and exciton binding energy (G_{ex}) of typical wide bandgap semiconductors

	crystal structure	bandgap energy E_g (eV)	binding energy of exciton G_{ex} (meV)
ZnSe	ZB	2.7	17
CdS	ZB	2.5	27
ZnS	ZB	3.7	37
ZnO	WZ	3.4	59
GaN	WZ	3.4	25
AlN	WZ	6.2	44[113]–80[114]

ZB denotes zinc blende and WZ denotes wurtzite crystal structures

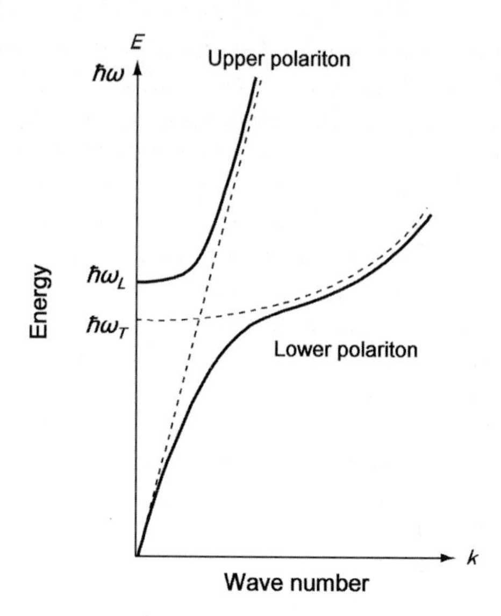

Fig. 2.22. Dispersion curve of excitonic polaritons

curve can be obtained by solving Maxwell's equations and consists of the upper polariton and lower polariton branches. These branches can be experimentally measured by resonant Brillouin scattering or two-photon Raman scattering through excitonic molecules as midstates. Figure 2.22 shows the dispersion curve of excitonic polaritons. Due to the small density of states of the upper branch, almost all the high-energy polaritons generated by optical absorption emit phonons along the lower branch and saturate near $\hbar\omega_T$. The electromagnetic wave component near the bottleneck $\hbar\omega_T$, and the upper branch (near $\hbar\omega_L$) is large, which implies a high probability of external photon emission outside the crystal. In high purity crystals, optical emission lines due to free excitons are observed as split exciton polaritons from both the upper and lower branches of the dispersion curve.

Free exciton emission lines are observed at energies that are less than the fundamental absorption edge by a value equal to the binding energy of the free excitons (G_{ex}). Exciton emission is usually observed from the ground state ($n = 1$). For high purity crystals, emission due to excited states ($n \geq 2$) is observed additionally. Under such conditions, it is possible to determine the binding energy of free excitons from the differences between the ground ($n = 1$) and excited ($n = 2$) exciton states.

Bound Excitons Emission

Excitons that are bound to impurities or lattice defects in semiconductors are called *bound excitons*. At low temperatures, emission lines due to bound excitons are observed at energies that are lower than free exciton emission. Bound exciton emission is represented by the following notation: neutral acceptor bound excitons (A^0, X) as the I_1 line; neutral donor (D^0, X) as the I_2 line; and ionized donor (D^+, X) as the I_3 line. Note that free excitons do not form bound states with ionized acceptors in wide bandgap semiconductors. For semiconductors in which the effective masses of electrons and holes are significantly different ($m_e^* << m_h^*$), the holes (the heavier of the two particles forming an exciton) move into close proximity to the ionized acceptors (A^-), with the result that the remaining electrons (lighter than the hole) only "see" the neutral impurities (A^0) and are not bound to the neutral impurities. According to theoretical calculations, the exciton-impurity binding energy depends on the ratio of the effective masses of electrons and holes ($\sigma = m_e^*/m_h^*$) [115, 116]. An exciton is not bound to ionized donors and acceptors simultaneously.

Haynes [117] experimentally observed the dissociation energy (D_0) of exciton-neutral impurity complexes to be approximately 10% of the impurity ionization energy. Thus according to the Haynes Law, $D_0(D^0, X) = 0.1E_D$ and $D_0(A^0, X) = 0.1E_A$, where E_D and E_A are the ionization energies of the donor and acceptor impurities, respectively. Emission lines from excitons bound to shallow donor and acceptor impurities (satisfying the conditions of the effective mass approximation) are observed at energies lower than the free exciton lines by an amount equal to the binding energy (D_0). The position of the emission line can be used in conjunction with Haynes Law to calculate the donor and acceptor levels. However, in actual crystals the binding energy depends on the chemical nature of the impurity. For example, in the case of excitons bound to donor impurities, the effect of short-range forces from donor impurities results in variations of the ground state (1s) for each chemical state. Thus it is necessary to correct the binding energy of (D^0, X) for each chemical state. The energy of the 1s ground state can be accurately obtained by central cell correction. The difference between the 1s and 2p states, $\{E_D(1s)–E_D(2p)\}$ can be deduced for each chemical state and the value of $E_D(2p)$ obtained using the effective mass approximation can be added. In photoluminescence, the value of $\{E_D(1s)–E_D(2p)\}$ can be determined by measurement of two-electron transitions. The two-electron transitions are observed when the final states of the optical transition for donor bound excitons are not 1s ground states but 2s or 2p excited states of the neutral donors. The binding energy of (A^0, X) can be analyzed in the same way as donor impurities by measurement of two-hole transitions.

Many-Body Effects of Excitons

In highly excited semiconductors, it is now well established that the interaction between two or more excitons causes the unique properties that are characterized by the formation of biexcitons (excitonic molecules) and the inelastic scattering of excitons [118]. Such characteristic phenomena of a dense excitonic system have been studied mainly in wide bandgap semiconductors because of the advantages of excitonic nature: excitons in wide bandgap semiconductors have relatively larger binding energies as well as smaller Bohr radii, so that the excitons exist stably at higher densities. As a result, the above phenomena are clearly observed in the dense excitonic systems of wide bandgap semiconductors.

A biexciton is a quasiparticle consisting of two excitons, i.e., two electrons and two holes in analogy to a hydrogen molecule. The optical transition responsible for the radiative recombination of biexcitons results from the disintegration of a biexciton into an exciton and a photon. The binding energy (G_m) of a biexciton is defined as the energy required for the dissociation of the complex into two excitons. Thus optical emission lines due to biexcitons appear at energies that are lower than free exciton emission by a magnitude equal to the binding energy of the biexciton. Table 2.7 shows the binding energy of biexcitons in wide bandgap semiconductors. The ratio of the biexciton binding energy relative to the exciton binding energy, G_m/G_{ex}, is found to be approximately 0.2 in these wide bandgap semiconductors. Theoretically, the value of G_m/G_{ex} depends only on the ratio of the effective masses of the electron and hole ($\sigma = m_e^*/m_h^*$) [119–121]. The magnitude of G_m/G_{ex} has been found to decrease with increasing σ and $G_m > 0$ in the range of $0 \leq \sigma \leq 1$. That is, stable biexcitons exist in semiconductors in which stable excitons are present. Many experimental reports show the energy of biexcitonic emission lines to be almost the same as that for neutral donor bound excitons. Measurement of two-photon biexciton resonance, which indicates the direct creation of biexcitons from a ground state by a two-photon absorption process, is used to identify emission lines due to biexcitons [122].

The radiative recombination processes of high density excitons are of interest in wide bandgap semiconductors for devices where the optical gain is important [123]. For example, the following advantages have been pointed out in the formation of optical gain due to radiative recombination of biexcitons [124]. First, as shown in Fig. 2.23, the energy level responsible for the radiative recombination of biexcitons essentially forms a four-level laser system, in which the population inversion is realized between biexciton and longitudinal exciton states. Also, polariton emission only occurs for excitons that have the same wavenumber vector of a photon. In contrast, it is possible for any biexciton above the dispersion curve to emit light by momentum transfer

Table 2.7. Binding energy of biexciton (G_m) and ratio of biexciton binding energy to exciton binding energy (G_m/G_{ex}) of typical wide bandgap semiconductors

	binding energy of biexciton G_m(meV)	G_m/G_{ex}
ZnSe	3.5	0.21
CdS	5.4	0.20
ZnS	8.0	0.22
ZnO	15	0.25
GaN	5.6	0.22

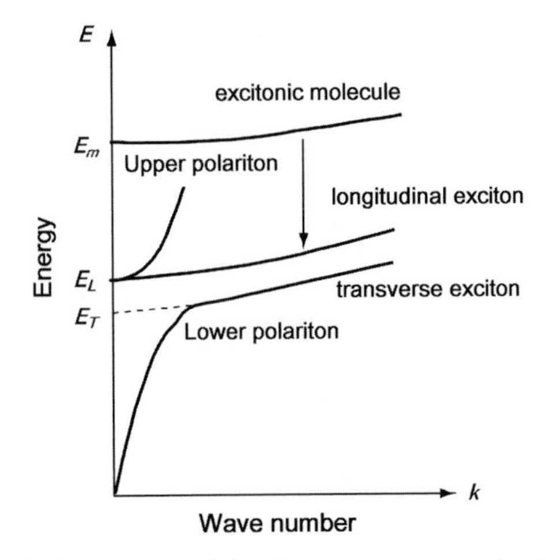

Fig. 2.23. Radiative recombination process of excitonic molecules

to one of the two excitons constituting the biexciton. This emission process has potential for generation of extremely large oscillator strengths, and as a result, it may be possible to fabricate lasers with threshold carrier densities of 2–3 orders of magnitude less than conventional devices based on the population inversion due to degenerate electron–hole plasma. Also, since free carriers are thought not to exist in "biexcitonic lasers," the differential quantum efficiency could be improved by eliminating losses due to the free-carrier absorption; there is thus potential for achieving an efficiency of 100%. To date, there have been no reports on the observation of optical gain due to radiative recombination processes of biexcitons at room temperature, although there have been reports of stimulated emission as a result of radiative recombination of biexcitons at cryogenic temperatures for ZnSe-based quantum-well structures at 120 K [125] and ZnS-based quantum-well structures at 250 K [126]. On the other hand, the optical gain due to inelastic scattering processes between excitons has been observed in ZnO (that has an extremely large exciton binding energy, $G_{ex} = 59\,\text{meV}$), with exciton scattering contributing to the stimulated emission at room temperature [127, 128].

Localized Exciton Emission

Spatial fluctuations in the composition of alloy semiconductors result in the inhomogeneous broadening of exciton density of states. Such an alloy broadening of exciton linewidth can be studied theoretically as a function of alloy composition [129]. The relatively small exciton volume (Bohr radius) in wide bandgap semiconductors leads to dramatic effects on the inhomogeneous broadening of the exciton linewidth [130, 131]. Excitons generated in such alloy semiconductors are localized at the low-energy side of the inhomogeneous broadening

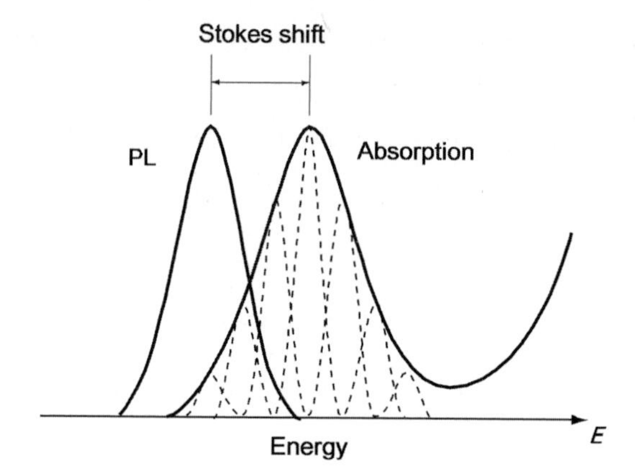

Fig. 2.24. Stokes shift and localized exciton emission

owing to excitonic energy relaxation processes. In such cases, as shown in Fig. 2.24, an energy difference arises between excitonic absorption and emission, which is defined as a Stokes shift of excitons. Both the inhomogeneous broadening and the Stokes shift of excitons are useful parameters in order to evaluate the degree of exciton localization.

The main characteristics of localized excitonic emission are seen in the emission energy dependence of the emission lifetime. At low temperatures, where excitons occupy localized states, the emission lifetime of the high-energy side of the localized excitonic emission band is short, and becomes longer as the emission band shifts to lower energies [130]. With increasing temperature, excitons undergo transitions from localized to delocalized states due to thermal energy, with decreasing differences of emission lifetimes in the low- and high-energy regions. At high temperatures the excitons become totally delocalized and the emission lifetimes are the same for all bands.

The first report on excitonic optical gain in quantum-well laser structures of wide bandgap semiconductors was based on the exciton localization in alloys [132]. Optical gain can be explained by considering a partial phase-space filling effect of localized excitons at the lower-energy states of inhomogeneously broadened exciton resonance.

Precise control of the bandgap energy is of vital importance for device applications. Almost all light-emitting devices incorporate active layers (emission layers) consisting of alloy semiconductors to control emission wavelengths. In light-emitting devices based on nitride semiconductors, $In_xGa_{1-x}N$ alloy thin films are used as active layers, where the emission wavelength is controlled by varying the alloy composition. It should be noted that the exciton localization contributes to the highly efficient light emission processes of the $In_xGa_{1-x}N$ alloy. However, in the case of $In_xGa_{1-x}N$ alloys, the localization mechanism of excitons cannot be explained by only considering alloy broadening, and more complicated models are required for an accurate understanding of highly efficient light emission processes. Details are presented in Chap. 3.

Methods for Measuring Refractive Indices and Dielectric Constants (Y. Kawakami)

The refractive index, dielectric constant, and other optical constants of semiconductors are usually expressed as complex numbers as shown below. These constants are important for the design of optical waveguides used in laser diodes (LDs).

The complex refractive index is defined by $n^* = n + i\kappa$, where the real part, n, is usually known as the refractive index. Also, the imaginary part, κ, is the extinction coefficient [133]. Also, if the complex dielectric constant is written as $\varepsilon^* = \varepsilon_\mathrm{r} + i\varepsilon_\mathrm{i}$, and the optical absorption coefficient as α, then the following equations are valid

$$\varepsilon_\mathrm{r} = n^2 - \kappa^2, \tag{2.6}$$

$$\varepsilon_\mathrm{i} = 2n\kappa, \tag{2.7}$$

$$\alpha = \frac{2\omega}{c}\kappa = \frac{\omega}{cn}\varepsilon_\mathrm{i}, \tag{2.8}$$

where ω is the angular frequency of light and c the velocity of light in vacuum. The important point is that since these optical constants appear as the total contributions of the various interband transitions, their magnitudes depend on the frequency. Since the Kramers–Kronig relationship is valid for the real and imaginary parts of the frequency dispersion of the complex refractive indices and dielectric constants, then if either one of them is known, it is possible to derive the other. That is, the following relationships hold:

$$\varepsilon_\mathrm{r} = 1 + \frac{\pi}{2}\int_0^\infty \frac{\omega'\varepsilon_\mathrm{i}(\omega')}{(\omega')^2 - \omega^2}d\omega', \tag{2.9}$$

$$\varepsilon_\mathrm{i} = -\frac{\pi}{2}\int_0^\infty \frac{\varepsilon_\mathrm{r}(\omega')}{(\omega')^2 - \omega^2}d\omega'. \tag{2.10}$$

Refractive indices can be experimentally determined using dispersion ellipsometry [134], transmission ellipsometry [135], and thermally induced methods [136]. Here dispersion ellipsometry is described where polarized light reflected from interfaces and surfaces is analyzed. If the parallel and perpendicular optoelectric reflection coefficients of polarized light are r_p, r_s, then the complex ratio ρ can be written as

$$\rho = \frac{r_\mathrm{p}}{r_\mathrm{s}} = \tan\psi \cdot e^{i\Delta} = \tan\psi(\cos\Delta + i\sin\Delta), \tag{2.11}$$

where tan ψ and cos Δ are known as ellipsometry parameters. Since dispersion ellipsometry is greatly affected by surface flatness and oxidation (contamination), the results must be analyzed accordingly. Figure 2.25 shows experimental fits to values of the high-energy refractive indices of $\mathrm{Al_xGa_{1-x}N}$ (x = 0–0.64) at room and high temperatures [137]. The refractive indices of

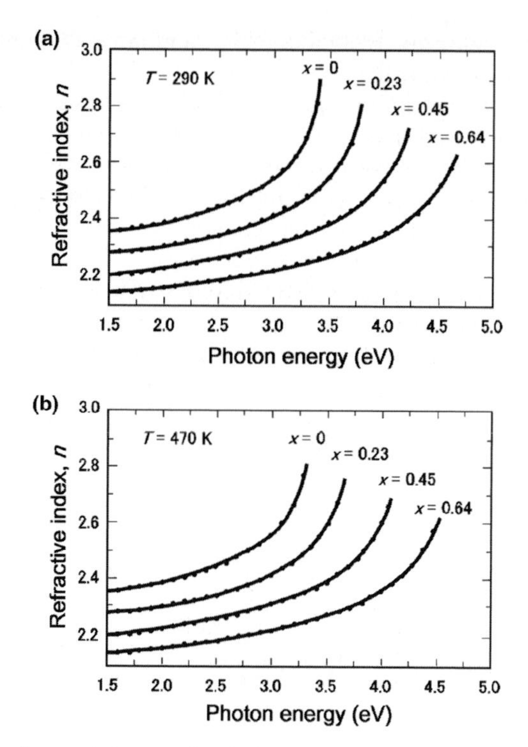

Fig. 2.25. Experimental fits to the high-energy refractive indices of $Al_xGa_{1-x}N$ (x = 0–0.64) at room (**a**) and high temperatures (**b**)

semiconductors show a tendency to increase with energy in the transparent region (where the extinction coefficient is close to zero) and this change is more pronounced near the absorption edge. Thus in the analysis of the wavelength separation between LD longitudinal modes, this dispersion relationship must be known accurately. Generally, dispersion characteristics of refractive indices can be approximated using Sellmeier's equation and many fitting functions have been reported based on considerations of alloy composition ratios and temperature dependence [133, 137, 138].

Strain and Polarization Effects in Semiconductors

Strain can be used to alleviate differences in lattice constants of heterostructures for the formation of coherent interfaces. The strain affects the semiconductor band structure and dispersion relationship of the bandgap width and the effective masses of carriers. Furthermore, electric fields are generated due to piezoelectric effects when hexagonal crystal structures (ionic crystals with low symmetry) are strained. For example, an $\mathrm{MV\,cm^{-1}}$ order electric field is induced in nitride semiconductor heterostructures grown in the c-axis direction [139]. This electric field affects the characteristics of high electron mobility transistors (HEMTs) and electron–hole recombination processes.

GaN-based semiconductors are four-coordinate, hexagonal crystals with bonding configurations as shown in Fig. 2.26. Here, Fig. 2.26a shows the case for unstrained regular tetrahedron crystals, where the net electric dipole moment is zero ($\mathbf{p}_1 + \mathbf{p}_2 + \mathbf{p}_3 + \mathbf{p}_4 = 0$). In contrast, in the presence of internal compressive strain, the Ga atom moves up and the N atom moves down changing the angle between the three base N atoms from θ to θ' ($\theta > \theta'$), the component of $\mathbf{p}_2 + \mathbf{p}_3 + \mathbf{p}_4$ in the x-direction increases, and $\mathbf{p}_1 + \mathbf{p}_2 + \mathbf{p}_3 + \mathbf{p}_4 \neq 0$. Thus the electric dipole causes the generation of a piezoelectric field from the surface toward the substrate direction (z-axis direction). Such directional electric fields due to a surface with Ga-polarity and compressive stress arise when InGaN/GaN quantum-well structures are grown on sapphire substrates by MOCVD. The piezoelectric polarization can be expressed by the following equation, where the electric field is seen to increase with increasing strain:

$$\begin{pmatrix} p_{\mathrm{x}} \\ p_{\mathrm{y}} \\ p_{\mathrm{z}} \end{pmatrix} = \begin{pmatrix} 0 & 0 & 0 & 0 & e_{15} & 0 \\ 0 & 0 & 0 & e_{15} & 0 & 0 \\ e_{31} & e_{31} & e_{33} & 0 & 0 & 0 \end{pmatrix} \begin{pmatrix} \varepsilon_{\mathrm{xx}} \\ \varepsilon_{\mathrm{yy}} \\ \varepsilon_{\mathrm{zz}} \\ \varepsilon_{\mathrm{yz}} \\ \varepsilon_{\mathrm{zx}} \\ \varepsilon_{\mathrm{xy}} \end{pmatrix} . \quad (2.12)$$

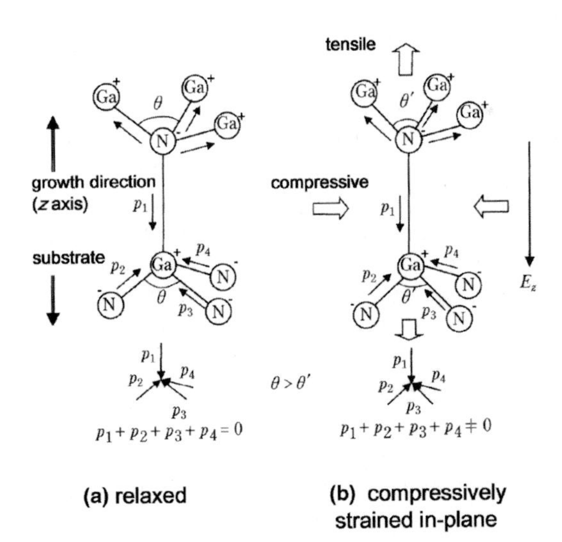

Fig. 2.26. Piezoelectric effects in GaN-based semiconductors. (**a**) In unstrained regular tetrahedron crystals the net electric dipole moment is zero. (**b**) With internal compressive strain a piezoelectric field results in a direction from the surface to the substrate (-z)

Here, p_i is the piezoelectric polarization, e_{ij} is the piezoelectric constant, and $\varepsilon_{\mathrm{ij}}$ is the magnitude of the strain (i, j = x, y, z). The following equations are valid for biaxial stress when the equivalent strain acts in the x- and y-directions:

$$\varepsilon_{\mathrm{xx}} = \varepsilon_{\mathrm{yy}} = -\frac{C_{33}}{2C_{13}}\varepsilon_{\mathrm{zz}}, \tag{2.13}$$

$$\varepsilon_{\mathrm{zz}} = \frac{c_S - c_0}{c_0}. \tag{2.14}$$

Here, C_{11} and C_{13} are elastic constants [140], and c_S and c_0 are the lattice constants along the c-axis for unstrained and biaxial strain (coherence length), respectively. From these relationships, the piezoelectric polarization, P_{z}, and the piezoelectric field along the c-axis, E_{z}, can be given by the following equations:

$$P_{\mathrm{z}} = e_{31}\varepsilon_{xx} + e_{31}\varepsilon_{\mathrm{yy}} + e_{33}\varepsilon_{\mathrm{zz}} = \left(e_{33} - \frac{C_{33}}{C_{13}}e_{31}\right)\varepsilon_{\mathrm{zz}}, \tag{2.15}$$

$$E_{\mathrm{z}} = -\frac{P_{\mathrm{z}}}{\varepsilon_{\mathrm{r}}\varepsilon_0}c. \tag{2.16}$$

Here, ε_r and ε_0 are the semiconductor relative dielectric constant and dielectric constant in a vacuum, respectively. The piezoelectric field generated in $\mathrm{In_xGa_{1-x}N}$ grown coherently on GaN can be approximated to be proportional to the In composition (x-value) assuming that the elastic and piezoelectric constants are invariant. The relationship $E_{\mathrm{z}} = k\mathrm{x}$ $(0 \leq \mathrm{x} \leq 1)$ is valid. There are many reports on piezoelectric coefficients, but due to large variations in the results [141,142], it can only be said that k is in the range of 5.6–16.9 MV $\mathrm{cm^{-1}}$ (Fig. 2.27) [143]. At the present time, it is difficult to determine more accurate values [144].

On the other hand, spontaneous polarization is known to arise in nonsymmetric crystals without strain. The magnitude of spontaneous polarizations in InN, GaN, and AlN have been reported to be 0.032, 0.029, 0.081 C $\mathrm{m^{-2}}$ [145], which is equivalent to electric fields of 2.5, 3.1, 8.5 MV $\mathrm{cm^{-1}}$, respectively. For Ga-polarized surfaces, the direction of spontaneous polarization in InN, GaN, and AlN is from the surface to the substrate, which is in the opposite direction to the aforementioned case in the presence compressive strain. In general, spontaneous polarization occurs uniformly within films. From film/substrate and air/film boundary conditions, the electric field occurs either at the surface or at interfaces. Thus, the effect of spontaneous polarization inside films is cancelled due to surface charge. In particular, the magnitude of the spontaneous polarization in GaN and InN is similar, which results in the spontaneous polarization being zero. In contrast, in compound structures such as AlGaN/GaN containing Al, spontaneous polarization has a major effect on the internal electric field.

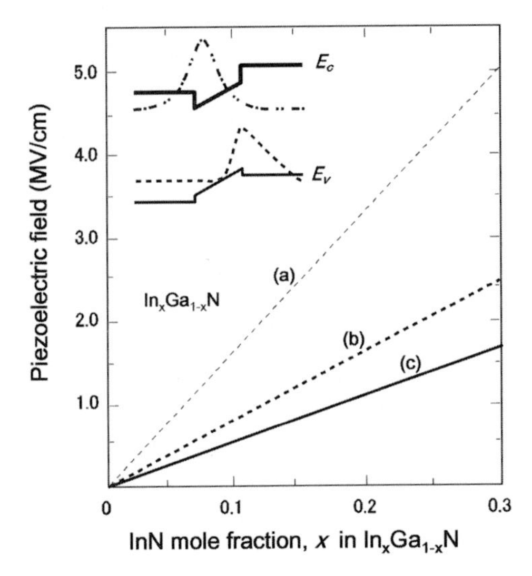

Fig. 2.27. In composition (x) dependence of the piezoelectric field generated in an $In_xGa_{1-x}N$ grown coherently on unstrained GaN

2.2.2 Quantum Structure of Wide Bandgap Semiconductors

Characteristics of Quantum Structures (I. Suemune)

Electrons are usually considered as having particle-like properties in bulk materials. As shown in Fig. 2.28a, this holds true when the size of the electron wave packet is much smaller than the dimensions of its surroundings. When the potential well formed between semiconductor heterojunctions narrows, the wave properties of electrons affect the microscopic properties of the semiconductor structure. For example, at room temperature, the electron wave packet has a spatial distribution of ~60 nm. When the relative width of a one-dimensional well becomes of a comparable size, the electron wave forms stationary waves in this direction and only its discrete states are allowed. This is called a "quantum–well" structure. Electrons can move freely along the two-dimensional directions perpendicular to the potential barriers.

A further dimensional constraint, as shown in Fig. 2.28c, allows electrons only to move in the one remaining dimension as the well width decreases. Such structures are called "quantum wires." Figure 2.28d shows the case for a "quantum dot" or "quantum box" structure, where the quantum-well width decreases in all three dimensions leading to the formation of 3-D standing waves and only discrete electron energy levels are allowed. Quantum dots are often referred to as artificial atoms because they consist of many atoms that resemble a quasiatom structure.

The electronic properties of a semiconductor depend on its density of states. As shown in Fig. 2.29a, the density of states in bulk semiconductors

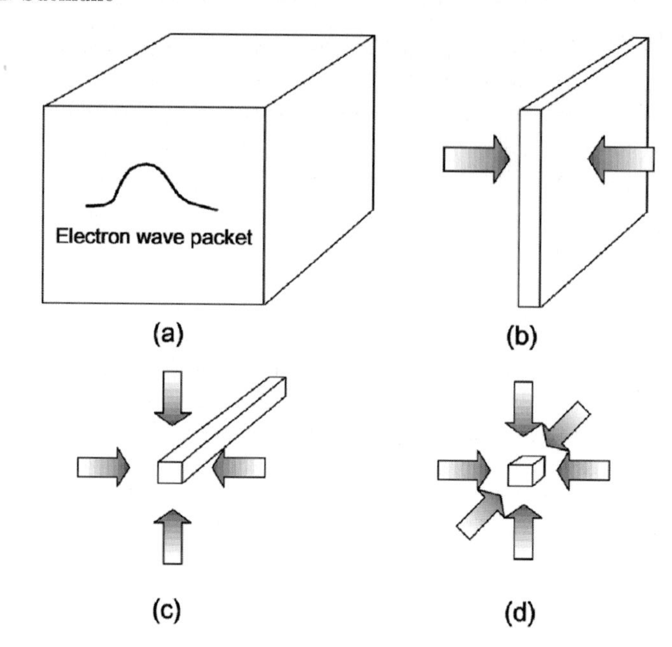

Fig. 2.28. Differing dimensions of semiconductor structures: (**a**) bulk structure, (**b**) quantum-well structure, (**c**) quantum wire structure, (**d**) quantum dot structure

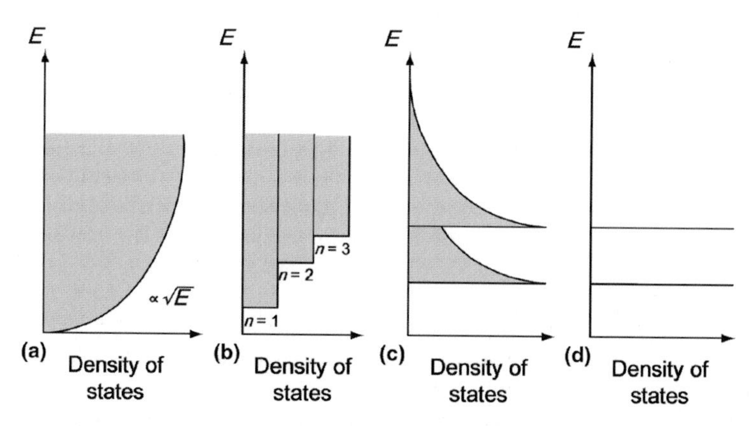

Fig. 2.29. Density of states of semiconductor structures with differing dimensions: (**a**) bulk, (**b**) quantum well, (**c**) quantum wire, (**d**) quantum dot

is proportional to $\sqrt{E}$, where E is the energy measured from the band edge. Thus the distribution broadens above room temperature and electrons with higher energy affect device operation.

As the degree of confinement increases from quantum wells (Fig. 2.29b) to quantum wires (Fig. 2.29c), the relative density of states in the high-energy region decreases, and the contribution of the electron density at the band edge increases. These effects enable a reduction in the electron density required for population inversion in devices such as lasers. However, since the position of the quantum level depends on the size of the quantum wells, the ability to control the width of the quantum-well precisely becomes a critical factor during the growth and fabrication of low-dimensional quantum structures.

Fabrication of Quantum Structures

Quantum Wells

The ideal means of controlling heterostructure growth is repeated, two-dimensional, atomic layer-by-layer deposition (Frank van der Merwe). Such two-dimensional growth can be monitored by the oscillations of the reflection high-energy electron diffraction (RHEED) signal measured during MBE process. RHEED has been used for monitoring the MBE growth of ZnO [146], laser ablation deposition of ZnO [147], and MBE growth of GaN, AlN [148].

Quantum Wires

Hexagonal wurtzite ZnO has been grown on submicron (50 nm) Au catalytic wires deposited on substrates at high temperatures of 800–900°C by the vapor–liquid solid (VLS) method [149]. However, 10 nm wide ZnO wires have been grown at low temperatures (400°C) without using such catalytic Au wires [150].

It is also possible to produce a wide range of "nanobelt" (ribbon) crystal structures by sublimation of oxide powders [151] without using catalysts. GaN columns have also been grown without catalysts [152].

Quantum Dots

Three-dimensional quantum dots are usually grown by self-assembly. The Stranski–Krastanov growth (S–K mode) has been used to grow III–V semiconductor quantum dots. In such structures, the initial two-dimensional (2-D) growth gives rise to 2-D compressive stress which leads to 3-D island growth and formation of dots on the surface. The formation of 3-D islands results in an increase in the area of stress-free surface, release of the accumulated strain energy, and minimization of the total energy of the semiconductor system. Many wide bandgap semiconductor dots such as CdSe [153] and GaN [154] are grown in the S–K mode.

Another method used for formation of quantum dots utilizes thin insulating films deposited on surfaces. GaN undergoes 2-D growth on AlGaN surfaces with low Al content but 3-D GaN dots grow on AlGaN surfaces covered in minute amounts of Si [155], where the Si acts as an "antisurfactant." ZnO has been reported to form islands on SiO_2 thin-film surfaces [156].

Properties of Quantum Structures and Applications

Quantum Structure, Excitons, and Molecular Exciton Optical Emission

Excitonic effects are very pronounced in wide bandgap semiconductors because of the large binding energy of excitons in these semiconductors (GaN = 28 meV, ZnO = 60 meV), the stability of excitons and the large oscillator strength of the optical transitions [157]. For example, room temperature excitonic emission has been observed in ZnO thin films as well as exciton–exciton scattering induced emission at 550 K [158].

Excitonic effects become even more pronounced in quantum structures. Upon reducing the width of ZnO/ZnMgO multiquantum wells to 1.8 nm, the exciton binding energy increases from 60 meV (bulk ZnO) to 105 meV, and the excitonic molecular binding energy from 15 meV (bulk) to 31 meV [159].

Optically excited emission at 385 nm has been observed at 300 K from ZnO nanowires with diameters of 70–100 nm and lengths of 2 μm [160]. Current-injection-induced laser operation was reported at room temperature for single CdS nanowires of 80–200 nm diameter [161]. The Fabry–Perot coupling mode was observed from the nanowire below the threshold as well as single mode oscillation above a threshold current of 200 μA.

Coupling of Semiconductor Quantum Structures and Optical Cavity

The 2001 Noble prize for physics was awarded for the verification of the Bose–Einstein Condensation (BEC) in alkaline atoms. This experiment conducted at an extremely low-temperature of 10^{-9} K confirmed Einstein's 1925 theory predicting condensation of Bose particles below a critical temperature. The increase of the critical temperature with decreasing particle mass has led to intense interest in the studies of excitons in semiconductors and polaritons in the quantum wells of ultra-small optical cavities. Especially the latter, which forms mixture states of excitons and photons due to the strong coupling of excitons and cavity modes, which can lead to the formation of polariton states with masses much smaller than that of excitons by a magnitude of 10^{-4}. This very small mass is expected to increase the critical temperature for the BEC up to the Kelvin level. Such strong couplings of excitons in quantum wells to microcavities based on CdTe material systems [162] and GaN material systems [163] have been observed. The strong coupling of CdS quantum dots to a microcavity was also reported recently [164].

Intersubband Transitions in Quantum Wells

Optical absorption in the 1.30–1.55 μm wavelength range has been observed from conduction band subband optical transitions in GaN/AlN multiquantum wells [165]. In nitrides, the short absorption recovery of ∼0.4 ps opens up possibilities for the realization of miniature ultra-high speed optical switches with pico-joule power dissipation.

Applications of Single Quantum Dots

GaN quantum dots grown on AlN are known to show a red shift up to the blue wavelength region (400 nm) due to the Stark effect [166] resulting from internal strain in the GaN dots. This internal field also influences the excitonic properties. Biexciton emissions are usually observed on the lower-energy side of exciton emissions. However, optical emission from excitonic molecules in single GaN quantum dots has been observed at higher energy (blue-shifted) than from excitonic emission [167]. This extraordinary phenomenon is explained by the reduced Coulomb attraction by the spatial separation of electrons and holes in a GaN dot and enhanced electron–electron and hole–hole Coulomb repulsion.

The lowest energy states in a quantum dot will be populated with two electrons in the conduction band and two holes in the valence band due to spin degeneracy. This leads to biexciton–exciton cascaded photon-pair generation. Antibunching photon characteristics have been observed up to room temperature with excitonic emissions by moderate excitation of CdSe-based quantum dots [168]. The generation of entangled photon pairs is expected with biexciton–exciton cascaded processes, and studies are in progress to solve the key issue of exciton energy splitting due to quantum dot anisotropies, which prevent the generation of entangled photon pairs [169]. The results of such investigations are expected to lead to advanced semiconductor quantum dot applications for quantum information processing, quantum encryption, and related optical communication technology.

2.2.3 Lattice Constants of Semiconductors (K. Hiramatsu)

Lattice Constants, Lattice Mismatch, and Lattice Relaxation of III–V and II–VI Type Semiconductors

Table 2.8 shows the lattice constants of III–V and II–VI compound semiconductors [170–172].

Lattice-Matched Systems

One example of a closely lattice-matched system is ZnSe grown on a GaAs substrate. The high vapor pressure of Se has made it difficult to produce bulk ZnSe wafers and so instead other substrates have been used as the base for growth of ZnSe thin films. The extremely small difference between the lattice constants of ZnSe (5.6681 Å) and GaAs (5.6533 Å), along with the fact that both materials have zinc blende crystal structures, has enabled the growth of ZnSe-based blue-emission optical devices on GaAs substrates.

Mismatched Systems

For lattice mismatched nitride semiconductors the emphasis is on problems related to lattice relaxation creating cracks, critical thicknesses, and the

Table 2.8. Lattice constants of wide bandgap semiconductors

family	materials	material structure	lattice constant (Å) a	c	reference
III–V	BN	h-BN	2.5040	6.6612	59
		R-BN	2.507	10.000	
		c-BN (ZB)	3.6160		
		W-BN (WZ)	2.5502	4.2131	
	BP	ZB	4.538		60
	BAs	ZB	4.777		
	AIN	WZ	3.112	4.982	59
		ZB	4.38		
	AIP	ZB	5.4625		60
	AIAs	ZB	5.6611		
	AISb	ZB	6.1355		
	GaN	WZ	3.8192	5.185	59
		ZB	4.52		
	GaP	ZB	5.4495		60
	GaAs	ZB	5.6533		
	GaSb	ZB	6.094		
	InN	WZ	3.548	5.70	59
		ZB	4.98		
	InP	ZB	5.8694		60
	InAs	ZB	6.058		
	InSb	ZB	6.478		
II–VI	ZnO	WZ	3.2407	5.1955	
	ZnS	WZ	3.82	6.26	
		ZB	5.4093		
	ZnSe	ZB	5.6681		
	ZnTe	ZB	6.1097		
	CdS	WZ	4.1368	6.7163	
		ZB	5.832		

differences in the thermal expansion coefficients between the substrate and epilayers.

Cracks and Critical Thickness

Sapphire substrates are widely used for the growth of nitride semiconductors. Early researchers found it difficult to obtain high-quality nitride films due to the large differences in lattice constant between sapphire and AlN (11.7%) or GaN (13.8%). However, the development of low-temperature AlN buffer layers enabled the growth of high-quality epitaxial films of GaN [173] and $Al_xGa_{1-x}N$ ($0 < x < 0.4$) [174] on sapphire substrates. Ito et al. [175] studied the growth of $Al_xGa_{1-x}N$ ($0 < x < 0.3$) on epitaxial GaN films. Figure 2.30 shows an SEM image of a 0.3 μm thick film of $Al_{0.1}Ga_{0.9}N$ grown on GaN at a growth rate of 0.1 μm min^{-1}. Cracks can be seen at the surface of the $Al_{0.1}Ga_{0.9}N$ film. Figure 2.31 shows cross-sectional SEM images of different

Fig. 2.30. SEM image of a 0.3 μm thick film of $Al_{0.1}Ga_{0.9}N$ grown on GaN at a growth rate of 0.1 μm min^{-1}

thickness $Al_{0.1}Ga_{0.9}N$ films grown on GaN substrates. The 0.15 μm films are free of cracks and have flat surfaces. Above 0.3 μm, cracks appear and above 1.5 μm v-shaped grooves are observed. These results demonstrate the existence of a critical thickness for AlGaN above which cracks appear. The cracks are thought to be due to the tensile stress created during growth because of the lattice constant of $Al_{0.1}Ga_{0.9}N$ being 0.25% smaller than GaN in the a-direction.

Lattice Relaxation and Differences in Coefficients of Thermal Expansion

Single-crystalline GaN grown by vapor phase epitaxy on sapphire substrates are known to contain cracks that arise due to stress that results from the large differences between the coefficients of thermal expansion of GaN and sapphire [176]. The coefficient of thermal expansion (CTE) of GaN is smaller than sapphire and as shown in Fig. 2.32, the film becomes warped with a convex contour. Further, as the GaN film becomes thicker, the action of the large tensile stress near the interface results in the generation of cracks that reduce the effects of residual thermal stress. The mechanism of stress relaxation at the GaN/sapphire interface was studied by Hiramatsu et al. [176, 177] who grew GaN films with the thickness in the range 1–1, 200 μm on sapphire substrates, using MOCVD and HVPE.

Figure 2.33 shows the thickness dependence of GaN's c-lattice constant. Changes due to thermal stress are seen in c for films that a few micrometers thick, but films thicker than 100 μm are almost completely relaxed with constant values of c. The lattice constant of strain free GaN is predicted to be $c_0 = 5.8150 \pm 0.0005$ Å. Figure 2.34 shows the relationship between the

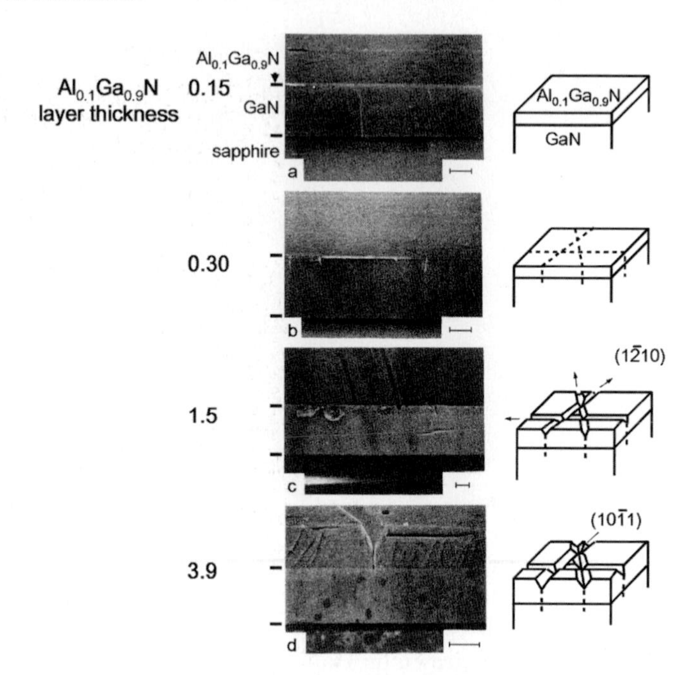

Fig. 2.31. Cross-sectional SEM images of different thickness $Al_{0.1}Ga_{0.9}N$ films grown on GaN substrates

lattice constants a and c. These results show that the GaN changes elastically with increasing layer thickness and that the lattice constant c decreases with increasing a. It can be assumed that under conditions of no strain, then the lattice constant a is $a_0 = 3.8192 \pm 0.0009$ Å.

2.2.4 Mechanical and Thermal Properties of Semiconductors (H. Kawanishi)

The four main nitride semiconductors – GaN, InN, AlN, and BN – have WZ and ZB crystal structures. BN also exists in rhombohedral (γ- BN) and hexagonal structures.

Epitaxial layers of these semiconductors are usually grown on sapphire, 4H- or 6H-SiC substrates at about 1,000°C using MOCVD or MBE. Lattice mismatches and differences in the thermal expansion coefficients lead to residual strain in the epilayers. Such strained layers show different electrical and optical properties than perfectly lattice-matched unstrained layers. Knowledge of the mechanical and thermal properties of these semiconductors is necessary for device design and fabrication.

Mechanical Properties

Table 2.9 shows the mechanical properties of nitride semiconductors.

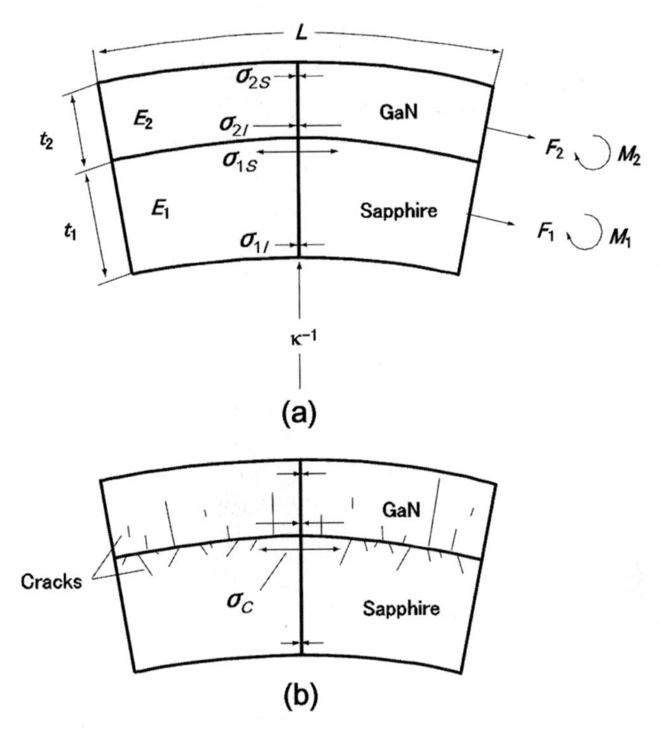

Fig. 2.32. Schematic diagram of the cross-section of GaN grown on sapphire substrates. The coefficient of thermal expansion of GaN is smaller than sapphire and the film becomes warped with a convex contour

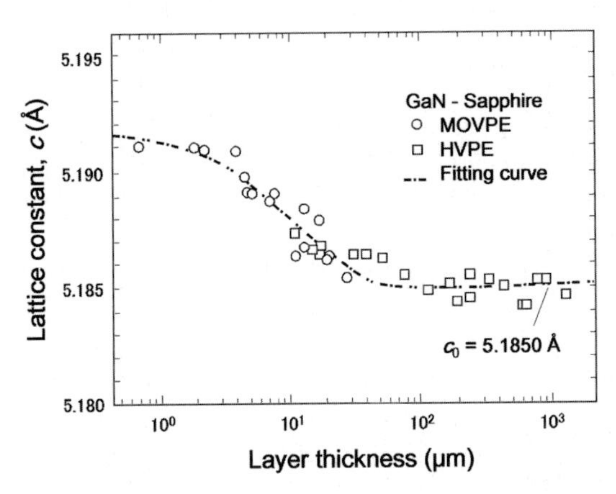

Fig. 2.33. Variation of the lattice constant of GaN (c) with its thickness

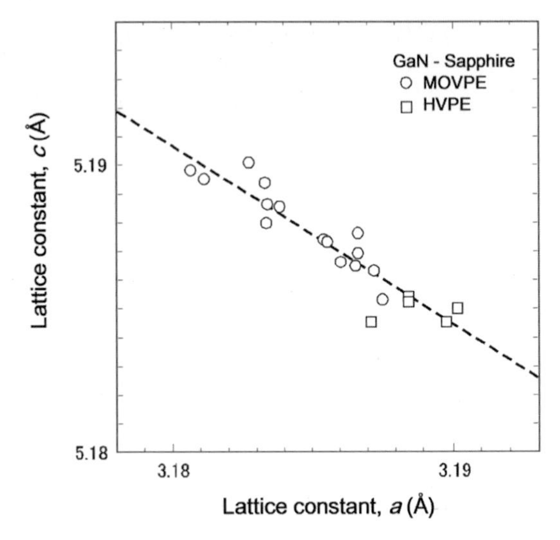

Fig. 2.34. Relationship between the lattice constants a and c of GaN

Constants of Elasticity (Elastic Stiffness Constants)

An elastic body deforms and becomes strained under the action of external forces (stress is defined as the force acting per unit cross-section area). Under such conditions the relationship between stress and strain is given by the elastic constant or elastic stiffness constant. Three-dimensional analysis of elastic bodies necessitates that the elastic constants be considered as "tensors." The relationship between stress, strain, and elastic constants is given by the equation

$$T_{ij} = C_{ijkl} e_{kl}. \tag{2.17}$$

Here, T_{ij}, e_{kl}, and C_{ijkl} are the stress (Pa), strain, and elastic tensor (Pa), respectively. The subscripts are indices with the values 1, 2, and 3 for the three Cartesian coordinate directions. Repeated indices are summed over. The stress and strain tensors are symmetric: $T_{ij} = T_{ji}$ and $e_{ij} = e_{ji}$.

In wurtzite structures the tensor notation is, c_{11}, c_{12}, c_{13}, c_{33}, c_{44}; in zinc blende it is c_{11}, c_{12}, c_{44}; and in hexagonal it is c_{11}, c_{12}, c_{33}, c_{44}.

Young's Modulus

Young's modulus is used to analyze elastic constants in linear, one-dimensional objects, such as wires and rods. Assume that, Δl is the change due to the action of a force, F, acting in the lengthwise direction of an elastic body with an original length, l. Then, if the cross-sectional area of the body is S, the stress is F/S, the strain $\Delta l/l$, and for small forces

Table 2.9. Mechanical properties of wide bandgap semiconductors

	crystal structure	elastic constant (GPa)	Young's modulus (GPa)	Poisson's ratio	bulk modulus (GPa)
GaN	WZ	c_{11} = 390 ± 15 (300 K)[68.75] c_{12} = 145 ± 20 (300 K)[68.75] c_{13} = 106 ± 20 (300 K)[68.75] c_{33} = 398 ± 20 (300 K)[68.75] c_{44} = 105 ± 10 (300 K)[68.75]	150[72.73]	0.20–0.30[74]	210 ± 10 (300 K)[71]
	ZB	c_{11} = 293 (300 K)[68] c_{12} = 159 (300 K)[68] c_{44} = 155 (300 K)[68]	181[71]	0.352[78]	204 (300 K)[71]
InN	WZ	c_{11} = 190[79] c_{12} = 104[79] c_{13} = 121[79] c_{33} = 182[79] c_{44} = 10[79]			1,651[74]
	ZB	c_{11} = 172[78] c_{12} = 119[78] c_{44} = 37[78]			
AlN	WZ	c_{11} = 410 ± 10 (300 K)[67.88] c_{12} = 149 ± 10 (300 K)[67.88] c_{13} = 99 ± 4 (300 K)[67.88] c_{33} = 389 ± 10 (300 K)[67.88] c_{44} = 125 ± 5 (300 K)[67.88]	308[69]	0.287 (0001)[70] 0.216 (1,120)[70]	210 (300 K)[69]
	ZB	c_{11} = 313[77] c_{12} = 168[77] c_{44} = 192[77]			216[78]
BN	h-BN	c_{11} = 750[76] c_{12} = 150[76] c_{33} = 18.7[76] c_{44} = 2.52[76]	22(//a)[80]		335[83]
	c-BN (ZB)	c_{11} = 819[77] c_{12} = 195[77] c_{44} = 475[77]	900[81]		382[85] 401[77]

Table 2.9. *Continued*

crystal structure	elastic constant (GPa)	Young's modulus (GPa)	Poisson's ratio	bulk modulus (GPa)
w-BN (WZ)	$c_{11} = 982$[77] $c_{12} = 134$[77] $c_{13} = 74$[77] $c_{33} = 1{,}077$[77] $c_{44} = 388$[77]	831[82]		392[86] 403[77]

$$\frac{F}{S} = \kappa \frac{\Delta l}{l}, \tag{2.18}$$

where κ is Young's modulus (Pa), which can be calculated using the elastic tensor, C_{ijkl}.

Poisson's Ratio

When an elastic body is stretched in one direction, for example in its lengthwise direction, the body is also deformed in the direction perpendicular to the direction of the imposed force. Poisson's ratio is defined as the ratio of the contraction strain normal to the applied load divided by the extension strain in the direction of the applied load. For an elastic body with an original length l and width w, the Poisson ratio σ is given by

$$\sigma = \frac{(\Delta w/w)}{(\Delta l/l)}, \tag{2.19}$$

where Δw is the change in the width, and Δl the change in the length, due to the action of a tensile force. Poisson's ratio can also be determined using the elastic tensor, C_{ijkl}.

Bulk Modulus

The bulk modulus, B, is defined as the relative change in volume of an elastic body due to a change in pressure

$$\Delta p = -B \frac{\Delta V}{V}. \tag{2.20}$$

Here, $\Delta V/V$ is the ratio of the change to the original volume and Δp is the change in applied pressure.

Thermal Properties

As described in Chap. 1 the growth of nitride semiconductors, such as GaN, InN, AlN, and BN, is carried out at high temperatures but the materials are used at room temperature, which is a temperature difference of ~1,000°C. For applications of nitrides as high-power optoelectronic devices, heat dissipation (thermal design) is an important aspect of device design and fabrication processes. Table 2.10 shows the thermal properties of a selection of nitride semiconductors.

Coefficient of Thermal Expansion

The change in volume of an elastic body when heated is expressed in terms of the material's CTE. The magnitude of CTE for nitride semiconductors is 10^{-5}–10^{-6} per °C. This value appears small but the effects are significant due to the high temperatures used for growing nitride semiconductors.

Surface cracks are known to result in epitaxial nitride films that contain residual strain due to tensile stress. On the other hand, cracks do not appear in cases of compression but piezoelectric effects give rise to carrier generation. The existence of such strain leads to warping and bowing of substrates, which affects device processing.

Differences of CTE also give rise to stress and strain between epitaxial nitride film layers having different compositions.

The CTE of hexagonal wurtzite semiconductors depends on whether the epitaxy occurs along the a- or c-axis.

Thermal Conductivity

The thermal conductivity of materials expresses the heat flux ($\mathrm{W\,m^{-1}\,K^{-1}}$) flowing through the material if a certain temperature gradient, ∇T ($\mathrm{K\,m^{-1}}$), exists in the material. The concept of thermal resistance, R_{th} ($\mathrm{K\,W^{-1}}$) is also used for the analysis of heat dissipation in optoelectronic devices.

Optoelectronic devices are usually mounted on heat sinks, which are materials – such as diamond, SiC, and c-BN – with small thermal resistances, i.e., large thermal conductivities.

Specific Heat

The specific heat is defined as the amount of heat per unit mass required to raise the temperature of a body by 1°C. Units for 1 g mass are $\mathrm{J\,K^{-1}\,g^{-1}}$ and for 1 mol are $\mathrm{J\,K^{-1}\,mol^{-1}}$.

Debye Temperature

In Debye's theory of the specific heat of solids, the Debye temperature (θ_D) is the temperature of a crystal's highest normal mode of vibration, i.e., the

Table 2.10. Thermal properties of nitride semiconductors

	crystal structure	coefficient of thermal expansion ($\times 10^{-6}\,\mathrm{K}^{-1}$)	thermal conductivity ($\mathrm{W\,m^{-1}\,K^{-1}}$)	molar specific heat ($\mathrm{J\,mol^{-1}\,K^{-1}}$)	Debye temperature (K)
GaN	WZ	$c = 3.1$[93]–3.7[93] (300 K) $a = 2.8$[93]–5.59[92] (300 K)	130[90]	38 (300 K)[94]	700 (T > 100K)[105]
InN	WZ	$c = 2.70$, $a = 3.40$ (190 K)[93] $c = 2.85$, $a = 3.75$ (260 K)[93] $c = 3.15$, $a = 4.20$ (360 K)[93] $c = 3.45$, $a = 4.80$ (460 K)[93] $c = 3.70$, $a = 5.70$ (560 K)[93]		41.87 (300 K)[102]	
AIN	WZ	$c = 5.27$ (20–800°C)[87,89] $a = 4.15$ (20–800°C)[87,89]	285[90]	34.87 (350 K)[105]	962 (T < 300K)[105]
BN	h-BN	$c = 40.5$ (300°C)[95] $a = -2.9$ (300°C)[95]	63[96]	19.92 (300 K)[101]	410[103]
	r-BN			20.71 (300 K)[101]	
	c-BN (ZB)	1.9 (300 K)[97]	700[99]	16.03 (300 K)[101]	1,610[104]
	w-BN (WZ)	$c = 2.6$ (300°C)[98] $a = 2.0$ (300°C)[98]	60[100]	16.51 (300 K)[101]	1,460[98]

highest temperature that can be achieved due to a single normal vibration. The Debye temperature is given by

$$\theta_D = \frac{h\nu_D}{k}, \tag{2.21}$$

where h is Planck's constant, k is Boltzmann's constant, and ν_D is the Debye frequency. The isovolumetric specific heat is proportional to the Debye temperature and is inversely proportional to the material's absolute temperature. The proportionality constant is $3Rf_D$, where R is the gas constant and f_D

the Debye function. Materials with high lattice oscillation frequencies, i.e., with strong interatomic bonding, are known to be hard materials with large Debye temperatures. Further, materials with short atomic radii and short interatomic distances are physically robust and have large Debye temperatures.

2.3 Electrical Properties of Wide Bandgap Semiconductors

2.3.1 Doping Technology (H. Amano)

Group III nitrides are extensively used for fabrication of commercial optoelectronic devices. This section describes (1) technology used for group III nitride semiconductors grown by MOVPE, and (2) the electrical properties of GaN and related semiconductors that are central to the development of devices.

n-Type Doping

In 1986 Sayyah et al. [217] reported on the use of SiH_4 for n-type doping of GaN and the effect of adding SiH_4 on MOVPE growth rates. The effect of doping on the electrical and optical properties was not discussed in the paper, probably because at that time the background donor concentration was higher than $10^{18}\,cm^{-3}$.

Recent studies show that the background donor concentration of GaN grown on sapphire substrates with low-temperature buffer layers is less than $10^{15}\,cm^{-3}$ [218]. In 1990, the first report was published on the use of SiH_4 for the controlled Si-doping of high-quality GaN crystals [219]. Germanium was also reported to act as a donor in AlGaN and GaInN [220, 221].

The activation energy of donors is given by the following equation

$$E_D = 13.56 \left(\frac{1}{\varepsilon_r}\right)^2 \frac{m_e}{m_0} [\mathrm{eV}],$$

where ε_r is the relative dielectric constant of GaN, and m_e is the effective mass of conduction band electrons. The activation energy of Si in GaN was determined, from Hall effect and PL measurements, to be 30 meV. This value is in good agreement with a hydrogen-like model.

p-Type Doping

The use of Mg for p-type doping of GaN was reported by Maruska et al. [222] for HVPE in 1972. In 1989, highly resistive Mg-doped GaN crystals were transformed into conducting p-type films by irradiation with a low-energy electron beam [223]. Afterward, hydrogen desorption was found to produce acceptor activation and p-type GaN films [224, 225].

The deactivating effects of hydrogen on accepter impurities in semiconductors are not limited to group III nitrides, Si- and GaInP-based semiconductors are all known to be affected [226]. Hydrogen forms a complex in semiconductor lattices with two stable positions being reported in GaN. Mg is known to act as an acceptor impurity in not only GaN but also in AlGaN and GaInN [227,228].

2.3.2 Mobility of Electrons and Holes

Carrier scattering in semiconductors occurs via different mechanisms with impurity scattering dominating at low temperatures and phonon scattering at high temperatures, and this also applies to the group III nitrides. Furthermore, dislocations in n-type GaN behave as acceptors and the resulting trapped-electron depletion region around the defect form carrier scattering centers. Figure 2.35 shows the variation of mobility with carrier concentration for a range of dislocation densities [229].

The most important issue in p-type GaN is to increase the concentration of holes. Figure 2.36 shows the relationship between Mg atomic concentration and room temperature hole concentration for p-GaN grown by MOVPE. An optimum Mg concentration is observed at $2 \times 10^{19}\ \mathrm{cm}^{-3}$, above which the concentration of holes decreases. Figure 2.37 shows a cross-sectional TEM image of GaN with twice the optimum concentration of Mg [230]. Reversed pyramidal defects were not observed for samples having Mg concentration of less than $2 \times 10^{19}\ \mathrm{cm}^{-3}$. These defects are thought to be due to surface segregation of Mg above certain doping levels. Surfaces covered with Mg are known to cause changes in the polarity of crystals [231].

Silicon is known to be an effective donor in n-type AlGaN and GaInN. The resistivity of Si-doped AlGaN increases with increasing AlN mole fraction although n-type AlGaN films with a high Al content have been produced for use as cladding layers in deep ultraviolet (~210 nm) LEDs.

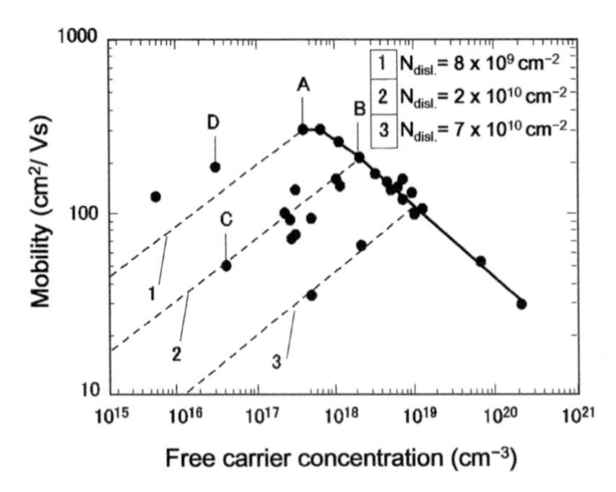

Fig. 2.35. Variation of mobility with carrier concentration of n-type GaN for a range of dislocation densities [229]

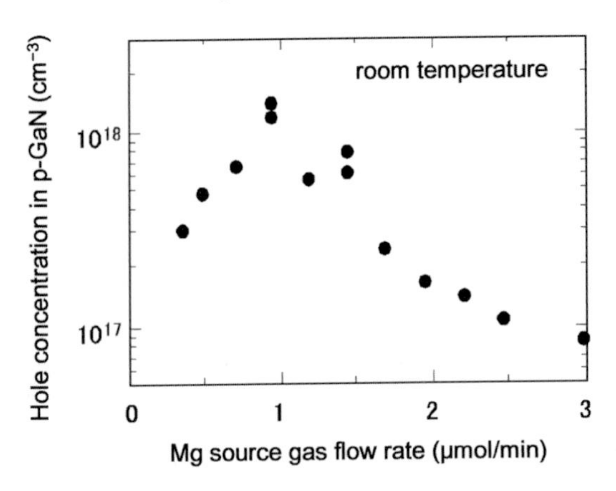

Fig. 2.36. Relationship between Mg atomic concentration and room temperature hole concentration for p-GaN grown by MOVPE

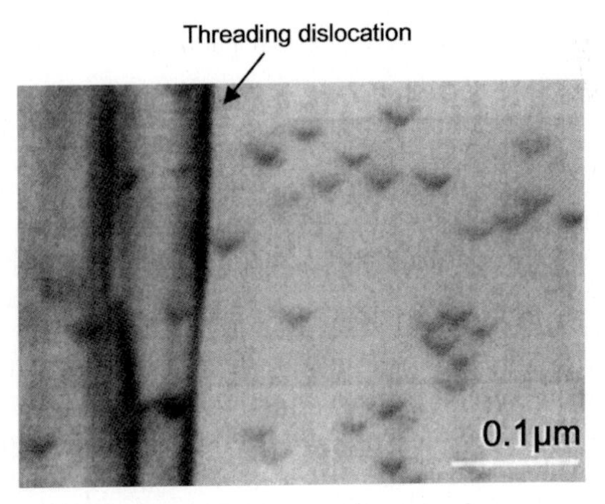

Fig. 2.37. Cross-sectional TEM image of GaN with twice ($4 \times 10^{19}\,\mathrm{cm}^{-3}$) the optimum concentration of Mg

For p-type AlGaN and GaInN films, Mg is known to be effective as an acceptor. For GaInN with low In contents of ~10%, the activation energy of Mg decreases with increasing InN mole fraction and hole concentrations greater than $10^{19}\,\mathrm{cm}^{-3}$ have been achieved [232]. However, for AlGaN, hole concentrations greater than $10^{17}\,\mathrm{cm}^{-3}$ (AlN mole fraction 0.25) are difficult to achieve because of a sudden increase of the activation energy with increasing AlN mole fraction.

Piezoelectric effects in strained Mg-doped AlGaN/GaN superlattice structures have been found to lead to reductions in the operating voltages of LEDs and LDs [233–237].

2.3.3 Electrical Properties of SiC (H. Okumura)

Silicon carbide is a IV–IV semiconductor that is expected to be used for fabricating high-power, high-frequency devices and large capacity switching devices. Figure 2.38 illustrates the operation margin of electron devices in terms of output power and operation frequency. In region 1, device performance is limited by heat generated during operation; in area 3, the device operation is limited by current gain. On the other hand, in region 2, the product of the output power and operation frequency is limited by the product of the saturation drift velocity of carriers and breakdown electric field. Extension of region 3 requires fine patterning of device structures and improvement of carrier mobility. GaAs HEMT, HBT, and InGaAs/InP devices satisfy these requirements. It is noteworthy that the properties of wide bandgap semiconductors become essential in region 2. The approach to use semiconductor materials having a large saturation drift velocity of carriers and high breakdown electric field will become increasingly important for the development of device functions with extended device operation margin. These electrical properties of SiC important for electron device applications are discussed in this section.

Breakdown Electric Field and Saturation Drift Velocity

In 1975, Ferry [238] theoretically analyzed carrier transport under high electric field in wide bandgap semiconductors and predicted saturation drift velocities above $2 \times 10^7\,\mathrm{cm\,s^{-1}}$. A detailed analysis of the electric field dependence of drift velocity can be found in a report by Trew et al. [239]. There are very few experimental studies on the breakdown field and saturation drift velocity of SiC. Meunch et al. found the breakdown field to be $2–3.7 \times 10^6\,\mathrm{V\,cm^{-1}}$ [240] and $2.0 \times 10^7\,\mathrm{cm\,s^{-1}}$ [241] for p-type layers of 6H-SiC with $N_A = 10^{17}–10^{18}\,\mathrm{cm^{-3}}$. Further, Baliga [242] reported the breakdown fields of n-type layers of 6H-SiC and 3C-SiC to vary as 10,640 ${N_D}^{0.142}$ ($\mathrm{V\,cm^{-1}}$) and 8,185 ${N_D}^{0.142}$ ($\mathrm{V\,cm^{-1}}$), respectively, with the donor concentration. More recently,

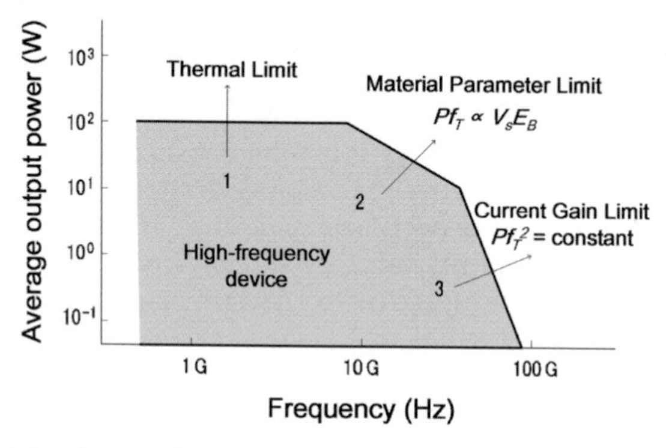

Fig. 2.38. Limiting factors for electron device operation. The operation margin is determined by thermal limit, material parameter limit or current gain limit.

Konstantinov et al. [243, 244] reported highly reliable results for 4H-SiC, the most important polytype in terms of device fabrication. They measured the electrical characteristics of a pin diode and empirically expressed the breakdown voltage (E_c) of the n-type layer as [245],

$$E_c = \frac{2.49 \times 10^6}{1 - \frac{1}{4}\log\left(\frac{N_D - N_A}{10^{16}}\right)} \; [\mathrm{V\,cm^{-1}}].$$

In addition to the existence of many polytypes of SiC as mentioned above, there have been several attempts to fabricate device structures on various kinds of its crystal surfaces. The breakdown electric field in the case on 4H-SiC< (0001) surface has been reported to be approximately 75% of that on other crystal surfaces [246]. Anyway, large drift velocity of $10^7\,\mathrm{cm^{-1}}$ at an electric field as high as 10^5–$10^6\,\mathrm{V\,cm}$ of SiC make this material extremely promising for high-power, high-frequency applications.

For actual device design, the breakdown voltage is usually calculated using a device simulator. For this purpose, the impact ionization coefficient is used rather than the breakdown electric field. Konstantinov et al. also analyzed the related issue in detail [245].

Effective Mass

The effective mass of electrons and holes in SiC has been determined using various methods, such as the Hall effect [247–251], Faraday rotation [252, 253] Zeeman splitting of photoluminescence emissions [254], cyclotron resonance [255–257], and infrared absorption [258–263].

Reliable reported values of the electron effective mass are in the range 0.45–0.48 for 6H-SiC, and 0.34–0.37 for 3C-SiC.

Also, infrared absorption measurements show the longitudinal and transverse effective masses to be 0.67 and 0.24–0.25 for 3C-SiC; 0.19–0.22 and 0.18–0.24 for 4H-SiC; and 1.3–1.7 and 0.24–0.26 for 6H-SiC, respectively.

On the other hand, there is a large discrepancy among reports on the effective mass of holes. Improvement in the accuracy of measurements is required.

Impurity Diffusion Coefficients and Oxidation Rates

The strong interatomic bonding, short bond lengths, and large bandgap of SiC result in a very small diffusion coefficient. Thus the diffusion of impurity atoms is practically negligible below 1,800°C. Several theoretical [264] and experimental [265] results have been reported: N $\sim 10^{-12}\,\mathrm{cm^2\,s^{-1}}$, B in the range 10^{-13}–$10^{-11}\,\mathrm{cm^2\,s^{-1}}$, and Al 10^{-14}–$10^{-12}\,\mathrm{cm^2\,s^{-1}}$.

The oxidation rate is extremely important in the fabrication of MOS devices that there have been many detailed studies [266–268]. The oxidation rate depends greatly on the crystal surface of the SiC crystal. The oxidation rate of the Si surface is lower than that of C surface with other surfaces having values between these two extremes [269].

2.3.4 Electrical Properties of GaN (T. Mizutani)

Electron Velocity-Electric Field Characteristics

Gallium nitride has a large effective electron mass of $0.19m_0$, which is approximately three times that of GaAs, resulting in an electron mobility at low field that is smaller than GaAs with a magnitude of only about 1,000–1,500 $\mathrm{cm^2\,V^{-1}\,s^{-1}}$ at room temperature even for HEMT structures.

However, at high electric fields, the performance of HEMT devices is governed not by the mobility but electron velocity. Figure 2.39 shows a comparison of the calculated electron velocities of GaN [270], Si, and GaAs. The peak velocity of GaN is $2.8 \times 10^7\,\mathrm{cm\,s^{-1}}$, a value that is 1.5 times higher than that of GaAs. However, the electric field at the maximum velocity is $180\,\mathrm{kV\,cm^{-1}}$ for GaN, a value much higher than for GaAs. This is because the energy gap between the Γ and heavy effective electron mass satellite valleys of the conduction band in GaN is 2.0 eV, a value much larger than for GaAs (0.33 eV).

Electron Velocity

The electron velocity has been estimated by analysis of the delay time obtained from measured cut-off frequency, f_T [271]. The channel effective electron velocity, v_{eff}, for HEMTs with gate lengths of 1.3 μm has been measured

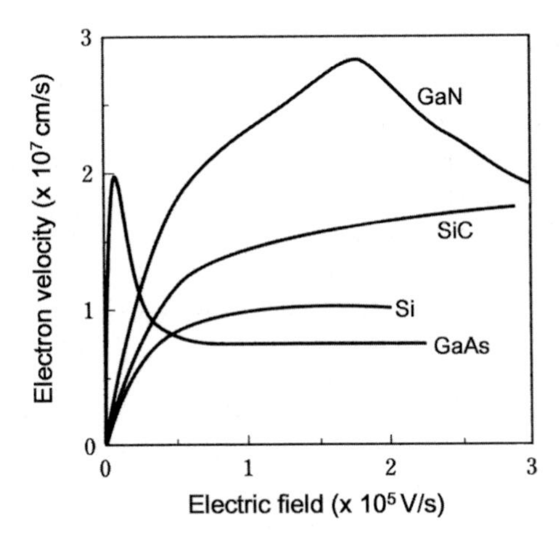

Fig. 2.39. Comparison of the calculated electron velocities of GaN, Si, SiC, and GaAs

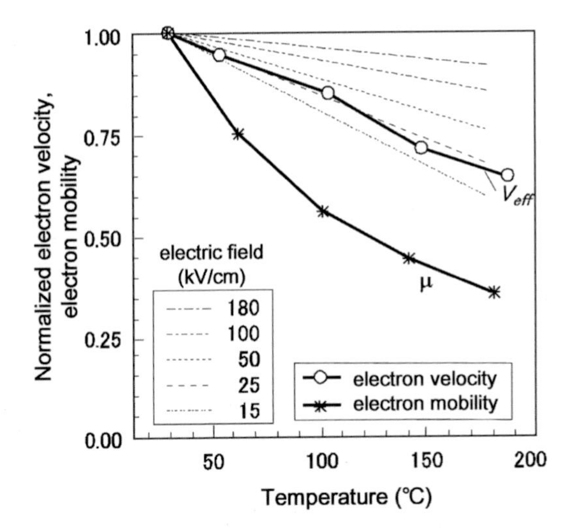

Fig. 2.40. Temperature dependence of the effective electron velocity and mobility. Measured effective electron velocity (*white circles*) shows a significantly larger temperature dependence than the calculated peak velocity at $E = 180\,\text{kV cm}^{-1}$

to be 1.24×10^7, 1.05×10^7, and $0.81 \times 10^7\,\text{cm s}^{-1}$ at 25, 103, and 187°C, respectively. These experimentally measured values for the effective electron velocities are half of the theoretical peak values, typically $\sim 2.8 \times 10^7\,\text{cm s}^{-1}$ at room temperature. The actual simulated electric field under the channel (excluding the edges) is approximately 15–20 kV cm^{-1} for devices with gate lengths of 1 μm, a value significantly smaller than the peak electric field.

These trends are confirmed by the results shown in Fig. 2.40, where the measured effective electron velocity (white circles) shows a significantly larger temperature dependence than the calculated peak velocity at $E = 180\,\text{kV cm}^{-1}$. The measured temperature dependence is similar to theoretical variation of the electron velocity at 25 kV cm^{-1}. This shows that the operation of the device is in an intermediate regime between mobility- and velocity-dominated modes.

Even higher effective electron velocities could be achieved by increasing the electric field under the gate region while reducing gate lengths. It is noteworthy that an effective electron velocity of $1.75 \times 10^7\,\text{cm s}^{-1}$ was reported for devices with short gate lengths of 0.25 μm, where delay times corresponding to the drain delay and mirror effects were subtracted [272]. High electric fields under the gates have been realized using inclined-gate-recess structures, which have experimentally been found to show improved transconductance and current gain cut-off frequencies [273].

On the other hand, in the case of ungated two terminal device structures, the electron velocity, as deduced from current–voltage characteristics, is found to be $2 \times 10^7\,\text{cm s}^{-1}$, a result which is less than the peak velocity of

$2.8 \times 10^7\,\mathrm{cm\,s^{-1}}$. The reason for the lower electron velocity was reported to be due to hot phonon effects [274]. Hot phonon effects are more pronounced when the lifetime of optical phonons is long and electron density is high.

Velocity Overshoot

Velocity overshoot is a phenomenon where the velocity of electrons increases above the steady state value during a short period immediately after the application of an electric field. This phenomenon occurs because the momentum relaxation time is shorter than the energy relaxation time. This effect is observable in III–V semiconductors that exhibit differential negative resistance. Overshoot can be expected at electric fields as low as $10\,\mathrm{kV\,cm^{-1}}$ in GaAs. In the case of GaN on the other hand, electric fields as high as $300\,\mathrm{kV\,cm^{-1}}$ are required in order to observe these phenomenon according to theoretical studies [275]. Even though no overshoot phenomena have been reported at present, they can be expected to be observed by decreasing the gate lengths to 30–100 nm.

Off-State Breakdown Voltage

Impact ionization is another phenomenon observed at high electric fields. The breakdown field of GaN is known to be high at approximately $3\,\mathrm{MV\,cm^{-1}}$. However, impact ionization causes voltage breakdown in device structures thus limiting high-power applications. This indicates the importance of implementing devices with a high breakdown voltage.

The off-state breakdown is more important than the on-state breakdown because the load line which determines the output power and the signal distortion is dominated by the off-state breakdown. It increases with increasing temperature and with decreasing gate leakage current [276]. Because of this behavior, it is thought that voltage breakdown occurs due to impact ionization that results from accelerating electrons that are injected from the gate through to the channel. Higher breakdown voltages can be achieved by using device structures with field plates [277] that prevent concentration of electric fields at the gate edge and reduce gate leakage currents. Surface passivation layers in HEMT structures can result either in increases or, in certain cases, decreases in the breakdown voltage depending on the surface state condition. This is probably due to the types and densities of surface states, which affect the electric field distribution in the channel, depend on the method used for depositing the passivation layer.

The experimentally measured breakdown electric field of HBT structures is $2.3\,\mathrm{MV\,cm^{-1}}$ [278], a value similar to theoretical predictions. An ionization coefficient of $2.9 \times 10^8 \exp\left(-3.4 \times 10^7/E\right)\,\mathrm{cm^{-1}}$ was obtained from measurements of the gate current of HEMT structures biased in the breakdown voltage region [279].

Electrical Transport at the Heterostructure Interface

The main scattering processes in two-dimensional heterostructures are (1) polar optical phonon scattering, (2) acoustic phonon scattering (deformation potential scattering and piezoelectric scattering), (3) ionized impurity scattering (remote impurity scattering and background impurity scattering), (4) interface roughness scattering, and (5) alloy scattering. Of these, (3)–(5) are independent of temperature to a first order approximation [280]. Further, the probability of interface roughness scattering and alloy scattering increases as the square of the 2DEG concentration at high electron concentrations [281]. Figure 2.41 shows the measured temperature dependence of electron mobility [282]. Below 10 K, the mobility is not dependent on the temperature and has a value of about $10^4\,\mathrm{cm^2\,V^{-1}\,s^{-1}}$ when the AlGaN barrier thickness is 10 nm. According to calculations, the mobility dominated by remote impurity scattering is $4.5 \times 10^5\,\mathrm{cm^2\,V^{-1}\,s^{-1}}$, which is larger than the measured mobility. This suggests that the mobility is dominated not by remote impurity scattering but by alloy scattering at these temperatures. For epitaxial layers of high-quality, acoustic phonon scattering dominates between 10 and 150 K and polar optical phonon scattering above 150 K.

Experiments show that alloy scattering contributes even at room temperature in AlGaN/GaN heterostructures due to the large 2DEG density. The inclusion of a 1 nm layer of AlN at the AlGaN/GaN interface has been found to reduce the effects of alloy scattering and improve the mobility of nitride HEMT structures [283]. Concerning the effects of dislocations on the electron transport, scattering by charged dislocation lines [284] and by the strain field surrounding edge dislocations [285] has been theoretically studied in a 2DEG system. It was found that such scattering mechanisms are not dominant when the 2DEG density is high.

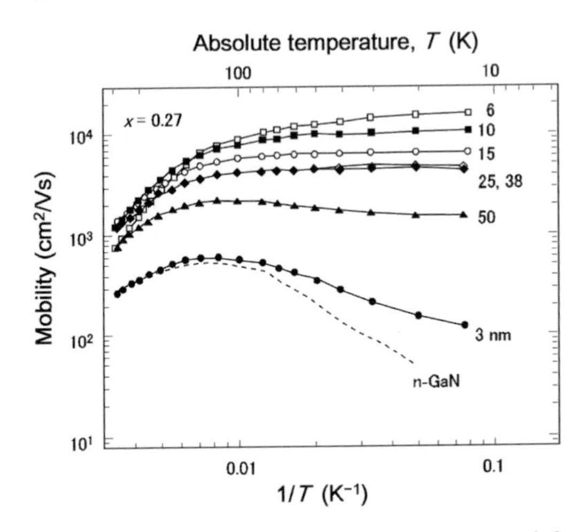

Fig. 2.41. Measured temperature dependence of electron mobility for a range of thicknesses of AlGaN

The discussion so far has focused on electron transport parallel to interfaces. In the case of vertical transport, band offsets rather than scattering are important. Due to the large conduction band offset of AlGaN/GaN (at $x_{\mathrm{Al}} = 0.25$, the offset is 0.36 eV) and its large effective electron mass (0.19), the heterostructure barrier acts as a large resistance. The use of GaN cap layers on modulation-doped AlGaN/GaN heterostructures to decrease the gate–source parasitic resistance is not suitable because it increases the barrier height due to polarization fields. Thus minimizing the source resistance is an important aspect in the design and fabrication of nitride-based HEMT devices. There have been reports on the use of recessed ohmic contacts and AlGaN/GaN superlattice cap layers [286].

References

1. L. Pauling: The Nature of the Chemical Bond, Cornell University Press, Ithaca, NY, 1960.
2. J.C. Phillips: Bonds and Bands in Semiconductors, Academic, New York, 1973.
3. W.R.L. Lambrecht, S. Limpijumnong, S.N. Rashkeev, and B. Segall: Phys. Status Solidi B: Basic Res. 202 (1997) 5.
4. W.J. Choyke, D.R. Hamilton, and L. Patrick: Phys. Rev. 133 (1964) A1163.
5. S. Yoshida: in Properties of Silicon Carbide (G.L. Harris, ed.), EMIS Datareview Series No. 13, INSPEC, London, 1995.
6. E. Biedermann: Solid State Commun. 3 (1965) 343.
7. W. van Haeringen, P.A. Bobbert, and W.H. Backes: Phys. Status Solidi B: Basic Res. 202 (1997) 63.
8. F. Bechstedt, P. Kackell, A. Zywietz, K. Karch, B. Adolph, K. Tenelsen, and J. Furthmüller: Phys. Status Solidi B - Basic Res. 202 (1997) 35.
9. S.G. Sridhara, F.H.C. Carlsson, J.P. Bergman, and E. Janzen: Appl. Phys. Lett. 79 (2001) 3944.
10. W.A. Harrison: Elementary Electronic Structure, World Scientific, Singapore, 1999.
11. J.E. Field (ed.): The Properties of Diamond, Academic, London, 1979; M.A. Prelas, G. Popovici, and L.K. Bigelow (eds.): Handbook of Industrial Diamonds and Diamond Films, Dekker, New York, 1998.
12. W. Saslow, T.K. Bergstresser, and M.L. Cohen: Phys. Rev. Lett. 16 (1966) 354; T. Watanabe, T. Teraji, T. Ito, Y. Kamakura, and K. Taniguchi: J. Appl. Phys. 95 (2004) 4866.
13. T. Teraji and T. Ito: J. Cryst. Growth 271 (2004) 409.
14. S. Koizumi, K. Watanabe, M. Hasegawa, and H. Kanda: Science 292 (2001) 1899.
15. H. Kato, S. Yamasaki, and H. Okushi: Appl. Phys. Lett. 86 (2005) 222111.
16. F. Maier, J. Ristein, and L. Ley: Phys. Rev. B 64 (2001) 165411.
17. F. Maier, M. Riedel, B. Mantel, J. Ristein, and L. Ley: Phys. Rev. Lett. 85 (2000) 3472.
18. H. Umezawa, K. Hirama, T. Arai, H. Hata, H. Takayanagi, T. Koshiba, K. Yohara, S. Mejima, M. Satoh, K.S. Song, and H. Kawarada: Jpn. J. Appl. Phys. 44(11) (2005) 7789.

19. M.C. Tamargo (ed.): II–VI Semiconductor Materials and Their Applications, Optoelectronic Properties of Semiconductors and Superlattices, Vol. 12, Taylor & Francis, Ann Arbor, 2002.
20. A. Mang, K. Reimann, and St. Rübenacke: Solid State Commun. 94 (1995) 251.
21. W.R.L. Lambrecht, S. Limpijumnong, and B. Segall: MRS Internet J. Nitride Semicond. Res. 4 (1999) art. G6.8.
22. K. Shindo, A. Morita, and H. Kaminuma: J. Phys. Soc. Jpn 20 (1965) 2054.
23. M. Cardona: Phys. Rev. 129 (1963) 69.
24. T. Makino, C.H. Chia, N.T. Tuan, Y. Segawa, M. Kawasaki, A. Ohtomo, K. Tamura, and H. Koinuma: Appl. Phys. Lett. 77 (2000) 1632.
25. K. Sakurai, T. Kubo, D. Kajita, T. Tanabe, H. Takasu, S. Fujita, and S. Fujita: Jpn J. Appl. Phys. Part 2: Lett. 39 (2000) L1146.
26. S. Shigemori, A. Nakamura, J. Ishihara, T. Aoki, and J. Temmyo: Jpn J. Appl. Phys. Part 2: Lett. 43 (2004) L1088.
27. T. Takagi, H. Tanaka, S. Fujita, and S. Fujita: Jpn J. Appl. Phys. Part 2: Lett. 42 (2003) L401.
28. K. Koike, K. Hama, I. Nakashima, S. Sasa, M. Inoue, and M. Yano: Jpn J. Appl. Phys. Part 1: Reg. Pap. 44 (2005) 3822.
29. K. Koike, I. Nakashima, K. Hashimoto, S. Sasa, M. Inoue, and M. Yano: Appl. Phys. Lett. 87 (2005) art. 112106.
30. H. Tanaka, S. Fujita, and S. Fujita: Appl. Phys. Lett. 86 (2005) art. 192911.
31. J.C. Phillips: Bonds and Bands in Semiconductors, Academic, New York, 1973.
32. E. Wimmer, H. Krakauer, M. Weinert, and A.J. Freeman: Phys. Rev. B 24 (1981) 864.
33. M. Suzuki, T. Uenoyama, and A. Yanase: Phys. Rev. B 52 (1995) 8132.
34. R.M. Chrenko: Solid State Commun. 14 (1974) 511.
35. B. Monemar: Phys. Rev. B 10 (1974) 676.
36. B. Gil (ed.): Group III Nitride Semiconductor Compounds: Physics and Applications. Oxford University Press, New York, 1998.
37. S. Chichibu, A. Shikanai, T. Azuhata, T. Sota, A. Kuramata, K. Horino, and S. Nakamura: Appl. Phys. Lett. 68 (1996) 3766.
38. A. Shikanai, T. Azuhata, T. Sota, S. Chichibu, A. Kuramata, K. Horino, and S. Nakamura: J. Appl. Phys. 81 (1997) 417.
39. G. Ramírez-Flores, H. Navarro-Contreras, A. Lastras-Martínez, R.C. Powell, and J.E. Greene: Phys. Rev. B 50 (1994) 8433.
40. H. Yamashita, K. Fukui, S. Misawa, and S. Yoshida: J. Appl. Phys. 50 (1979) 896.
41. W.R.L. Lambrecht and B. Segall: Phys. Rev. B 43 (1991) 7070.
42. M.P. Thompson, G.W. Auner, T.S. Zheleva, K.A. Jones, S.J. Simko, and J.N. Hilfiker: J. Appl. Phys. 89 (2001) 3331.
43. I. Vurgaftman, J.R. Meyer, and L.R. Ram-Mohan: J. Appl. Phys. 89 (2001) 5815.
44. F. Bechstedt and J. Furthmüller: J. Cryst. Growth 246 (2002) 315.
45. D. Bagayoko, L. Franklin, and G.L. Zhao: J. Appl. Phys. 96 (2004) 4297.
46. J. Wu, W. Walukiewicz, W. Shan, K.M. Yu, J.W. Ager, S.X. Li, E.E. Haller, H. Lu, and W.J. Schaff: J. Appl. Phys. 94 (2003) 4457.
47. Y. Ishitani, H. Masuyama, W. Terashima, M. Yoshitani, N. Hashimoto, S.B. Che, and A. Yoshikawa: Phys. Status Solidi C 2 (2005) 2276.

48. S.H. Wei, X.L. Nie, I.G. Batyrev, and S.B. Zhang: Phys. Rev. B 67 (2003).
49. T. Inushima, N. Kato, Y. Sasaki, T. Takenobu, and M. Motokawa: Phys. Status Solidi C 2 (2005) 2271.
50. R. Goldhahn, A.T. Winzer, V. Cimalla, O. Ambacher, C. Cobet, W. Richter, N. Esser, J. Furthmuller, F. Bechstedt, H. Lu, and W.J. Schaff: Superlattices Microstruct. 36 (2004) 591.
51. Y. Ishitani et al.: Proceedings of 6th International Conference on the Physics of Nitride Semiconductors, Bremen, 2005.
52. Y.P. Varshni: Physica 34 (1967) 149.
53. P.B. Allen and M. Cardona: Phys. Rev. B 23 (1981) 1495.
54. Y. Ishitani, H. Yaguchi, and Y. Shiraki: Jpn J. Appl. Phys. Part 1: Reg. Pap. 40 (2001) 1183.
55. Y. Ishitani, H. Hamada, S. Minagawa, H. Yaguchi, and Y. Shiraki: Jpn J. Appl. Phys. Part 1: Reg. Pap. 36 (1997) 6607.
56. Y. Koide, H. Itoh, M.R.H. Khan, K. Hiramatu, N. Sawaki, and I. Akasaki: J. Appl. Phys. 61 (1987) 4540.
57. T. Takeuchi, H. Takeuchi, S. Sota, H. Sakai, H. Amano, and I. Akasaki: Jpn J. Appl. Phys. Part 2: Lett. 36 (1997) L177.
58. T. Onuma, S.F. Chichibu, A. Uedono, T. Sota, P. Cantu, T.M. Katona, J.F. Keading, S. Keller, U.K. Mishra, S. Nakamura, and S.P. DenBaars: J. Appl. Phys. 95 (2004) 2495.
59. W. Walukiewicz, S.X. Li, J. Wu, K.M. Yu, J.W. Ager, E.E. Haller, H. Lu, and W.J. Schaff: J. Cryst. Growth 269 (2004) 119.
60. T. Onuma, S. Chichibu, Y. Uchinuma, T. Sota, S. Yamaguchi, S. Kamiyama, H. Amano, and I. Akasaki: J. Appl. Phys. 94 (2003) 2449.
61. W. Terashima et al.: Proceedings of 47th TMS 2005 Electronic Materials Conference, Santa Barbara, CL, 2005.
62. V.Y. Davydov, A.A. Klochikhin, V.V. Emtsev, S.V. Ivanov, V.V. Vekshin, F. Bechstedt, J. Furthmüller, H. Harima, A.V. Mudryi, A. Hashimoto, A. Yamamoto, J. Aderhold, J. Graul, and E.E. Haller: Phys. Status Solidi B 230 (2002) R4.
63. T. Matsuoka, M. Nakao, H. Okamoto, H. Harima, and E. Kurimoto: Jpn J. Appl. Phys. Part 1: Reg. Pap. 42 (2003) 2288.
64. J.N. Baillargeon, K.Y. Cheng, G.E. Hofler, P.J. Pearah, and K.C. Hsieh: Appl. Phys. Lett. 60 (1992) 2540.
65. W.G. Bi and C.W. Tu: Appl. Phys. Lett. 70 (1997) 1608.
66. S. Miyoshi and K. Onabe: Jpn J. Appl. Phys. Part 1: Reg. Pap. 37 (1998) 4680.
67. W. Shan, W. Walukiewicz, K.M. Yu, J. Wu, J.W. Ager, E.E. Haller, H.P. Xin, and C.W. Tu: Appl. Phys. Lett. 76 (2000) 3251.
68. P.J. Klar, H. Gruning, W. Heimbrodt, J. Koch, F. Hohnsdorf, W. Stolz, P.M.A. Vicente, and J. Camassel: Appl. Phys. Lett. 76 (2000) 3439.
69. L. Bellaiche, S.H. Wei, and A. Zunger: Phys. Rev. B 54 (1996) 17568.
70. S.H. Wei and A. Zunger: Phys. Rev. Lett. 76 (1996) 664.
71. J. Salzman and H. Temkin: Mater. Sci. Eng. B: Solid State Mater. Adv. Technol. 50 (1997) 148.
72. W.A. Harrison and J. Tersoff: J. Vac. Sci. Technol. 14 (1977) 1016.
73. S.W. King, C. Ronning, R.F. Davis, M.C. Benjamin, and R.J. Nemanich: J. Appl. Phys. 84 (1998) 2086.
74. F. Herman and S. Skillman: Atomic Structure Calculation, Prentice-Hall, New Jersey, 1963.

75. S. Sakai, Y. Ueta, Y. Terauchi: Jpn J. Appl. Phys. Part 1: Reg. Pap. 32 (1993) 4413.
76. Y. Ueta, S. Sakai, Y. Kamiyama, and H. Sato: in MRS Symposium D "Diamond SiC and Nitride Wide-Bandgap Semiconductors" Proceedings, Vol. 339 (C.H. Carter, Jr, G. Gildenblat, S. Nakamura, and R.J. Nemanich, eds.), Materials Research Society, San Francisco, CA, 1994, p. 459.
77. G. Martin, A. Botchkarev, A. Rockett, and H. Morkoç: Appl. Phys. Lett. 68 (1996) 2541.
78. D.R. Hang, C.H. Chen, Y.F. Chen, H.X. Jiang, and J.Y. Lin: J. Appl. Phys. 90 (2001) 1887.
79. H. Morkoç: Nitride Semiconductors and Devices, 1st ed. Springer, Berlin Heidelberg New York, 1999.
80. H. Sato, H.X. Wang, D. Sato, R. Takaki, N. Wada, T. Tanahashi, K. Yamashita, S. Kawano, T. Mizobuchi, A. Dempo, K. Morioka, M. Kimura, S. Nohda, T. Sugahara, and S. Sakai: Phys. Status Solidi A: Appl. Res. 200 (2003) 102.
81. J.P. Ibbetson, P.T. Fini, K.D. Ness, S.P. DenBaars, J.S. Speck, and U.K. Mishra: Appl. Phys. Lett. 77 (2000) 250.
82. S. Rajan, H.L. Xing, S. DenBaars, U.K. Mishra, and D. Jena: Appl. Phys. Lett. 84 (2004) 1591.
83. S.J. Pearton, J.C. Zolper, R.J. Shul, and F. Ren: J. Appl. Phys. 86 (1999) 1.
84. B.P. Luther, S.E. Mohney, T.N. Jackson, M.A. Khan, Q. Chen, and J.W. Yang: Appl. Phys. Lett. 70 (1997) 57.
85. H. Kim, J. Lee, D.M. Liu, and W. Lu: Appl. Phys. Lett. 86 (2005) art. 143505.
86. L.C. Chen, J.K. Ho, C.S. Jong, C.C. Chiu, K.K. Shih, F.R. Chen, J.J. Kai. and L. Chang: Appl. Phys. Lett. 76 (2000) 3703.
87. Y.J. Lin and K.C. Wu: Appl. Phys. Lett. 84 (2004) 1501.
88. R.H. Horng, C.C. Yang, J.Y. Wu, S.H. Huang, C.E. Lee, and D.S. Wuu: Appl. Phys. Lett. 86 (2005) art. 221101.
89. J.W. Park, J.O. Song, D.S. Leem, and T.Y. Seong: Electrochem. Solid State Lett. 8 (2005) G17.
90. H.W. Jang and J.L. Lee: Appl. Phys. Lett. 85 (2004) 4421.
91. S. Chichibu, T. Azuhata, T. Sota, H. Amano, and I. Akasaki: Appl. Phys. Lett. 70 (1997) 2085.
92. S. Yoshida, S. Misawa, and S. Gonda: J. Appl. Phys. 53 (1982) 6844.
93. Y. Koide, H. Itoh, M.R.H. Khan, K. Hiramatu, S. Sawaki, and I. Akasaki: J. Appl. Phys. 61 (1987) 4540.
94. T. Takeuchi, H. Takeuchi, S. Sota, H. Sakai, H. Amano, and I. Akasaki: Jpn J. Appl. Phys. 36 (1997) L177.
95. M.D. McCluskey, C.G. Van de Walle, L.T. Romano, B.S. Krusor, and N.M. Johnson: J. Appl. Phys. 93 (2003) 4340.
96. C.A. Parker, J.C. Roberts, S.M. Bedair, M.J. Reed, S.X. Liu, N.A. El-Masry, and L.H. Robins: Appl. Phys. Lett. 75 (1999) 2566.
97. H. Amano, N. Sawaki, I. Akasaki, and Y. Toyoda: Appl. Phys. Lett. 48 (1986) 353.
98. S. Nakamura: Jpn J. Appl. Phys. Part 2: Lett. 30 (1991) L1705.
99. S.D. Lester, F.A. Ponce, M.G. Craford, and D.A. Steigerwald: Appl. Phys. Lett. 66 (1995) 1249.
100. W. Qian, M. Skowronski, M.D. Graef, K. Doverspike, L.B. Rowland, and D.K. Gaskill: Appl. Phys. Lett. 66 (1995) 1252.

101. X.H. Wu, L.M. Brown, D. Kapolnek, S. Keller, B. Keller, S.P. Denbaars, and J.S. Speck: J. Appl. Phys. 80 (1996) 3228.
102. T. Hino, S. Tomiya, T. Miyajima, K. Yanashima, S. Hashimoto, and M. Ikeda: Appl. Phys. Lett. 76 (2000) 3421.
103. T. Miyajima, T. Hino, S. Tomiya, A. Satake, E. Tokunaga, Y. Masumoto, T. Maruyama, M. Ikeya, S.-I. Morishima, K. Akimoto, K. Yanashima, S. Hashimoto, T. Kobayashi, and M. Ikeda: International Workshop on Nitride Semiconductors (IWN2000), Vol. 1, IPAP, Nagoya, 2000, pp. 536–539.
104. T. Miyajima, T. Hino, S. Tomiya, K. Yanashima, H. Nakajima, T. Araki, Y. Nanishi, A. Satake, Y. Masumoto, K. Akimoto, T. Kobayashi, and M. Ikeda: Phys. Status Solidi B: Basic Res. 228 (2001) 395.
105. A. Kikuchi, T. Yamada, S. Nakamura, K. Kusakabe, D. Sugihara, and K. Kishino: Jpn J. Appl. Phys. Part 2: Lett. 39 (2000) L330.
106. D. Sugihara, A. Kikuchi, K. Kusakabe, S. Nakamura, Y. Toyoura, T. Yamada, and K. Kishino: Phys. Status Solidi A: Appl. Res. 180 (2000) 65.
107. A. Sakai, H. Sunakawa, and A. Usui: Appl. Phys. Lett. 71 (1997) 2259.
108. K. Yanashima, S. Hashimoto, T. Hino, K. Funato, T. Kobayashi, K. Naganuma, T. Tojyo, T. Asano, T. Asatsuma, T. Miyajima, and M. Ikeda: J. Electron. Mater. 28 (1999) 287.
109. J. Elsner, R. Jones, P.K. Sitch, V.D. Porezag, M. Elstner, T. Frauenheim, M.I. Heggie, S. Öberg, and P.R. Briddon: Phys. Rev. Lett. 79 (1997) 3672.
110. C.J. Fall, R. Jones, P.R. Briddon, A.T. Blumenau, T. Frauenheim, and M.I. Heggie: Phys. Rev. B 65 (2002) art. 245304.
111. J.E. Northrup: Appl. Phys. Lett. 78 (2001) 2288.
112. N. Peyghambarian, S.W. Koch, and A. Mysyrowicz: Introduction to Semiconductor Optics, Prentice Hall Series in Solid State Physical Electronics, Prentice Hall, Englewood Cliffs, NJ, 1993 (Chap. 5).
113. T. Onuma, S.F. Chichibu, T. Sota, K. Asai, S. Sumiya, T. Shibata, and M. Tanaka: Appl. Phys. Lett. 81 (2002) 652.
114. K.B. Nam, J. Li, M.L. Nakarmi, J.Y. Lin, and H.X. Jiang: Appl. Phys. Lett. 82 (2003) 1694.
115. J.J. Hopfield: in Proceedings of the Seventh International Conference on the Physics of Semiconductors (M. Hulin, ed.), Dunod Cie & Academic, Paris, 1964, p. 725.
116. R.R. Sharma and S. Rodriguez: Phys. Rev. 159 (1967) 649.
117. J.R. Haynes: Phys. Rev. Lett. 4 (1960) 361.
118. C. Klingshirn and H. Haug: Phys. Rep. 70 (1981) 315.
119. O. Akimoto and E. Hanamura: J. Phys. Soc. Jpn 33 (1972) 1537.
120. W.F. Brinkman, T.M. Rice, and B. Bell: Phys. Rev. B 8 (1973) 1570.
121. W.-T. Huang: Phys. Status Solidi B: Basic Res. 60 (1973) 309.
122. Y. Yamada, T. Sakashita, H. Watanabe, H. Kugimiya, S. Nakamura, and T. Taguchi: Phys. Rev. B 61 (2000) 8363.
123. S.W. Koch, H. Haug, G. Schmieder, W. Bohnert, and C. Klingshirn: Phys. Status Solidi B: Basic Res. 89 (1978) 431.
124. S. Shionoya: Kotai Butsuri 30 (1995) 438 (in Japanese).
125. F. Kreller, J. Puls, and F. Henneberger: Appl. Phys. Lett. 69 (1996) 2406.
126. Y. Yamada: J. Lumin. 87–89 (2000) 140.
127. P. Yu, Z.K. Tang, G.K. Wong, M. Kawasaki, A. Ohtomo, H. Koinuma, and Y. Segawa: in Proceedings of the 23rd International Conference on the Physics of Semiconductors (M. Scheffler and R. Zimmermann, eds.), Vol. 2, World Scientific, Singapore, 21–26 July 1996, pp. 1453–1456.

128. D.M. Bagnall, Y.F. Chen, Z. Zhu, T. Yao, S. Koyama, M.Y. Shen, and T. Goto: Appl. Phys. Lett. 70 (1997) 2230.
129. R. Zimmermann: J. Cryst. Growth 101 (1990) 346.
130. Y. Kawakami, M. Funato, Sz. Fujita, Sg. Fujita, Y. Yamada, and Y. Masumoto: Phys. Rev. B 50 (1994) 14655.
131. G. Coli, K.K. Bajaj, J. Li, J.Y. Lin, and H.X. Jiang: Appl. Phys. Lett. 78 (2001) 1829.
132. J. Ding, H. Jeon, T. Ishihara, M. Hagerott, A.V. Nurmikko, H. Luo, N. Samarth, and J. Furdyna: Phys. Rev. Lett. 69 (1992) 1707.
133. S. Adachi: Properties of Group-IV, III–V and II–VI Semiconductors, Wiley, New York, 2005.
134. T. Kawashima, H. Yoshikawa, S. Adachi, S. Fuke, and K. Ohtsuka: J. Appl. Phys. 82 (1997) 3528.
135. D. Brunner, H. Angerer, E. Bustarret, F. Freudenberg, R. Hopler, R. Dimitrov, O. Ambacher, and M. Stutzmann: J. Appl. Phys. 82 (1997) 5090.
136. A.C. Boccara, D. Fournier, W. Jackson, and N.M. Amer: Opt. Lett. 5 (1980) 377.
137. U. Tisch, B. Meyler, O. Katz, E. Finkman, and J. Salzman: J. Appl. Phys. 89 (2001) 2676.
138. A.B. Djurisic and E.H. Li: J. Appl. Phys. 85 (1999) 2848.
139. T. Takeuchi, S. Sota, M. Katsuragawa, M. Komori, H. Takeuchi, H. Amano, and I. Akasaki: Jpn. J. Appl. Phys. P.2 236 (1997) L382.
140. A.F. Wright: J. Appl. Phys. 82 (1997) 2833.
141. F. Bernardini, V. Fiorentini, and D. Vanderbilt: Phys. Rev. B 56 (1997) 10024.
142. A.D. Bykhovski, V.V. Kaminski, M.S. Shur, Q.C. Chen, and M.A. Khan: Appl. Phys. Lett. 68 (1996) 818.
143. Y. Kawakami: in "Low Dimensional Nitride Semiconductors" (B. Gil, ed.), Series on Semiconductor Science and Technology, Oxford University Press, New York City, 2002, p. 233–256.
144. S.F. Chichibu, T. Sota, K. Wada, O. Brandt, K.H. Ploog, S.P. DenBaars, and S. Nakamura: Phys. Status Solidi A - Appl. Res. 183 (2001) 91.
145. O. Ambacher, R. Dimitrov, M. Stutzmann, B.E. Foutz, M.J. Murphy, J.A. Smart, J.R. Shealy, N.G. Weimann, K. Chu, M. Chumbes, B. Green, A.J. Sierakowski, W.J. Schaff, and L.F. Eastman: Phys. Status Solidi B - Basic Res. 216 (1999) 381.
146. Y. Chen, H.-J. Ko, S.-K. Hong, and T. Yao: Appl. Phys. Lett. 76 (2000) 559.
147. A. Tsukazaki, A. Ohtomo, S. Yoshida, M. Kawasaki, C.H. Chia, T. Makino, Y. Segawa, T. Koida, S.F. Chichibu, and H. Koinuma: Appl. Phys. Lett. 83 (2003) 2784.
148. N. Grandjean and J. Massies: Appl. Phys. Lett. 71 (1997) 1816.
149. H.T. Ng, B. Chen, J. Li, J. Han, M. Meyyappan, J. Wu, S.X. Li, and E.E. Haller: Appl. Phys. Lett. 82 (2003) 2023.
150. W.I. Park, D.H. Kim, S.-W. Jung, and G.-C. Yi: Appl. Phys. Lett. 80 (2002) 4232.
151. Z.W. Pan, Z.R. Dai, and Z.L. Wang: Science 291 (2001) 1947.
152. K. Kusakabe, A. Kikuchi, and K. Kishino: J. Cryst. Growth 237 (2002) 988.
153. M. Arita, A. Avramescu, K. Uesugi, and I. Suemune: Jpn J. Appl. Phys. 36 (1997) 4097.
154. B. Daudin, F. Widmann, G. Feuillet, Y. Samson, M. Arlery, and J.L. Rouviere: Phys. Rev. B 56 (1997) R7069.

155. S. Tanaka, S. Iwai, and Y. Aoyagi: Appl. Phys. Lett. 69 (1996) 4096.
156. S.W. Kim, S. Fujita, and S. Fujita: Jpn J. Appl. Phys. 41 (2002) L543.
157. I. Suemune, K. Yoshida, H. Kumano, T. Tawara, A. Ueta, and S. Tanaka: J. Cryst. Growth 248 (2003) 301.
158. D.M. Bagnall, Y.F. Chen, Z. Zhu, T. Yao, M.Y. Shen, and T. Goto: Appl. Phys. Lett. 73 (1998) 1038.
159. C.H. Chia, T. Makino, K. Tamura, Y. Segawa, M. Kawasaki, A. Ohtomo, and H. Koinuma: Appl. Phys. Lett. 82 (2003) 1848.
160. M.H. Huang, S. Mao, H. Feick, H.Q. Yan, Y.Y. Wu, H. Kind, E. Weber, R. Russo, and P.D. Yang: Science 292 (2001) 1897.
161. X. Duan, Y. Huang, R. Agarwal, and C.M. Lieber: Nature 421 (2003) 241.
162. L.S. Dang, D. Heger, R. Andre, F. Boeuf, and R. Romestain: Phys. Rev. Lett. 81 (1998) 3920.
163. T. Tawara, H. Gotoh, T. Akasaka, N. Kobayashi, and T. Saitoh: Phys. Rev. Lett. 92 (2004) art. 256402.
164. T. Tawara, I. Suemune, and H. Kumano: Physica E 13 (2002) 403.
165. N. Iizuka, K. Kaneko, and N. Suzuki: Appl. Phys. Lett. 81 (2002) 1803.
166. F. Widmann, J. Simon, B. Daudin, G. Feuillet, J.L. Rouviere, N.T. Pelekanos, and G. Fishman: Phys. Rev. B 58 (1998) R15989.
167. S. Kako, K. Hoshino, S. Iwamoto, S. Ishida, and Y. Arakawa: Appl. Phys. Lett. 85 (2004) 64.
168. P. Michler, A. Imamoglu, M.D. Mason, P.J. Carson, G.F. Strouse, and S.K. Buratto: Nature 406 (2000) 968.
169. V.D. Kulakovskii, G. Bacher, R. Weigand, T. Kummell, A. Forchel, E. Borovitskaya, K. Leonardi, and D. Hommel: Phys. Rev. Lett. 82 (1999) 1780.
170. I. Akasaki (ed.): Group III Nitride Semiconductors – Category 1: Electronic Materials, Physical Properties, Devices. Advanced Electronics Series, Vol. I-21, Baifukan, Tokyo, 1999 (in Japanese).
171. I. Akasaki (ed.): Groups III–V Compound Semiconductors. Advanced Electronics Series, Baifukan, Tokyo, 1994 (in Japanese).
172. K. Iga (ed.): Semiconductor Laser. The Japan Society of Applied Physics Series, Ohmsha, Tokyo, 1994 (in Japanese).
173. H. Amano, N. Sawaki, I. Akasaki, and Y. Toyoda: Appl. Phys. Lett. 48 (1986) 353.
174. Y. Koide, N. Itoh, K. Itoh, N. Sawaki, and I. Akasaki: Jpn J. Appl. Phys. Part 1: Reg. Pap. 27 (1988) 1156.
175. K. Ito, K. Hiramatsu, H. Amano, and I. Akasaki: J. Cryst. Growth 104 (1990) 533.
176. T. Detchprohm, K. Hiramatsu, K. Itoh, and I. Akasaki: Jpn J. Appl. Phys. Part 2: Lett. 31 (1992) L1454.
177. K. Hiramatsu, T. Detchprohm, and I. Akasaki: Jpn J. Appl. Phys. Part 1: Reg. Pap. 32 (1993) 1528.
178. L.E. McNail, M. Grimsditch, and R.H. French: J. Am. Ceram. Soc. 76(5) (1993) 1132–1136.
179. A.F. Wright: Elastic properties of zinc-blende and wurtzite AlN, GaN, and InN. J. Appl. Phys. 82(6) (1997) 2833–2839.
180. D. Gerlich, S.L. Dole, and G.A. Slack: J. Phys. Chem. Solids 47(5) (1986) 437–441.
181. R. Thokala and J. Chaudhuri: Thin Solid Films 266(2) (1995) 189–191.

182. R. Truell, C. Elbaum, and B.B. Chick: Ultrasonic Methods in Solid State Physics, Academic, New York, 1969.
183. I. Akasaki and H. Amano: in Properties of Group III Nitrides (J.H. Edgar, ed.), EMIS Datareviews Series, IEE, London, 1994, p. 222.
184. T. Detchprohm, K. Hiramatsu, K. Itoh, and I. Akasaki: Jpn J. Appl. Phys. 31 (1992) L1454.
185. C. Kisielowski et al.: Phys. Rev. B 54 (1996) 17745.
186. A. Polian, M. Grimsditch, and I. Grzegory: Elastic constants of gallium nitride. J. Appl. Phys. 79(6) (1996) 3343–3344.
187. L. Duclaux, B. Nysten, J.-P. Issi, and A.W. Moore: Phys. Rev. B 46 (1992) 3362.
188. K. Shimada, T. Sota, and K. Suzuki: J. Appl. Phys. 84 (1998) 4951.
189. M.E. Sherwin and T.J. Drummond: J. Appl. Phys. 69 (1991) 8423.
190. V.A. Savastenko and A.U. Sheleg: Phys. Status Solidi A 48 (1978) K135.
191. A.W. Moore: J. Cryst. Growth 106 (1990) 6.
192. V.B. Shipilo, N.A. Shishonok, and A.V. Mazovko: Inorg. Mater. 26 (1990) 1401.
193. V.A. Pesin: Sverktverd. Mater. 6 (1980) 5.
194. Y.N. Xu and W.Y. Ching: Phys. Rev. B 44 (1991) 7787.
195. P.E. Camp, V.E. Van Doren, and J.T. Devreese: Phys. Rev. B 41 (1990) 1598.
196. E.V. Yakovenko, I.V. Aleksandrov, A.F. Goncharov, and S.M. Stishov: Sov. Phys. JETP 68 (1989) 1213.
197. Y.N. Xu and W.Y. Ching: Phys. Rev. B 48 (1993) 4335.
198. N.N. Sirota and V.Z. Golodushko: Tezisy Dokl., Vses Konf. Khi., Svyazi Poluprovdn. Polumetallakh 5 (1974) 98.
199. G.A. Slack, R.A. Tanzilli, R.O. Pohl, and J.W. Vandersande: J. Phys. Chem. Solids 48(7) (1987) 641–647.
200. W. Qian, M. Skowronski, and G.R. Rohrer: Structural defects and their relationship to nucleation of GaN thin films. in III-Nitride, SiC, and Diamond Materials for Electronic Devices (D.K. Gaskill, C.D. Brandt, and R.J. Nemanich, eds.), Vol. 423, Material Research Society Symposium Proceedings, Pittsburgh, PA, 1996, pp. 475–486.
201. E.K. Sichel and J.I. Pankove: Thermal conductivity of GaN.25–360 K. J. Phys. Chem. Solids 38(3) (1997) 330.
202. M. Leszczynski, T. Suski, H. Teisseyre, P. Perlin, I. Grzegory, J. Jun, S. Porowski, and T.D. Moustakas: J. Appl. Phys. 76 (1994) 4909.
203. H.P. Maruska and J.J. Tietjen: Appl. Phys. Lett. 15 (1969) 327.
204. A.U. Sheleg and V.A. Savastenko: Vestsi AN BSSR, Ser. Fiz.-Mat. Navuk 126 (1976).
205. J.C. Nipko, C.-K. Loong, C.M. Balkas, and R.F. Davis: Appl. Phys. Lett. 73 (1998) 34.
206. R.S. Peace: Acta Crystallogr. 5 (1952) 356.
207. A.W. Moore: J. Cryst. Growth 106 (1990) 6.
208. V.B. Shipilo, N.A. Shishonok, and A.V. Mazovko: Inorg. Mater. 26 (1990) 1401.
209. Z.I. Kolupayeva, M. Ya Fuks, L.I. Gladkikl, A.V. Arinkin, and S.V. Malikhin: J. Less-Common Met. 117 (1986) 259.
210. H. Sumiya, K. Tsuji, and S. Yazu: Proceedings of 2nd International Conference on New Diamond Science Technology, Washington, DC, 23–27 September 1990, MRS, Pittsburgh, PA, 1991, p. 1063.
211. D.V. Fedoseev, A.V. Lavrent'ev, I.G. Varshavskaya, A.V. Bochko, and G.G. Karyuk: Poroshk. Metall. 3 (1978) 92.

212. K.S. Gavrichev, V.L. Solozhenko, V.E. Gorbunov, L.N. Golushina, G.A. Totrova, and V.B. Lazarev: Thermochim. Acta 217 (1993) 77.
213. I. Barin, O. Knacke, and O. Kubashewski: Thermochemical Properties of Inorganic Substances, Springer, Berlin Heidelberg New York, 1977.
214. E.K. Sichel, R.E. Miller, M.S. Abrahams, and C.J. Buiocchi: Phys. Rev. B 13 (1976) 4607.
215. T. Atake, S. Takai, A. Honda, Y. Saito, and K. Saito: Rep. Res. Lab. Eng. Mater. Tokyo Inst. Technol. 16 (1991) 15.
216. J.H. Edgar, S. Strite, I. Akasaki, H. Amana, and C. Wetzel: Gallium Nitride and Related Semiconductors, A1.4, emis, Data Reviews Series, No. 23.
217. K. Sayyah, B.-C. Chung, and M. Gershenzon: J. Cryst. Growth 77 (1986) 424.
218. H. Amano, N. Sawaki, I. Akasaki, and Y. Toyoda: Appl. Phys. Lett. 48 (1986) 353.
219. H. Amano and I. Akasaki: MRS Symposium Extended Abstract, Vol. 165 (1990), Materials Research Society, Boston, MA, November 27–December 2, 1989, p. EA-21.
220. I. Akasaki and H. Amano: in Proceedings of 1991 MRS Fall Meeting Symposium – Wide-Bandgap Semiconductors (T.D. Moustakas, J.I. Pankove, and Y. Hamakawa, eds.), MRS Symposium Proceedings, Vol. 242, Materials Research Society, Boston, MA, 2–6 December 1991, p. 383.
221. S. Nakamura, T. Mukai, and M. Senoh: Jpn J. Appl. Phys. Part 2: Lett. 32 (1993) L16.
222. H.P. Maruska, W.C. Rhines, and D.A. Stevenson: Mater. Res. Bull. 7 (1972) 777.
223. H. Amano, M. Kito, K. Hiramatsu, and I. Akasaki: Jpn J. Appl. Phys. 28 (1989) L2112.
224. S. Nakamura, N. Iwasa, M. Senoh, and T. Mukai: Jpn J. Appl. Phys. Part 1: Reg. Pap. 31 (1992) 1258.
225. J.A. Van Vechten, J.D. Zook, R.D. Horning, and B. Goldenberg: Jpn J. Appl. Phys. Part 1: Reg. Pap. 31 (1992) 3662.
226. J.I. Pankove and N.M. Johnson (eds.): Hydrogen in silicon. Semiconductors and Semimetals, Vol. 34, Academic, San Diego, 1991.
227. S. Yamasaki, S. Asami, N. Shibata, M. Koike, K. Manabe, T. Tanaka, H. Amano, and I. Akasaki: Appl. Phys. Lett. 66 (1995) 1112.
228. I. Akasaki, H. Amano, H. Murakami, M. Sassa, H. Kato, and K. Manabe: J. Cryst. Growth 128 (1993) 379.
229. H.M. Ng, D. Doppalapudi, T.D. Moustakas, N.G. Weimann, and L.F. Eastman: Appl. Phys. Lett. 73 (1998) 821.
230. D. Cherns, M.Q. Baines, Y.Q. Wang, R. Liu, F.A. Ponce, H. Amano, and I. Akasaki: Phys. Status Solidi B - Basic Res. 234 (2002) 850.
231. Z. Liliental-Weber, M. Benamara, W. Swider, J. Washburn, I. Grzegory, S. Porowski, and R.D. Dupuis, in Proceedings of MRS Fall Meeting – Symposium W: GaN and Related Alloys (R. Feenstra, T. Myers, M.S. Shur, and H. Amano, eds.), MRS Symposium Proceedings, Vol. 595, Materials Research Society, Boston, MA, November 29–December 3, 1999, p. W9.7.1.
232. K. Kumakura, T. Makimoto, and N. Kobayashi: J. Cryst. Growth 221 (2000) 267.
233. S. Nakamura, M. Senoh, S.-I. Nagahama, N. Iwasa, T. Yamada, T. Matsushita, H. Kiyoku, Y. Sugimoto, T. Kozaki, H. Umemoto, M. Sano, and K. Chocho: Jpn J. Appl. Phys. Part 2: Lett. 36 (1997) L1568.

234. K. Kumakura, T. Makimoto, and N. Kobayashi: in Proceedings of MRS Spring Meeting – Symposium T: Wide-Bandgap Electronic Devices (R.J. Shul, F. Ren, M. Murakami, and W. Pletschen, eds.), MRS Symposium Proceedings, Vol. 622, Materials Research Society, San Francisco, CA, 24–28 April 2000, p. T5.11.
235. E.L. Waldron, J.W. Graff, and E.F. Schubert: Appl. Phys. Lett. 79 (2001) 2737.
236. M.L. Nakarmi, K.H. Kim, J. Li, J.Y. Lin, and H.X. Jiang: Appl. Phys. Lett. 82 (2003) 3041.
237. M.Z. Kauser, A. Osinsky, A.M. Dabiran, and P.P. Chow: Appl. Phys. Lett. 85 (2004) 5275.
238. D.K. Ferry: Phys. Rev. B 12 (1975) 2361.
239. R.J. Trew, J.-B. Yan, and P.M. Mock: Proc. IEEE 79 (1991) 598.
240. W.V. Meunch and I. Pfaffeneder: J. Appl. Phys. 48 (1977) 4831.
241. W.V. Meunch and E. Pettenpaul: J. Appl. Phys. 48 (1977) 4823.
242. B.J. Baliga: Springer Proc. Phys. 71 (1992) 305.
243. A.O. Konstantinov, Q. Wahab, N. Nordell, and U. Lindefelt: J. Electron. Mater. 27 (1998) 335.
244. A.P. Dmitriev, A.O. Konstantinov, D.P. Litvin, and V.I. Sankin: Sov. Phys. Semicond. 17 (1983) 686.
245. A.O. Konstantinov, Q. Wahab, N. Nordell, and U. Lindefelt: Appl. Phys. Lett. 71 (1997) 90.
246. S. Nakamura, H. Kumagai, T. Kimoto, and H. Matsunami: Mater. Sci. Forum 389–393 (2002) 651.
247. G.N. Violina, Y. Liang-hsiu, and G.F. Kholuyanov: Sov. Phys. Sollid State 5 (1964) 2500.
248. G.A. Lomakina and Yu.A. Vodakov: Sov. Phys. Solid State 15 (1973) 83.
249. B.W. Wessels and J.C. Gatos: J. Phys. Chem. Solid 38 (1977) 345.
250. L.S. Aivazova, S.N. Gorin, V.G. Sidyakin, and I.M. Shvarts: Sov. Phys. Semicond. 11 (1978) 1069.
251. A. Suzuki, A. Ogura, K. Furukawa, Y. Fujii, M. Shigeta, and S. Nakajima: J. Appl. Phys. 64 (1988) 2818.
252. B. Ellis and T.S. Moss: Proc. R. Soc. Lond. A 299 (1967) 383.
253. B. Ellis and T.S. Moss: Proc. R. Soc. Lond. A 299 (1967) 393.
254. P.J. Dean, W.J. Choyke and L. Patrick: J. Lumin. 15 (1977) 299.
255. R. Kaplan, R.J. Wagner, H.J. Kim, and R.F. Davis: Solid State Commun. 55 (1985) 67.
256. J. Kono, S. Takeyama, H. Yokoi, N. Miura, M. Yamanaka, M. Shinohara, and K. Ikoma: Phys. Rev. B 48 (1993) 10909.
257. W.M. Chen, N.T. Son, E. Janzén, D.M. Hofmann, and B.K. Meyer: Phys. Status Solidi A 162 (1997) 79.
258. M.A. Il'in, A.A. Kukharskii, E.P. Rashevskaya, and V.K. Subashiev: Sov. Phys. Solid State 13 (1972) 2078.
259. S.A. Geidur, V.T. Prokopenko, and A.D. Yas'kov: Sov. Phys. Solid State 20 (1978) 1654.
260. R. Helbig, C. Haberstroh, T. Lauterbach, and S. Leibenzeder: Abs. Electrochem. Soc. Conf. 477 (1989) 695.
261. A.V. Mel'nichuk and Yu.A. Pasechnik: Sov. Phys. Solid State 34 (1992) 227.
262. W. Suttrop, G. Pensl, W.J. Choyke, R. Stein, and S. Leibenzeder: J. Appl. Phys. 72 (1992) 3708.

263. W. Götz: J. Appl. Phys. 72 (1993) 3332.
264. J. Bernhok, S.A. Kajihra, C. Wang, A. Antonelli, and R.F. Davis: Mater. Sci. Eng. B 11 (1992) 265.
265. A.G. Zubatov, V.G. Stepanov, Yu.A. Vodakov, and E.N. Mokhov: Sov. Tech. Phys. Lett. 8 (1982) 120.
266. A. Suzuki, H. Ashida, N. Furui, K. Mameno, and H. Matsunami: Jpn J. Appl. Phys. 21 (1982) 579.
267. J. Schmitt and R. Helbig: J. Electrochem. Soc. 141 (1994) 2262.
268. A. Rys, N. Singh, and M. Cameron: J. Electrochem. Soc. 142 (1995) 1318.
269. K. Ueno: Phys. Status Solidi A 162 (1997) 299.
270. D. Albrecht, R.P. Wang, P.P. Ruden, M. Farahmand, and K.F. Brennan: J. Appl. Phys. 83 (1998) 4777.
271. M. Akita, K. Kishimoto, and T. Mizutani: Phys. Status Solidi A: Appl. Res. 188 (2001) 207.
272. T. Inoue, Y. Ando, K. Kasahara, Y. Okamoto, T. Nakayama, H. Miyamoto, and M. Kuzuhara: IEICE Trans. Electronics E86-C (2003) 2065.
273. Y. Aoi, Y. Ohno, S. Kishimoto, K. Maezawa, and T. Mizutani: in Extended Abstracts of International Conference on Solid State Devices and Materials, Kobe, Japan, 13–15 September 2005, p. I-5–3.
274. L. Ardaravicius, A. Matulionis, J. Liberis, O. Kiprijanovic, M. Ramonas, L.F. Eastman, J.R. Shealy, and A. Vertiatchikh: Appl. Phys. Lett. 83 (2003) 4038.
275. B.E. Foutz, L.F. Eastman, U.V. Bhapkar, and M.S. Shur: Appl. Phys. Lett. 70 (1997) 2849.
276. Y. Ohno, T. Nakao, S. Kishimoto, K. Maezawa, and T. Mizutani: Appl. Phys. Lett. 84 (2004) 2184.
277. Y. Ando, Y. Okamoto, H. Miyamoto, N. Hayama, T. Nakayama, K. Kasahara, and M. Kuzuhara: in Proceedings of 2001 International Electron Devices Meeting, IEDM Technical Digest, Washington, DC, 2–5 December 2001, pp. 17.3.1–17.3.4.
278. T. Makimoto, K. Kumakura, and N. Kobayashi: Phys. Status Solidi C 0 (2002) 95.
279. K. Kunihiro, K. Kasahara, Y. Takahashi, and Y. Ohno: IEEE Electron Device Lett. 20 (1999) 608.
280. L. Hsu and W. Walukiewicz: J. Appl. Phys. 89 (2001) 1783.
281. T. Ando: J. Phys. Soc. Jpn 51 (1982) 3900.
282. I.P. Smorchkova, C.R. Elsass, J.P. Ibbetson, R. Vetury, B. Heying, P. Fini, E. Haus, S.P. Denbaars, J.S. Speck, and U.K. Mishra: J. Appl. Phys. 86 (1999) 4520.
283. L. Shen, S. Heikman, B. Moran, R. Coffie, N.Q. Zhang, D. Buttari, I.P. Smorchkova, S. Keller, S.P. Denbaars, and U.K. Mishra: IEEE Electron Device Lett. 22 (2001) 457.
284. D. Jena, A.C. Gossard, and U.K. Mishra: Appl. Phys. Lett. 76 (2000) 1707.
285. D. Jena and U.K. Mishra: Appl. Phys. Lett. 80 (2002) 64.
286. T. Murata, M. Hikita, Y. Hirose, Y. Uemoto, K. Inoue, T. Tanaka, and D. Ueda: IEEE Trans. Electron Devices 52 (2005) 1042.

3

Photonic Devices

3.1 Physical Properties

3.1.1 Fundamental Properties of Optical Devices (Y. Kawakami)

Structure and Quantum Efficiency

Relationship Between Emission, Power, and External and Internal Quantum Efficiency

Improving the emission efficiency of LEDs and other such optical devices is of paramount importance for use in the illumination and display industry. The definition and factors affecting emission efficiency will be discussed here.

In illumination engineering, the optical emission efficiency is defined as the total luminous flux that can be extracted from an optical source for a given input power. The units of efficiency are $\mathrm{lm\,W^{-1}}$.

Since the human eye is sensitive to light only in the 380–780 nm wavelength range, the emission efficiency with respect to infrared and ultraviolet light is 1 lm/W. Table 3.1 shows the relative visible sensitivity (V_λ) and spectroscopic emission efficiency (K_λ) per 1 W optical power. In practice, in order to produce an output of 1 W, it is necessary to inject a much larger power. This efficiency is known as the power efficiency (η_P) or "wall-plug" efficiency. The aforementioned emission efficiency in illumination engineering can be written as the product, $\eta_\mathrm{P} \cdot K_\lambda$. However, the physical properties of materials are more accurately understood by defining the optical emission efficiency in terms of the fraction of photons generated by the injection of electron–hole pairs into the active layer of the device. This is called the external quantum efficiency, η_ex.

If the light output from an LED is P_out, and the input power is IV, then the various efficiencies can be expressed by the following equations:

$$\eta_\mathrm{P} = \frac{P_\mathrm{out}}{IV} = \eta_\mathrm{v}\eta_\mathrm{ex} \tag{3.1}$$

Table 3.1. Wavelength dependence of visible sensitivity and spectroluminous efficiency

wavelength (nm)	visible sensitivity (V_λ)	spectroluminous efficiency K_λ (lm W^{-1})
380	0.0000	0.00
400	0.0004	0.27
420	0.0040	2.7
440	0.023	15.6
460	0.060	40.8
480	0.139	94.5
500	0.323	219.6
520	0.710	482.8
540	0.954	648.7
555	1.000	683.0
560	0.995	676.6
580	0.870	591.6
600	0.631	429.1
620	0.381	259.1
640	0.175	119.0
660	0.061	41.5
680	0.017	11.6
700	0.0041	2.8
720	0.00105	0.71
740	0.00025	0.17
760	0.00006	0.04
780	0.000015	0.01

Maximum is achieved at 555 nm

$$\eta_{\mathrm{ex}} = \frac{P_{\mathrm{out}}}{IV_{\mathrm{g}}} = \eta_{\mathrm{int}}\eta_{\mathrm{extr}} \tag{3.2}$$

Here, η_{v} is the voltage efficiency, V_{g} bandgap voltage of the active layer, η_{int} the internal quantum efficiency, and η_{extr} the optical extraction efficiency. The magnitudes of η_{v} and η_{int} are give by the following equations:

$$\eta_{\mathrm{v}} = \frac{V_{\mathrm{g}}}{V_{\mathrm{g}} + IR_{\mathrm{S}}} \tag{3.3}$$

$$\eta_{\mathrm{int}} = \frac{I_{\mathrm{rad}}}{I_{\mathrm{rad}} + I_{\mathrm{non-rad}} + I_{\mathrm{of}}} \tag{3.4}$$

Here, R_{S} is the total series resistance comprised of the resistive components due to the various device layers and the electrode contact resistances. Further, I_{rad} is the current due to radiative recombination, $I_{\mathrm{non-rad}}$ is the current component due to the non-radiative recombination (heat dissipation due to phonons), I_{of} is the over flow current from the active layer to the cladding layer that becomes significant in heterostructures and quantum well structures with insufficient carrier confinement or at high temperatures and conditions of high current injection.

For high efficiency LEDs, the most important parameters are η_{extr} and η_{int}. Optical extraction is limited by light generated within the semiconductor being totally reflected at the semiconductor surface. Light that is totally reflected is lost due to absorption by semiconducting and electrode layers of the device. These problems are minimized by (a) the use of a current diffusion layer for reducing optical recombination under the electrodes [1], (b) improvement of surface reflection efficiency by use of distributed Bragg reflectors [2], (c) use of surface texture structures for reducing total reflection [3], (d) use of reverse tapered device structures [4], and recently, the use of periodic structures, such as photonic crystal lattices [5, 6] and (e) surface plasmon coupling [7, 8].

Emission Lifetime and Nonemission Lifetimes

The recombination of carriers or excitons (electrons and holes) injected by optical excitation or forward biased pn junctions occurs by the two competing processes of radiative or non-radiative recombination. When a "pulse" of carriers is injected into a semiconductor, then the emission intensity after excitation is given by,

$$I(t) = I_0 \exp\left(-\frac{t}{\tau_{\text{L}}(T)}\right) \tag{3.5}$$

where $\tau_{\text{L}}(T)$ is the emission lifetime at temperature T. Here, the magnitude of $\tau_{\text{L}}(T)$ is the sum of the following three different recombination times,

$$\frac{1}{\tau_{\text{L}}(T)} = \frac{1}{\tau_{\text{rad}}(T)} + \frac{1}{\tau_{\text{non-rad}}(T)} + \frac{1}{\tau_{\text{trans}}(T)} \tag{3.6}$$

where $\tau_{\text{rad}}(T)$ and $\tau_{\text{non-rad}}(T)$ are the radiative and non-radiative recombination lifetimes, respectively; $\tau_{\text{trans}}(T)$ represents the time constant, where excitations distributed at higher energy band relax to lower energy bands. Assuming that recombination processes occur at the ground levels of the energy band then the following equation is valid,

$$\frac{1}{\tau_{\text{L}}(T)} = \frac{1}{\tau_{\text{rad}}(T)} + \frac{1}{\tau_{\text{non-rad}}(T)} \tag{3.7}$$

Also, assuming that overflow effects are negligible, then the internal efficiency of optical emission, $\eta_{\text{int}}(T)$, is given by,

$$\eta_{\text{int}}(T) = \frac{1/\tau_{\text{rad}}(T)}{1/\tau_{\text{rad}}(T) + 1/\tau_{\text{non-rad}}(T)} = \frac{\tau_{\text{non-rad}}(T)}{\tau_{\text{rad}}(T) + \tau_{\text{non-rad}}(T)} \tag{3.8}$$

Or, if such effects cannot be neglected, then effects due to overflow can be assumed to be incorporated in the term, $\tau_{\text{non-rad}}(T)$. Thus, if it is possible to measure $\tau_{\text{L}}(T)$ and $\eta_{\text{int}}(T)$ simultaneously, then it is possible to analyze the temperature dependence of both radiative and non-radiative processes. Further, $\eta_{\text{int}}(T)$ can be determined by simultaneously measuring the external quantum and extraction efficiencies. However, the optical extraction efficiency varies with sample structure and surface flatness, thus it is difficult to determine precise values for these parameters. Also, since these processes occur inside semiconductors, the magnitude of $\eta_{\mathrm{int}}(T)$ cannot be measured from outside the semiconductor. However, as described below, $\eta_{\mathrm{int}}(T)$ can be determined if certain assumptions are valid.

That is, assuming that the density of non-radiative centers is $N_{\mathrm{non-rad}}$, the carrier capture cross-section of non-radiative centers is σ_{t}, the carrier thermal velocity is v_{th}, then the non-radiative lifetime, $\tau_{\mathrm{non-rad}}(T)$, is given by,

$$\frac{1}{\tau_{\mathrm{non-rad}}(T)} = N_{\mathrm{non-rad}} \times \sigma_{\mathrm{t}} \times v_{\mathrm{th}} \tag{3.9}$$

Thus it is possible to neglect non-radiative emission processes at low temperatures in the case of high quality crystals with a small $N_{\mathrm{non-rad}}$ and small values of σ_{t} and v_{th}. Figure 3.1 shows the temperature dependence of the photoluminescence (PL) from InGaN single quantum wells. The PL intensity is constant below 50 K. Thus the internal emission quantum efficiency can be approximated as being 1 within this temperature range. If the emission intensity is I_{C}, and the temperature range over thermal quenching occurs is $I(T)$, then the radiative and non-radiative lifetimes can be written as [9],

$$\tau_{\mathrm{rad}}(T) = \tau_{\mathrm{L}}(T) \times \frac{I_{\mathrm{C}}}{I(T)} \tag{3.10}$$

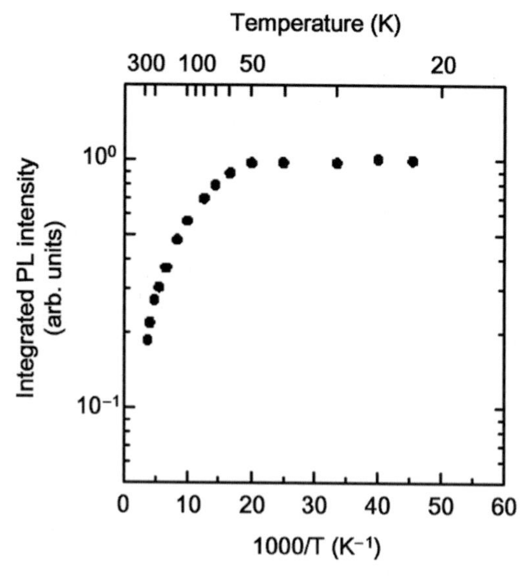

Fig. 3.1. Temperature dependence of the photoluminescence (PL) from a single InGaN quantum well

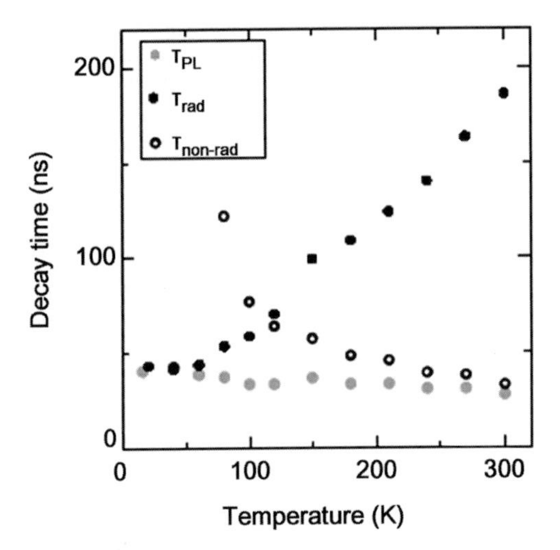

Fig. 3.2. Dependence of the PL lifetime, radiative, and non-radiative lifetimes for blue optical emission from single InGaN quantum well

$$\tau_{\mathrm{non-rad}}(T) = \tau_{\mathrm{L}}(T) \times \frac{I_{\mathrm{C}}}{I_{\mathrm{C}} - I(T)} \tag{3.11}$$

Figure 3.2 shows the temperature dependence of the PL lifetime, radiative, and non-radiative lifetimes for blue optical emission from InGaN single quantum wells.

Recombination Dynamics of Localized Excitons

In quantum structures incorporating alloys as active layers there are localized nonuniformities due to fluctuations in quantum well widths and alloy composition. These effects result in potential fluctuations with associated localization of carriers and excitons. Localization effects are particularly pronounced in wide-bandgap semiconductors such as InGaN due to the following properties of the GaN and InN binary compounds (a) large difference between the bandgap energies, (b) relatively easy generation of composition modulation due to the large differences in the lattice constants [10], and (c) generation of differences of the alloy composition of the volumes occupied by excitons due to their small radii [11, 12]. Narrow localized centers have been reported to have a positive effect on the high optical efficiency of LEDs since such centers reduce the probability of excitons being trapped by non-radiative centers. However, the gain of lasers is reduced by such effects as a result of increases in the nonuniformity of the width of optical emission. Analysis of the energy emission peaks observed from time resolved luminescence can be used to determine emission lifetimes, $\tau_L(E)$ of localized structures. An example is shown in Fig. 3.3, where analysis of the shift of the emission energy to lower energies is used to study the localization process and the lifetime of localized excitons [14]. If localized energy levels, as shown in Fig. 3.4,

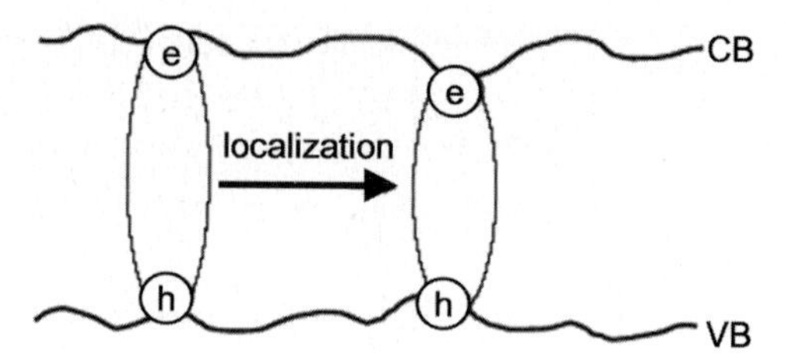

Fig. 3.3. Illustration of the model used to study the localization process and the lifetime of localized excitons

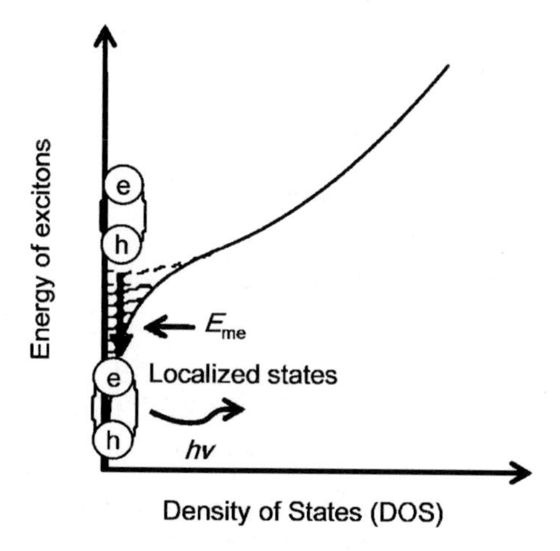

Fig. 3.4. Illustration of the localized energy levels of excitons

can be approximated by the equation $g(E) = A\exp(-E/E_0)$, then E_0 represents the degree of localization. Further, assuming that non-radiative recombination is negligible and that the emission lifetime, τ_{rad} is energy independent at the energy band being considered, then the exciton density, $n(E,t)$, at an energy, E, and a time, t, following an excitation energy pulse, can be expressed by the following rate equations [14],

$$\frac{\mathrm{d}n(E,t)}{\mathrm{d}t} = -\frac{n(E,t)}{\tau_{\mathrm{rad}}} - \sum_{E'<E} W(E \to E')n(E,t) + \sum_{E'>E} W(E' \to E)n(E',\, t) \tag{3.12}$$

$$W_{\mathrm{tr}}(E) \propto \int_E^{\infty} g(E')\mathrm{d}E' \propto \exp(-E/E_0) \tag{3.13}$$

where $W(E \to E')$, is the probability of transitions from energy, E to E', $W_{\mathrm{tr}}(E)$ is the sum of all the probabilities of transitions from levels at energies less than E. The solutions of these equations leads to the following expressions for $n(E,t)$ and $\tau_L(E)$,

$$n(E,t) = n_0 \exp\{-t/\tau_{\mathrm{L}}(E)\} \tag{3.14}$$

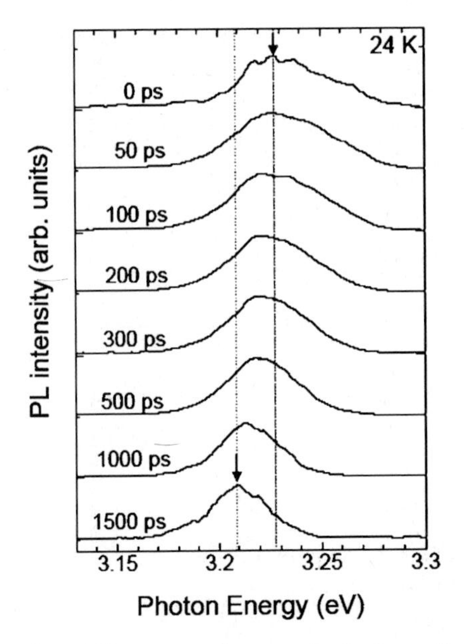

Fig. 3.5. Time-dependent PL spectra from a single $In_{0.08}Ga_{0.92}N$ epitaxial layer with a thickness of 0.1 μm

$$\tau_L(E) = \frac{\tau_{rad}}{1 + \exp\{(E - E_{me})/E_0\}} \tag{3.15}$$

where E_{me} is the energy level corresponding to the mobility edge.

Figure 3.5 shows the time-dependent PL spectra from a single $In_{0.08}Ga_{0.92}N$ epitaxial layer with a thickness of 0.1 μm. The PL peak immediately after excitation is at 3.227 eV, but shifts to lower energies with time and is 3.208 eV after 1,500 ps [15]. These results show the classical dynamics of localized excitons. Figure 3.6 shows the time-integrated PL spectra and $\tau_L(E)$ for this sample. The dotted line shows a fitting of the experimental results using τ_{rad}, E_0, and E_{me} [15]. It is important to note that these results represent the physical parameters of the average degree of localization inside the optical excitation spot of light. Scanning near-field optical microscopy (SNOM) measurements show that the emission lifetime and depth of localization have a submicron spatial distribution [16, 17]. Again, it should be noted that the aforementioned analytical process is based on simplistic approximations. For example, in semiconductor systems with significant nonuniformity, the time dependence of the emission strength cannot be fitted by simple single exponentials. Multiexponential and stretched exponential treatments have been reported [18, 19]. Further, as shown in Fig. 3.7, the quantum confinement Stark effect (QCSE) is observed in systems such as InGaN/GaN quantum wells to exhibit strong piezoelectric effects. In such systems, the internal electric field screening due to optically generated carriers, the transition energy, and optical emission probability are large immediately after excitation but

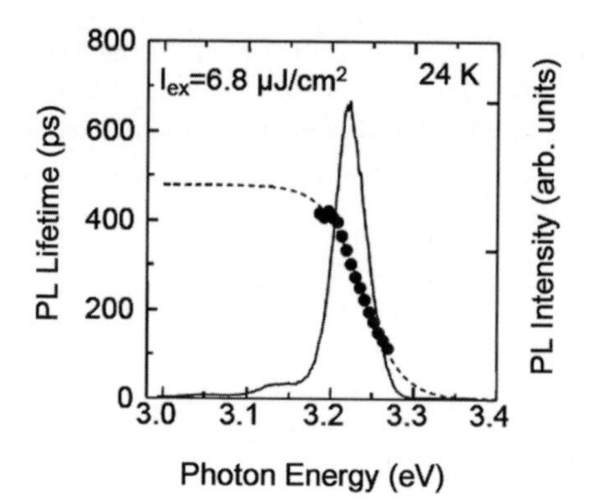

Fig. 3.6. Time-integrated PL spectra and PL lifetime for the same sample shown in Fig. 3.5

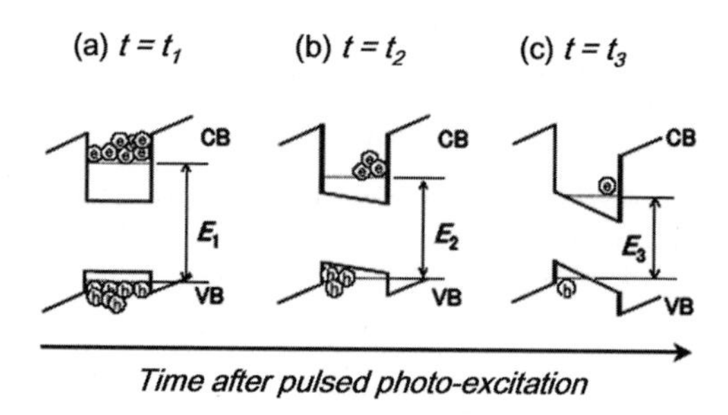

Fig. 3.7. Band structure and transition energy of a quantum well irradiated with a pulse excitation under a piezoelectric field

decrease with decreasing carrier concentration as time elapses. This process appears to be the same as the localized dynamics of excitons but a careful analysis is required for a deeper understanding of the actual processes leading to the observations [20].

Emission Lifetime and Degrees of Exciton Confinement

Quantum well layers are used in many optical device structures. Ideally, the QW structures should be two dimensional but in reality potential fluctuations occur in most cases. Thus localized centers can be considered as being like quantum dots and zero dimensional. Here, the term "zero dimensional" means that excitons are localized at the ground state and excited to higher localized levels by thermal energy, but the excitons do not move to continuous levels in nonlocalized states. Thus in weakly localized systems, zero-dimensional effects are only observed at ultra-low temperatures. In order to realize zero-dimensional systems at room temperature, it is necessary that the energy difference between the ground and first excited states should be approximately two times the thermal energy at room temperature on top of the requirements of sufficient potential confinement and a small structure. Sugawara et al. [21] derived the following equation to calculate the emission lifetime, $\tau_{\mathrm{rad,0D}}$ required for the observation of zero-dimensional localized excitons,

$$\frac{1}{\tau_{\mathrm{rad,0D}}} = \frac{e^2 n_{\mathrm{r}} \omega_{\mathrm{ex}} \mu_{\mathrm{cv}}^2 F^2 \left|\phi(0)\right|^2}{\pi m_0^2 \varepsilon_0 \hbar c^3} \times 2\pi\beta^2 \tag{3.16}$$

Here, $e, m_0, \varepsilon_0, \hbar, c$ are the elementary charge, the mass of a stationary electron, dielectric constant in a vacuum, Planck's constant, and velocity of light, respectively. Further, $n_{\mathrm{r}}, \omega_{\mathrm{ex}}, \mu_{\mathrm{cv}}, F, \phi$ are the refractive index, the angular frequency of optical transition, the moment matrix of the conduction band–valence band transition, the overlap integral of the electron and hole wavefunctions in the z-direction, and the wavefunction for in-plane exciton motion, respectively. In addition, β represents the degree of expansion of the exciton wave function inside the surface, $\sqrt{2}\beta$ is the radius of the localized center. Here, the effect of QCSE as described in "Recombination Dynamics of Localized Excitons" is given by F.

Further, it is known that as the localized center decreases in size, that is the zero dimensionality becomes more pronounced, the optical lifetime increases. However, in order for (3.1)–(3.16) to be valid, the thermal excitation of excitons must be negligible. In cases where thermal excitation occurs, the optical emission lifetime increases with increasing temperature due to the generation of excitons possessing large barycentric coordinate momentum, which do not satisfy the momentum conservation laws for light emission [22]. Thus, from measurements of the temperature dependence of emission lifetimes it is possible to calculate the β, a parameter that indicates the localization of excitons [23–25].

3.1.2 Optical Gain and Laser Action (S. Kamiyama)

Inherent Properties of the Valence Band of Nitride Semiconductor Systems

The band structure and optical gain of nitride semiconductors that have a wurtzite crystal structure are different than conventional semiconductors with zinc blende structures. Figure 3.8 shows the energy dispersion curves of GaN near its band edge [26, 27]. There are three energy states at $k = 0$ of the valence band edge due to the low symmetry of the crystal lattice. The heavy hole (HH), $\Gamma_9^{3/2}$ is located at the valence band edge, the light hole (LH) is in the middle at $\Gamma_7^{3/2}$, and the deepest energy level is called the crystal field split hole (CH), $\Gamma_7^{1/2}$. The Bloch functions of these states are given as,

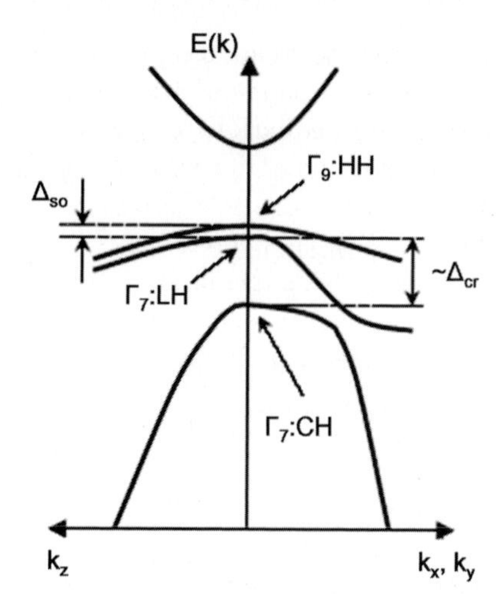

Fig. 3.8. Energy dispersion curves of GaN near its zone center

Table 3.2. Effective mass of GaN

electron		HH		LH		CH	
$M_{e\perp}$	$m_{e//}$	$m_{h\perp}$	$m_{h//}$	$m_{h\perp}$	$m_{h//}$	$m_{h\perp}$	$m_{h//}$
$0.23m_0$	$0.20m_0$	$1.61m_0$	$1.76m_0$	$0.14m_0$	$1.76m_0$	$1.04m_0$	$0.16m_0$

x-, y-axis is ⊥, z-axis is || corresponding to the normal orientation of c-axis

$$\begin{aligned}
\left|\Gamma_9^{3/2}\right\rangle &= \frac{1}{\sqrt{2}}(X+iY)\uparrow, \frac{1}{\sqrt{2}}(X-iY)\downarrow \\
\left|\Gamma_7^{3/2}\right\rangle &= \frac{1}{\sqrt{2}}(X+iY)\downarrow, \frac{1}{\sqrt{2}}(X-iY)\uparrow \\
\left|\Gamma_7^{1/2}\right\rangle &= Z\uparrow, Z\downarrow
\end{aligned} \tag{3.17}$$

The X, Y, Z in (3.1)–(3.17) are the unit vectors in Cartesian coordinates, where the Z component is the *c*-direction of the crystal lattice and the XY components correspond to in-plane directions. The HH and LH splitting energy shown by Δ_{so}, is due to spin–orbit interaction and the splitting energy between LH and CH, shown as Δ_{cr}, is mainly due to the crystal field. On the other hand, the conduction band can be thought of being an s-like state with only weak anisotropy in the *c*-direction and directions perpendicular to c-axis – the same as for zinc blende crystal structures.

Table 3.2 shows the effective masses of various states near the zone center. The effective masses in both the conduction and valence bands are seen to be extremely large compared with conventional semiconductors used for fabricating lasers. Thus high carrier densities are necessary in order to obtain population inversion for laser action. Further, nitride semiconductors exhibit changes in densities of states due to the dimensional confinement of carriers, but not as large as in zinc blende structures, where large changes in the densities of states occur due to reductions of mixing in the valence bands. This is because the HH and LH bands located at the band edge have almost the same effective masses in the z-direction, which is quantized, and the energy difference between these bands does not change due to quantum size effects. Only the CH band has a different effective mass in the z-direction with the result that the relative energy does change but since the CH is energetically separated from the band edge, laser action is not significantly affected. Furthermore, the results are the same in the case of the biaxial strain present in quantum wells.

The above discussion has been about GaN, but the same features are seen in the case of GaInN, which is generally used as an active layer in nitride devices. The addition of In results in a slight decrease in the effective mass of GaInN but the mole fraction of InN is only ~10%, which is not too different from GaN, and the effective mass is similar to GaN.

Thus as described earlier, the effective masses of electrons and holes in nitride semiconductors are large and, due to the large density of states near the band edges, the use of these semiconductors leads to large threshold current densities.

Optical Gain in Nitride Semiconductors

The optical gain of quantum well structures forming the active layers of semiconductor lasers can be derived using

$$g(E) = \frac{\pi q^2 \hbar}{c\bar{n}\varepsilon_0 m_0^2 E} \sum_{k_{||}} \frac{1}{L_z} \left|M_{cv}\left(k_{||}\right)\right|^2 \left[f_c\left(k_{||}\right) + f_v\left(k_{||}\right) - 1\right] \times \frac{1}{\pi} \frac{\hbar/\tau_{in}}{\left[E_{cv}\left(k_{||}\right) - E\right]^2 + \left(\hbar/\tau_{in}\right)^2} \tag{3.18}$$

Here, the inherent characteristics of GaN result from the dipole matrix elements, $|M_{cv}(k_{||})|^2$, the density of bonds, $\sum_{k_{||}}$, and relaxation time inside the band, τ_{in}. The dipole matrix element, $|M_{cv}(k_{||})|^2$, includes the square of the overlap integral of the electron and hole wavefunctions, and as shown by (3.1)–(3.17), it includes the overlap of the Bloch functions of the hole and the s-like Bloch function of electrons and the electron and the hole envelope functions determined by the shape of the quantum well potential.

In cases where the well width is relatively large, there is a drastic reduction of the overlap between the envelope functions of the electrons and holes due to an effect of the internal electric field that arises as a result of the inherent properties of the wurtzite structure. In actual device structures, thin quantum wells with widths of less than 4 nm are used to reduce such effects [29]. Further, the polarization of photons emitted during recombination depends on the

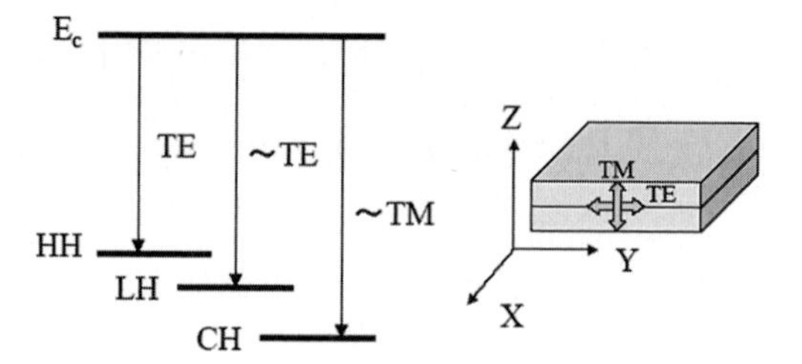

Fig. 3.9. Polarization allowed for photons emitted during recombination

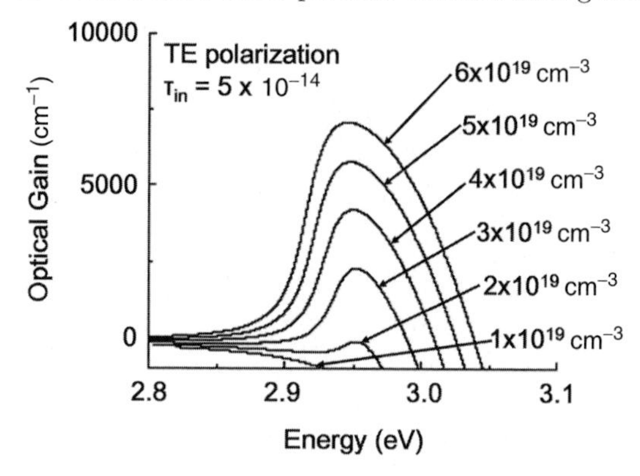

Fig. 3.10. Gain spectrum of GaInN quantum wells calculated using this parameter for TE polarization

dipole anisotropy (which is included in the hole Bloch wavefunctions), and only the polarizations shown in Fig. 3.9 are allowed. The density of bonds near the band edge depends on the effective mass of electrons and holes, which is large in GaN-based semiconductors. However, a large bond density is not necessarily an advantage for the optical gain since a reduction of the Fermi distribution function $f_c(k_{||})$ or $f_v(k_{||})$ can lead to decreases in the optical gain.

Further, the intraband relaxation time, τ_{in}, corresponds to a carrier collision time of 10^{-13} s. This parameter is the main cause of broadening of the gain spectrum. However, in GaInN alloys the broadening is also affected by nonuniformities in the alloy composition, and the intraband relaxation time is estimated at 5×10^{-14} s, after accounting for this new parameter. This magnitude agrees with the experimental results [30].

Figure 3.10 shows the gain spectrum of GaInN quantum wells calculated using this parameter for TE polarization. The maximum gain corresponds to transitions between the ground states of the conduction band and the HH level, where only the TE polarization is allowed. A closer examination of the variation of the gain with peak transition energy shows an inversion from absorption to gain above an excited carrier density of 10^{19} cm^{-3}. The carrier density for the population inversion is approximately one order of magnitude larger than for GaAs, thus necessitating the injection of an extremely high density of carriers to produce laser oscillation. The population inversion is difficult to achieve in GaN-based materials due to the large density of bond states in these materials. As described later, these physical properties of nitride semiconductors are the reason for the large threshold currents required for inducing the laser action.

Device Structures

Figure 3.11 shows two typical device structures for nitride semiconductor lasers. In Fig. 3.10a, a blue-violet semiconductor laser is fabricated on a sapphire substrate, with the growth carried out in two steps in order to reduce the dislocation density. The growth and fabrication is carried out as follows (1) growth of a low temperature buffer layer on the sapphire substrate by MOVPE, (2) GaN layer deposition, (3) formation of striped dielectric mask, (4) growth of the remaining semiconducting layers including the GaInN/GaN multiquantum well structures [31]. Semiconductor crystals do not grow on the dielectric masked areas. Semiconductor layers formed in the mask opening grow laterally, covering the dielectric mask and producing a flat surface. The use of lateral growth enables the deposition of semiconductor layers with extremely low dislocation densities (apart from the center of the mask opening) and the fabrication of high performance devices. It is noteworthy that since sapphire substrates do not conduct electricity, it is necessary to form n-type electrodes on the surface of the semiconductor layers. Thus it is necessary to etch the device structure to expose the n-GaN layers, as shown in Fig. 3.11a.

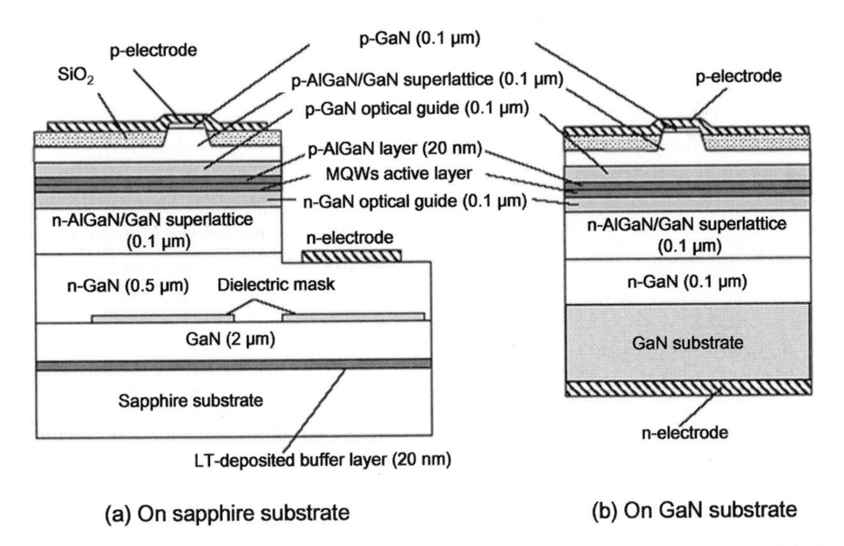

Fig. 3.11. Two typical device structures for nitride semiconductor lasers. (**a**) Sapphire substrate and (**b**) GaN substrate

Figure 3.11b shows a homoepitaxial laser structure grown on a conducting GaN single crystal substrate [32], where a low temperature buffer is not necessary. Further, the almost negligible dislocation density of structures grown homoepitaxially on GaN substrates enables the fabrication of laser devices in a single step [33]. Also, the n-type contact can be produced from the back surface of the substrate, without etching the n-GaN layer as in the processing of lasers fabricated using conventional III–V semiconductors. The high costs of using GaN substrates are tempered by the simplicity of the process used for device fabrication.

The "ridge-type" optical waveguides in both of the structures shown in Fig. 3.11a, b are produced by buried dielectric layers. The GaInN quantum well active layers are thermally unstable because they result from re-growth of the semiconductor layer, a process also used in GaAs- and AlGaInP-based lasers. However, the fabrication of ridge-type optical waveguides necessitates extremely precise and reproducible etching to produce structures of less than 2 μm width, in order to stabilize the in-plane lateral mode emission. High precision technology for fabricating these structures is mature but the use of extremely narrow stripes results in p-type electrodes with small surface areas which in turn result in high contact resistances of several tens of ohms. Thus the operating voltage of blue-violet semiconductor lasers is high and improvements in device structures are desirable.

Lasing Characteristics

The characteristics of nitride blue-violet semiconducting lasers are good enough for use as light sources in commercial high density optical storage systems. Table 3.3 shows a comparison of nitride and AlGaInP-based red lasers used in DVD systems. It can be seen that the threshold currents, the slope efficiency, the beam spreading angle, the aspect ratio, and the astigmatic difference of nitride lasers is comparable with conventional red lasers. The

Table 3.3. Characteristics of nitride based semiconductor laser

items	AlGalnP-based red semiconductor laser	Nitride-based violet semiconductor laser
threshold current 1_{th}	50 mA	45 mA
slope efficiency Se	$1\,W\,A^{-1}$	$1.2\,W\,A^{-1}$
operation current 1_{op}	120 mA (70 mW)	100 mA (60 mW)
operation voltage V_{op}	2.6 V (70 mW)	5.0 V (60 mW)
specific temperature T_0	130 K	210 K
spreading $\theta_{//}$	10°	8°
angle $\theta_{\perp}$	17°	20°
aspect ratio	1.7	2.5
difference of non interstice ΔZ	2 μm	2 μm
oscillation wavelength λ	650 nm	405 nm
relative intensity of noise RIN	$-135\,dB\,Hz^{-1}$	$-125\,dB\,Hz^{-1}$

blue-violet semiconductor laser incorporates a simple ridge stripe structure (dielectric) to control the characteristics of the lateral mode and beam. In the buried ridge stripe used in infrared or red semiconductor lasers, stable operation is possible even for relatively large stripe widths, but a very narrow stripe of 1.5–2 μm is required in blue-violet semiconductor lasers. However, the threshold current density at $\sim$2.4 kA cm^{-2} in nitride devices is several times higher than for conventional lasers due to the large contact resistance of the p-electrode and p-type layer in these structures. Further, the relative noise intensity (RNI) during high frequency superposition is an important parameter for optical disk applications, and again this factor is larger in nitride lasers than conventional red semiconducting lasers. Further, the temperature dependence of the threshold current, T_0, is extremely large for blue-violet lasers.

3.2 Visible LEDs

3.2.1 Visible AlGaInP LEDs (G. Hatakoshi)

Properties of AlGaInP

AlGaInP-based materials have direct bandgaps and emit in the red to green spectral region. Red laser diodes used as light sources in the DVD system are made with this semiconductor. The material is also extremely useful for fabrication of LEDs.

An attractive technique for metalorganic chemical vapor deposition (MOCVD) of AlGaInP material system is epitaxial growth on off-angle substrates. The bandgap energy of AlGaInP alloys depends on the presence or absence of natural superlattices that can be formed under certain growth conditions, where disordered structures are known to have larger bandgaps. The disordered state is enhanced by using off-angle substrates with crystal planes orientated from the (100) to the [011] direction. This technique is effective for shortening the emission wavelength without increasing the alloy's Al content, which has a significant effect on the formation of undesirable non-radiative recombination centers associated with incorporation of oxygen impurities. Short-wavelength and high-efficiency AlGaInP LEDs on off-angle substrates have been reported [34].

AlGaInP LEDs with wavelengths in the range of 640–560 nm are commercially available. As shown in Fig. 3.12, the emission spectra of AlGaInP LEDs are narrower than those for the GaP and GaPAs LEDs with indirect bandgaps. The fabrication of short-wavelength and high-efficiency devices necessitates use of highly doped AlGaInP cladding layers with high Al content. The use of off-angle substrates enables incorporation of p-type impurities (Zn) at a high density, limiting oxygen contamination as well [35,36]. Carrier concentrations of more than 10^{18} cm^{-3} have been reported for AlInP alloys grown on off-angle substrates [37].

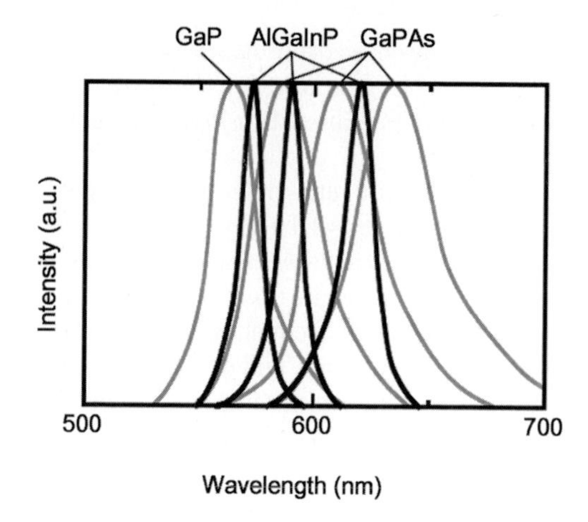

Fig. 3.12. Comparison of the emission spectra of AlGaInP, GaP, and GaPAs LEDs

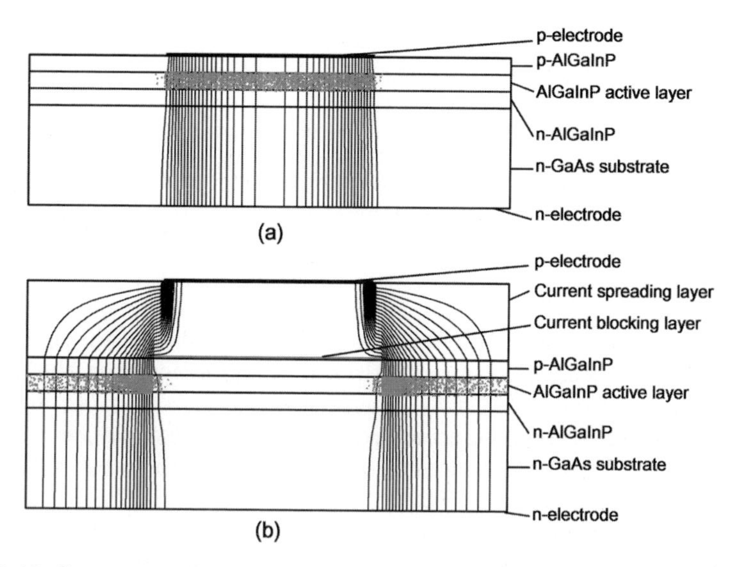

Fig. 3.13. Structure and current distribution of AlGaInP-based LEDs. (**a**) Conventional LED and (**b**) a structure incorporating a current spreading layer and current blocking layer

Device Design for High-Efficiency LEDs

In the initial stage of the development of AlGaInP LEDs, the optical extraction efficiency was extremely low due to difficulties in the growth of thick AlGaInP layers and low-resistivity p-type cladding layers. Thus, as shown in Fig. 3.13a, in conventional LED structures, the current in the cladding layer does not spread laterally and optical emission only occurs immediately under

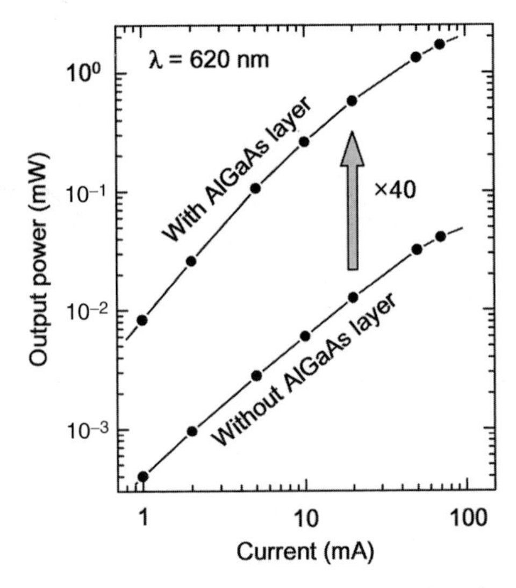

Fig. 3.14. Effect of the current spreading layer

the electrode, which blocks the passage of light. This problem was resolved by fabricating the structure shown in Fig. 3.13b, where a low-resistivity p-AlGaAs "current-spreading layer" and a "current-blocking layer" were grown on the p-type cladding layer, thereby reducing the current flowing directly below the electrode [34, 38]. As shown in Fig. 3.14, the light output increased 40-fold due to insertion of the current-spreading layer [38–40].

GaAs, used as a substrate for the growth of AlGaInP material system, has a large absorption coefficient at the emission wavelength of AlGaInP LEDs, and thus the light directed at the substrate would be absorbed and not emitted externally. This problem is circumvented by use of multilayered distributed Bragg reflectors (DBR) [41–43] and transparent GaP substrates [44].

DBRs are widely used in vertical cavity surface-emitting lasers (VCSELs), where a high reflectivity of near 100% is required. In the case of LEDs, such high reflectivity is not always necessary, but it is important to broaden the reflection bandwidth in order to maintain a realistic fabrication tolerance. DBRs for AlGaInP LEDs can be constructed by AlInP/AlGaInP and AlInP/GaAs. The former is transparent to the light of AlGaInP LEDs, but the refractive index difference between AlInP and AlGaInP is not very large and it is necessary to increase the number of layers to improve reflection efficiency, with the result that the reflection bandwidth becomes extremely narrow. On the other hand, the use of GaAs results in some absorption losses but the refractive index difference between AlInP and GaAs is large and it is possible to attain high reflection efficiency using a small number of layers, and hence obtain a large reflection bandwidth. DBR structures incorporating both materials have been reported [43], enabling the possibility of a twofold increase in the extraction efficiency.

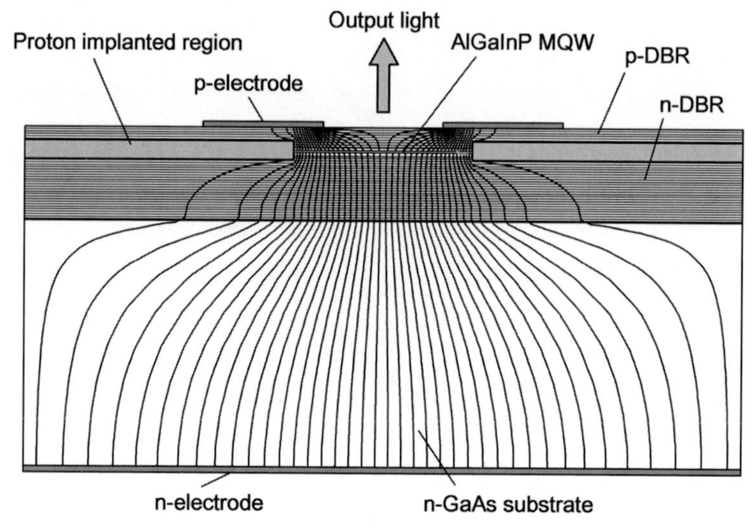

Fig. 3.15. Structure of a AlGaInP-based resonant cavity LED

Resonant cavity LEDs (RC-LEDs), as shown in Fig. 3.15, have also been reported [45,46], where the light extraction efficiency is improved by the effect of emission angle selectivity of resonant cavity and photon recycling.

Generally, the light extraction efficiency in semiconductor LEDs is limited by the total reflection when light passes from the semiconductor to the air. This problem has been resolved by the introduction of structure reshaping [47], random surface texturing, and photonic crystals [48–50].

Device simulation of LEDs [39, 51, 52] including the analysis of the electrical, optical, and thermal properties has become more important due to recent progress in the development of high-efficiency LEDs as light sources for display and illumination. AlGaInP LEDs have played an important role in the development of these technologies.

3.2.2 Visible GaN LEDs (T. Mukai, Y. Narukawa)

Development of Visible GaN LEDs

In a series of pioneering experiments on the growth of GaN, Ammano and Akasaki reported (a) on the use AlN low temperature buffer layers to dramatically improve the crystalline quality of GaN thin films [53], (b) the control of n-type conductivity GaN by Si doping [54], (c) p-type doping of GaN by Mg doping [55] and electron beam irradiation [56]. In 1989, based on these initial findings, Ammano and Akasaki [56] fabricated the first GaN pn junction light-emitting diode. Then in 1991, Yoshimoto et al. [57] produced high quality InGaN ternary alloys which made possible the fabrication of double heterostructure LEDs and realization of GaN-based visible LEDs. Then in

1993 Nichia Corporation announced the sale of high brightness blue LEDs incorporating InGaN emission layers [58]. After this, the emission output power of GaN-based LEDs has improved on a yearly basis [59–61] and now, in 2006, the emission output power of commercial LEDs is approximately 20 mW, a figure of merit that is ten times higher than the first commercial LEDs. Further, green, yellow, red, and ultraviolet LEDs are available [62–64].

Characteristics of GaN LEDs

One of the unique properties of (Al,Ga,In)N group III nitride semiconductors is the existence of a direct bandgap ranging from 0.6 eV for InN to 6.2 eV in AlN, enabling the use of these materials for fabricating devices emitting from the visible to vacuum ultraviolet (UV) region. That is, group III nitrides can be used for generating all wavelengths of light.

High efficiency GaN-based LEDs consume less energy and have longer life times compared with conventional incandescent lighting and they are being increasingly used for industrial and household illumination.

History of the Development of GaN LEDs

High brightness GaN-based LEDs were first manufactured for commercial sales in 1993 when technology for increasing the In composition of the InGaN emitting layer had not yet been established. And as a result, donor–acceptor pair recombination from a double heterostructure doped with Zn and Si was used since it was difficult to obtain blue emission via band to band recombination. These structures enabled fabrication of LEDs emitting blue light at ∼450 nm with high intensity of ∼1 cd. Figure 3.16 shows the structure of a

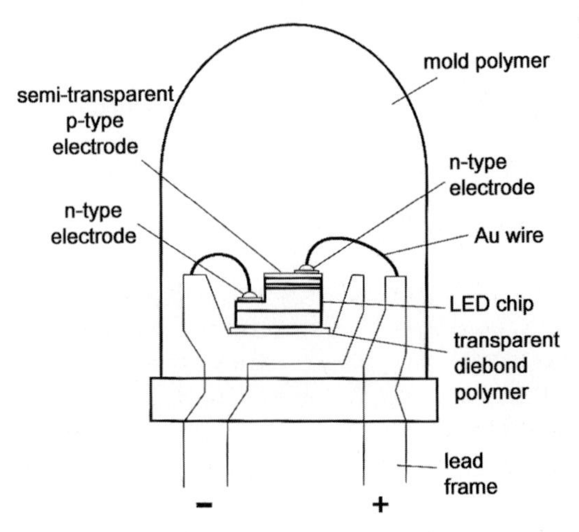

Fig. 3.16. Structure of a typical double heterostructure LED (DH-LED)

typical double heterostructure LED (DH-LED). Due to the insulating sapphire substrate, the p- and n-type electrical contacts are made from the surface of the device structure as opposed to the use of substrates in conventional red-LED device structures that employ conducting substrates. The use of the semitransparent p-type electrodes improves the light extraction efficiency as well as increasing the diffusion currents and thereby reducing the operating voltage. This innovative device structure represents a major milestone in the history of the development of group III nitride LEDs. Also, compared with conventional red LEDs that rely on band edge emission, blue LEDs fabricated using group III nitrides exploit impurity related recombination from the InGaN layer doped with Zn and Si. Thus GaN-based blue LEDs have not just replaced conventional red LEDs. They can be considered as being a unique technology.

Figure 3.17 shows the emission spectrum of a blue LED at an operating current of 20 mA, that has a peak wavelength at 450 nm, a full width half maximum of 70 nm and an output power of 1.5 mW. These LEDs have are extensively used in traffic lights as shown in Fig. 3.18. LEDs installed in such

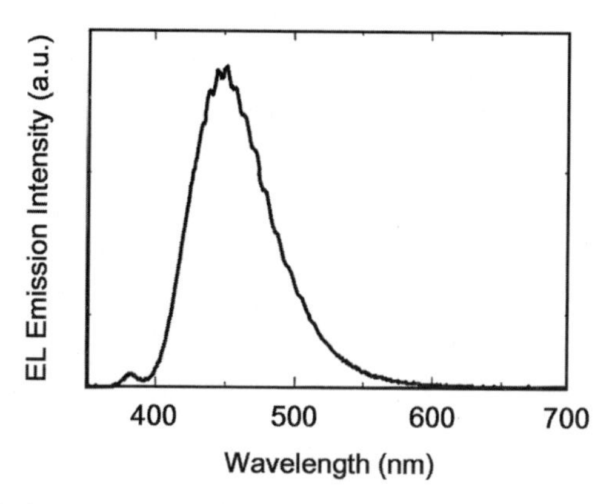

Fig. 3.17. Emission spectrum of a blue LED at an operating current of 20 mA

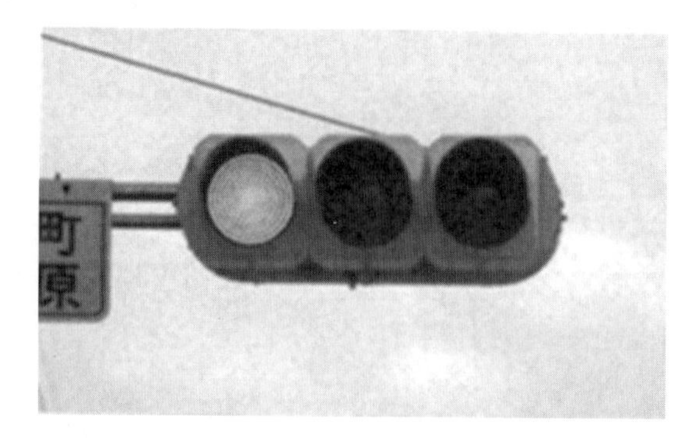

Fig. 3.18. Blue LEDs used in traffic lights

equipment 10 years ago are still operating today underscoring the long life time of these devices. However, in spite of the long lifetimes, as can be seen in the emission spectrum, the full width at half maximum of the emission peak is large and there is a small "white" component due to the poor chromatic purity of the light. This makes the use of the LEDs in displays difficult. Further, as shown in Fig. 3.19, as the operating current is increased, the impurity related emission starts to saturate and band edge related emission appears at 380 nm. Furthermore, the mold plastic is damaged at high operating currents. These problems were resolved by increasing the In content of the InGaN emission layer and using blue band edge emission from this layer as opposed to the impurity related emission in earlier structures. The major result of such experiments was the development of group III nitride LEDs incorporating quantum well (QW) structures. The use of QW structures enabled dramatic improvement of the efficiency of electron–hole recombination in the well layer. Further, since an extremely thin layer of InGaN is used, it became possible to use InGaN layers with large In contents. As a result, in 1995, the first single QW blue LEDs (SQW blue LED) were commercialized. The emission spectra of typical SQW-LED and DH-LED operating at 20 mA, are shown in Fig. 3.20. The FWHM of the SQW-LED is 30 nm, less than half of the 70 nm for the DH-LED. The use of quantum wells in LED device structures resolved the problems of chromatic purity of the light emission, degradation of the mold plastic, and enabled the production of high brightness LEDs with long lifetimes. Furthermore the output power of the SQW-LED was 3 mW, which is approximately double that of DH-LEDs, and by increasing the In content of the emission layer, pure green LEDs (SQW-green LED) of high brightness (6 cd) were also realized; at that time this brightness was 60 times the value achieved in GaP LEDs. The new blue and green LEDs were combined with

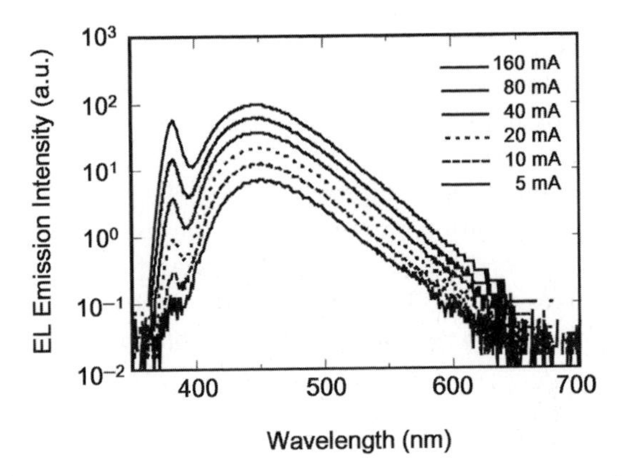

Fig. 3.19. Operating current dependence of the emission spectra of a DH structure blue LED

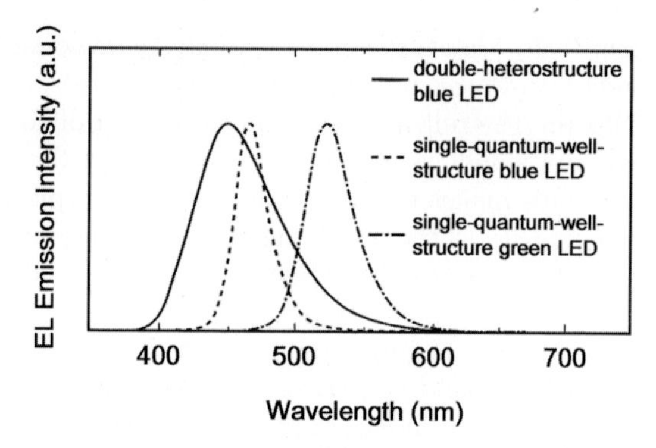

Fig. 3.20. Emission spectra of typical SQW-LED and DH-LED operating at 20 mA

Fig. 3.21. Blue, green, and red LEDs used for a full color display

conventional red emitters for the development of full color displays as shown in Fig. 3.21, which can now be seen in many street corners.

In 1996, blue LEDs and YAG phosphors were combined to produce unique white LED structures which are now used as back lights in the compact full color displays that are an integral component of mobile telephones. At the present time, approximately 70% of GaN-based LEDs are used for fabricating white LED light sources.

Recently, a combination of improvements in the crystalline quality of epilayers, electrodes, fabrication processes, and the use of multiquantum wells in the active layers has enabled the manufacture of blue LEDs with an output power of 20 mW at 20 mA operating currents; a ten times improvement compared with the figures of merit of the early blue LEDs.

Improvements of the Lifetime of GaN-Based LEDs

InGaN ternary alloy layers can be used to cover the whole visible spectrum from GaN (3.4 eV) to InN (0.6 eV). However, as the In content of InGaN is increased, the large differences between the lattice constants of InGaN and

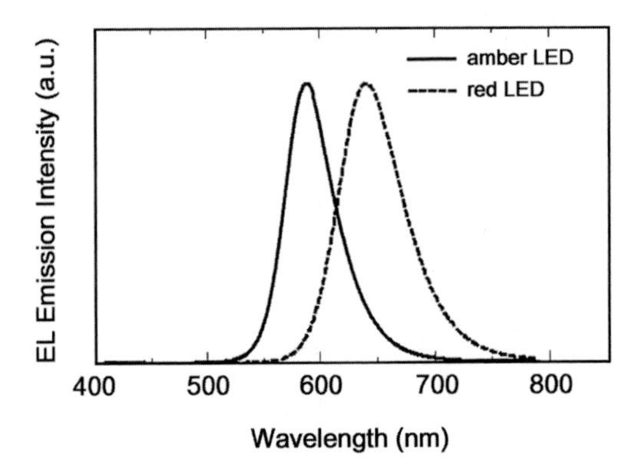

Fig. 3.22. Emission spectrum of recent InGaN-based long wavelength LEDs

GaN leads to increases in crystalline defects and related problems such as the formation of In droplets. Also, at large In compositions, the piezoelectric effect induces increased spatial separation of electrons and holes in the quantum well layers which leads to a reduction of the optical recombination efficiency.

Figure 3.22 shows the emission spectrum of a recent InGaN-based long wavelength LED. The peak wavelengths and FWHM of the amber LED are 585 nm and 49 nm, respectively, and for the red-LED they are 645 nm and 66 nm. For comparison, it is noteworthy that the FWHM of conventional AlInGaP-based LEDs is extremely sharp at 20 nm. Thus the color purity of group III nitride LEDs is not as good as conventional LEDs. Also, the output power InGaN-based LEDs is 1 mW at an operating current of 20 mA, which is 1/10 that of AlInGaP-based devices. For these reasons, in the red spectral region, InGaN-based LEDs exhibit inferior device characteristics compared with AlInGaP devices. However, the optical efficiencies of InGaN and AlInGaP structures are comparable in the amber to yellow region. The emission wavelengths of AlInGaP devices increase, and output powers decrease, if the ambient temperature increases. In contrast, the peak wavelength and output power of InGaN-based LEDs are not affected by the surrounding temperature. Thus InGaN-based LEDs are used for outdoor applications.

Future Trends of GaN-Based LEDs

A number of improvements and challenges in InGaN-based LEDs are expected in the coming years, including:

- Reduction of operating voltage
- Problems related to optical output decrease in the short wavelength region due to changes in the color of the mold plastic
- Improvement of internal and external quantum efficiencies
- Reduction of the dislocation density

In summary, over the last decade, the optical efficiency of blue LEDs has increased tenfold. It is possible to use GaN-based LEDs to produce light with wavelengths over the range from UV to red. Further technological improvements are necessary to realize the full potential of InGaN-based LEDs.

3.2.3 BeZnSeTe-Based Green LEDs (I. Nomura, K. Kishino)

Expectations of II–VI Based Optical Devices

To-date, red semiconductor laser diodes (650 nm) have been fabricated using AlGaInP-based III–V compound semiconductors and blue devices (400 nm) using AlGaInN-based nitride semiconductors. However, laser diodes covering the green to yellow wavelengths are difficult to fabricate using such semiconductors. The development of green laser diodes would enable the use of semiconductors for fabricating light sources covering all three primary colors. II–VI semiconductors show promise for covering the green to yellow wavelength range.

Concerted efforts to develop ZnSe-based II–VI laser diodes started in 1991 and the room temperature (CW; lifetime of 400 h) operation of ZnCdSe/MgZnSSe blue-green laser diodes grown on GaAs substrates has been demonstrated [65]. However, it has not been possible to fabricate devices with lifetimes longer than 400 h and commercial devices are not available. The main reasons for this short lifetime are thought to be (1) defects that arise due to the nitrogen (N) doping of the p-type cladding layer, (2) defects due to strain energy in the strained ZnCdSe active layer, and (3) the brittleness of ZnCdSe/MgZnSSe-based II–VI semiconductor crystals.

II–VI Semiconductors on InP Substrates for the Visible Wavelength Range

Attempts at resolving the aforementioned problems include using InP substrates [66]. MgZnCdSe alloys can be grown lattice matched on InP substrates with direct bandgap energies in the visible region ranging from 2.1 to 3.8 eV (590–326 nm). This range enables fabrication of semiconductor light sources for the green to yellow spectrum [67]. The good optical properties of MgZnCdSe epitaxial layers grown on InP are promising for use of this semiconductor as the active layer in LEDs and laser diodes. In spite of the large bandgap energy ($\sim$3 eV) of MgZnCdSe system, it can be doped with chlorine (Cl) to electron concentrations of $\sim 5 \times 10^{17}\,\mathrm{cm}^{-3}$, thus enabling its use as a highly effective n-type cladding layer. But it is difficult to grow p-type cladding layers thus limiting its use.

BeZnTe has potential as a wide-bandgap semiconductor [68]. This material shows potential as a p-type cladding layer because it can be grown lattice matched to InP, doped p-type, and has a wide bandgap of $\sim$2.8 eV. However, the straightforward use of BeZnTe as a p-type cladding layer has been elusive because the BeZnTe and ZnCdSe active layers form type II heterojunctions.

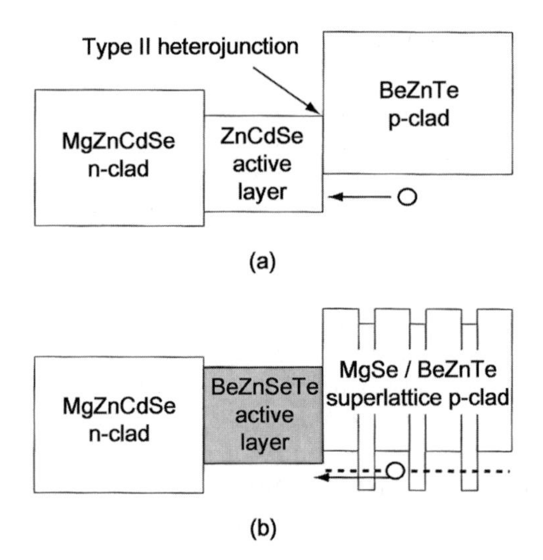

Fig. 3.23. Energy band line up of BeZnSeTe-based DH laser structures

Figure 3.23a shows the energy band line up for a structure consisting of a ZnCdSe active, MgZnCdSe n-cladding, and BeZnTe p-cladding layers. The energy of the valence band edge of BeZnTe is ~0.77 eV higher than the valence band edge of ZnCdSe which results in a large heterojunction barrier in the valence band. This functions as a potential barrier for the injection of holes from the p-cladding layer into the active layer. Holes trapped at the ZnCdSe/BeZnTe type-II heterointerface recombine with electrons in the ZnCdSe active layer and produce radiative or non-radiative transitions. This dramatically degrades the emission properties from the active layer.

Two novel materials have been proposed to resolve these problems. The first is MgSe/BeZnTe superlattice p-cladding layer [69], and the other BeZnSeTe active layer [70]. Figure 3.23b shows the band line up of a double heterostructure formed using these materials. MgSe has a large bandgap energy of 4.0 eV and the edge of the valence bandgap is 1.27 eV lower than BeZnTe. The subband inside an MgSe/BeZnTe superlattice has shifted to a lower energy, thus lowering the potential barrier with ZnCdSe. The other characteristics of MgSe/BeZnTe superlattices are (1) it is possible to dope BeZnTe layers to high hole concentrations and (2) the refractive index difference with MgSe layers is large thus increasing optical confinement. In addition, the lattice constant of MgSe is similar to InP thus the lattice mismatch between MgSe/BeZnTe superlattices and InP is small for any thickness of films.

Thus MgSe/BeZnTe superlattices show promise as the p-cladding layers of LEDs and laser diodes because in these materials the type-II heterostructure barrier is low and p-type doping is readily achieved. However, the use of this superlattice does not completely solve the type II problem of ZnCdSe and use of BeZnSeTe as active layers has been proposed.

As shown in Fig. 3.23b, the band lineup of BeZnSeTe is at a higher energy than ZnCdSe and forms a type-I junction with MgSe/BeZnTe superlattices. BeZnSeTe can be lattice matched to InP and the bandgap energy controlled between 2.1 and 2.8 eV and shows promise for the 590–440 nm emission band. Furthermore, incorporation of BeSe and BeTe into the active layers enhances the crystal bond strength due to their high cohesive energy, thereby preventing the diffusion of defects.

BeZnSeTe LEDs

Figure 3.24 shows a structure of an LED with a BeZnSeTe active layer. The device structure consists of a 7.5 nm thick BeZnSeTe active layer sandwiched between MgSe/ZnCdSe superlattice barrier, a Cl doped, n-type MgSe/ZnCdSe superlattice, and a nitrogen doped, p-type MgSe/BeZnTe superlattice cladding layers. Figure 3.25 shows examples of the emission spectra for this LED. LEDs for various (Be:Se) compositions of the BeZnSeTe active layer yielded the following wavelength emissions: for (0.015:0.52), $\lambda = 594$ nm; (0.055:0.47), $\lambda = 575$ nm; and (0.12:0.40), $\lambda = 542$ nm.

Figure 3.26 shows the room temperature time dependence of the normalized optical output at 16 mA (130 A cm^{-2}) of a yellow (575 nm) LED with an electrode stripe width of 20 μm bonded to a heat sink. There is no rapid degradation of optical output. The device operated continuously for more than 5,000 h. This performance is remarkable compared with conventional II–VI LED structures grown on GaAs substrates that only operate for ~100 h.

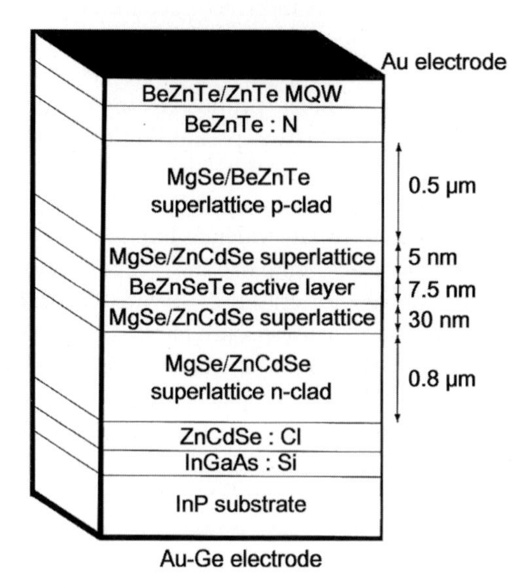

Fig. 3.24. Structure of an LED with a BeZnSeTe active layer

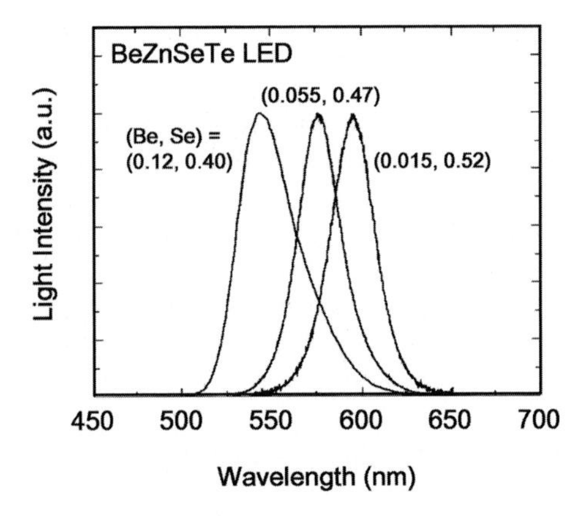

Fig. 3.25. Composition dependence of emission spectra of LEDs with BeZnSeTe active layers

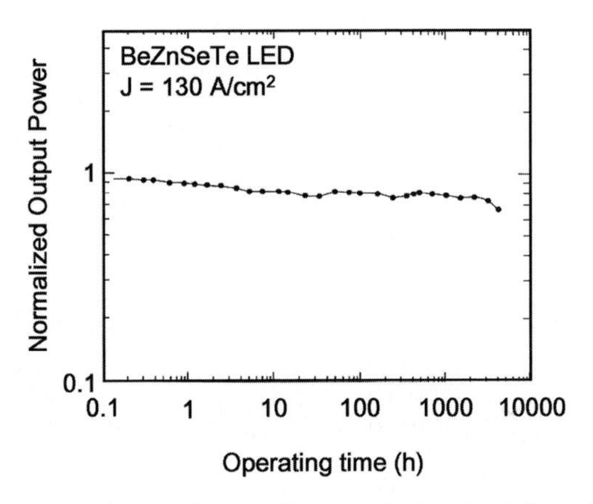

Fig. 3.26. Time dependence of normalized optical output from BeZnSeTe LED

As mentioned above, the reasons for the short lifetimes of above conventional II–VI devices are defects due to the N doping of p-type cladding layers; defects due to strain energy in the strained active layer; and the brittleness of the conventional materials. These problems could be resolved using the BeZnSeTe LEDs proposed here. Because the BeZnSeTe active layer is lattice matched to the substrates, and the p-cladding layer is made of not conventional MgZnSSe but MgSe/BeZnTe superlattices. In addition, including BeSe and BeTe into the active layers was effective in enhancing the crystal bond strength due to their higher cohesive energy, as mentioned before. These factors are effective for suppressing defect generation and diffusion during operation, and for extending lifetimes. These material and device properties of BeZnSeTe LEDs are promising for the realization of high reliability green light emitting devices.

3.3 Ultraviolet Devices

3.3.1 Issues Related to the Development of High Efficiency and Short Wavelength Ultraviolet LEDs (H. Hirayama)

Expectations of UV LEDs

Semiconductor laser diodes and light-emitting diodes with emission wavelengths in the range 250–350 nm have a wide range of potential applications [71–73]. Figure 3.27 shows the main applications of UV light sources. The main markets are white light illumination, sterilization and water purification, medicine and biochemistry, light sources for high density optical recording, fluorescence analytical systems and related information sensing fields, air purification equipment, and zero emission automobiles.

Figure 3.28 shows the relationship between the lattice constants and bandgap of nitride semiconductors with wurtzite structures and the emission wavelengths of several gas lasers. AlGaN has a direct bandgap between 3.4 and 6.2 eV and covers the spectrum possible using gas lasers. The main advantages of using nitride semiconductors for UV light sources are (1) the possibility of obtaining high efficiency optical emission from quantum wells, (2) the possibility of producing both p- and n-type semiconductors in the wide-bandgap spectral region, (3) the nitrides are mechanically hard and devices

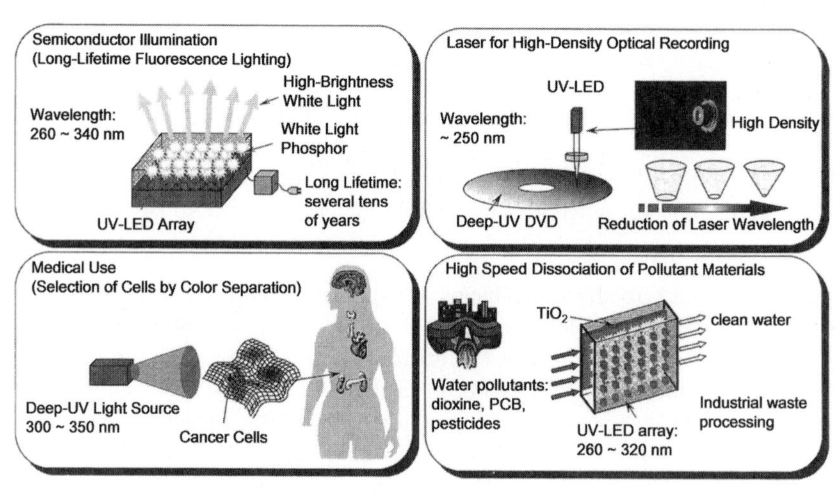

Fig. 3.27. Potential applications of UV LEDs and LDs

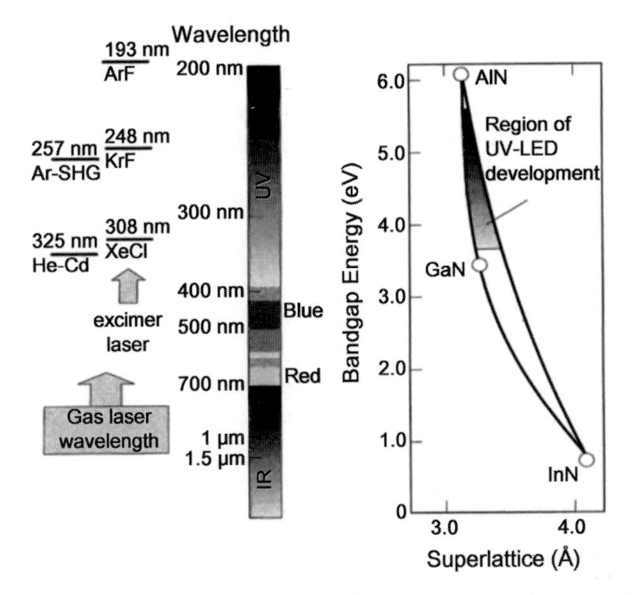

Fig. 3.28. Lattice constants and bandgap of nitride semiconductors with wurtzite structures and the emission wavelengths of several gas lasers

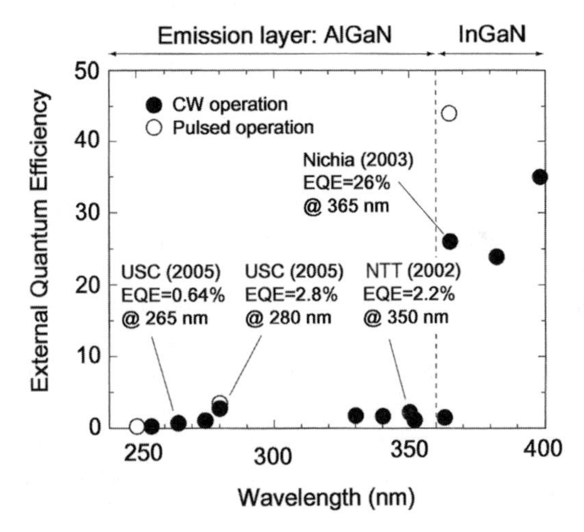

Fig. 3.29. Wavelength dependence of the external quantum efficiency (EQE) of nitride UV LEDs measured at room temperature under CW operation

have long lifetimes, and (4) the materials are free of harmful arsenic. The development of high efficiency, short wavelength UV LEDs and LDs is becoming extremely competitive [74–78]. Figure 3.29 shows the wavelength dependence of the external quantum efficiency (EQE) of nitride UV LEDs measured at room temperature under CW operation. Blue LEDs with EQEs of more than 40% have already been reported but in 2004 Nichia Corporation announced a 365 nm UV InGaN LED with an output of 1 W and an EQE of 26% at room temperature and CW operation [74]. However, optical emission efficiency degrades dramatically for wavelengths of less than 360 nm, with EQEs dropping to 1–2%. Thus the fabrication of high efficiency, short wavelength LEDs is still a major topic of research.

Fabrication of High Efficiency, Short Wavelength UV LEDs

The observed sudden drop in efficiency of UV LEDs below 360 nm is due to three main factors:

1. The intensity of light emitted from the AlGaN layer depends on the density of dislocations penetrating this layer.
2. The p-type doping of AlGaN is difficult.
3. GaN substrates cannot be used because they absorb strongly in the UV region. High quality, single crystal AlN substrates, and AlGaN layers with low dislocation densities are still under development.

The dislocation density of AlGaN buffer layers deposited on sapphire substrates is 10^9–$10^{10}\,cm^{-2}$ and it is difficult to obtain high efficiency optical emission from AlGaN-based quantum wells grown on such buffer layers. In order to fabricate high efficiency UV LEDs using AlGaN-based emission layers it is necessary to reduce the dislocation density to $\sim 10^7\,cm^{-2}$. Thus in devices incorporating an AlGaN emission layer, it is thought that the use of AlN single substrates or epitaxial lateral overgrowth (ELO) are effective methods for producing high quality AlGaN buffer layers with low dislocation densities. But such substrates are expensive and their use may be limited to high end UV LDs as opposed to mass produced devices.

The mass production of low cost, high efficiency UV LEDs could be achieved by using materials, like InGaN, that emit light in spite of the presence of a high density of dislocations. Quaternary InAlGaN alloys are wide-bandgap semiconductors that have the possibility of high efficiency emission at room temperature and show tremendous potential for use as emission layers in UV LEDs [72, 73]. There have been reports on growth conditions [79, 80] and emission from quantum wells using high quality InAlGaN alloys [73]. Internal quantum efficiencies of ∼40% have been obtained even with the presence of a high dislocation density for LEDs emitting in the 305–370 nm band. Further, the optical efficiencies in the 350 nm band for InAlGaN LEDs were reported to be superior to AlGaN-based devices [78]. Thus InAlGaN is of great interest for mass producing low cost, high efficiency UV LEDs emitting in the 280–380 nm wavelength range.

There is another major problem to resolve in the quest for short wavelength nitride UV LEDs, namely, the difficulties in obtaining p-type doping in AlGaN. As the Al content of p-type AlGaN increases, the activation energy of acceptors also increases with the result that it becomes difficult to produce highly doped p-type AlGaN. Actual experiments show that it is possible to make electrical measurements of Mg doped AlGaN with Al content of 30–40%, but such samples are difficult to measure for Al compositions greater than 40%. In cases where the hole carrier density is low, the electron injection efficiency into the emission layer decreases markedly due to electron overflow into the p-type layer, and as a result the p-type layer becomes highly resistive and the optical efficiency decreases due to sample heating. The efficiency is expected to be particularly low for wavelengths less than 300 nm.

Recently, the reports of a 250–300 nm band, short wavelength LED have attracted a lot of attention [76,77]. Khan et al. from South Carolina, reported the pulsed operation of 250 nm band LED with a AlGaN multiquantum well grown on a sapphire substrate [76]. Devices operating on the 280 nm waveband, with single peak emission and 0.94% external quantum efficiencies were reported [77]. In these experiments, flip-chip structures were used and UV light was emitted from the back side of the substrate. However the low efficiency of less than 1% implies that there are still significant technological issues to resolve.

There have been no reports on solutions to the problems of producing p-type AlGaN for high Al compositions. Co-doping has been proposed but so far no successful technique has been found for high Al composition AlGaN.

Quaternary InAlGaN Alloys for High Efficiency UV Emission Devices

By introducing a few percent of In into AlGaN, high efficiency UV emission has been observed even in cases where there was a high dislocation density [79,80]. The introduction of In has the effect of localizing electron–hole pairs due to In composition modulation and, as a result, electron–hole recombination occurs before the excited carriers are trapped by lattice defects. Hence high efficiency emission is possible even for layers with large densities of crystalline defects.

Figure 3.30 shows the room temperature photoluminescence of quaternary InAlGaN alloys and InGaN quantum wells for a range of compositions. Table 3.4 shows the parameters of the quantum well structures corresponding to the peaks (a) to (f). In spite of a dislocation density of $10^{10}\,\mathrm{cm}^{-2}$ in these structures, there is strong emission from InAlGaN quantum wells at room temperature in the 280–380 nm wavelength range. The emission is of a similar strength as that from InGaN quantum wells in the 320–350 nm region.

Modulation of the In composition in quaternary InAlGaN alloys has been observed as spatial fluctuations in cathode luminescence images [79]. Also, the room temperature intensity of the PL from InGaN and InAlGaN is up to two orders of magnitude stronger than that from GaN and AlGaN wells [80]. The internal quantum efficiency of InGaN quantum wells has been measured at room temperature to be 40% even for structures with a high density of dislocations. These results show that InAlGaN alloy quantum wells are promising for the fabrication of UV optical devices emitting in the 280–380 nm wavelength range.

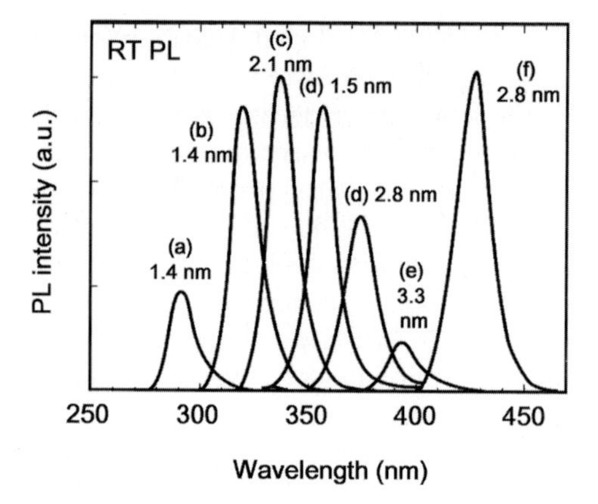

Fig. 3.30. Room temperature photoluminescence of quaternary InAlGaN alloys and InGaN quantum wells for a range of compositions

Table 3.4. Structural parameters of fabricated InAlGaN and InGaN quantum wells

quantum well structure (well/barrier)	composition wavelength		optimum thickness (nm)	wavelength spectrum of high brightness emission
	well (nm)	barrier (nm)		
(a) $In_{0.03}Al_{0.05}Ga_{0.47}N$/ $In_{0.01}Al_{0.06}Ga_{0.39}N$-3MQW	308	270	2.1	290–313
(b) $In_{0.05}Al_{0.34}Ga_{0.61}N$/ $In_{0.02}Al_{0.06}Ga_{0.38}N$-3MQW	340	300	1.4	318–338
(c) $In_{0.06}Al_{0.08}Ga_{0.86}N$/ $In_{0.02}Al_{0.52}Ga_{0.46}N$-3MQW	364	310	2.1	325–355
(d) $In_{0.06}Al_{0.09}Ga_{0.85}N$/ $In_{0.03}Al_{0.18}Ga_{0.79}N$-SQW	362	345	1.5	358–375
(e) $In_{015}Ga_{0.85}N$/ $In_{0.02}Al_{0.15}Ga_{0.83}N$-3MQW	408	334	(3.3)	371–395
(e) $In_{0.2}Ga_{0.8}N$/ $In_{0.02}Ga_{0.98}N$-SQW	430	370	2.8	405–448

InAlGaN Alloys for 310–350 nm UV LEDs

The acceptor activation energy in Mg doped AlGaN with an Al composition of ~30% is greater than 300 meV and the hole activation is less than 1%. Thus high doping levels of $\sim 10^{20}$ cm^{-3} are necessary. A hole density of 2×10^{18} cm^{-3} was reported for Mg doped AlGaN with an Al composition of 32% [81]. Hall Effect measurements for Al compositions greater than 40% have not been

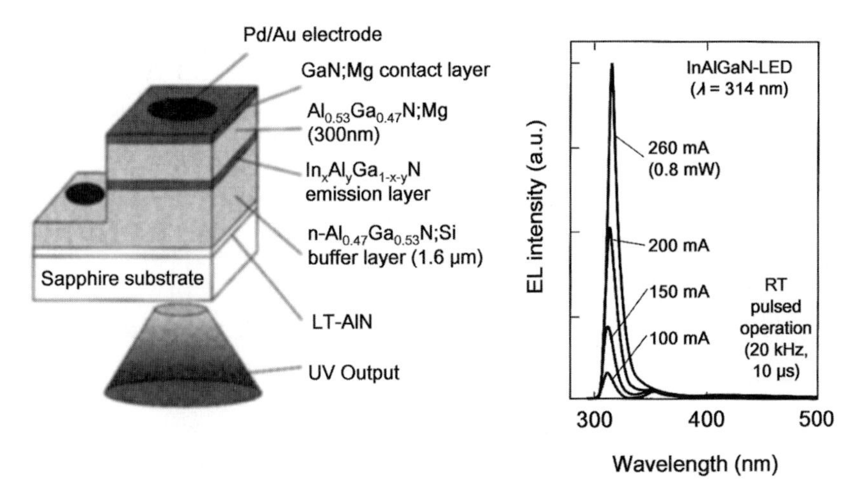

Fig. 3.31. Structure and optical spectrum of a 310 nm LED incorporating an InAlGaN emission layer

reported. However, p-type AlGaN with Al compositions of 70%, for 250 nm wavelength LEDs, has been fabricated [76].

Figure 3.31 shows the structure and optical spectrum of a 310 nm LED incorporating an InAlGaN emission layer [72]. The LED structure is grown on a sapphire substrate and consists of n-AlGaN (Al = 47%), InAlGaN emission, p-AlGaN (Al = 53%), and a p-GaN contact layer. The p-AlGaN layer was grown using an alternate supply of gases. Under high current injection conditions, there is only a single peak. The output power at wavelengths of 308 nm and 314 nm was 0.4 mW and 0.8 mW, respectively. The external quantum efficiency of both LEDs is low at 0.1%. Further increases in the doping of the p-layers and optimization of the quantum well optical layers are required for higher optical emission efficiency.

Figure 3.32 shows the structure and CW spectrum of an LED fabricated on a GaN substrate using a quaternary InAlGaN alloy [78]. The use of GaN substrates is expected to reduce the dislocation density. The LED structure consists of n-GaN, an InGaN layer to prevent cracking, n-AlGaN (Al = 18%), an InAlGaN/InAlGaN double quantum well, a p-AlGaN (Al = 28%) electron blocking layer, and a p-AlGaN/AlGaN superlattice. Semitransparent, Ni/Au electrodes are formed on the front surface and Ti/Al on the back. LEDs have also been fabricated using a p-AlGaN (Al = 18%) layer instead of the p-AlGaN/AlGaN superlattice. The substrate was made by Sumitomo Electric Industries, LTD. and yields a dislocation density of less than 10^6 cm^{-2}. The emission peak was at 352 nm and had a FWHM of 9 nm. The applied voltage was 4.4 V at an injection current of 100 mA.

Figure 3.33 shows the current-output power and external quantum efficiency characteristics. The maximum output power at an injection current

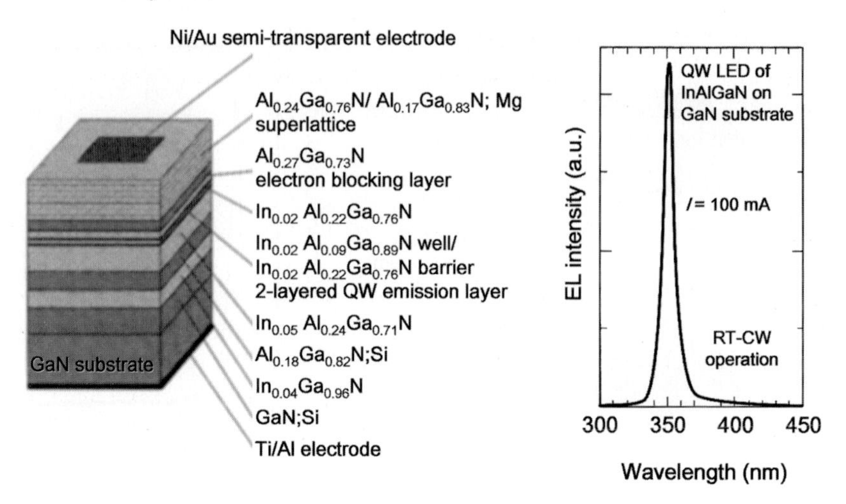

Fig. 3.32. Structure and CW spectrum of an LED fabricated on a GaN substrate with a quaternary InAlGaN alloy

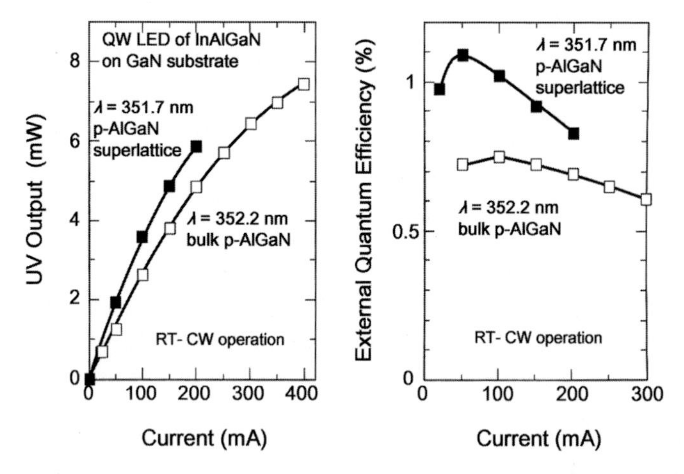

Fig. 3.33. Current–output power and external quantum efficiency characteristics of InAlGaN LEDs grown on GaN substrates

of 400 mA was 7.4 mW. The output power at 100 mA for a LED with a p-AlGaN/AlGaN superlattice was 3.9 mW. This power figure is higher than the 3.0 mW obtained from a 350 nm AlGaN QW LED grown on a GaN substrate [75]. A maximum value of 1.1% was obtained for the external quantum efficiency of surface-emitting LED structures operating at 350 nm, confirming the superiority of quaternary InAlGaN alloy LEDs over AlGaN-based devices.

Further, quaternary alloy structures have recently been found to show higher efficiencies than AlGaN LEDs grown on GaN substrates even in the presence of a high density of dislocations. The external quantum efficiency is limited to 1% because the GaN substrate absorbs UV light. The light extraction efficiency could be improved by using GaN free structures and extracting the UV light from the backside of the substrate.

3.3.2 ZnO LEDs (M. Kawasaki, A. Ohtomo)

Expectations of ZnO

Zinc (Zn) is the 23rd most plentiful element in the earth's crust. Zinc oxide (ZnO) is used as an additive for hardening gum of automobile tires and glass (~10,000 tonnes are consumed in Japan alone), and as a whitener in cosmetics and baby powder. In the electronics industry, sintered ZnO is used in varistors for absorbing voltage surges and in conducting lightning rods. As a thin film, ZnO shows promise as an alternative to In–Sn–O (ITO) for transparent electrodes and as a material for gas sensors exploiting the change of the electrical resistance of ZnO particles in a large range of gases. Honeywell manufactures ZnO oxygen sensors used in the Space Shuttle. Also, ZnO has tremendous potential for fabrication of UV LED and LD devices.

ZnO

Table 3.5 shows a comparison of the figures of merit for ZnO and GaN. ZnO is a semiconductor of a wurtzite structure, with bandgap and lattice constant almost the same as GaN. The exciton binding energy is extremely large at 60 meV and is almost three times that of GaN (24 meV). This exciton binding energy is larger than the room temperature thermal energy (25 meV), which

Table 3.5. Comparison of ZnO and GaN for use as LED materials

		GaN	ZnO	Improvements
bandgap (eV)		3.43	3.37	–
exciton binding energy		26	60	–
lattice constant	a-axis (Å)	3.19	3.25	–
	c-axis (Å)	5.19	5.20	–
substrate	lattice mismatch	large	none	crystallinity
	electrical conductivity	none	exist	cooling
electrode		one side	both sides	resistivity
refractive index		2.5	2.2	optical extraction and emission efficiency
construction		bridge-type	die wire bonding	manufacturing

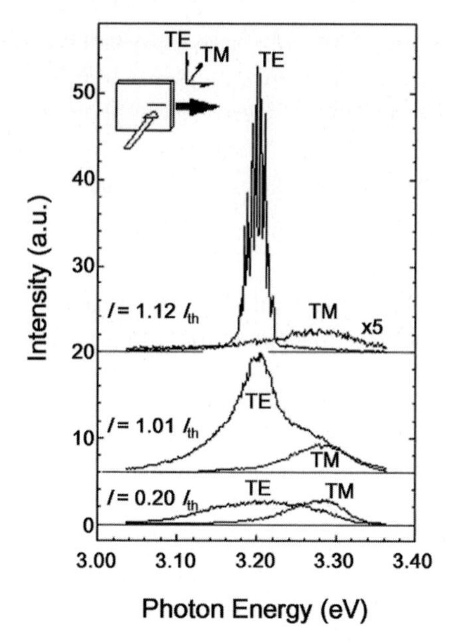

Fig. 3.34. Polarized spectra observed from the edges of ZnO films irradiated with Nd:YAG laser light (355 nm)

leads to the possibility of achieving high efficiency exciton emission at room temperature. In 1996, Tang et al. reported room temperature of laser emission in near-UV (390 nm) from high quality ZnO thin films upon optical excitation [82–85]. Figure 3.34 shows the polarized spectra observed from the edges of ZnO films irradiated with Nd:YAG laser light (355 nm). The TE and Fabry Perot modes are clearly observed above the extremely low threshold excitation intensity of $I_{th} = 24\,\mathrm{kW\,cm^{-2}}$. It is easy to produce n-type ZnO films. However, device applications also require the fabrication of p-type films [86–88].

Comparison of GaN and ZnO LED Structures

The main difference is the substrate. GaN devices are grown on insulating sapphire substrates, whereas it is possible to grow ZnO structures on conducting ZnO substrates. The growth of bulk GaN crystals is still expensive but it is possible to growth 3 in. ZnO crystals by the hydrothermal synthesis (Fig. 3.35) and mass production will enable reductions in cost [89]. Figure 3.36 shows an actual structure for GaN LEDs and a realistic structure for ZnO LEDs.

Control of the Bandgap of ZnO and Superlattices

The crucial technology for fabrication of ZnO-based LEDs is the control of the bandgap by use of appropriate alloys and p-type doping. For the former

Fig. 3.35. Large substrates of bulk GaN crystals

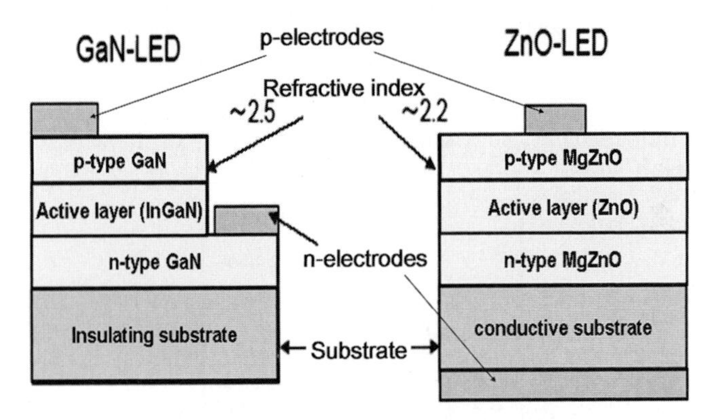

Fig. 3.36. Realistic structures for ZnO and GaN LEDs

case, Ohtomo et al. demonstrated the use of Mg and Cd doping for bandgap control [90, 91]. Figure 3.37 shows a comparison of the bandgap and a-axis length of the wurtzite structures of ZnO and GaN alloys. Addition of 7% Cd enables variation of the ZnO bandgap from 3.3 to 3.0 eV and the addition of 30% Mg produces an increase to 4.0 eV. The point to notice is that the length of the a-axis only changes by ~0.5%. This is an important fact for the growth of heterojunctions. The solubility limit of Mg has been found to be ~10% under thermal equilibrium conditions by annealing experiments. Optimizing growth conditions leads to metastable solubility of 50% and the bandgap can be increased to 4.5 eV [92].

The properties of superlattice structures employing ZnO quantum well layer and $Mg_xZn_{1-x}O$ barrier layer have been extensively studied. In a particularly notable series of experiments, approximately ten superlattice structures, with different well widths, were grown on a single substrate using an innovative masking mechanism, referred to as the combinatorial method [93–95]. Figure 3.38 shows the absorption and photoluminescence spectra of $ZnO/Mg_xZn_{1-x}O$ superlattices with $x = 0.12$. In the ZnO thin film, phonon replicas are clearly visible in addition to excitonic absorption peaks A and B.

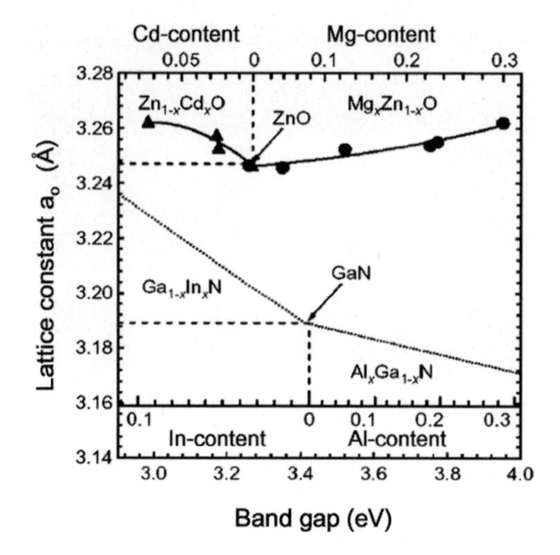

Fig. 3.37. Comparison of the bandgap and a-axis length of the wurtzite structures of ZnO and GaN alloys

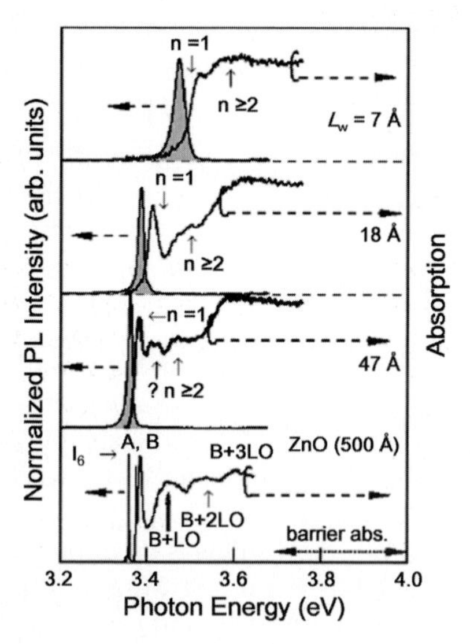

Fig. 3.38. Absorption and photoluminescence spectra of $ZnO/Mg_xZn_{1-x}O$ superlattices with $x = 0.12$

For the superlattices, the peaks of the absorption and emission spectra are clearly seen to shift to lower wavelengths due to quantum size effects and quantum levels are clearly visible in the absorption spectrum. A detailed analysis of these results was reported by Makino et al. [96,97]. The most notable result was the observed increase in the exciton binding energy, as shown in Fig. 3.39. For $x = 0.12$ and $x = 0.27$, at the well widths of 1.5 nm, the exciton binding

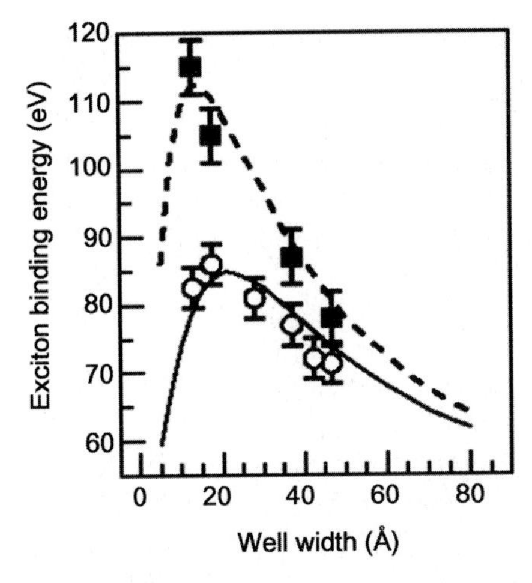

Fig. 3.39. Variation of the exciton binding energy with quantum well width

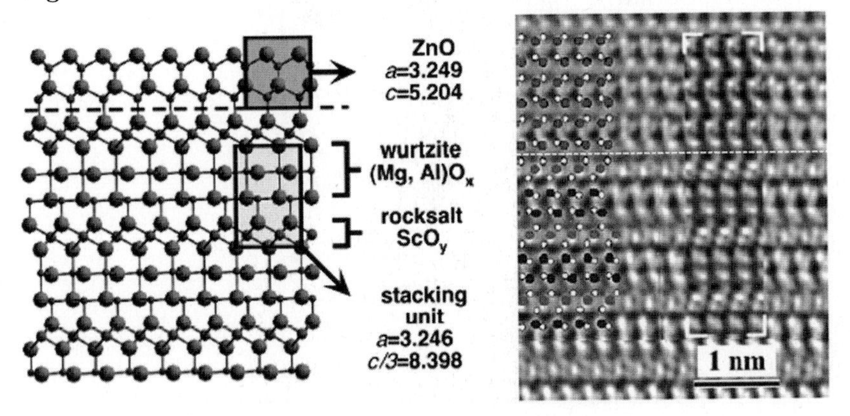

Fig. 3.40. Plausible atomic arrangement of a ZnO/$ScAlMgO_4$ heterojunction

energies were 85 meV and 115 meV, respectively. These structures would produce exciton emission at room temperature in LED and laser devices.

Growth of High Quality ZnO Thin Films

There have been reports on the growth of ZnO on sapphire substrates using low temperature ZnO or ultra-thin buffer layers, as well as homoepitaxial growth on single crystal ZnO substrates [98]. A novel approach was reported by Ohtomo et al. using single crystal $ScAlMgO_4$ substrates [99]. These substrates were developed at Bell Labs for the growth of GaN, but heteroepitaxy on these substrates did not yield favorable results and it was concluded that oxides and nitrides do not make a good combination [100]. Figure 3.40 shows a plausible atomic arrangement of a ZnO/$ScAlMgO_4$

heterojunction [101]. $ScAlMgO_4$ is considered as a natural superlattice composed of (0001) slab of wurtzite $(MgAl)O_x$ and the (111) slab of rock salt ScO_x. The lattice mismatch of ZnO and these substrates is only 0.09%, and it is much easier to grow high quality crystalline films on $ScAlMgO_4$ substrates than on sapphire. Figure 3.41 shows the optical spectrum of a ZnO thin film. The separation of the peaks A and B is not clearly resolved for films grown on sapphire substrates but the difference is clear for the

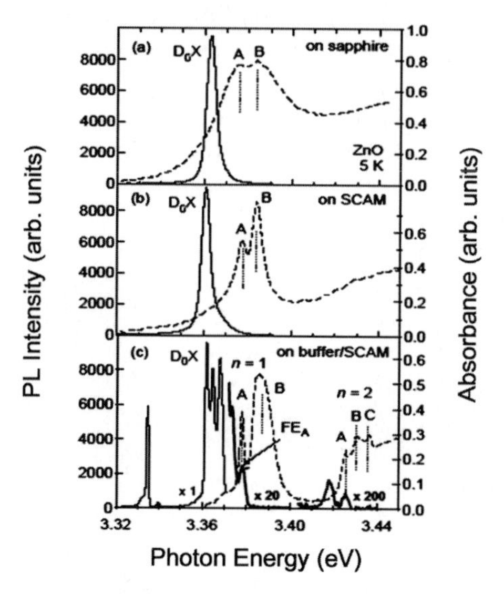

Fig. 3.41. Emission and absorption spectra of ZnO thin films grown on various substrates

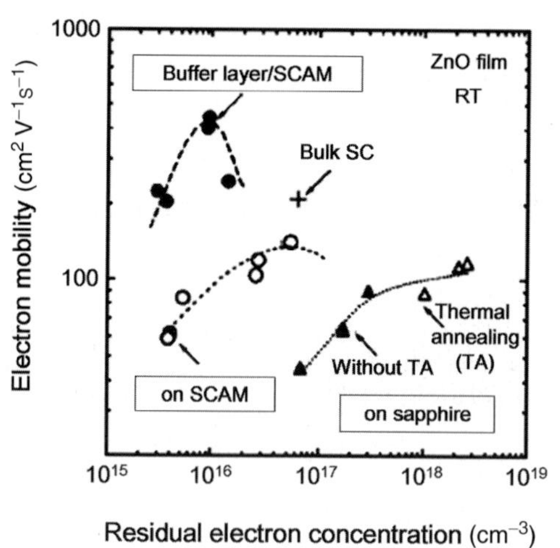

Fig. 3.42. Mobility and residual electron density of undoped ZnO thin films grown on various substrates

$ScAlMgO_4$ substrate. Figure 3.42 shows the mobility and density of electrons of undoped ZnO thin films. On sapphire, the background density is $\sim 10^{17}\,cm^{-3}$ and the mobility increases to $\sim 100\,cm^2\,V^{-1}\,s^{-1}$ with increasing electron density. High temperature annealing causes an increase of the size of crystals and the electron density increases to $10^{18}\,cm^{-3}$, but the mobility saturates at $100\,cm^2\,V^{-1}\,s^{-1}$. However, for the film grown on the $ScAlMgO_4$ substrate, the mobility is still $\sim 100\,cm^2\,V^{-1}\,s^{-1}$, although the background electron density is reduced to $10^{16}\,cm^{-3}$. In the p-type doping of ZnO, it is important to maintain a high mobility and a low background electron density. $ScAlMgO_4$ substrates offer the possibility of producing intrinsic ZnO layers.

Further, a buffer layer annealed at high temperature was developed for growth of ZnO layers with low background electron densities and high mobilities [102]. A ZnO layer with a thickness of ~ 100 nm was deposited on a $ScAlMgO_4$ substrate, annealed at 1,000°C and, as shown in Fig. 3.43a, the surface was found to be atomically flat [103, 104]. Then, the RHEED intensity oscillations persist while ZnO is grown on this surface (Fig. 3.43b). The half-width of the X-ray rocking curve was less than 18 arcsec.

Figure 3.41(c) shows the optical and electrical properties of ZnO films grown on semi-insulating MgZnO buffer layers. High order ($n = 2$) excitonic absorption peaks, A, B, and C are clearly seen in the absorption spectrum. Further, the room temperature mobility is $440\,cm^2\,V^{-1}\,s^{-1}$ and background electron density is reduced to less $\sim 10^{15}\,cm^{-3}$. Figure 3.44 shows a comparison of these results wit the Eagle Pitcher data referred as the best data for intrinsic ZnO grown by a vapor-transport method (dotted line). The background electron density is one order of magnitude lower and mobility several times higher. A carrier mobility of $5{,}000\,cm^2\,V^{-1}$ was recorded at 100 K, just prior to carrier freeze-out [105]. The lifetime of free excitons from this film was determined to be 2.5 ns, an extremely large value that shows the high quantum efficiency of this material [106]. Figure 3.44 also shows a comparison of Ga doped n-type ZnO and N doped p-type ZnO to be described later.

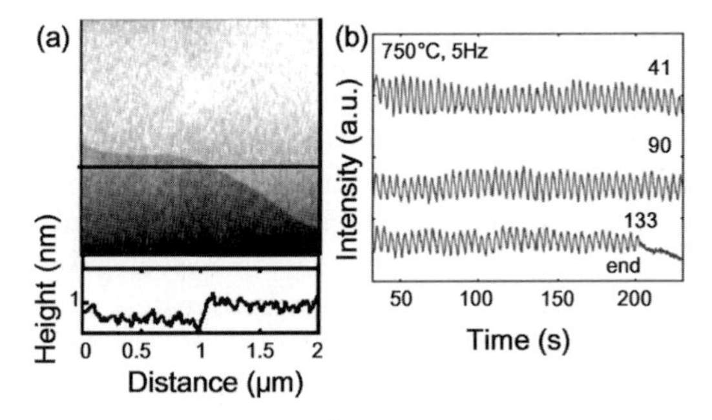

Fig. 3.43. (**a**) AFM images of buffer layers annealed at high temperature and (**b**) RHEED oscillations during growth of ZnO grown on the same buffer layers

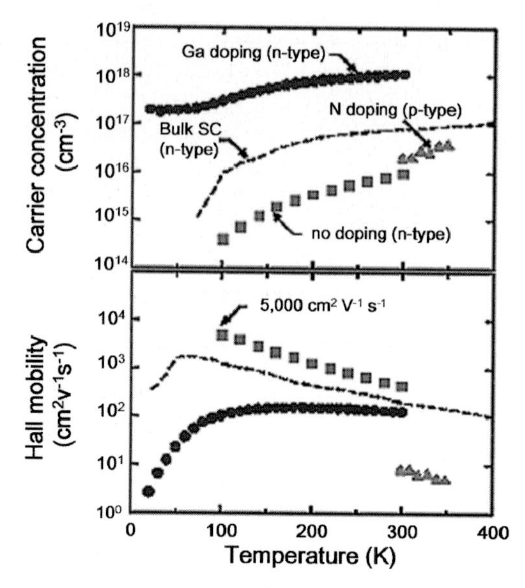

Fig. 3.44. Temperature dependence of the hole and carrier density of Ga doped n-type ZnO and N doped p-type ZnO

p-type Doping of ZnO and LEDs

Zinc oxide must be doped with acceptor impurities to produce p-type films. Possibilities include Li, Cu, N, P, and As. But nitrogen has the most potential because the radius of the ion is similar to O and it will result in less compensation effects. However, the major problem with nitrogen is that densities of $10^{20}\,\mathrm{cm}^{-3}$ can only be achieved for growth temperatures of less than 500°C, a temperature that is too low for growth of high quality ZnO films, which require growth temperatures of more than 800°C. High N doping can be achieved even at high growth temperatures on Zn-surface of ZnO, but the film on sapphire and $ScAlMgO_4$ substrates yields in O-surface [98] and cannot incorporate N at high temperatures.

A solution to this dilemma is the repeated temperature modulation method, shown in Fig. 3.45 [88]. A thin (15 nm) N-doped layer is grown at 400°C and then the substrate temperature is suddenly increased to 1,000°C. The effect of this annealing is to produce an extremely flat surface. Additional deposition of a layer with a thickness of ~1 nm is made to ensure surface flatness. Then the substrate temperature is reduced again and the low temperature growth resumed. This process is repeated.

By use of this method, it is possible to obtain surfaces that are atomically flat and ZnO films with N densities of $\sim 10^{20}\,\mathrm{cm}^{-3}$. Optimization of the growth process enables fabrication of p-type ZnO films with room temperature hole concentrations of $\sim 10^{16}\,\mathrm{cm}^{-3}$ and mobilities of $8\,\mathrm{cm}^2\,\mathrm{V}^{-1}\,\mathrm{s}^{-1}$. Figure 3.44 shows the hole concentration and mobility of p-ZnO films determined by the

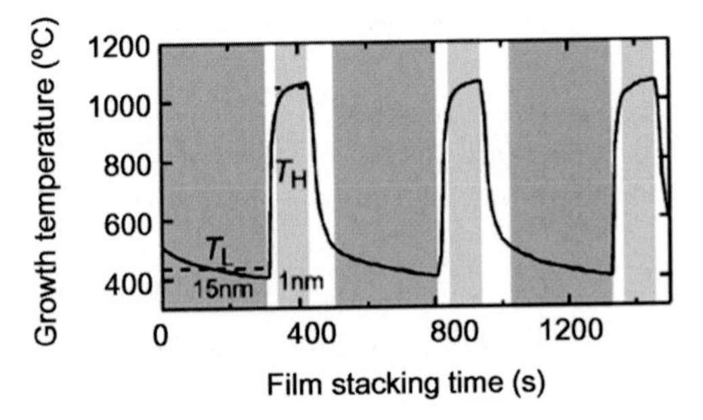

Fig. 3.45. Temperature profile of the repeated temperature modulation method

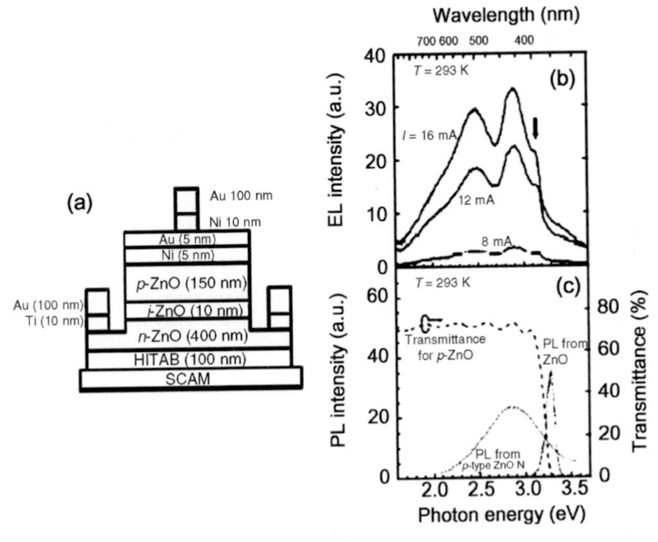

Fig. 3.46. Structure and optical spectrum of an LED fabricated using p-ZnO films

Hall measurements for microfabricated Hall bars with five proves. From these results, the compensation ratio (N_A/N_D) and acceptor activation energy were calculated to be 0.8 and 100 meV, respectively.

Figure 3.46 shows the structure and optical spectrum of an LED fabricated using the aforementioned p-ZnO films [107,108]. The pn junction on insulating $ScAlMgO_4$ substrate was patterned into a 300 μm mesa, and the electrode on the p-ZnO is formed through a semitransparent metallization. Application of a forward bias of more than 5 V results in clear blue emission. The brightness of the emission varies linearly with the drive current. The subpeak seen in the spectrum is due to optical interference as the light travels thought the multilayered device structure. The LED emission is almost the same as the PL from p-ZnO films. In this pn junction the p-type side is lightly doped, which results in a depletion region on this side of the junction where the majority of the electron–hole recombination occurs.

Future

The growth of p-type ZnO is a major step in the development of ZnO light-emitting devices. It is still necessary to increase the hole concentration of ZnO films above $10^{18}\,cm^{-3}$ and to produce p-type (MgZn)O films. The development of homoepitaxial technology for growing ZnO on ZnO substrates is also an important area of research. ZnO-based, high-performance phosphors have been demonstrated [109] and white LEDs using ZnO UV LEDs are under investigation.

3.3.3 Diamond Light Emitting Devices (H. Okushi)

Free-Exciton Emissions in Diamond

Like silicon, diamond is an indirect semiconductor. However, it is known that diamond exhibits strong excitonic emissions even at room temperature. Figure 3.47 shows near-band-edge CL spectra at room temperature, where (a) is the spectrum for high quality diamond film grown by homoepitaxial CVD growth and (b) is that for Ib diamond substrates used in homoepitaxial CVD growth [110]. In the figure, strong sharp emission lines due to the free-exciton were observed at 235 nm (5.27 eV) and 242 nm (5.12 eV). These lines are associated with the transverse optical phonon (FE_{TO}) and its replica, respectively. The excitons in diamond have a large binding energy (80 meV) and a small Bohr radius (1.57 nm) because of diamond's low dielectric constant [111]. Therefore, a high density of excitons can be generated even at room temperature, and one expected application of such high density excitons (more than $10^{19}\,cm^{-3}$)

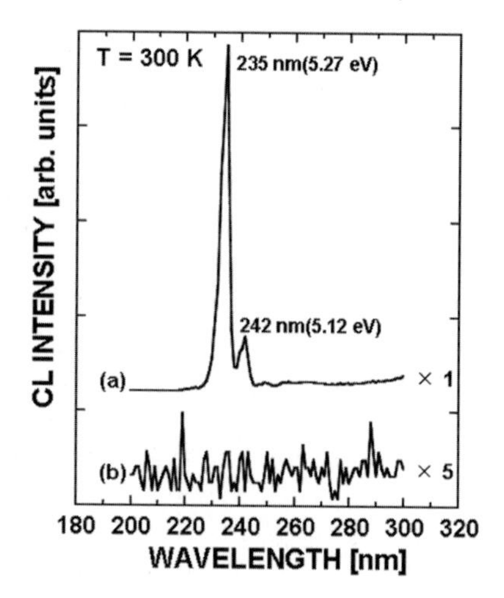

Fig. 3.47. Near-band-edge CL spectra at room temperature: (**a**) the spectrum for high quality diamond film grown by homoepitaxial CVD growth and (**b**) Ib diamond substrates used in homoepitaxial CVD growth

has been as a light source for deep-ultraviolet (deep-UV) light-emitting diodes (LEDs). In this section, recent achievements in diamond deep-UV LED are reviewed in terms free-exciton emission characteristics in diamond, potential for practical exciton LEDs, p–n junction properties and LED performance.

Characteristics of Excitonic Emission in Semiconductors

Here, the common characteristics of excitonic emission in semiconductors will be first discussed before a description of a diamond exciton deep-UV LED. It is necessary to note that excitons are generated by an intrinsic property of the material and not from extrinsic features such as the presence of impurity atoms. In general, electron–hole pairs generated in semiconductors exist as free electrons and holes at high temperature. However, at low temperature, a low energy, spatially intimate electron–hole pair formed by the Coulomb interaction is more stable than the free electron–hole state. Such a weakly bound state of electron–hole pair is called an exciton. The exciton can exist regardless of whether the semiconductor has a direct or indirect bandgap.

However, whether excitons are actually stable or not is dependent on the relationship between the binding energy of the exciton and the thermal dissociation energy. Therefore, excitons in conventional narrow-gap semiconductors exist only at low density at low temperature because of their low-biding energy as shown in Table 3.6. On the other hand, the binding energy of excitons in wide-gap semiconductors is generally large. In particular, the binding energy of excitons in diamond is as large as 80 meV and then a high exciton density (more than 10^{19} cm^{-3}) is attained even at room temperature, as mentioned above.

As described in Sect. 3.3.1, the emission efficiency of LEDs is determined by many factors such as internal quantum efficiency, external efficiency, and voltage loss. Among them, the theoretical possibility of high-efficiency LEDs based on particular semiconducting material or light emission mechanisms is usually judged based on the internal quantum efficiency (η). If overflow currents are neglected, η can be expressed as follows:

$$\eta = \frac{\tau_R^{-1}}{\tau_R^{-1} + \tau_{NR}^{-1}} = \left[1 + \frac{\tau_R}{\tau_{NR}}\right]^{-1} \tag{3.19}$$

Table 3.6. Binding energy of exciton and Mott density

	binding energy (meV)	Bohr radius(nm)	Mott density (cm^{-3})
Si[a]	14.7	5	1×10^{18}
GaAs	4.2	14	1×10^{16}
ZnSe	17	3.5	6×10^{17}
GaN	25	2.9	1×10^{18}
ZnO	60	1.4	1×10^{19}
Diamond	80	1.5	6×10^{19}

[a]Ref. Introduction to Solid State Physics, eds. C. Kittel (Wiley, New York)

where τ_R is the radiative recombination time of the electron–hole pair and τ_{NR} is the non-radiative recombination time of the electron–hole pair via defect or impurity states.

In direct semiconductors, the free electron–hole pair exists at the same symmetry point (Γ point) of a crystal where they can also recombine. Therefore, τ_R is very short (for example, less than 10^{-9} s) and the relation $\tau_R < \tau_{NR}$ holds in the usual case and η can take a value near the theoretical maximum of 1 in (3.19). On the other hand, in indirect semiconductors, the free electron and holes are located at different symmetry points of the crystal from each other and phonons are needed to assist radiative recombination of electrons and holes. Therefore, τ_R becomes longer by three to six orders in magnitude than τ_R in direct semiconductors and the lifetime of free electrons and holes is governed via defect or impurity states, namely the relation of $\tau_R \gg \tau_{NR}$ holds in the usual case, and η takes on only very small values, i.e., $\eta \ll 1$ (see Table 3.7).

This large difference between the direct and indirect semiconductors for the value of η is reduced by using the exciton emission process. Since the excitons consist of spatially intimate electron–hole pairs, the direct recombination rate of electron–hole pairs becomes larger. Therefore, as shown in Table 3.7, even in the case of indirect semiconductors, it theoretically enables η to approach 1. It should be noted that if the density of excitons generated in the steady state is more than the density of recombination centers due to defect and impurity states, the effective recombination rate via recombination centers is decreased and τ_{NR} becomes even longer and consequently it accelerates the approach of η toward 1. From the viewpoint of high density exciton generation, realization is much easier in indirect semiconductors than that in direct semiconductors because the lifetime of excitons is much longer in the former case.

Table 3.7. Internal quantum efficiency for direct and indirect transitions

recombination of free electron and hole pair	
direct transition	indirect transition
τ_R: short (recombination at Γ point)	τ_R: long (phonon assisted recombination)
$\tau_R \leq \tau_{NR}$ or $\tau_R \ll \tau_{NR} \rightarrow \eta \cong 1$	$\tau_R \gg \tau_{NR} \rightarrow \eta \cong 0$
recombination of exciton (τ_R becomes shorter by spatially intimate electron–hole pair)	
direct transition	indirect transition
$\tau_R \leq \tau_{NR}$ or $\tau_R \ll \tau_{NR} \rightarrow \eta \cong 1$	$\tau_R \geq \tau_{NR}$ or $\tau_R \leq \tau_{NR}$ it is also possible to realize $\tau_R \ll \tau_{NR}$ condition (see in the text) $\rightarrow \eta \cong 1$

[a] $\eta = \dfrac{\tau_R^{-1}}{\tau_R^{-1} + \tau_{NR}^{-1}} = \left[1 + \dfrac{\tau_R}{\tau_{NR}}\right]^{-1}$

Thus, η in indirect semiconductors can be expected to equal that in direct semiconductors by using excitonic emission. However, the density of excitons generated at steady state needs to be more than the density of recombination centers in order to obtain a comparable η. From the argument above, it can be seen that high quality CVD diamond is among the few candidates for a high-efficiency exciton LED based on indirect semiconductors.

Diamond Exciton Deep-UV LED

The first report of diamond deep-UV LED was presented by Koizumi et al. (NIMS group) in 2001 [112]. They succeeded in making n-type diamond films using homoepitaxial CVD growth at high pressure and high temperature (HPHT). The synthesized diamond (111) substrate used phosphorus atoms as dopant [113]. Based on this result, they made a diamond p–n junction diode and succeeded in observing electroluminescence from free excitons at room temperature [112]. The structure of this LED consists of three components (1) a low-resistance B-doped HPHT synthesized diamond (111) substrate; (2) a B-doped p-layer CVD diamond film; and (3) a P-doped n-layer deposited on the previous layers. However, as will be seen later, the structure of this LED using a diamond (111) substrate is unsuitable for commercial device fabrication, and the development of this LED has not progressed.

Horiuchi et al. (Tokyo Gas group) developed the research and development of a diamond-based LED in a new positive direction [114]. They synthesized a B-doped p-type layer by homoepitaxial CVD growth on a low-resistance B-doped HPHT (100) substrate and further synthesized an S-doped n-type layer to make diamond p–n junction LEDs. Although Hall effect measurements do not show a clear evidence of n-type conduction for the S-doped CVD layer, the junction characteristics of the diode shows a rectification ratio of three orders of magnitude at $\pm 10\,\mathrm{V}$ under room temperature conditions. A relatively high performance of electroluminescence was obtained by a current injection into these p–n junctions. Best performance values reported for this LED structure up to the present time include a 0.027% external quantum efficiency and $7\,\mu\mathrm{W}$ output power at $10\,\mathrm{mA}$ [115]. In order to fabricate higher efficiency and further improve this LED, the S-doped "n-type layer" – which does not show an n-type Hall effect – needs to be improved.

The AIST group has continued research of diamond exciton deep-UV LEDs [116–118]. When we consider utilization of diamond LEDs as practical devices, the structure using a diamond (111) substrate is unsuitable. This is because the hardness of the relevant crystal face makes it difficult to polish to a flat surface. Therefore, n-type control of homoepitaxial growth of CVD diamond films on a diamond (001) substrate is needed for machine polishing, micro fabrication, and etching steps that correspond to commercial device fabrication processes. Horiuchi et al. adopted the S-doped n-type CVD diamond films in order to make use of a diamond (001) substrate [114], but, as

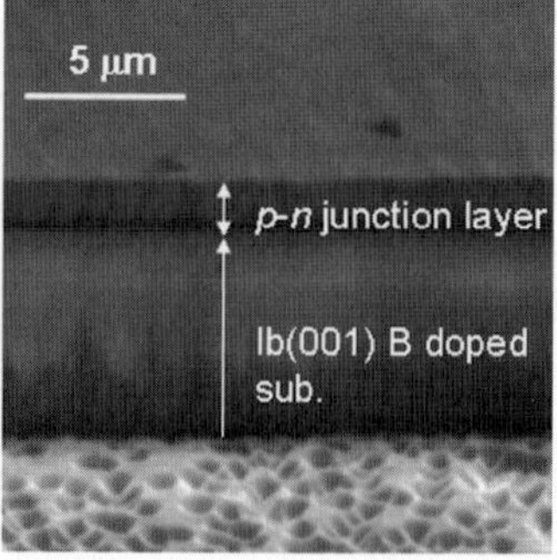

Fig. 3.48. A scanning electron microscope (SEM) image of a typical (001) p–n junction diode

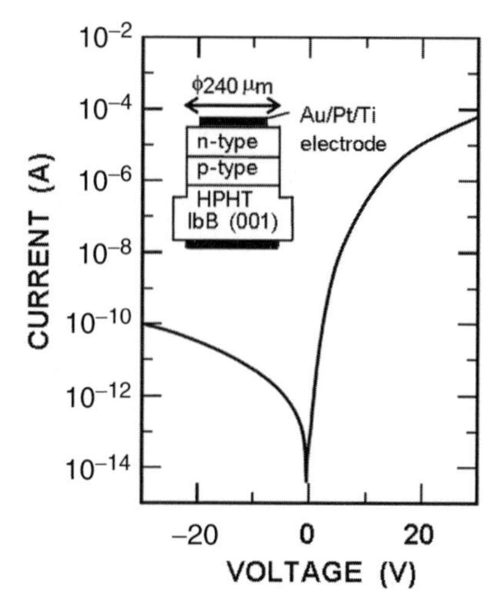

Fig. 3.49. *I–V* characteristic of diamond p–n junction diode. The inset shows a schematic presentation of the p–n junction diode

mentioned above, the S-doped n-type layer needs to be improved. Kato et al. (AIST group) succeeded in making n-type homoepitaxial diamond CVD films on an HPHT synthesized diamond (001) substrate using phosphorus as the dopant by optimizing and extending the range of synthetic conditions [114]. In their P-doped n-type CVD film, a relatively high electron mobility of about $400\,\mathrm{cm^2\,V^{-1}\,s^{-1}}$ was obtained, as determined by Hall-effect measurements.

Makino et al. (AIST group) are fabricating p–n junction diodes based on the P-doped n-type layer on diamond (001) substrate [118]. Figure 3.48 shows a scanning electron microscope (SEM) image of a typical (001) p–n junction diode. The current–voltage (*I–V*) characteristic of this p–n junction diode has the rectification ratio of six orders of magnitude at ±30 V, as shown in Fig. 3.49. The capacitance–voltage characteristics also show the standard behavior of conventional p–n junctions fabricated with other semiconductors. Great improvements in diamond p–n junction characteristics are expected from research and development based on data accumulation and analysis.

Light Emission Characteristics

Figure 3.50 shows two electroluminescence spectra (A and B) of the p–n junction whose I–V characteristic is shown in Fig. 3.49, where the spectra A and B were observed under forward bias voltages of 43 and 47 V and currents of 23 and 35 mA, respectively [118]. The Fig. 3.50 inset graph shows the strong emission at 235 nm from free excitons associated with transverse optical phonons (FE_{TO}) and its replica ($FE_{TO+O\Gamma}$). Broadband emissions were also observed at the peak energy around 350 nm (3.5 eV) and around 500 nm (2.5 eV). The origin of the 3.5 eV emissions has not yet been clarified, although it is often observed from cathodoluminescence (CL) of B-doped CVD diamond films. The broad emission at around 500 nm is attributed to the well-known H3 center.

It has been also observed that the ratio between the free exciton and defect-related emission intensities ($I_{\text{exciton}}/I_{\text{defect}}$) increases as the forward current is increased. This tendency was also observed in diamond LEDs with S-doped n-type layer. These results strongly suggest that the recombination rate of the electron–hole pair via the recombination center due to defect impurity states decreases effectively as the density of excitons increases. This phenomenon has been also observed from the exciton emission spectrum of CVD diamond films by CL when the density of excitons exceeds $10^{18}\,\text{cm}^{-3}$ [119]. This result means that large internal quantum efficiencies can be obtained in diamond even if there is a relatively high density of recombination centers due to defects or impurity states in the LED material. This point is an advantage

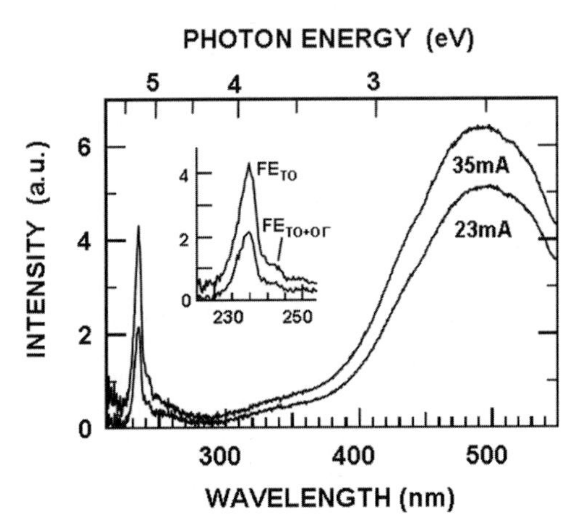

Fig. 3.50. Current-injected emission spectra of p–n junction diode under forward bias at room temperature. The forward currents are (**A**) 23 mA and (**B**) 35 mA. The inset expands the spectra around 235 nm

for indirect diamond LEDs in comparison with those fabricated from direct bandgap semiconductors.

Fabrication of diamond LEDs based on a P-doped n-type layer using a diamond (001) substrate has just started and performance data are scanty. Very recently, however, a high performance LED fabricated with the P-doped n-type layer has been obtained by the AIST group [120], where the defect-related emission intensity is greatly reduced compared to the case of Fig. 4. Judging from this result, the performance of this LED is much better than that of LED fabricated with the S-doped n-type layer mentioned above.

High-Efficiency LED Manufacture

For practical diamond LEDs, we have to solve both scientific and technical problems. Although recent diamond semiconductor R&D has been progressing rapidly, it is still far behind other semiconductors. Since diamond is a single element semiconductor like silicon, it is theoretically possible to make a higher quality single crystal compared with compound semiconductors. High quality diamond thin films are obtained by homoepitaxial CVD growth using a low growth rate method [119, 121]. However, in order to utilize diamond for practical devices, a synthetic technology for high quality diamond with a high growth rate needs to be developed. This is one of the future big subjects for diamond R&D. In terms of doping technology, a low resistance n-type diamond is required.

Concerning ohmic electrode formation, another important factor for high efficiency LEDs, a low contact resistance of less than $10^{-5}\,\Omega\,\mathrm{cm}$ has been realized on high quality p-type films [122], which is sufficient for diamond devices. However, ohmic electrodes are more problematic for n-type diamond at present. Although there is a report of ohmic electrodes for n-type diamond [123], there are still no reliable data for contact resistances. In the diamond LED shown in Figs. 3.48 and 3.49, ohmic electrodes for the n-type layer could not be used and the voltage losses are large because of the high resistance of the electrodes. The subject of ohmic electrodes for n-type diamond is an important issue in improving diamond LED performance.

3.3.4 AlN LEDs (Y. Taniyasu, M. Kasu, T. Makimoto)

First synthesis of aluminum nitride (AlN) was reported in 1907 [124]. AlN has a direct bandgap energy of 6 eV at room temperature, the largest among available semiconductors. Therefore, AlN-based light emitters are expected to emit deep ultraviolet (deep-UV) light with a wavelength $\lambda \sim 210\,\mathrm{nm}$ [125]. Deep-UV light sources have attracted considerable attention because of their potential uses in high-density optical data storage, biomedical research, water and air purification, and sterilization [126]. In addition, nanofabrication technology and environmental science both demand light sources with shorter emission wavelength: the former for improved resolution in photolithography and the latter for sensors that can detect minute particles such as cancer-causing and toxic substances. The deep-UV light sources available at present are only gas sources, such as mercury lamps or gas lasers. Therefore, AlN-based deep-UV solid-state light sources, such as light-emitting diodes (LEDs) and laser diode (LDs), are of great technological interest as an alternative to large toxic low-efficiency gas sources.

Epitaxial Growth of AlN

AlN layers are commonly grown by low-pressure metalorganic vapor phase epitaxy (MOVPE). The sources are trimethylaluminum (TMA) and ammonia (NH_3). A parasitic reaction of the TMA and NH_3 in the gas phase easily occurs [127]. This reaction decreases the growth efficiency and strongly affects the quality of AlN layer. In addition, AlN layers are heteroepitaxially grown on a lattice-mismatched substrate, such as sapphire or SiC, because large bulk AlN substrates are not available. As a result, the heteroepitaxial AlN layers contain a large number of threading dislocations, with the density of 10^{10}–10^{11} cm^{-2}.

The threading dislocations strongly degrade the optical and electrical properties in AlN. The dislocations act as non-radiative recombination centers and limit the internal quantum efficiency in nitride semiconductors [128]. In addition, the dislocations act as compensating and scattering centers in n-type AlN and thereby decrease doping efficiency and electron mobility [129]. By suppressing the parasitic reaction of the sources and by increasing growth temperature, the dislocation density was decreased to 10^8–10^9 cm^{-2} [130]. In addition, the concentrations of unintentional impurities, like C and O, were decreased to levels below the detection limits of SIMS ([C] $<5 \times 10^{16}$ cm^{-3} and [O] $<5 \times 10^{17}$ cm^{-3}).

Doping Control in AlN

n-type Doping

n-type AlN layers were obtained by Si doping using silane (SiH_4) [129,131,132]. For Si-doped AlN with dislocation densities of 2×10^9–1×10^{10} cm^{-2}, n-type conduction was observed for Si doping concentrations between 2×10^{17} cm^{-3} (lower doping limit) and 8×10^{19} cm^{-3} (upper doping limit). When the Si doping concentration is below the lower doping limit, the Si-doped AlN becomes highly resistive because Si donors are fully compensated by dislocation-related defects. Thus, the lower doping limit depends on the quality of the epitaxial layer, and it can be extended by decreasing the dislocation density. On the other hand, when Si doping concentrations exceed the upper doping limit, the Si-doped AlN becomes highly resistive again because a self-compensation effect of Si occurs [131]. A theoretical calculation

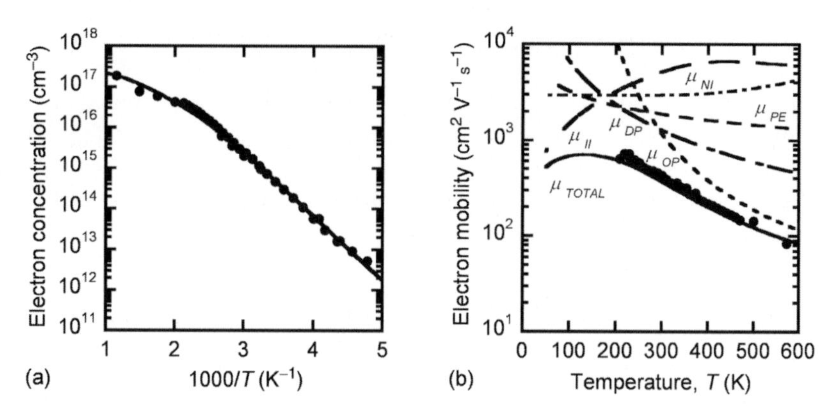

Fig. 3.51. Temperature dependence of (**a**) electron concentration and (**b**) electron mobility for an n-type Si-doped AlN

has predicted that the acceptor-like defects created by the self-compensation effect are cation vacancies for n-type AlN [133]. Therefore, precise control of Si doping concentration is important for obtaining n-type conduction.

Figure 3.51 shows the temperature dependence of (a) electron concentration and (b) electron mobility for an n-type Si-doped AlN layer grown on bulk AlN (0001) substrate. The Si doping concentration was 3.5×10^{17} cm^{-3}. At 300 K, the electron concentration and mobility were 7.3×10^{14} cm^{-3} and 426 cm^2 V^{-1} s^{-1}, respectively. As the temperature increased, the electron concentration exponentially increased and then saturated at high temperature. From the temperature dependence, the donor concentration N_D, the acceptor concentration N_A, and the donor ionization energy E_D can be obtained by assuming the charge neutrality equation for n-type semiconductor with a shallow donor level and a deep acceptor level [131]. The best-fit values are $N_D = 3.0 \times 10^{17}$ cm^{-3}, $N_A = 2.6 \times 10^{16}$ cm^{-3}, and $E_D = 282$ meV. The N_D of 3.0×10^{17} cm^{-3} agreed well with the Si doping concentration of 3.5×10^{17} cm^{-3}, indicating that almost all of the Si atoms act as donors in AlN. The compensation ratio N_A/N_D was about 0.1.

On the other hand, the electron mobility increased monotonically with decreasing temperature. As the temperature decreased to 220 K, the mobility increased to 730 cm^2 V^{-1} s^{-1}, the highest value ever reported for n-type AlN. In calculating the mobility, scatterings of neutral impurities (μ_{NI}), ionized impurities (μ_{II}), polar optical phonons (μ_{OP}), acoustic deformation potential (μ_{DP}), and piezoelectric potential (μ_{PE}) were considered, Matthiesen's rule was assumed, and the best-fit values of N_D, N_A, and E_D were used [132]. The calculated electron mobility (solid line) agreed very well with the measured one.

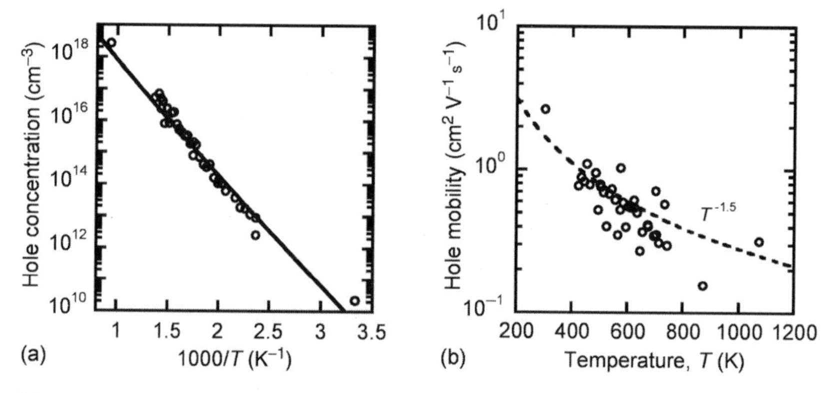

Fig. 3.52. Temperature dependence of (**a**) hole concentration and (**b**) hole mobility for a p-type Mg-doped AlN

p-type Doping

p-type AlN layers were obtained by Mg doping using bis-cyclopentadienyl-magnesium (Cp_2Mg) [125]. When the Mg doping concentration exceeded the upper doping limit of 2×10^{20} cm^{-3}, the p-type conduction was not observed. This is similar to the tendency observed in the n-type Si-doped AlN and probably results from a self-compensation effect of Mg in p-type AlN. This also indicates that control of the Mg doping concentration is important for obtaining p-type conduction. In addition, the as-grown Mg-doped AlN was highly resistive and became p-type conductive after thermal annealing in N_2 ambient at 800°C for 10 min. For the highly resistive as-grown Mg-doped AlN, the H concentration was in the same range as the Mg concentration, and after annealing the H concentration decreased. This means that, in the as-grown AlN, H passivated Mg dopant, but after annealing H was desorbed and Mg was activated, like in GaN [134].

Figure 3.52 shows the temperature dependence of (a) hole concentration and (b) hole mobility for an AlN layer doped with Mg to a concentration of 2×10^{19} cm^{-3}. As the measurement temperature increased, the hole concentration increased exponentially but did not saturated even at a high temperature of around 1,000 K. Thus, from the present data, only the acceptor ionization energy E_A was obtained by assuming that the hole concentration follows $\exp(-E_A/k_BT)$, where k_B is the Boltzmann constant and T is the sample temperature. The E_A was estimated to be 630 meV, which is in good agreement with an optically obtained one (510 meV) [135]. Due to the high acceptor ionization energy, the room-temperature hole concentration is as low as about 10^{10} cm^{-3}. On the other hand, the hole mobility decreased as temperature increased. The temperature dependence was well fitted to $\mu \propto T^{-1.5}$ (broken line). This indicates that the phonon scattering dominates the hole mobility.

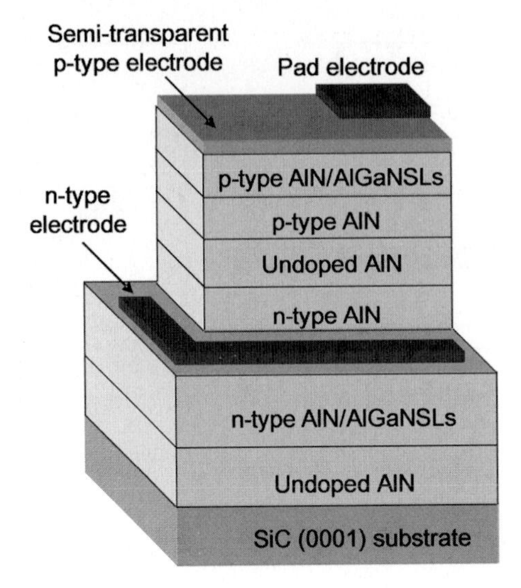

Fig. 3.53. Schematic of a device structure of an AlN LED grown on SiC (0001) substrate

AlN Homojunction Light-Emitting Diode

Figure 3.53 shows device structures of a p-type AlN/undoped AlN/n-type AlN homojunction LED grown on SiC (0001) substrate [125]. The device consisted of a 750-nm-thick undoped AlN layer, n-type uniformly Si-doped 100-period AlN/AlGaN (1.2 nm/3 nm) superlattices (SLs) with a Si doping concentration *[Si]* of $3 \times 10^{19}\,\text{cm}^{-3}$, a 200-nm-thick n-type Si-doped AlN layer with *[Si]* of $2 \times 10^{18}\,\text{cm}^{-3}$, a 60-nm-thick undoped AlN layer, a 20-nm-thick p-type Mg-doped AlN layer with a Mg doping concentration *[Mg]* of $4 \times 10^{19}\,\text{cm}^{-3}$, p-type uniformly Mg-doped 3-period AlN/AlGaN (1.2 nm/3 nm) SLs with *[Mg]* of $4 \times 10^{19}\,\text{cm}^{-3}$. The average carrier concentration of the SLs is as high as $10^{18}\,\text{cm}^{-3}$, because the doping efficiency of the SLs is much higher than that of the AlN layers [136]. Therefore, the SLs were used to improve the lateral conduction in the n-type region and reduce the contact resistance of the electrodes and thereby lower the driving voltage of the LED. A mesa structure $200\,\mu\text{m} \times 200\,\mu\text{m}$ in size was formed by dry etching down to the n-type SLs. An n-type Ti/Al/Ti/Au electrode was formed on the n-type SLs, while a semitransparent p-type Pd/Au electrode was formed on the p-type SLs. A pad electrode was then formed on the Pd/Au electrode.

The characterization of the electrical and optical properties was performed on-wafer under a direct current (DC) bias condition at room temperature. The current–voltage (I–V) characteristics showed a rectifying property with rectification ratios of higher than 10^4 at $\pm 40\,\text{V}$ as shown in Fig. 3.54. The current remarkably increased above the voltages V_D of 5 V. The voltage is close to the diffusion (built-in) potential of the p–n junctions. The current

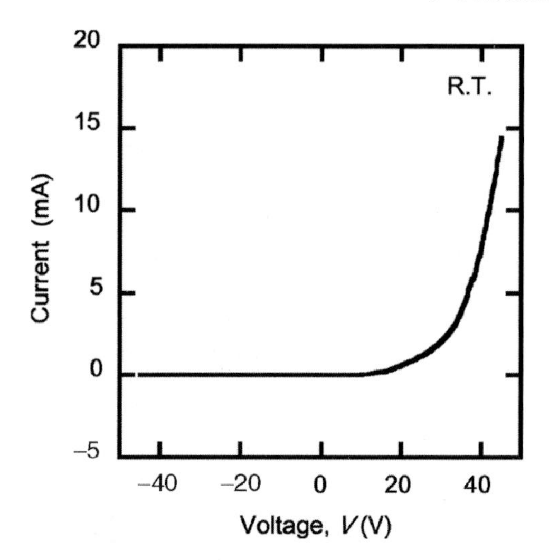

Fig. 3.54. Current–voltage characteristics of the AlN LEDs

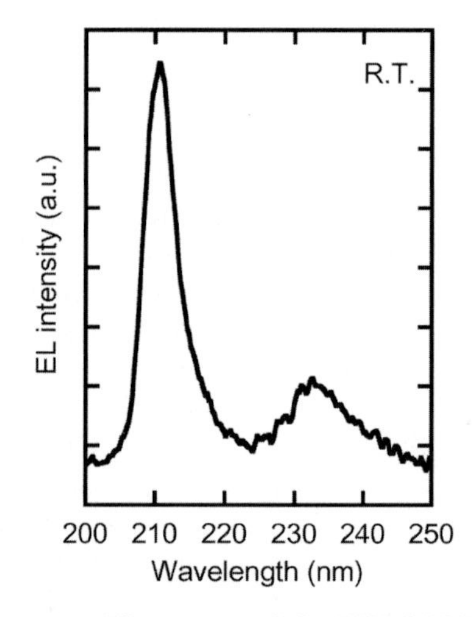

Fig. 3.55. EL spectra of the PIN AlN LEDs

was limited by a high series resistance, which originates from the p-type AlN layer. Consequently, the operating voltages at 20 mA were about 45 V, and were much higher than the voltages estimated from the diffusion potentials because of the high series resistance.

Figure 3.55 shows electroluminescence (EL) spectra of the PIN LED. The EL was observed at a wavelength of approximately 210 nm, which is the shortest ever reported among semiconductors. This emission wavelength was almost the same as that of the free-exciton recombination in AlN layers measured by photoluminescence at room temperature. This indicates that the origin of the EL is unambiguously the near-band-edge recombination in the AlN. The output power of the near-band-edge emission at $\lambda = 210\,\text{nm}$ was measured to be about $0.02\,\mu\text{W}$ at a DC current of 40 mA. The external quantum efficiency η_{ext} of the on-wafer device was estimated to be on the order of $10^{-6}\%$, which is still lower than that for typical packaged commercial visible LEDs ($\eta_{ext} = 1$–10%).

Future Prospects

For the practical use of these AlN LEDs, it is necessary to decrease the driving voltage and to increase the emission efficiency. The keys to improving the performance are to (1) increase p-type doping efficiency, (2) decrease dislocation density, (3) introduce carrier confinement structures, and (4) increase extraction efficiency.

1. Increasing the hole concentration in p-type AlN will enhance internal quantum efficiency because the hole concentration injected into the AlN active layer will increase. This will also decrease the driving voltage because the series resistance will decrease. Mg is commonly used as a p-type dopant for nitride semiconductors, but Mg's doping efficiency in AlN is quite low. Alternative elements for acceptor with lower ionization energy must be explored.
2. Our device still has a high dislocation density of $10^9\,\text{cm}^{-2}$ as a result of the lattice mismatch and the heterovalency between the SiC substrate and AlN layer. The use of AlN substrate will decrease the dislocation density in the device and thereby increase the internal quantum efficiency because the dislocations act as non-radiative recombination centers.
3. Introducing carrier confinement structures (double heterostructures or quantum well structures) will increase the internal quantum efficiency. In addition, using an AlGaN/AlN quantum well structure, the emission wavelength can cover the region from deep-UV (210 nm) to near-UV (364 nm).
4. The light extraction efficiency will be increased by using transparent substrates like sapphire. In addition, the near-band edge emission parallel to the *c*-axis is much weaker than that perpendicular to it because optical transition between the conduction band and top most valence band is allowed for $E||c$ but almost completely prohibited for $E\perp c$ in AlN. Therefore, the light extraction efficiency will be increased by using m- or a-plane substrate and by designing an appropriate device structure.

3.3.5 Ultra Violet Photodetectors (A. Hirano)

Applications and Demand for UV Photodetectors

Silicon photodetectors operate in the UV waveband, thus detectors fabricated using other compounds must have distinct advantages to compete with silicon devices. For example, InGaAs photodetectors are used in the 1.3–1.55 μm wavelength range, where the sensitivity of Si detectors is poor. Further, GaAs photodetectors operating in the visible spectrum exhibit much higher response times than Si devices.

The lifetime of Si photodetectors is limited by damage due to exposure to UV light. This deficiency in Si devices can be overcome by the use of wide-bandgap compound semiconductors that are not damaged by UV radiation and could be used for fabricating UV photodetectors operating over wavelengths ranging from 280 to 400 nm. The stable operation of waveband gap semiconductors at high temperatures enables their use for applications such as flame detectors. Table 3.8 shows some applications of UV photodetectors.

Suitable materials include compound semiconductors belonging to II–VI (SnO, ZnO, SrS, CaS), III–V (In)AlGaN(P), and IV (SiC, C*[diamond]*) groups. The ionic bonding of II–VI compounds leads to large electron and hole effective masses, which makes these semiconductors unsuitable for conventional electron devices. Group VI, indirect bandgap semiconductors are stable at high temperatures but their detection wavelength shows a red shift at high temperatures, which limits their applications for wavelength selective optical detectors. From the viewpoint of peripheral technology, group III–V (In)AlGaN UV detectors are promising.

Development and Current Status of Nitride Compound Semiconductors

The fabrication of optical detectors necessitates crystals with lower dislocation densities than for light-emitting devices. Dislocations in emission devices behave as non-radiative recombination centers and in detectors they become photocurrent traps. Wide-bandgap semiconductors are not damaged at high temperatures and can thus be used for fabrication of devices operating at high current densities. Thus the adverse effects of non-radiative recombination centers can be reduced in light-emitting devices. However, in the case of optical detectors, the density of dislocations determines the detection sensitivity and limits the response time.

Optical detectors require large detection areas and their chip size is larger than emission devices. In the case of nitride compound semiconductors, the required surface area of optical detectors is not readily obtained using the ELOG method developed by Nakamura et al. [137]. Also, dislocations limit high temperature operation because electrons and holes are thermally trapped or emitted from trap centers.

Development of Absorption Layers with Low Dislocation Densities

The optical properties of absorption layers are evaluated by measuring the photoconductivity of single absorption layers. Although blue LEDs are grown on single buffer layers, the density of dislocations penetrating the GaN layers ranges from 8×10^{8}–1×10^{9} cm^{-2}. Amano et al. developed interlayer technology

Table 3.8. Types and applications of the main types of photodetectors

	wavelength range(nm)	application	disadvantages of conventional devices	advantages of nitrides
UV-A photodetector and wide spectrum UV photodetector	400–315 400–high end	personal UV radiation meter/ health care instruments/ i-radiation meter/laser power meter (excimer HeCd)	Si is degraded by UV radiation	possible to make highly accurate measurements and nitrides are not degraded by UV radiation
UV-B photodiode	315–280	meteorological UV radiation measurements	limits in accuracy in discrimination between UV-A and visible radiation	possible to make highly accurate measurements and nitrides are not degraded by UV radiation
UV-C photodetector	280–200	flame sensor/ozone hole observation/ sterilization lamp meter	limits to heat resistance; almost impossible to discriminate stray light less than 280 nm	heat resistant (flame sensor), use of band edge and filters enables discrimination up to five orders of magnitude

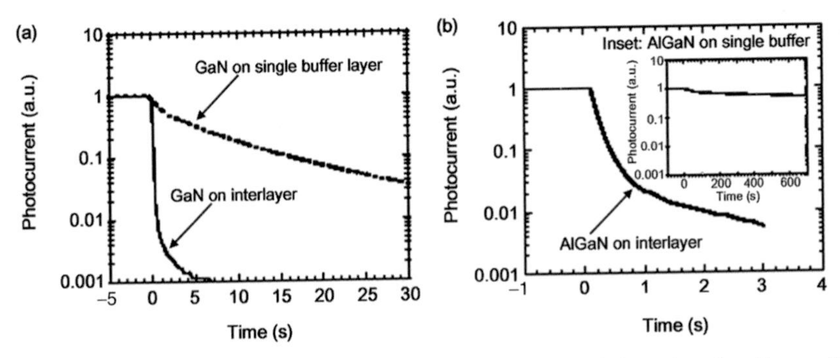

Fig. 3.56. (**a**) and (**b**) are a comparison of the effects of dislocation density on the photoconduction response times of GaN and AlGaN absorption layers, respectively

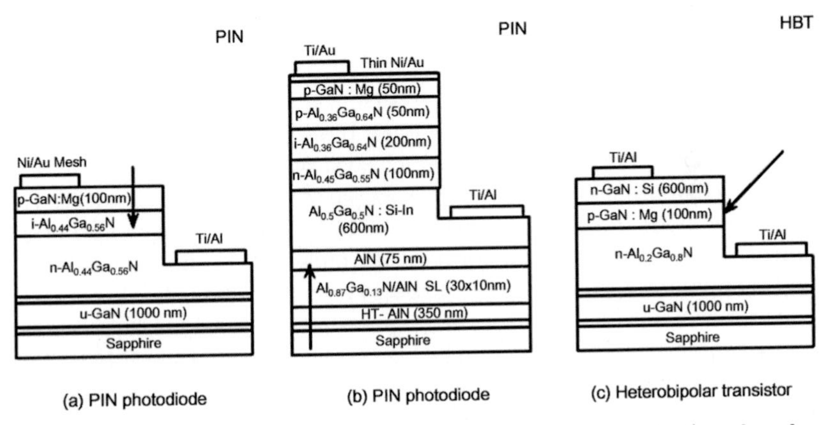

Fig. 3.57. Typical structures for a PIN diode optical detector (*left*) and a phototransistor (HBT)

to reduce the density of dislocations over the wafer diameter [138], where the dislocation density of GaN was reduced to $\sim 10^8\,\mathrm{cm}^{-2}$ and that of AlGaN (AlN $\sim$20%) was reported to be $6 \times 10^9\,\mathrm{cm}^{-2}$, compared with $10^{11}\,\mathrm{cm}^{-2}$ for conventional growth. Figure 3.56a, b show a comparison of the effects of dislocation density on the photoconduction response times of GaN and AlGaN absorption layers, respectively. The response time is improved by reduction of dislocation density.

Examples of Optical Detectors

Compared with photoconductors, a strong electric field exists in the absorption layer of junction-type optical detectors resulting in a significant reduction of the adverse effects of recombination centers. Figure 3.57 shows typical structures for a PIN diode optical detector (left) and a phototransistor (HBT). In the PIN diode, the GaN layer on the sapphire substrate is

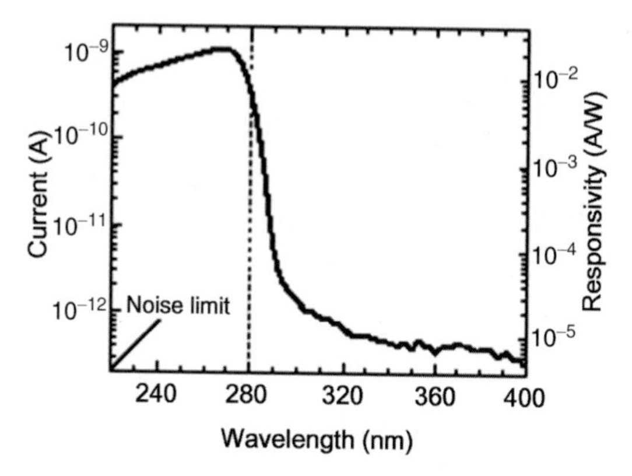

Fig. 3.58. Example of the wavelength sensitivity of a PIN photodiode incorporating AlN (x = 44%)

opaque over a wide range of the UV spectrum. Because of this, although it is necessary to detect light traveling through the top layer, it is not possible to prevent absorption in the p-GaN contact layer which limits the sensitivity of the device [139]. Figure 3.58 shows an example of the wavelength sensitivity of a PIN photodiode incorporating AlN (x = 44%). Solutions to the absorption problem are structures that do not use GaN buffer layers and light irradiation from the sapphire substrate surface has also been reported [140]. In the case of the phototransistor (right), electrons trapped in the absorption layer are emitted and amplified, resulting in a reduction of the response time [141]. Internal amplification requires further reduction of the dislocation density. There have been reports on structures without p-GaN layers in Schottky diodes [142] and photo-FETs [143]. In Schottky diodes, the detection sensitivity is limited by the transparency of the electrodes and absorption in n-type semiconductors. The main problems encountered in FET structures are dark currents and there are no obvious advantages in using such devices compared with junction detectors.

The dark current determines the lower limit of detection. Dark currents result from tunneling through the junctions and carrier flow around the periphery of mesa areas. The tunnel currents through a GaN pn junction grown on a single buffer layer are large although dark currents around mesa areas cannot be ignored [144]. A passivation layer deposited over the mesa reduces the dark currents around these structures as reported by Monroy et al. [142], who deposited a SiO_2 dielectric by electron beam evaporation. Figure 3.59 shows a comparison of the dark currents observed in Schottky diodes with and without passivation layers.

UV-C Selective Optical Detectors and Flame Sensors

Emission from flames containing UV-C, sunlight and household illumination lighting cover wavelengths down to 280 nm. As described in "Examples of

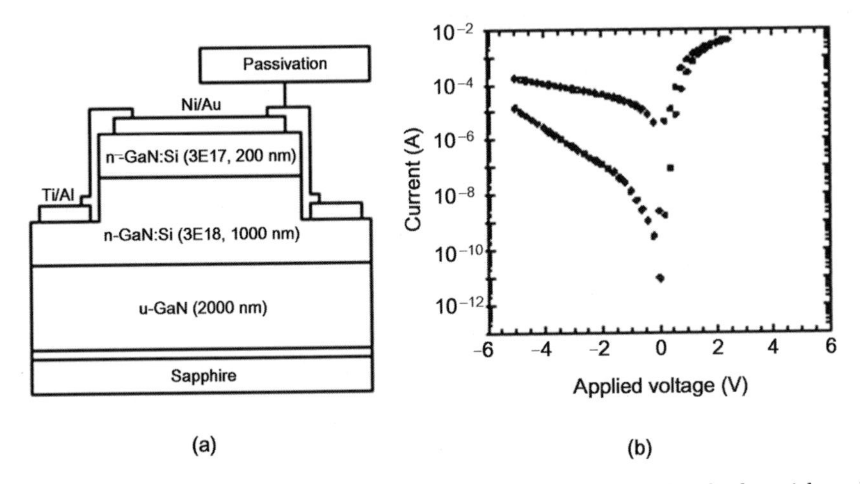

Fig. 3.59. Comparison of the dark currents observed in Schottky diodes with and without passivation layers

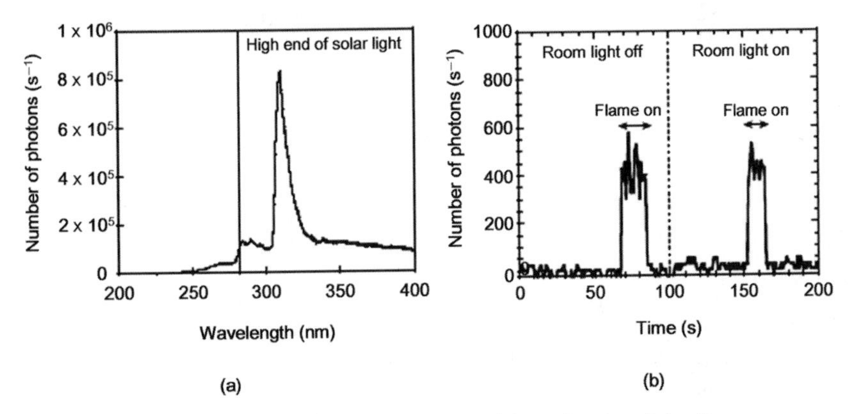

Fig. 3.60. Spectrum of a flame and its selective detection

Optical Detectors", an AlGaN detector fitted with a glass filter that cuts visible light, but transmits UV radiation, will not respond to household lighting but will react to a flame [145]. Figure 3.60 shows the spectrum of a flame and its selective detection.

Currently, flames in industrial furnaces are monitored using UV-C selective optical detection tubes but the lifetime of these devices is only about 1 year and the maximum operating temperature is $\sim 120°C$. The heat resistance of nitride semiconductors makes them promising for fabrication of wavelength selective flame detectors.

3.4 White Light Emitting Devices (K. Kohmoto)

3.4.1 White LEDs for Illumination

Comparison of Vacuum and Solid-State Light Sources

The first shipment of commercial white LEDs in Japan was in 1996. Improvements in the power output per device over the last decade have resulted in even greater expectations for white LED illumination.

The main light sources currently used for illumination are incandescent bulbs, fluorescent lamps, and high intensity discharge (HID) lamps, which operate by energy conversion in a vacuum and are called "vacuum system sources," compared with the "solid-state lighting sources" from LED-based technology.

Table 3.9 is a comparison of the main features of vacuum and solid-state-based light sources. As shown in Table 3.10, solid-state sources have many advantages over vacuum systems.

Requirements of Light Sources for Illumination

Table 3.11 shows the main requirements for the light sources used for illuminating environments where people live and work. The main requirements are:

(a) Development of white LEDs. The light emitted from LEDs must be similar to that from the sun; that is, the artificial light from white LEDs must have wavelengths covering the whole visible spectrum. Table 3.12 shows the range of white LEDs developed for illumination to date.

(b) Color rendering. The use of color rendering indices (CRI) were devised by the International Commission on Illumination (CIE) as a means of mathematically expressing the ability of artificial light sources to reproduce colors of objects being illuminated by a given source. Monochromatic sources, such as halogen lamps, have a CRI close to zero, and blackbody radiators, like vacuum light bulbs, have a CRI close to 100, which is the maximum of the scale.

(c) Emission efficiency (energy conversion efficiency). The emission efficiency of light sources is an extremely important characteristic. Energy conversion in LEDs occurs via one or two steps in LED+phosphor type configurations, thus high conversion efficiencies are possible. Table 3.13 shows the experimentally measured results for energy conversion efficiencies for two-color white LEDs. For comparison, Table 3.13 also includes the efficiencies for vacuum system light sources. It can be seen that further improvements are necessary for further industrialization of white LEDs.

Table 3.9. Comparison of vacuum and solid-state optical sources

comparison	vacuum system optical source	solid-state optical source
characteristics of energy conversion	light generated by energy conversion inside a specific gaseous vapor isolated from vacuum or air (thermal radiation or discharge plasma luminescence)	form a semiconductor pn junction, pass a current in the forward direction and light is generated due to the recombination of electron–hole pairs (recombination luminescence)
size of emission element	basic element consists of an optically transparent and air tight container; difficult to reduce the size below certain dimensions	1. the size of one device is small thus it is possible to split the light into small units
		2. by dividing light into small parts it is possible to add value to the light source system
energy conversion efficiency	there are two to four steps in the conversion process from electrical energy accompanied by large losses	electrical energy is directly converted to optical radiation; only one step in the conversion process; possibility of further improving efficiency
mechanical strength	transparent airtight vessels (glass) are fragile	compound semiconductors are mechanically strong
durability and lifetime	1. not possible to use broken vessels	1. semiconductors are conductors and do not become insulators
	2. heat resistant materials are not sufficiently durable	2. it is possible that it will not be necessary to replace light source sections
	3. thus vacuum system optical sources are replaceable components	
output power of an individual element of emission section	size of an element required for discharge plasma becomes large due to use of heat resistant materials inside the airtight vessel; input power also becomes large	not possible to increase dimensions because increasing the thickness of compound semiconductor layers leads to nonuniform flow of current through the device; the input power is also small
price of each element of emission section	well established manufacturing processes mean reasonably priced elements	currently, the price is greater than vacuum based optical sources
standardization	vacuum systems have been developed over the last 60–120 years; extensive international standardization of related circuits and equipment	no standardized products on the market yet; device specifications depend on the manufacturers

Table 3.10. The strongpoint of solid illuminant in comparison with the vacuum system illuminant

1. possible for light fractionation	easy to digitize the illumination
2. easy to fabricate and compound	easy for the system interaction
3. multifunction	easy for multifunctional illumination
4. less energy-exchanging steps	high efficiency of energy-exchanging
5. high mechanical intensity	it is not necessary to use a vacuum vessel
6. long lifetime	solder jointing
7. process and materials are completely different	possible for the new creation from new domains and employment

White LEDs as Light Sources for Illumination

Table 3.14 is a comparison of the relevant properties of vacuum and white LEDs as light sources for general illumination based on the roadmaps published by LED manufactures. The data for LEDs is that for two-color white LEDs.

Future Trends

The following conditions must be fulfilled for the proliferation of white light LEDs into everyday illumination.

1. Improvement of energy conversion efficiency
2. Increase of the optical flux emitted per device
3. Standardization of LED devices and product specifications
4. Development of new environmental illumination systems and standardization of their specifications
5. Improvement of device fabrication process to increase yield
6. Reduction of optical source costs

3.4.2 Development Trends of Phosphors (H. Kanie)

The emission bands of blue, green, and red LEDs are extremely narrow thus a simple combination of these three LEDs does not produce high quality white light for illumination. In order to increase the bandwidth of emitted light the following combinations of LEDs and phosphors have been reported for producing white light.

(a) Blue LED light source for exciting a complementary color yellow phosphor: complementary (B+Y) type LED
(b) Addition of a red phosphor to complement red wavelengths and improve color rendering properties: three-color (B+YR) type LED

Table 3.11. Requirement of optical sources for illumination

subject	details of subject	requirements	considerations for LED applications
1. light color	the appearance of the color of the light	to obtain white light emission	possible to produce white light by combining monochromatic LEDs
2. color rendering properties	appearance of illuminated objects	natural appearance	the validity of using the current CIE rendering evaluation method for LEDs has been questioned
3. emission efficiency	conversion efficiency of input power into light	greater than 30%	devise methods for reducing heat losses
4. lifetime	total time until the device fails to switch on a act as a light source	greater than 20,000 h	absolute lifetime: from infinite to light source replacement not necessary effective lifetime: future improvements are imperative
5. variation of specifications between products	variations between products used in the same areas	within ±10% of middle values	indispensable that many LEDs are used at the same place
6. power source circuit	requirements of lighting power supplies and circuits	function using commercially available power supplies	development of suitable lighting circuits
7. safety	risk to person being illuminated	not affect the health of a person under standard illumination conditions	evaluate risks using CIE standardization

Table 3.12. Types and configurations of white LEDs

type of LED used	related phosphor	structure and principle of light emission	example of actual products	
			emission efficiency (lm/W)	average rendering value (Ra)
blue + green + red	(none)	white light produced interconnecting B,G,R LEDs near each other and switching them all on simultaneously	30	80
blue + yellow, or turquoise + amber	(none)	white light produced when two complimentary color LEDs (B + Y or BG + Am) are inter-connecting near each other and switching them all on simultaneously	25	60
blue LED	YAG phosphor (yellow emission)	produce light by mixing blue and yellow emission: the yellow light is produced by using a blue LED to irradiate a YAG phosphor which yield yellow emission	40	60
ultra-violet LED	three wavelength phosphor (B + G + R)	produce white light by irradiating a three wavelength phosphor with light from a UV LED	40	80

(c) Excitation of green and red phosphors for improving color rendering: three-color type (B+GR) type LED

Furthermore, a three color (Pu+BGR) LED has been proposed for using a 380–410 nm near ultraviolet LED light source for exciting red, blue, and green phosphors.

Requirements of Phosphors for Use in White LED Sources

(a) *Emission efficiency/strength.* In optically excited phosphors, the wavelength of the emitted light is longer than that of the excitation light

Table 3.13. Energy exchanging ratio of different illuminants

illuminant	UV radiation	visible radiation	infrared radiation	convection loss
Sun's rays	2–5	40–50	50–55	–
incandescent lamp	0–0.2	8–14	80–85	5–8
fluorescent lamp	0.5–1	25	30	44
high-pressure mercury lamp	2–4	13–16	60	16–22
metal halide lamp	2–7	20–40	50–67	7–20
high-pressure sodium lamp	0.3	27–30	47–63	10–23
white LED (two-color type)	0	10–20	0–0.2	80–90

Table 3.14. Comparison between LED illuminant and previous illuminants

item	previous vacuum system illuminant			white LED illuminant	
	incandescent lamp	fluorescent lamp	compact fluorescent lamp	forecast for 2005	forecast for 2006
longest dimension of one unit (mm)	60–250	150–2,400	90–900	3–12	< 20
electric power for one unit (W)	10–1,500	4–110	4–96	0.1–10	1–50
luminous flux for one unit (lm)	75–32,000	100–10,000	200–9,000	1.5–120	10–600
lamp efficiency ($\mathrm{lm\,W^{-1}}$)	8–20	60–100	50–80	35	> 100
integrated efficiency ($\mathrm{lm\,W^{-1}}$)	8–200	55–90	45–75	30	> 90
absolute lifetime (h)	1,000–2,000	5,000–20,000	3,000–9,000	> 100, 000	> 100, 000
effective lifetime (%lm = 50) (h)	>absolute lifetime	>absolute lifetime	>absolute lifetime	< 10, 000	< 10, 000
illuminant price per luminous flux	< 0.4	< 0.4	0.4–4.0	20–40	< 4.0

(Stokes shift) and the energy loss increases with increasing differences in the wavelengths of the excitation and emitted light. Thus improvements in the emission efficiency at long (red) wavelengths are required.

(b) *Wide optical emission spectrum.* The strength of light emitted in the visible spectrum must be uniform in order to obtain high color rendering indices.

(c) *Small temperature dependence of optical emission.* The power dissipation of GaN-based LEDs is high with chip temperatures reaching 150°C. Thus it is necessary to ensure temperature-independent emission spectra.

(d) *Robustness.* Phosphors in lamp type LEDs are used interfused with epoxy resin and must be resistant to humidity, high temperatures (150°C), and water. Further, phosphors with near ultraviolet LED light sources must be robust and resistant to oxidation under UV and water conditions.

Phosphors made from sulfide and oxysulfide compounds are widely used but do not satisfy all the aforementioned conditions. Thus, in Japan, there are concerted efforts to develop alternative phosphor materials based on oxides, nitrides and nitride matrices.

Yellow Phosphors Using Blue LED Excitation Sources

(a) *Ce Doped YAG ($Y_3Al_5O_{12}$:Ce) [146].* White LEDs based on blue LEDs and yellow phosphors are well known and widely used. The color rendering is poor but these white LEDs are used as backlights for LCD panels used in mobile telephones and displays.

(b) *Eu doped α-sialon (CaSiAlON:Eu) [147, 148].* The yellow emission from Eu is used to fabricate complementary white LEDs. The α-sialon is a silicon nitride ceramic containing a small percentage of aluminum oxide and has a α-silicon nitride crystal structure. A typical example is $M_x(Si,Al)_{12}(O,N)_{16}$ (M = Ca, Sr, Ba). Here M is a metallic ion and x the solubility limit. To date, α-sialon has been studied as a material having good heat resistance properties. However, a tremendous new interest in this material has been sparked by the use of M = Ca that was found to produce optical absorption at wavelengths corresponding to the emission from blue LEDs. Figure 3.61 shows the photoluminescence excitation (PLE) and PL spectra of CaSiAlON:Eu where strong emission is observed at 580 nm due to optical excitation by LEDs emitting in the near ultraviolet and blue wavelengths.

Green Phosphors Using Blue LED Excitation Sources

(a) *Eu doped β-sialon ($Si_{6-z}Al_zO_zN_{8-z}$:Eu) [149].* β-sialon has a silicon nitride crystal structure: $Si_{6-z}Al_zO_zN_{8-z}$. Here, z is the solubility quantity of the element. Figure 3.62 shows the excitation and emission spectra of $Si_{6-z}Al_zO_zN_{8-z}$:Eu. Emission is observed at 540 nm when excited with near UV or blue LEDs.

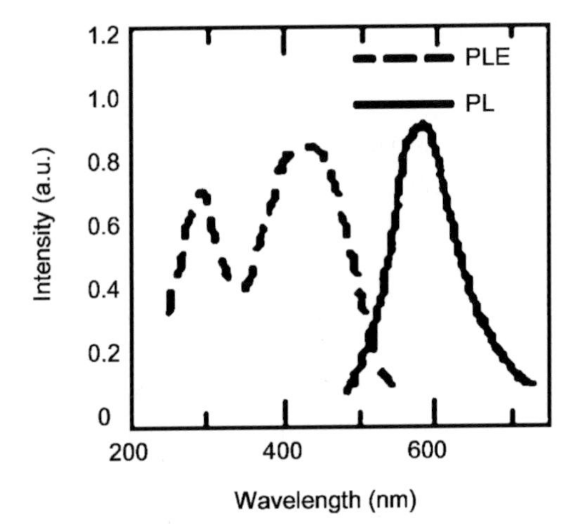

Fig. 3.61. Photoluminescence excitation spectrum and photoluminescence spectrum of CaSiAlON:Eu

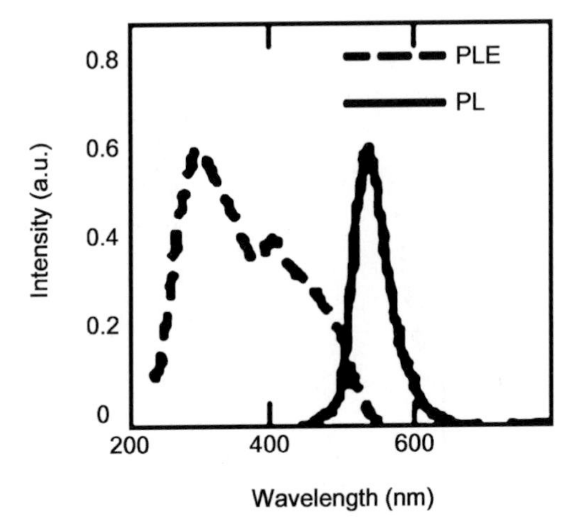

Fig. 3.62. Photoluminescence excitation spectrum and photoluminescence spectrum of $Si_{6-z}Al_zO_zN_{8-z}$:Eu

(b) *Ce doped CSSO* ($Ca_3Sc_2Si_3O_{12}$*:Ce*) *[150].* This is a completely new phosphor and has the absorption and emission spectra shown in Fig. 3.63. Excitation with a blue LED produces emission at 508 nm. Being an oxide, this material is highly robust and has temperature-independent optical emission properties. The main issue is the price of Sc.

Red Phosphors for Blue LED Excitation Sources

Eu doped $CaAlSiN_3$ ($CaAlSiN_3$*:Eu*) *[149, 151].* The optical characteristics are shown in Fig. 3.64. This phosphor has a wide excitation range from 300

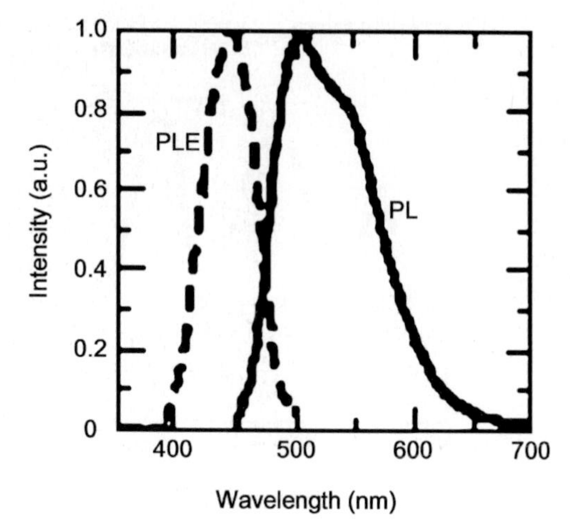

Fig. 3.63. Photoluminescence excitation spectrum and photoluminescence spectrum of $Ca_3Sc_2Si_3O_{12}$:Ce

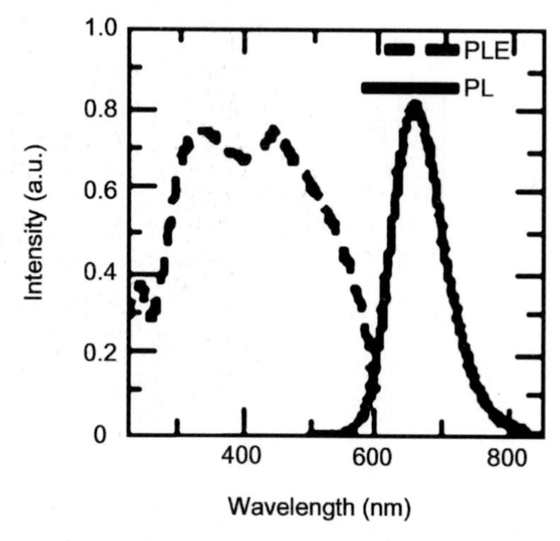

Fig. 3.64. Photoluminescence excitation spectrum and photoluminescence spectrum of $CaAlSiN_3$:Eu

to 550 nm and emits in the red spectrum region between 600 and 700 nm. LEDs emitting at less than 400 nm (UV and 405 nm violet LEDs; 450–480 nm blue LEDs; 500–550 nm green LEDs) can be used for excitation and since the strength of emitted light is independent of temperature from –240°C to +100°C, there is very little change in the color of emitted light when incorporated into LEDs.

3.4.3 Improvements of the Color Rendering of GaN-Based White LEDs (N. Shibata)

Introduction

The emission wavelength of LEDs depends on the bandgap energy of the semiconductor used for fabricating the devices. The bandgap energy is an inherent property of the particular semiconductor.

Gallium nitride is a direct bandgap semiconductor with an energy bandgap of 3.4 eV (365 nm). These properties enable the use of GaN-based III–V nitride semiconductors for fabricating optical devices emitting from the UV to blue and green wavelength regions.

The major breakthroughs in the development of high quality III–V nitride semiconductors are

1. Use of a low temperature buffer layer between the GaN and sapphire substrate [152]
2. Control of n-type conductivity by silicon doping [153, 154]
3. Use of low energy electron beam irradiation (LEEBI) for activation of Mg to control p-type conductivity [155]

The collaboration initiated in 1986 between Toyoda Gosei Co., Ltd and Akasaki's group led to the development of the blue LED in 1991 [156]. This LED, where Zn was used for the p-type doping of GaN, exhibited a luminous intensity of 200 mcd at a wavelength of 490 nm. This was the highest output from a blue LED at the time. Then in 1995, double heterostructure LEDs with luminous intensities of 1 cd were manufactured [157, 158]. Current blue LEDs show outputs of 10 mW at applied currents of 20 mA.

Applications of white LEDs can be divided into three main areas. (1) Light sources placed behind filters for applications such as back lights in LCDs used in mobile phones. In such cases it is necessary to match the white light to the characteristics of the color filter itself. (2) Indicators and displays seen directly by people where it is sufficient to have just white light without too many quality requirements on the light. (3) White LEDs for lighting objects such as in general illumination. For such applications, the light source must emit all wavelengths of the spectrum in order to light up all colors of the object.

Thus the required quality and properties of the light emitted by white light LEDs depends on specific applications.

Operating Principles of White Light LEDs and High Color Rendering

White light LEDs are classified according to their operating principles. Figure 3.65 shows the three main types of white LED.

The three color LEDs shown in the left-hand side column indicate an LED package containing red, green, and blue LEDs. The white light is produced by

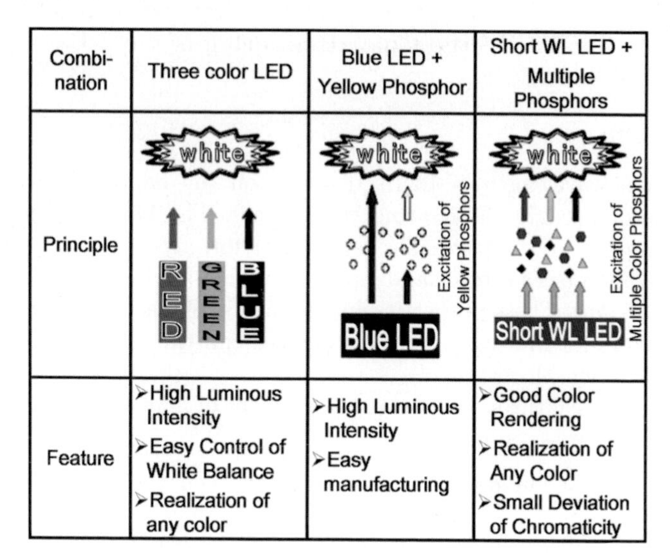

Fig. 3.65. The operation principles of the three main types of white LED

controlling the intensity of the optical output of the individual LEDs. However, since three LEDs are placed into a single package, the size and cost of the package may be unfavorable for applications requiring extremely small and inexpensive devices. The middle column shows a blue LED + yellow phosphor arrangement, where typically a YAG phosphor produces yellow light upon excitation by the blue LED. The main feature of this configuration is that high intensity white light can be produced using a single chip. However, the quality of color rendering is an issue since white light is produced by mixing only blue and yellow light. The right-hand column shows a short wavelength LED + multiple phosphor configuration, where a near infrared LED (wavelength of less than 400 nm) light source is combined with several phosphors to produce white light. Here again, white light can be produced using a single chip with the added advantage of high quality color rendering because of the use of multiple phosphors emitting at the desired wavelengths. Here, the main issue is that the emission efficiency is poor compared with the blue LED + yellow phosphor configuration.

Table 3.15 shows the combinations of LED light sources and phosphors used to produce white light.

The first example is a combination of a 380–390 nm near infrared LED ("TG Purple") and tricolor (blue, green, red) phosphors. This is shown as Type A in Table 3.15. The luminous intensity and general color rendering index (40) of this LED are low. It is clear from Fig. 3.66 that the emission spectrum is split into the blue-green and green-red regions and that this LED does not emit at wavelengths in the middle of these two regions. These optical properties are the reason for the low CRI value. Furthermore, excitation of

Table 3.15. Comparison of various white LEDs

light source of phosphors excitation	phosphors blue 450 nm	green 515 nm	yellow (New) 575 nm	red 625 nm	general color rendering index	relative luminous intensity	x, y coordinates	
TG purple	○	○		○	40	100	0.32, 0.32	type A
TG purple	○	○	○	○	91	240	0.32, 0.33	type B
TG purple	○		○	○	88	280	0.33, 0.33	type C
TG purple	○		○		78	300	0.31, 0.32	type D
blue LED			YAG		82	300	0.32, 0.32	type E

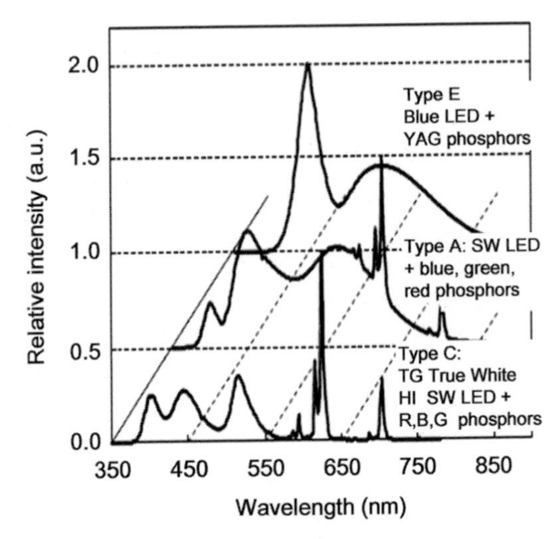

Fig. 3.66. Luminescence spectra from Type-A, Type-C, and Type-E white LEDs

red phosphors with short-wavelength light sources results in a relatively large energy conversion loss, leading to a low luminous intensity.

The quantity of phosphor added to the white LED is an extremely important factor in determining the quality of white light. If too much phosphor is added, then the phosphor itself can block light traveling through the white LED structure. Thus, in cases when multiple phosphors are used, it is important to optimize the composition and total volume of phosphors used.

The second example (Type B) is a combination consisting of "TG-Purple" and four phosphors, as shown in Table 3.15. Use of a phosphor with an emission spectrum with a wide half-width produces white LEDs with significantly better color rendering characteristics. In this Type-B LED, the CRI is very high with a value of 91.

The third example (Type-C) is similar to Type-B, but without the use of the green phosphor. The CRI of such white LEDs is 88, a value that is slightly less than Type-B LEDs. On the other hand, the luminous intensity of Type-C LEDs is 280, far greater than the 240 for Type-B LEDs. This shows that in cases where three or four types of phosphors are used, and where the emission spectra overlap, it is imperative that the quantity and ratios of the multiple phosphors be optimized in order to obtain white light. The emission spectrum is shown in Fig. 3.66.

Type-C white LEDs are known as "TG True White Hi" because of their high CRI, proximity to natural sunlight and their ability to simultaneously produce CRI and luminous intensity at a high level. However, Table 3.15 shows that there is a trade-off between CRI and luminous intensity. Further developments to increase the conversion efficiency and improve the emission spectrum of the phosphors are still necessary.

The fourth example is that of white LEDs where both red and green phosphors are not used (Type-D). The use of red phosphors results in low luminous intensity thus removing them leads to a significant improvement of light intensity to the detriment of the CRI, which decreases to 78.

Type-E is the fifth example, which is a combination of a blue LED and yellow phosphor. The emission spectrum is shown in Fig. 3.65. This type of white LED has a high luminous intensity but its CRI (82) is lower than Type B and Type C structures. The lower CRI results from the production of white light through mixing the blue emission from the LED and yellow emission from the phosphor, but there is a deficiency of red and green components.

Summary

White light LEDs with high CRI and luminous intensity can be produced by combining "TG Purple" LEDs with appropriate phosphors. A typical example of a high performance white LED is the "TG True White Hi."

In order for white LEDs to gain wider acceptance as a light source for general illumination (where the light source is used to illuminate an object and the human eye views the light reflected from the object), progress on improving the CRI is of utmost importance.

3.4.4 High Efficiency and High-Power LEDs (T. Mukai, Y. Narukawa)

Advances in White LEDs

The invention of blue LEDs in 1993 [159] and pure green LEDs in 1995 [160] has enabled fabrication of white LEDs by combining the three primary colors: red, green, and blue. However, the use of this configuration is complicated because each of the operating voltages and output intensities of each of the individual LEDs must be controlled individually. Thus there is a demand for a simpler means of using nitride LEDs for producing white light. In 1996, Nichia Corp. reported a unique solution to this problem where a blue LED was used to excite a YAG phosphor to produce commercial white light LEDs [161]. These lightweight and compact white light LEDs are used as backlighting of LCDs in mobile phones and other such products. Further, the emission efficiency (η_L) of commercial white LEDs currently on the market is 50 lm W^{-1}, which is three times larger than incandescent bulbs (15 lm W^{-1}). However, since the emission efficiencies of fluorescent tubes are 75–100 lm W^{-1}, it is still necessary to improve the performance of white LEDs by further improvements in the emission efficiency of blue LEDs. Currently, the total luminous flux (Φ) of a white LED at a constant drive current of 20 mA is only 4 lm. This output is much less than the 1,000 lm required for illumination and it would require 250 white LEDs to meet such requirements. There are demands to improve the emission output of current single white LEDs by a factor of ten.

Structure and Operating Principle of White LEDs

As shown in Fig. 3.67, the structure of a typical white LED consists of a blue LED incorporating an InGaN emission layer that is covered with a yellow phosphor. White light comes from the conversion of blue light from the LED into yellow light by the phosphor. Since blue and yellow are complementary colors, they mix to produce light that appears white to the human eye. A typical yellow phosphor is $(Y_{1-a}Gd_a)_3(Al_{1-b}Ga_b)_5O_{12}$: Ce^{3+}, a YAG phosphor that has an extremely large emission conversion efficiency and a negligible temperature dependence of the emission wavelength over a wide temperature range. Also, this material is extremely stable, easy to manufacture, and of low cost.

Figure 3.68 shows the typical emission spectrum of a white LED with a correlated color temperature (T_{cp}) and CRI (R_a) of 6,500 K and 85, respectively.

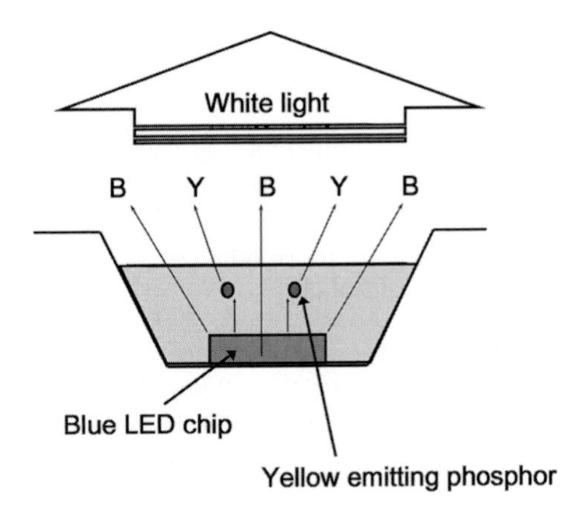

Fig. 3.67. Typical white LED consisting of a blue LED integrated in an InGaN emission layer covered with a yellow phosphor

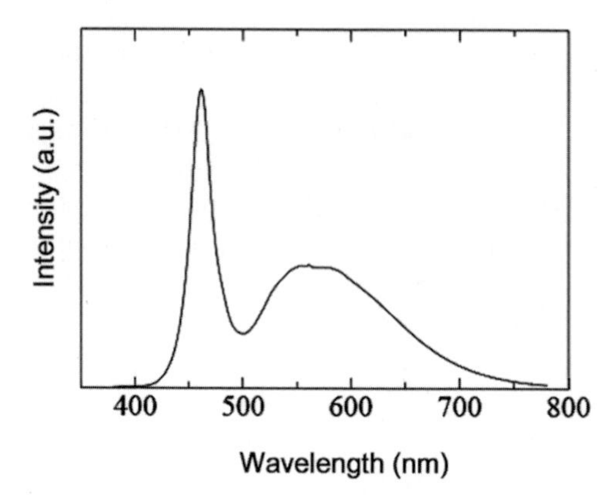

Fig. 3.68. Emission spectrum of a white LED with a correlated color temperature (T_{cp}) and CRI (R_a) of 6,500 K and 85, respectively.

The R_a of this white LED is much higher than the value of 67 for fluorescent tubes. Also since these sources do not contain UV or infrared light, objects illuminated by white LEDs are not adversely affected by optical irradiation. This is a unique property of white LEDs which will increase the range of applications of these devices.

Improvement of the Emission Efficiency

The emission efficiency of white LEDs is larger than incandescent lamps but not as good as fluorescent tubes. The following factors must be considered in order to improve the emission efficiency of white LEDs:

1. Improvement of the emission efficiency of the excitation source (blue LED chip)
2. Improvement of the emission efficiency of the phosphor
3. Optimization of the method used to coat the blue LED chip with the phosphor material
4. Development of packaging for more efficient use of emitted light

Advances in (1) have been reported by (a) improvements of internal quantum efficiency by growth of high quality epitaxial crystals; (b) reduction of operating voltage (V_f) due to implementation of the results of (a); and (c) modification of device processes for improving the efficiency of light extraction from LED chips.

In 2002, epitaxial layers used for commercial blue LEDs were grown on sapphire substrates using p-type Ni/Au semitransparent electrodes. Since the refractive index of nitride semiconductors is larger than both sapphire substrates and the mold resin, approximately 60% of the light generated in the active layer travels along the nitride semiconductor layer. Further, the transmission coefficient of Ni/Au semitransparent electrodes is 40% and the optical

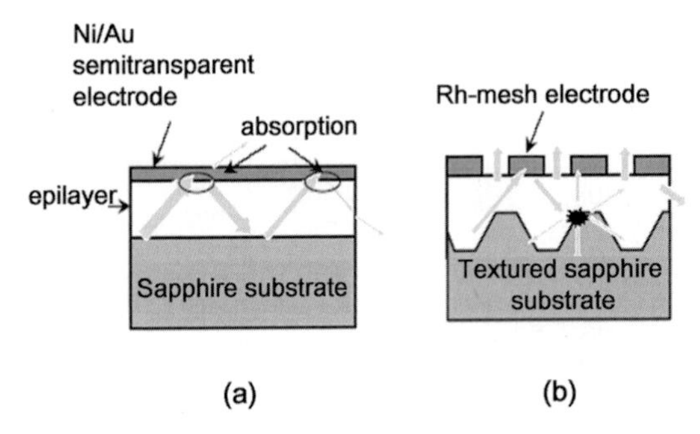

Fig. 3.69. (**a**) Cross-section a conventional blue LED and (**b**) the new more efficient blue LED chip

absorption is relatively large. Thus, as shown in Fig. 3.69, when the light traveling along the nitride semiconductor layer is reflected at the p-type electrode, its amplitude decreases. Thus it is necessary to decrease the light absorbed at electrodes.

This problem was solved using two methods [162]. The first involves the use of rhodium (Rh), which has a high reflection coefficient, as the p-type electrode, as well as the use of a mesh-shaped structure to reduce the area of the p-type electrode covering the epilayer surface. The use of this method reduces the probability of the light being incident on the p-type electrode and less of the light is absorbed even if it strikes the p-type electrode. In the other method, undulating structures are produced in the sapphire substrate so as to scatter light at the substrate/epilayer interface, increasing the efficiency of extracting light from the structure. Figure 3.69b shows an implementation of these methods where a blue LED driven by a 20 mA current produced an optical output power (Po) and external quantum efficiency (η_{ex}) of 19.3 mW and 35.8%, respectively. The operating voltage (V_f), Φ, η_L, and T_{cp} of a white LED with this blue LED light source (drive current = 20 mA) were 3.4 V, 4.2 lm, 62 lm W^{-1}, and 5,500 K, respectively. The magnitudes of Φ and η_L are double those of commercial LEDs sold in 2002 and closer to the 75 lm W^{-1} of conventional fluorescent tubes. However, the use of undulated substrates results in a reduction of device yield during manufacture.

This latter problem was resolved in 2004 by replacing the p-type Ni/Au electrode with a material with a transparency of more than 90%. White LEDs fabricated by this method had an emission efficiency that was increased by 1.7 times without the necessity for undulations in the substrate. The precise figures of merit at a drive current of 20 mA are a V_f of 3.4 V, $\Phi = 3.6$ lm, $\eta_L = 54$ lm W^{-1}, x = 0.31, and y = 0.32 in the CIE chromaticity diagram. Also, in 2005, the V_f was further reduced by modifications to the epitaxial structure, and at a drive the current of 20 mA, Vf = 3.0 V, $\Phi = 4.2$ lm, $\eta_L = 70$ lm W^{-1}, x = 0.31, and y = 0.32. These white LEDs are commercially available and have almost the same emission efficiency as fluorescent tubes.

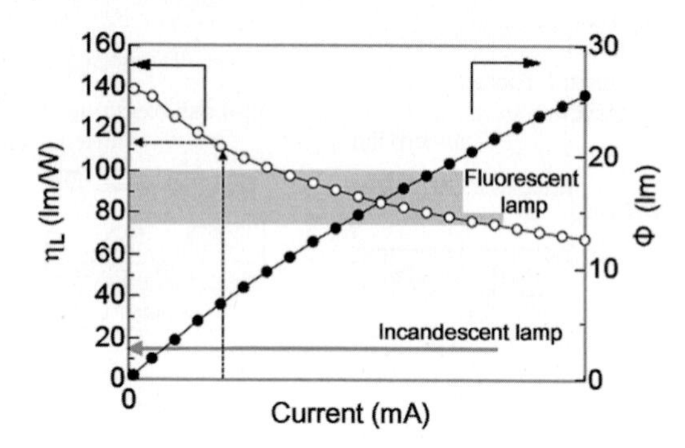

Fig. 3.70. The I–Φ and I–η_L characteristics in the range from 1 to 100 mA

White LEDs with even higher figures of merit have been reported with a drive current = 20 mA, $V_f = 3.0$ V, $\Phi = 6.8$ lm, $\eta_L = 113\,\mathrm{lm\,W^{-1}}$, x = 0.34, and y = 0.35. The 320 × 320 μm chip size of these LEDs is the same as commercial devices. Figure 3.70 shows the I–Φ and I–η_L characteristics for 1–100 mA for these devices. The η_L is seen to decrease with increasing current but the emission efficiency ($75\,\mathrm{lm\,W^{-1}}$) is comparable to fluorescent tubes up to currents of 80 mA. Also, it is notable that the emission efficiency at 1 mA is higher than at 20 mA. It is known that UV LEDs are less prone to such decreases in the emission efficiency with increasing drive currents [163, 164] and resolving this problem in blue LEDs is expected to enable emission efficiencies exceeding $150\,\mathrm{lm\,W^{-1}}$.

High-Power White LEDs

As described above, the luminous flux (Φ) of commercially available white LEDs is ∼4 ml per device and about 250 LEDs would be required for use as illumination sources that typically have $\Phi \sim 1{,}000$ lm. Assembling hundreds of LEDs would mean an increase of the surface area of the equipment and result in a decrease in the luminous flux per unit area. Thus, in order to resolve this problem, it is necessary to increase the luminous flux per device by a factor of ten, which would require increasing the input power per white LED. However, high input powers result in increased heat generation and the chip package must be cooled for stable device operation. Further, even if suitable methods for heat dissipation could be devised, the size of the blue LED chip will have to be increased to accommodate the large input power requirements. In response to these needs, Nichia Corp have developed a new, extremely compact package with a thermal resistance of $8\,^{\circ}\mathrm{C\,W^{-1}}$ and a blue LED chip with a size of 1 × 1 mm, which is ten times larger than that of conventional devices. These two developments have enabled the realization of compact white LEDs with power outputs of more than one order of magnitude larger than conventional devices with a drive current = 350 mA, $V_f = 3.8$ V, $\Phi = 42$ lm, $\eta_L = 32\,\mathrm{lm\,W^{-1}}$, x = 0.31, and y = 0.32. Figure 3.71 shows the I–Φ

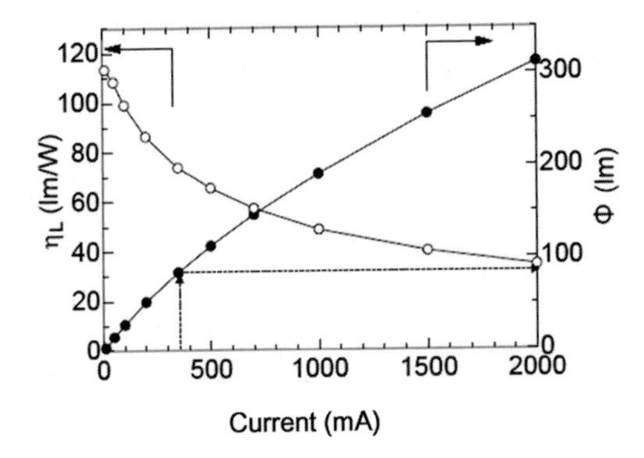

Fig. 3.71. The I–Φ and I–η_L characteristics of high-power output white LED

and I–η_L characteristics of a white LED with the following, even better, figures of merit: drive current = 350 mA, $V_f = 3.3$ V, $\Phi = 85$ lm, and $\eta_L = 74\,\mathrm{lm\,W^{-1}}$. At 2 A, the maximum power output is 312 lm, which is equivalent to a 20 W incandescent bulb. Combining a few tens of these new high-power white LEDs would enable brightness similar to that of HID sources used for automobile headlights.

The color rendering of these white LEDs is comparable or better than fluorescent tubes. Recently, even higher power test devices (240 × 420 μm) have been fabricated with an output of $125\,\mathrm{lm\,W^{-1}}$ [165]; similar to fluorescent tubes. Finally, a 1 × 1 mm chip operating at 350 mA has been found to have a total luminescence flux, Φ, of 85 lm and emission efficiency, η_L, of $74\,\mathrm{lm\,W^{-1}}$.

3.4.5 ZnSe-Based White LEDs (T. Nakamura)

The use of ZnSe-based semiconductors for fabricating white LEDs was reported in 1998 [166]. These unique white LEDs consisted of ZnSe-based LEDs grown homoepitaxially on ZnSe substrates. Emission in the range 480–490 nm (blue/blue-green) from the ZnSe-based LEDs excited the substrate material, producing further light emission of ∼585 nm (green to red). The two wavebands of emitted light were mixed which resulted in white light from a simple solid-state device. Here, the operation principle of ZnSe-based white LEDs and the methods used for suppressing degradation are described.

Operational Principle and Main Features of ZnSe-Based White LEDs

The operation of white light ZnSe LEDs exploits the band tailing and self-activated (SA) emission centers of group II–VI semiconductors. Conducting ZnSe substrates are produced by doping bulk ZnSe and by appropriate thermal processing. The homoepitaxial growth of LED structures on ZnSe substrates yields the following results (1) emission of blue light from the active layer;

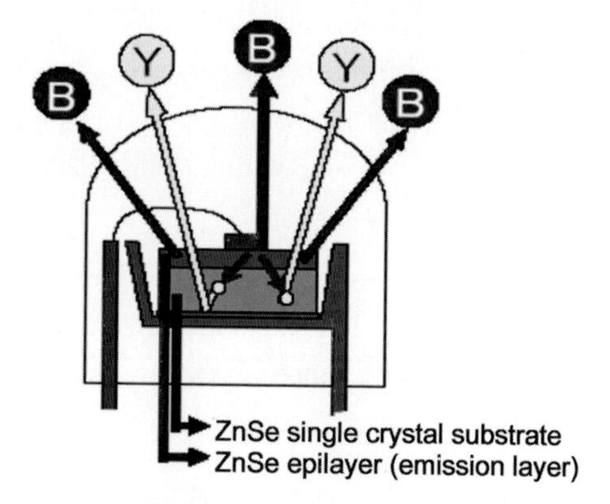

Fig. 3.72. Cross-section of a typical ZnSe-based white LED

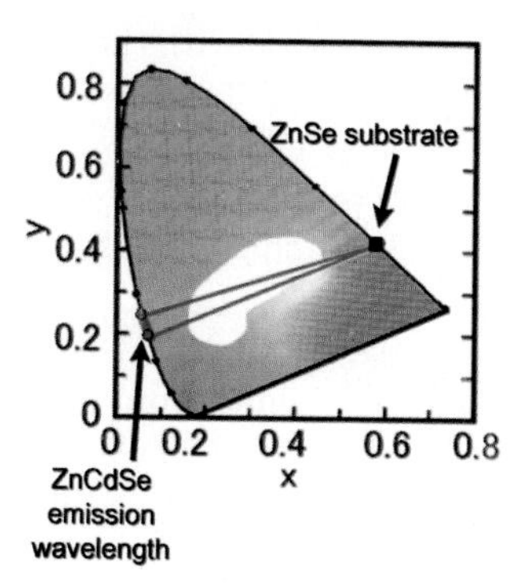

Fig. 3.73. Control of the chromaticity of ZnSe-based LEDs

(2) injection of the blue light into the substrate produces emission of yellow light; (3) blue and yellow light emitted from the LED and substrate mix and produce white light (Fig. 3.72).

It is also possible to control the chromaticity of the light emitted from such LEDs as shown in Fig. 3.73. The chromaticity of light from iodine doped ZnSe substrates is almost the same as 585 nm light, indicated by the black square (■) in Fig. 3.73. In the white spectrum, it is possible to change the emission wavelength by using complementary light in the 480–490 nm range with the 585 nm from the substrate. Furthermore, the intensity of light from the substrate can be changed by appropriate selection of the dopants and annealing conditions. Thus the properties of light from white light ZnSe-based LEDs can be changed by suitable structural changes of the substrate and the active LED in the homoepitaxial layer. The main features of white light ZnSe-based LEDs are (1) phosphors are not necessary and the fabrication process is simple because the electrodes are on the surface of the substrate (backside); (2) compared with white light GaN LEDs, the ZnSe-based structures operate at a low voltage of 2.5 V; (3) it is possible to control the wavelength with negligible fluctuations of the color tone; (4) the use of a conducting substrate reduces the possibility of damage due to electrostatic discharge.

These excellent characteristics have raised interest in the use of white light ZnSe LEDs in mobile telephones, but the lifetime of these devices is still requires further improvements.

Lifetime of White Light ZnSe LEDs

Figure 3.74 shows the time dependence of the optical power output at high temperature of a white light ZnSe LED. Generally, the lifetime of LEDs is specified by the time at which the device output decreases to half of the initial value, but the point at which there is a sudden decrease (sudden degradation) determines the actual usefulness of the device [167]. The origin of slow mode degradation has been proposed as being related defects induced by electrons overflowing into the nitrogen doped ZnMgSSe cladding layer [168, 169].

Rapid degradation has been studied by scanning spreading resistance microscope (SSRM) measurements of device cross-sections before and after degradation [170, 171]. Figure 3.75 shows the resistance distribution of the cladding layer of a device before and after degradation. These measurements show that (1) the conductivity decreases only at the p-cladding region and (2) the conductivity is particularly rapid on the active layer side which suggests the following failure progression: leakage of electrons into the p-cladding layer → propagation of defects due to recombination at N-defect sites of the p-cladding layer near the active layer → reduction of the effective carrier density → and finally reduction of the injection efficiency.

The overflow of electrons from the active layer into the p-cladding layer can be suppressed by using a large energy barrier (heterojunction barrier: ΔE_c) ZnMgBeSe for the cladding layer, a material having a large covalency.

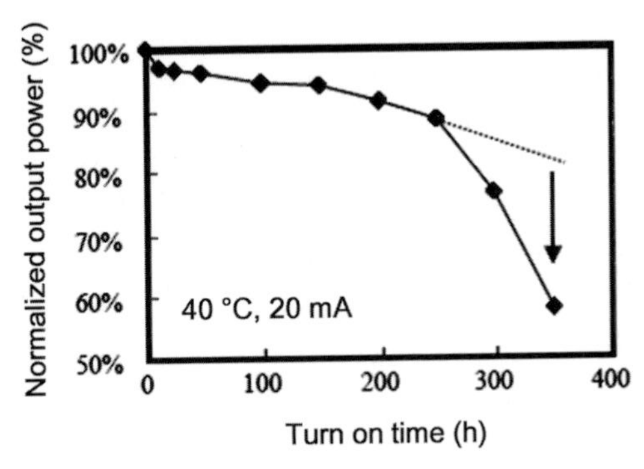

Fig. 3.74. Time dependence of the optical power output at high temperature of a white light ZnSe LED

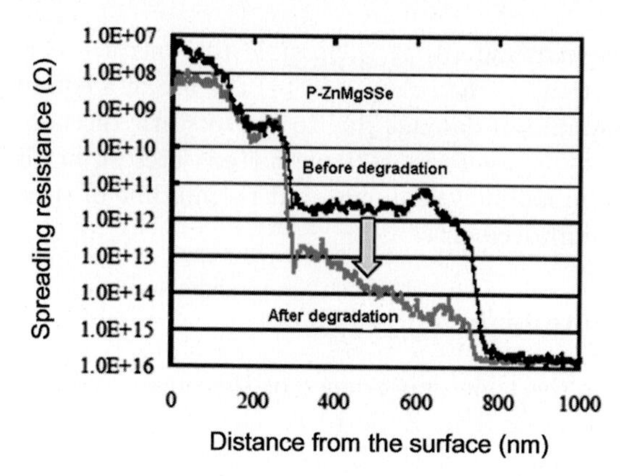

Fig. 3.75. Resistance distribution of the cladding layer of a device before and after degradation

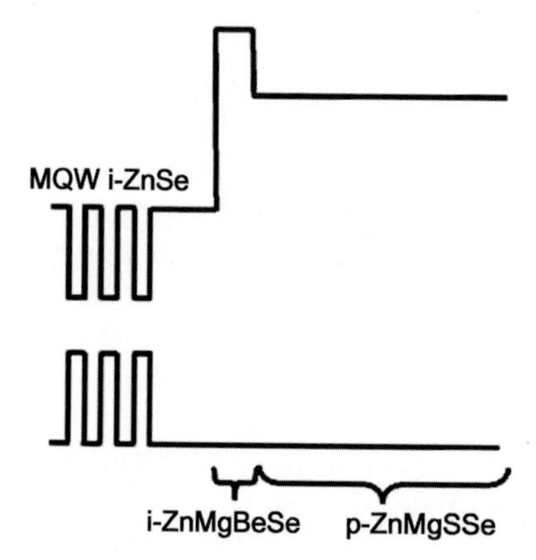

Fig. 3.76. Energy band structure for a double cladding i-ZnMgBeSe/p-ZnMgSSe arrangement

However, difficulties in the p-type doping have led to a proposal of a double cladding structures consisting of ZnMgSSe and ZnMgBeSe.

Figure 3.76 shows the energy band structure for a double cladding i-ZnMgBeSe/p-ZnMgSSe arrangement. The ZnMgBeSe layer is not doped. The main feature of this structure is that of the two roles played by the p-cladding layer, that is, injection of holes into the active layer, and suppression of electron overflow; the latter task is undertaken by the i-ZnMgBeSe layer [172].

Figure 3.77 shows the test results at 70°C of conventional LEDs (15 mA) with a ZnMgSSe cladding layer and LEDs (15 mA) with an optimized double cladding structure. It should be noted that at 70°C (15 mA) the time taken for the output to decrease to 50% is 400 h which corresponds to 10,000 h for ZnSe-based white LEDs at room temperature (activation energy of 0.6 eV).

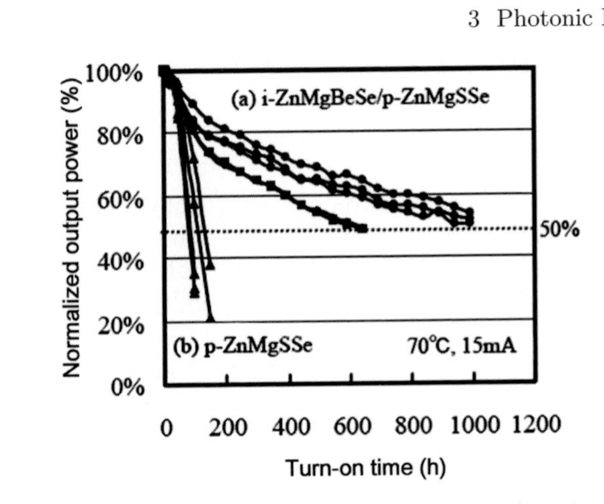

Fig. 3.77. Results at 70°C of a conventional LED (15 mA) with a ZnMgSSe clad layer and an optimized double cladding structure

Table 3.16. Comparison between ZnSe-based white LEDs and GaN-based white LEDs

white LEDs	driving voltage	color dispersion	electric static discharge (kV)	lifetime at RT (h)
ZnSe-based	2.5 V	$\sigma = 0.015$	>8	>10,000
GaN-based	3.5 V	$\sigma = 0.15$	~0.5	>10,000

The lifetime of the p-ZnMgSSe system was 130 h at 70°C, which corresponds to ~3,000 h at room temperature. On the other hand the double cladding structure does not show sudden degradation and the minimum output was measured to occur at 600 h, which corresponds to a lifetime of 15,000 h at room temperature.

These results show that (1) the lifetime characteristics are greatly affected by the characteristics of the i-ZnMgBeSe layer and (2) compared with the single cladding layer, the use of the two-layered structure results in improved carrier confinement and suppression of electron overflow, and consequently long lifetimes.

Comparison of ZnSe and GaN white LEDs (Table 3.16)

ZnSe white light LEDs using substrate emissions offer the unique advantage of producing white light using a single discrete device, but the lifetime must still be extended. The use of double cladding structures enables lifetimes that are comparable with GaN-based LEDs.

3.4.6 Applications and Prospects of LEDs (K. Tadatomo)

Introduction

Recent improvements in the luminous efficiency ($\mathrm{lm\,W^{-1}}$) of nitride semiconductor LEDs has enabled new fields of applications including traffic signals, large displays, and the illumination industry. The LED industry has evolved from simple discrete devices used in indicator panels to the illumination use of white LEDs. Development issues have evolved from luminous intensity (cd) or luminance ($\mathrm{cd\,m^{-2}}$) to luminous flux (lm) or illuminance ($\mathrm{lx = lm/m^2}$) of LEDs [173, 174]. This section is a review of the potential for LEDs fabricated by using InGaN semiconductor structures.

Table 3.17 shows potential applications of LEDs. The increased use of mobile communications terminals has led to increased demand for double wavelength-type white LEDs (produced by combining blue LEDs with yellow-YAG:Ce phosphors) for use as backlight optical sources of liquid crystal displays (LCD) in portable telephones and personal digital assistant (PDA) hand sets. Surface mounted white LEDs are often optically connected to light guiding plates for use as backlighting in LCDs. Further, the negligible self-heating of nitride semiconductor LEDs enables their safe household use, and their small size, high-energy efficiency, and battery-driven operation are other features of merit.

Further, the high brightness and visibility of nitride semiconductor LEDs has many potential outdoor applications such as street lighting, large area full color displays, road and railroad signals, information boards at railway stations, and advertisements in shopping arcades.

Further improvements in the luminous efficiency will lead to illumination applications competing with incandescent light sources. Table 3.18 is a comparison of the main characteristics of white LEDs, incandescent bulbs, and fluorescent lamp [175]. The luminous efficiency of white LEDs has surpassed that of incandescent bulbs and is approaching that of fluorescent lamp. The two main impediments to the proliferation of white LEDs in the illumination industry are the small total luminous flux emitted from a single source and the low cost efficiency. At present, white LEDs are mainly used as supplementary lighting such as foot lighting and scenic illumination, road signs, and internal lighting in cars. There have also been feasibility studies on their use in the agricultural and fisheries industries.

The total market for incandescent bulbs and illumination equipment is about 1 trillion Yen in Japan and 4 trillion worldwide. Thus there is potentially a huge market if white LEDs were to replace even only one-kind of conventional light source. But as shown in Table 3.18, the cost efficiency of white LEDs is still smaller than that of fluorescent lamp, thus more technological innovations are required before LEDs are more widely used in lighting.

Table 3.17. Some examples of high efficient LED application

area/function	applications	examples
indicator/display	portable equipment	backlight of LCD of mobile telephones and PDA
	display equipment	large area full color displays used in shopping arcades
	decorations	Christmas tree lights
	amusements	backlight for slot machines
	cars, planes and other vehicles	backlight for measurement instrumentation, direction indicators, stop lamps, rear lights, DRL (daytime running light)
	traffic (road and rail)	displays for road information, traffic signals, information displays inside railway stations, rail signals
illumination/lighting	subsidiary lighting for inside of buildings, background scene lighting	emergency lighting, foot lighting, background stage lighting
	mobile equipment	pocket flashlight, flash lamp for digital cameras
	traffic and vehicles	traffic information display boards, internal lighting for cars, street lighting
	agriculture and fishing	plant growth-enhancement and harvesting, fishing illumination
other	electronic equipment	light source for optical readers of paper money, printers
	optical catalyst	excitation light source for optical catalysts

Table 3.18. Comparison of different illuminants

	white LED	incandescent bulb (54 W)	fluorescent lamp (38 + 30 + 28 W)
total luminous flux (lm)	2	810	7,880
emission efficiency (lm W^{-1})	30	15	80
average color rendering index (Ra)	80–90	100	84
lifetime (h)	~10,000	1,000	6,000
initial cost (JPY)	100	160	4,300
cost efficiency (lm/JPY)	ca. 0.02	5	1.8

New potential markets include medical equipment, automobiles, backlights for large liquid crystal TVs, and public lighting such as street lamps. These areas will now be considered in more detail.

Medical Equipment

The HID (high intensity discharged) lamp such as metal halide lamps are constructed so as to illuminate a surgical operation without shadows obscuring the procedure in a hospital. However, in practice, a surgeon's head can block the light and reduce the field of view. The use of lamps worn around the head (headlamps) can improve the field of view but surgeons experience increased fatigue when wearing headlamps during long surgical procedures. The negligible self-heating of white LEDs enables the construction of compact solid-state light sources. Furthermore, white LEDs are also suitable for compact light sources because reflectors and lenses can be integrated into the light source, thus enabling precise focusing without bulky peripheral equipment.

Members from Kyoto University and Kyoto Prefecture University Hospital recently reported on the development and use of surgical goggles incorporating white LEDs as light sources. Figure 3.78 shows the goggles along with the world's first operation being carried out using white LED goggles [176, 177]. The goggles consist of two-wavelength type white LEDs (Nichia Chemical) with an optical emission efficiency of 25 lm W^{-1} and a beam divergence angle of 60°, under standard operation conditions of 3.5 V and 20 mA. There are 56 LEDs attached on each side of the goggles. The surgery was conducted at Kyoto Prefectural Hospital in September 2000 on a patient suffering from chronic renal failure using a procedure known as "intra-shunt." Assuming that an area of 15×15 cm (0.0225 m^2) is illuminated under standard operating conditions, then an illuminance of 8,700 lx is attained. Recently, goggles have

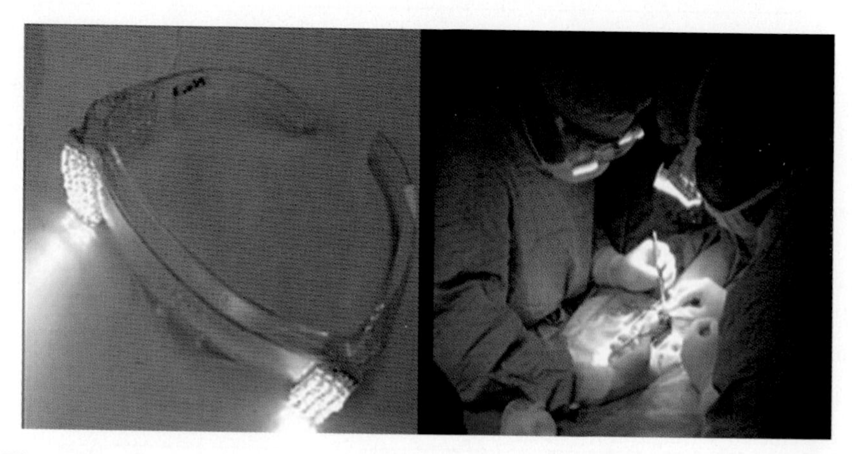

Fig. 3.78. Goggles and the world's first operation being carried out using white LED goggles

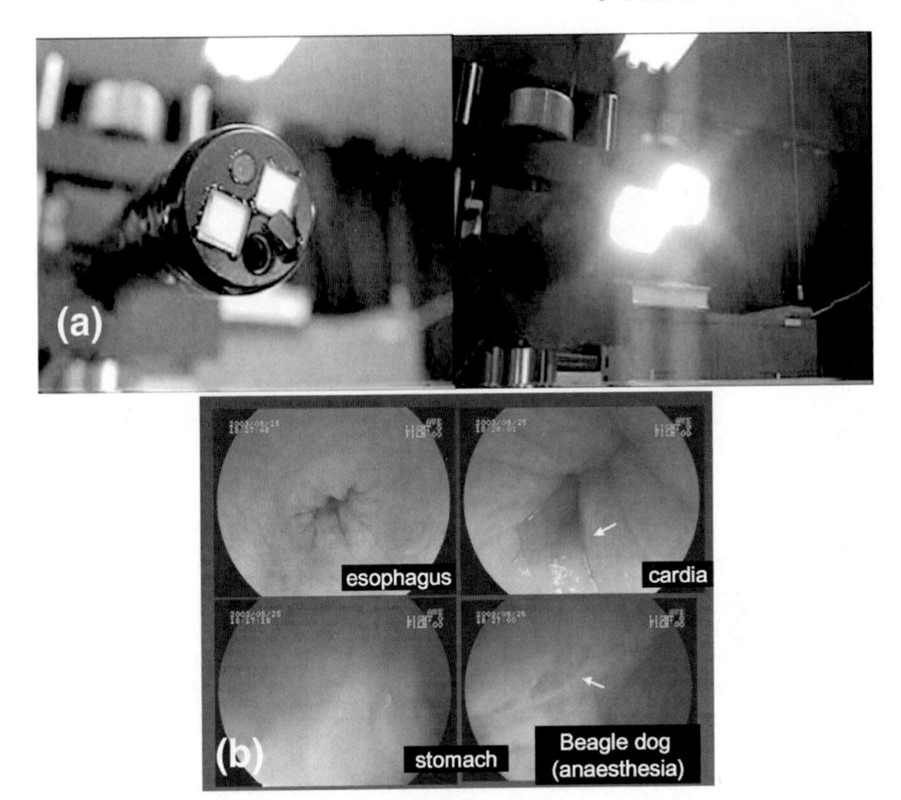

Fig. 3.79. (**a**) Two high color-rendering white LEDs attached to the tip of an endoscope. (**b**) Images of the upper digestive tract of a dog (Beagle) taken with an endoscope with the aforementioned white LEDs

been constructed incorporating 17 white LEDs each with a power of 47 lm (350 mA and 3.77 V) and an illuminance of 20,000 lx has been achieved, which is the minimum illuminance required for use in Japanese hospitals.

Researchers at Yamaguchi University are studying the possibility of using white LEDs for gastro-endoscopy as part of the "Knowledge Cluster Initiative" [178]. Figure 3.79a shows two high color-rendering white LEDs attached to the tip of an endoscope. The devices shown are three-wavelength fluorescent phosphor-type white LEDs (total luminous flux of 10 lm, average color-rendering index of 94, correlated color temperature of 6,470 K, beam divergence of 130°) excited by LEDs with central emission wavelengths in the near ultraviolet at 395 nm [179, 180]. Figure 3.79b shows images of the upper digestive tract of a dog (Beagle) taken with an endoscope with the aforementioned white LEDs. There are differences between the images taken with white LEDs and conventional endoscopes using Xenon lamps [181]. But, the total luminous flux from white LEDs is still insufficient for

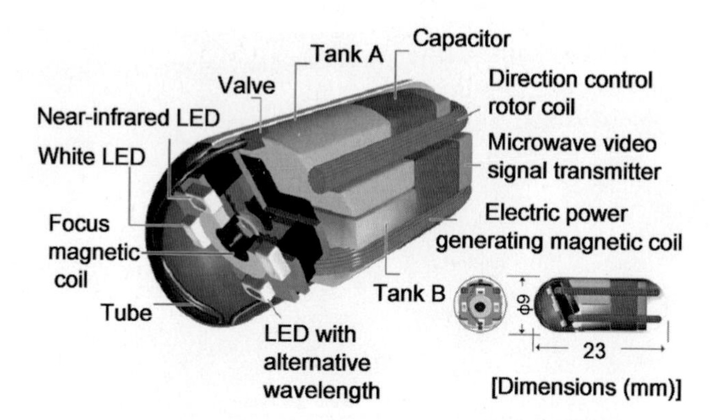

Fig. 3.80. Main components of the latest version of such an endoscopic capsule, the NORIKA-III, developed by RF Ltd of Japan

imaging the entire stomach and the diameter of the endoscope, at 12 mm, is still too large. Thus, there is still further study required to produce lighter endoscopes with greater illumination.

The fictional world in Hollywood films like "Fantastic Voyage" (1966) came a step closer to reality in 2000 when "Given Imaging Ltd" (Israel) announced the world's first endoscopic capsule [182, 183]. Figure 3.80 shows the cross-section of the main components of the latest version of such an endoscopic capsule, the NORIKA-III, developed by RF Ltd of Japan. The capsule is 23 mm in length and has a diameter of 9 mm. The optics consists of a CCD camera with a focusing mechanism and white and near infrared LEDs for horizontal and vertical illumination, respectively. Three-dimensional imaging is possible with the white LEDs and the near infrared LEDs enable observation of the inside of the digestive canal. The capsule is powered by wireless technology and the 30 images per second taken by the 410,000 pixel CCD camera are transmitted by microwaves to external image storage equipment. In conventional endoscopic capsules, it is not possible to control or specify the location of imaging since the capsules simply send images of their surroundings during their passage through the digestive system. However, the combination of a microcoil inside the NORIKA-III capsule and magnetic fields generated by external coils placed close to the body of the patient enables rotation of the capsule, permitting imaging with positional control of 5 mm and viewing angles of up to 15°. Studies are in progress for extending the functions of this technology for biopsies and drug delivery [184].

Automobiles and Transport

There is increasing interest in the use of white LEDs to replace incandescent lighting used in the automobile industry. White LEDs are increasingly being used as daytime running lights (DRL) as well as car interior light. Figure 3.81 shows an example of a headlight under development by

Fig. 3.81. Headlight under development by Stanley Electric Co Ltd (Japan) that incorporates ten white LEDs per lamp

Stanley Electric Co. Ltd. (Japan) that incorporates ten white LEDs per lamp. The total power consumption is 60 W with a maximum luminous intensity of 27,500 cd and a luminous flux of 900 lm.

Backlights for Large Liquid Crystal Displays

White LEDs are widely used as backlights in liquid crystal displays of portable telephones, but large liquid crystal TVs still employ cold cathode fluorescent lamps (CCFLs). However, in 2004, SONY announced the use of LEDs (Luxeon series made by Lumileds Ltd; ratio of 1 red:2 green:1 blue) for the "Toriluminos backlighting system" of their digital high vision LED televisions (QUALIA 005) [185,186]. The use of LEDs enabled the realization of lifelike-colors (chromaticity diagram, Fig. 3.82) with a 150% increase in the range of color reproduction compared with the CCFL configuration. The lower power consumption of LEDs compared with CCFLs is another factor that will lead to the increasing use of LED backlighting for large displays.

General Illumination

There are demands for low cost lighting to increase street lighting and to reduce maintenance costs of lighting systems placed at high altitude. Figure 3.83 shows a prototype street light (42 white LEDs) with a total power consumption of 200 W (50% that of sodium street lamps), maximum luminous intensity 3,800 cd, luminous flux 3,600 lm and average illuminance of 15 lx. Industrial roadmaps forecast that the emission efficiency will surpass 100 lm W^{-1} by 2010, thus making it compatible with fluorescent lamps. The day when solid-state lighting will replace Edison's incandescent bulb is not far in the future. The last major issue is reduction of manufacturing costs. Technological developments in materials and fabrication procedures together with the benefits of economies of scale due to mass production should resolve cost related issues.

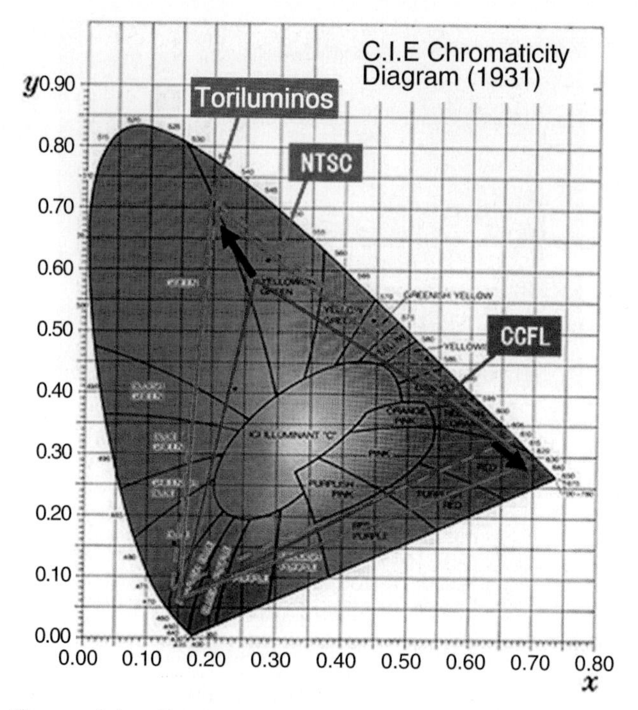

Fig. 3.82. Chromaticity diagram with a comparison of CCFL, NTSC, and Sony's Toriluminos backlighting system for their digital high vision LED televisions (QUALIA 005) as announced in 2004

3.5 Laser Diodes (LDs)

3.5.1 Blue/Violet Laser Diodes for Next Generation DVD Applications (M. Kuramoto)

Laser diodes for optical memory applications must have the following characteristics: high output power, high reliability, low threshold currents/high emission efficiency, low noise, and single transverse mode control. The development of low dislocation GaN substrates has advanced efforts in the high yield, mass production of these devices. This section describes the growth of low dislocation LD structures on sapphire substrates, design of high-power laser structures, growth of high-power LD structures on GaN substrates, and three wavelength lasers for optical disks.

Advances in the growth of nitride semiconductor lasers on sapphire substrates

Reduction of Dislocation Density and Reliability

Many advances in nitride laser structures have been reported since the first report of pulsed operation of blue/violet nitride semiconductor lasers in 1995

Fig. 3.83. Prototype street light (42 white LEDs) with a total power consumption of 200 W

[187, 188]. The main breakthroughs leading to this invention were growth of high quality GaN crystals by the use of low temperature buffer layers [189], formation of low p-type resistance by Mg doping with subsequent electron irradiation and thermal annealing [190,191], and growth of high quality GaInN crystals [192]. The first continuous wave laser oscillation was reported in 1996 [193] but the most important issue in the development of lasers is device reliability. At the time, lack of GaN substrates led engineers to grow GaN on sapphire substrates (13.8% lattice mismatch with GaN) that resulted in the presence of 10^8–10^{10} dislocations per cm^2 [194, 195]. The presence of such a large density of dislocations resulted in laser structures with lifetimes of ∼40 min at low power outputs of 1 mW [196]. The development of epitaxial lateral overgrowth (ELO) enabled growth of epitaxial GaN layers with dislocation densities as low as 10^5–$10^6\,cm^{-2}$ [197–203].

Figure 3.84 shows an example of a laser structure grown on a sapphire substrate using ELO growth. MOCVD was used to grow a GaN layer on a sapphire substrate with an intermediate GaN buffer layer. Next, an SiO_2 layer was deposited onto the GaN layer followed by patterning to define a 18 μm wide window and subsequent reactive ion etching (RIE) down to the sapphire substrate to produce a 6 μm wide GaN seed crystal. Next, the seed layer above the SiO_2 was removed and an n-type GaN ELO layer formed by MOCVD regrowth.

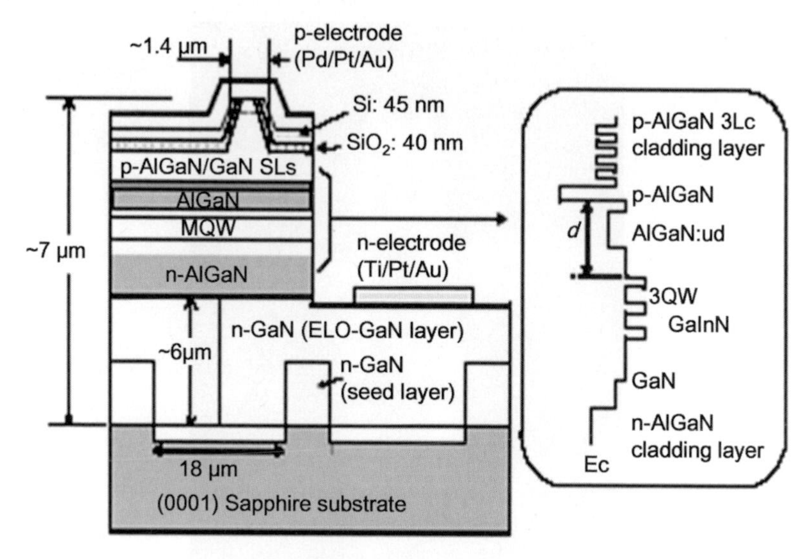

Fig. 3.84. Example of a laser structure grown on a sapphire substrate using ELO growth

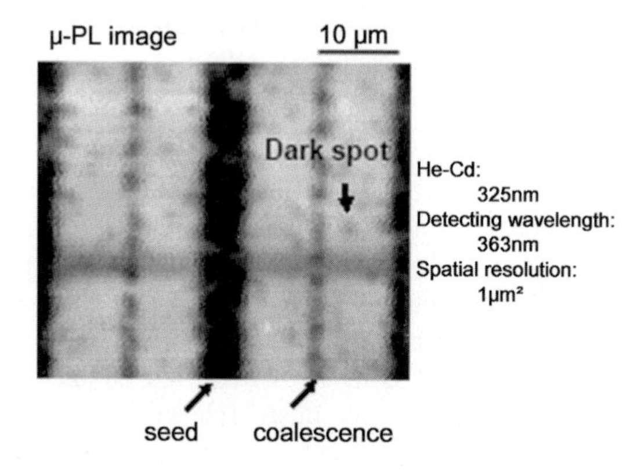

Fig. 3.85. Micro-PL image of the surface of ELO-GaN

Figure 3.85 is a scanned microphotoluminescence image (resolution = $1\,\mu m^2$; excitation source = 325 nm He–Cd laser; detection wavelength = 363 nm) from the surface of an ELO layer where the emission intensity at the surface of dislocations is less (dark points) than surrounding areas. Many dark points can be seen at coalescence regions.

Figure 3.86 shows the relationship between the dark spot density and reliability. The mean time to failure (MTTF) was defined as the time at which the current increased by 20% at 50 mW and 70°C. A clear relationship is observed between the MTTF and density of dark spots, which indicates that the dislocation density must be reduced for reliable devices. Thus, in cases

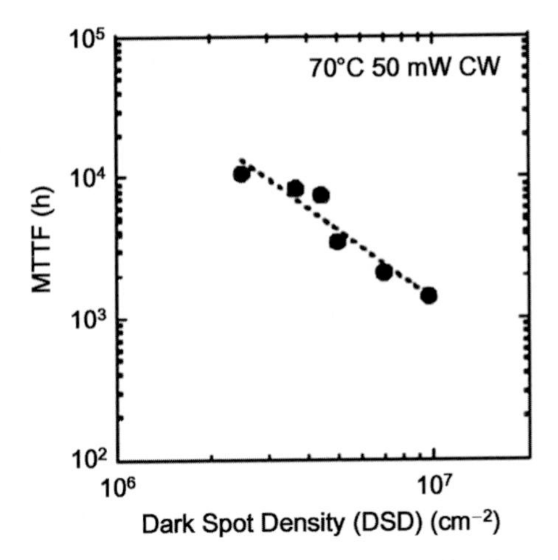

Fig. 3.86. Relationship between the dark spot density and reliability

when ELO technology is used for fabricating lasers on sapphire substrates, it is necessary to position the laser stripe in between the seed and coalescence regions, as shown in Fig. 3.84.

Laser Structures for High-Power Devices

Design of the LD Layer Structure

The main technological issues for the fabrication of high-power nitride lasers for next generation DVD applications are low threshold current density, high slope efficiency, and improvement of temperature and noise characteristics.

The layer structure shown in Fig. 3.84 consists of ELO-GaN, n-AlGaN cladding, an n-GaN guide, GaInN multiquantum well (MQW) region, a GaInN and AlGaN intermediate layer, p-AlGaN electron barrier, p-GaN/AlGaN superlattice clad, and p-GaN contact. The basic design of the laser consists of a separated confinement multiquantum well structure which is modified by the inclusion of a p-GaN/AlGaN superlattice clad layer [204] for low voltage operation and high Al composition p-AlGaN electron barrier [187] to reduce carrier overflow.

The main differences of the recent laser development designs and those of the early days are a decrease in the number of MQWs in the active layer from 26 to 3 layers and the use of an intermediate layer between the active layer and p-AlGaN electron barrier. Decreasing the number of quantum wells in conventional semiconductor lasers has been studied as a means of reducing the volume of the active layer in order to reduce the threshold current. In nitride laser structures, the overflow current increases if the number of quantum wells is large because of nonuniform hole injection due to the large effective mass of holes and difficulties in producing high p-type materials [205, 206].

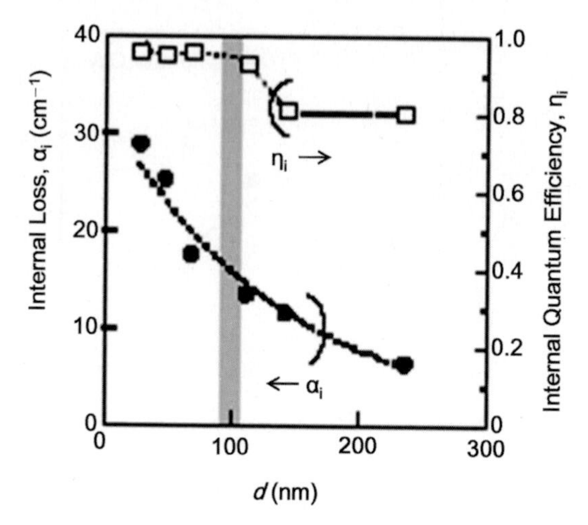

Fig. 3.87. Variation of internal losses and internal quantum efficiency with the distance (d) from the end of the active layer to the Mg doped layer

Large internal losses due to nonuniform hole injection greatly affect the threshold current density and slope efficiency. The use of two to three layers of quantum wells has been reported to result in a decrease of the threshold current density to 1.9–2.5 kA cm^{-2} [207], but the internal loss was still high at between 29 and 43 cm^{-1} [208, 209]. The origin of the large internal loss in nitride semiconductors was reported to be due to the large absorption coefficient of the Mg doped layer [210]. Shifting the light distribution peak to the n-type side or, as shown in Fig. 3.87, use of an intermediate layer of GaInN or AlGaN between the active and p-clad layers was reported to be effective in separating the Mg doped layer from the high light density of the active layer [211].

Figure 3.87 shows the variation of internal losses and internal quantum efficiency with the distance (d) from the end of the active layer to the Mg doped layer. The internal loss (α_i) decreases with increasing d to values of $\sim$5 cm^{-1}, which are comparable with conventional lasers. However, the internal quantum efficiency only decreases for d greater than 100 nm. Thus laser structures are usually designed with d $\sim$ 100 nm.

Figure 3.88 shows the variation of the threshold current and slope efficiency with the position of Mg doping. The internal loss decreases, as the Mg doping position is moved further from the active layer. Further, the threshold current decreases, and the slope efficiency improves. Introduction of the AlGaN and GaInN intermediate layers also leads to an improvement in temperature characteristics [203] due to suppression of carrier overflow.

Figure 3.89 shows the relationship between the threshold current and relative intensity noise (RIN) at a laser power of 2.5 mW when the high frequency superposition has not been added. It can be seen that for a laser with an

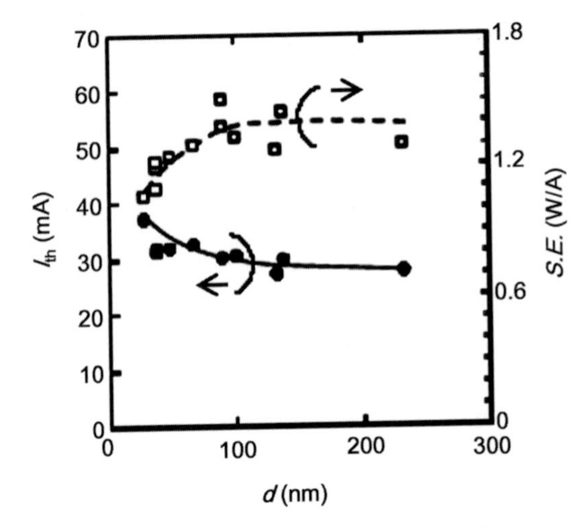

Fig. 3.88. Variation of the threshold current and slope efficiency with the position of Mg doping

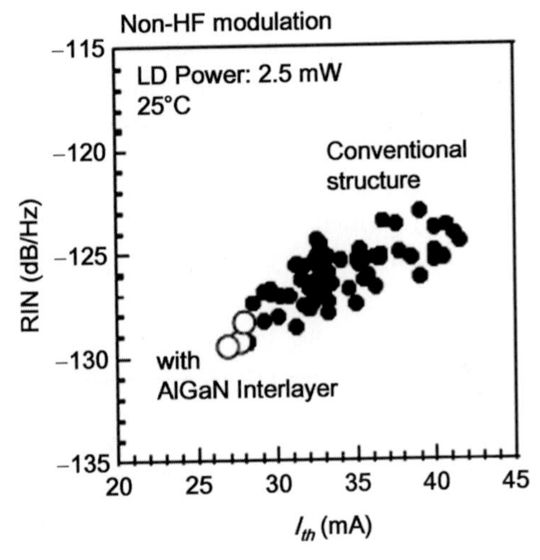

Fig. 3.89. Relationship between the threshold current and relative intensity noise (RIN) at a laser power of 2.5 mW

AlGaN intermediate layer the RIN at 2.5 mW is $-130\,\mathrm{dB\,Hz^{-1}}$, sufficient for next generation DVD applications.

Current Constriction and Transverse Mode Control

Control of the transverse modes in lasers is extremely important for optical disk applications. Current constriction structures have major effects on the stability of the transverse mode and the reduction of the threshold current.

Laser oscillation in nitride semiconductor lasers was first observed in a gain guide type structure. Later, alternatives – such as ridge structures fabricated by dry etching [196, 201, 203] and selective epitaxial growth [210], and inner stripes [212, 213] – were developed.

High-kink levels are extremely important in the design of high-power lasers. As the difference in the horizontal refractive index increases, the angle of the far-field radiation pattern ($\theta_{//}$) increases and at the same time the kink level decreases. The majority of kinks are a hybrid of the fundamental and first modes of the transverse direction [214]. The value of $\theta_{//}$ is designed to be 8–9°, but, if $\theta_{//}$ is decreased, then the aspect ratio increases and there is a deterioration of the optical tolerance. It is necessary to develop means of suppressing secondary modes for $\theta_{//}$ values of the size described above.

Control of the transverse mode of the laser shown in Fig. 3.84 is usually carried out by the use of dry etching to form ridge structures with SiO_2 and Si layers deposited on both sides of the ridge [215]. Since the Si layer absorbs light in the 400 nm waveband, it suppresses the generation of the first-order mode, resulting in an increase in the kink level. Figure 3.90 shows the relationship between the thickness of the SiO_2 layer and the theoretical absorption coefficients for the fundamental and first modes. The absorption coefficients of both modes are seen to increase with increasing SiO_2 layer thickness with the absorption coefficient of the first mode increasing more dramatically than the fundamental mode. This difference in absorption coefficient can be used to suppress the first mode. If the SiO_2 layer is too thin, then absorption of the fundamental mode increases, which leads to increase in the threshold current and reduction of the slope efficiency. Typically, for a ridge width of 1.7 μm the SiO_2 is 40 nm thick.

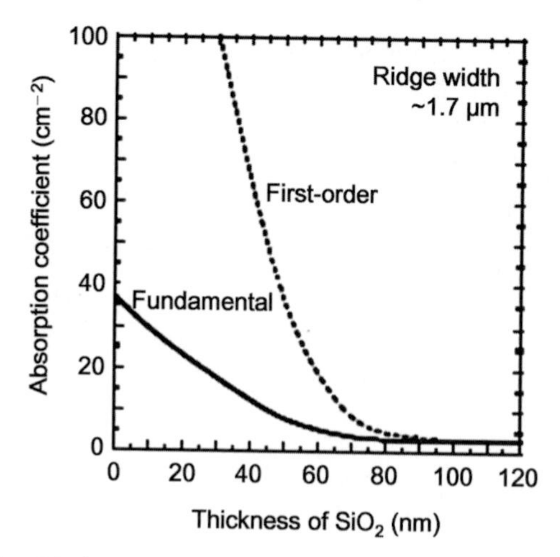

Fig. 3.90. Relationship between the thickness of the SiO_2 layer and the theoretical absorption coefficients for the fundamental and first modes

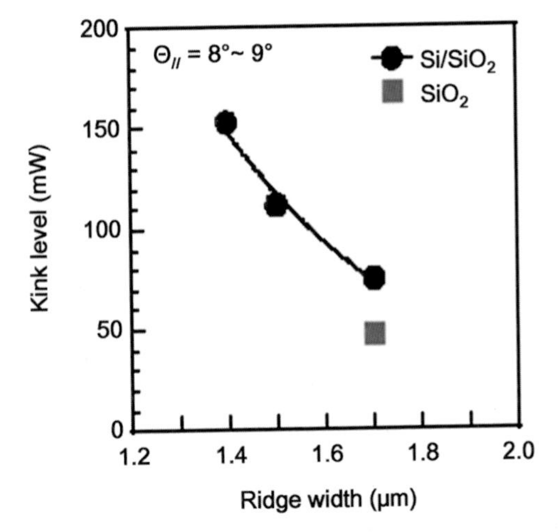

Fig. 3.91. Variation of the kink level with the ridge width under continuous wave conditions

Further, reducing the width of the ridge is expected to lead to major improvements in the kink level. Figure 3.91 shows the variation of the kink level with the ridge width under continuous wave conditions. The kink level increases to 30 mW and 150 mW for ridge widths of 1.7 μm and 1.4 μm, respectively.

Blue/Violet LDs on GaN Substrates

Laser Structure

GaN substrates were not available in the early 1990s and LD structures were grown on sapphire, SiC [216, 217] and spinel substrates [218]. In 1998, blue LDs were fabricated on high resistivity free standing GaN substrates [200] with electrodes formed on the upper surface of the LD structures in the same way as in sapphire substrates. In 1999, LD structures were developed with electrodes on the reverse side of the GaN substrates [219], but, as shown in Fig. 3.84, it was not possible to produce reliable LDs due to the high dislocation density ($2–4 \times 10^7$ dislocations per cm^2) in the substrates. In 2001, the first practical, high quality blue LDs were reported due to the development of low defect density ($\sim 10^5\,cm^{-2}$) GaN substrates [220].

Figure 3.92 is an illustration of a laser structure grown on a GaN substrate. The laser structure itself is similar to that shown in Fig. 3.84 but the fabrication process has been dramatically simplified. Table 3.19 is a summary of the differences in the procedures used to process LDs on ELO/sapphire and GaN substrates.

The width of low dislocation density regions of LDs grown on sapphire substrates is ~6 μm [202, 203] and that on GaN substrates is ~150 μm (defect

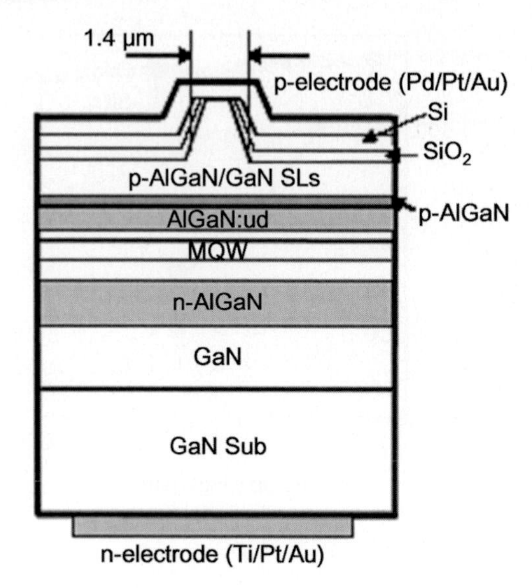

Fig. 3.92. Illustration of a GaN-based blue laser structure grown on a GaN substrate

Table 3.19. Comparison of LD on sapphire and GaN substrates

	ELO/sapphire substrate	GaN substrate
dislocation density (cm^2)	$\sim 10^6$	$\sim 10^5$
low dislocation region (μm)	∼6	150
growth cycles (time)	2	1
formation of n-type contacts	complex	easy
chip width	large	small
ease of cleavage	difficult	easy
constriction	junction-down (difficult)	junction-up/down (easy)

density less than 1×10^6 cm^{-2}) [221]. Thus it is easier to fabricate stripe regions in LD structures grown on GaN than ELO/sapphire substrates. Also, the use of conducting GaN substrates enables formation of electrodes on the reverse side of the substrate and consequently reduction of the chip size. Further, mirror facets are easily formed by cleaving techniques. LD on GaN substrates are easier to cleave than that on sapphire substrates. The aforementioned figures of merit, in addition to the favorable thermal conductivity of GaN compared with sapphire, have been important factors in the development of high performance LDs to-date.

Characteristics of Blue/Violet Lasers Grown on GaN Substrates

Figure 3.93 shows the current–output power (a) and current–voltage (b) characteristics of a laser structures grown on a GaN substrate. The output power

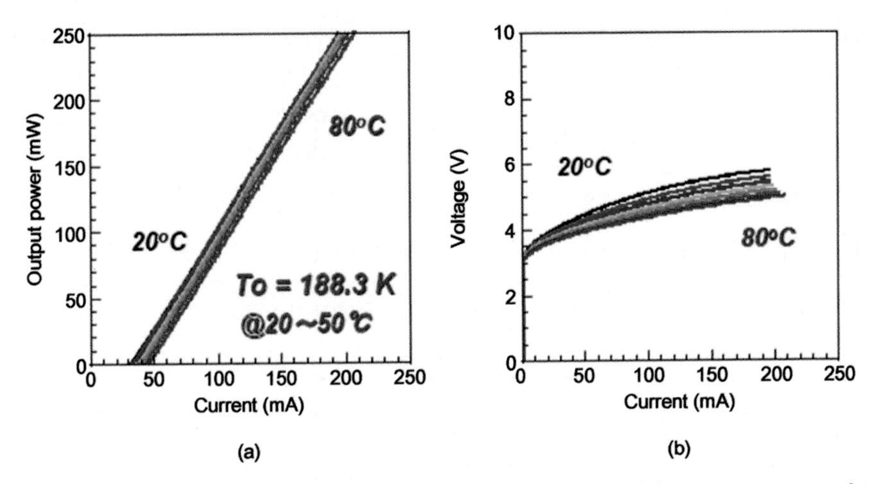

Fig. 3.93. Current–output power (**a**) and current–voltage (**b**) characteristics of a laser structures grown on a GaN substrate

was 250 mW between 20 and 80°C. The threshold current and voltage at 20°C were 31 mA and 4.2 V, respectively. At 80°C, these figures were 46 mA and 3.9 V.

The power output characteristics, shown in Fig. 3.93, are high at 1.6 W A^{-1}, with good linearity and high kink level. Further, from the fact that the power output is observed up to 250 mW, it can also be assumed that the catastrophic optical damage (COD) power is also above 250 mW. The power density at the front edge, estimated from the COD level, is 50 MW cm^{-2}, which is more than one order of magnitude greater than the COD critical density, 2–4 MW cm^{-2}, of AlGaInP and AlGaAs-based lasers [222, 223]. Thus lasers with COD levels of 250 mW can be fabricated using GaN-based materials without the need for window structures.

Figure 3.94(a), (b) shows the far-field patterns of blue/violet lasers fabricated on sapphire and GaN substrates, respectively. As shown, a ripple pattern is observed in the in the $\theta_{\perp}$ direction for lasers grown on sapphire substrates due to optical interference at the sapphire/GaN interface. However, such a ripple is not observed for lasers grown on GaN substrates.

Reliability Tests

Figure 3.95 shows the results of aging tests of blue/violet lasers grown on GaN substrates at 75°C and 120 mW pulse currents (pulse width = 50 nm, duty cycle = 50%). Stable operation was observed even after 500 h. The estimated lifetime where the current decreases to 40% is 9,000 h. The operating current degradation is proportional to the root of the time (same as for sapphire substrates) and the failure mechanism is thought to be due to diffusion related phenomena.

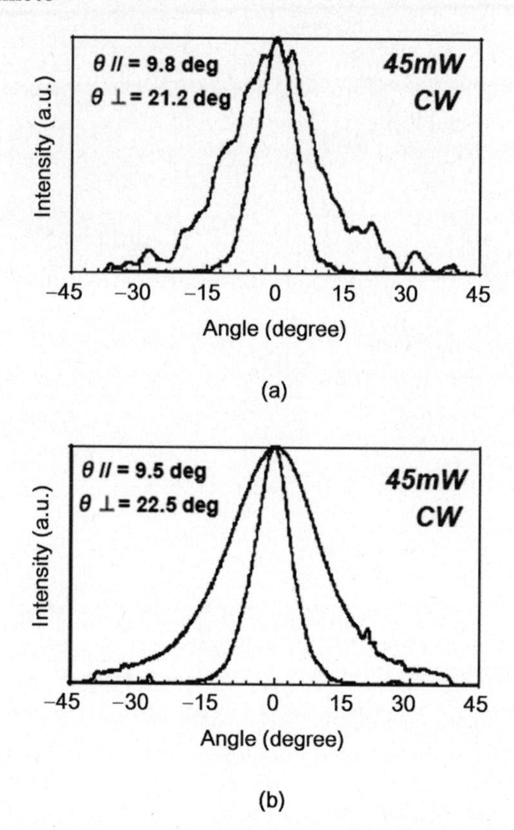

Fig. 3.94. (**a**) and (**b**) show the far-field patterns of blue/violet lasers fabricated on sapphire and GaN substrates, respectively

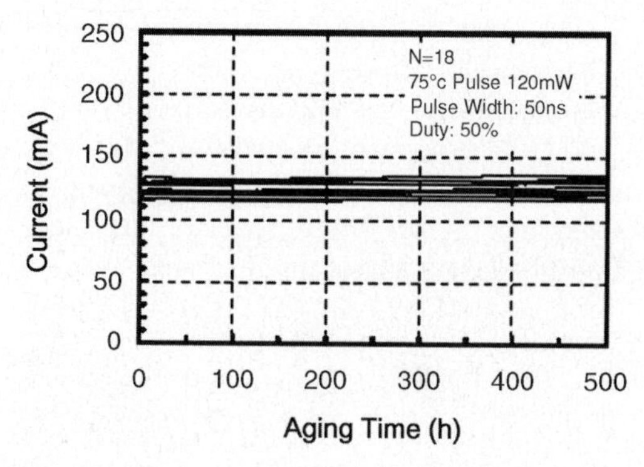

Fig. 3.95. Results of aging tests of blue/violet lasers grown on GaN substrates at 75°C and 120 mW pulse currents (pulse width = 50 nm, duty cycle = 50%)

Further, TEM observations of structures after failure show there to be no increase in the dislocation density, although such increases are found in GaAs-based lasers [224]. There is a clear correlation between reliability and dislocation density. The observed increase of the operating current is thought to result from an increase in the non-radiative capture cross-section due to (1) Mg related defects causing the motion of dislocations or (2) the diffusion of point defects (due to nitrogen vacancies) in the crystal lattice toward dislocations [225]. However, further clarification of the degradation mechanism of high-power GaN-based lasers will have to include greater consideration of degradation at the edge face.

The activation energy estimated from temperature dependent lifetime measurements is estimated to range between 0.32 eV [226] and 0.5 eV [227]. The values are extremely small compared with the 0.7–0.9 eV for AlGaAs [228] and AlGaInP [229] based lasers.

Three Wavelength Laser Systems

Modern optical recording systems employ both 780 nm and 650 nm lasers to enable readout from both CD and DVD-based media. There is strong demand for blue laser (405 nm) based systems capable of being able to read both CD and DVD media. The structure and main features of such three wavelength optical recording systems is described.

Structure of three Wavelength Lasers

Commercially available two wavelength lasers used in CD/DVD optical recording systems contain an AlGaAs-based laser for CDs and AlGaInP-based laser for DVDs. Both laser structures are grown on the same GaAs substrate as a single monolithic chip. It is extremely difficult to fabricate monolithic chips using GaN-based materials. The approach in GaN lasers is to use a two stage hybrid structure consisting of a two wavelength CD/DVD laser mounted on a blue/violet laser grown on a GaN substrate (Fig. 3.96).

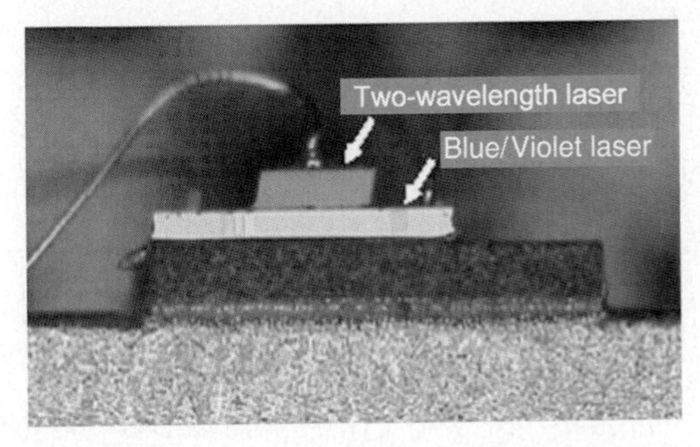

Fig. 3.96. Hybrid type three wavelength laser

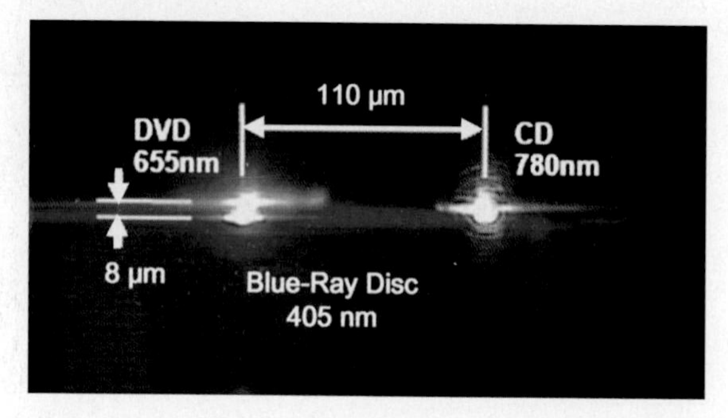

Fig. 3.97. Emission from a three wavelength laser

In this structure, the use of the high thermal conductivity GaN substrate (200 W m^{-1} K^{-1}) below the two wavelength laser enables efficient dissipation of heat generated during the operation of the two wavelength laser. The thermal conductivity of GaN is comparable to Si (150 W m^{-1} K^{-1}) and AlN (230 W m^{-1} K^{-1}) that are normally used for mounting two wavelength lasers. Another advantage of the two stage configuration is that the emission points can be positioned in very close proximity to the optical axis of the objective lens.

Figure 3.97 is a photograph of the emission points of a hybrid three wavelength laser system. The vertical distance between the emission points of the blue/violet and two wavelength lasers is 5–10 μm. The horizontal distance between the emission points for the CD and DVD lasers is 110 μm. This horizontal positioning of the emission points can be adjusted to match the location and structure of the optical pickups.

Fundamental Characteristics of Three Wavelength Lasers

Figure 3.98 shows the current–voltage and current–output power characteristics of a three wavelength laser (all with a resonator cavity length of 400 μm) for optical play back. The threshold voltage increases in the order CD, DVD, and blue/violet laser in correspondence to the increases of the bandgap energy of the semiconductors used. Also, the electrical resistance of the blue/violet laser is the largest due to the difficulties in forming p-type low resistance to wide-bandgap semiconductors. However, from the current–power output characteristics, the operating current at 10 mW for all three wavelengths is the same at less than 35 mA. Such characteristics are important for the low power, high performance operation of laser devices.

3.5.2 Long Wavelength GaN-Based Laser Diodes (T. Mukai, T. Yanamoto)

There have been extensive reports on the fabrication and characterization of high brightness blue/green and yellow GaN-based light-emitting diodes [230, 231]. Similarly, laser diodes with emission in the 365–482 nm wavelength range [232–235] show tremendous potential for use in fields including medicine,

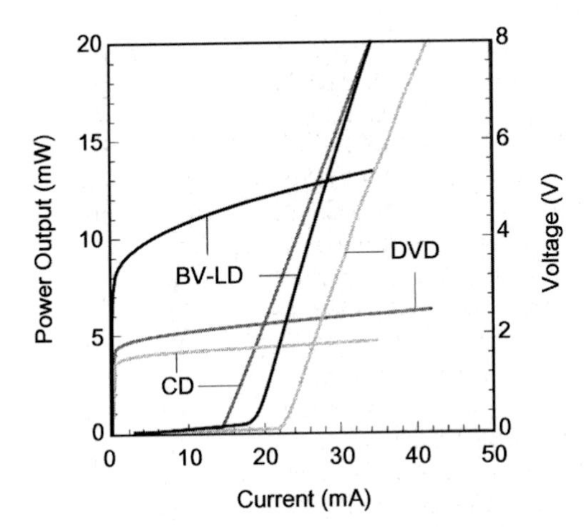

Fig. 3.98. Current–voltage and current–output power characteristics of a three wavelength laser (all with a resonator cavity length of 400 μm) for optical play back

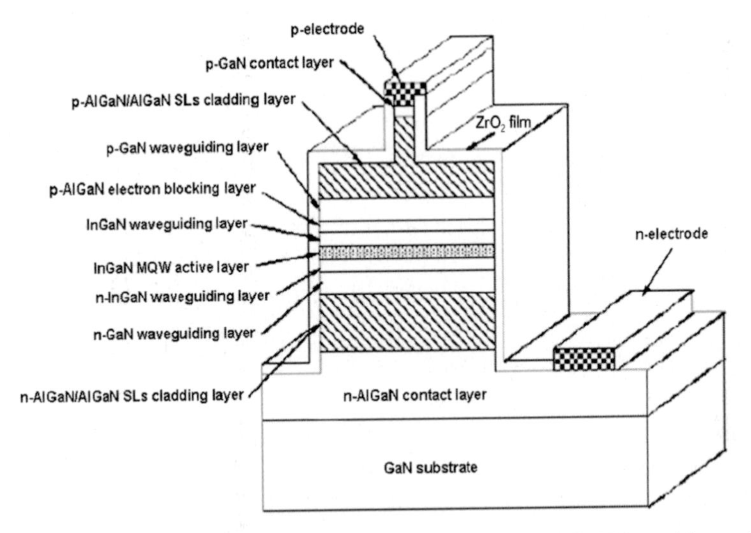

Fig. 3.99. Structure used for fabrication of long wavelength GaN-based laser diodes

biotechnology related equipment, and full-color displays. In the case of color displays, there is strong demand for high-power laser diodes in the 440–460 nm range. This section describes long wavelength GaN-based laser diodes (LDs) and in particular high-power blue LDs.

Long Wavelength GaN Laser Diodes

Figure 3.99 shows the structure used for fabrication of long wavelength GaN-based laser diodes. The active layer is an InGaN-MQW (multiquantum well) structure. Long wavelength oscillation was achieved by reducing the growth

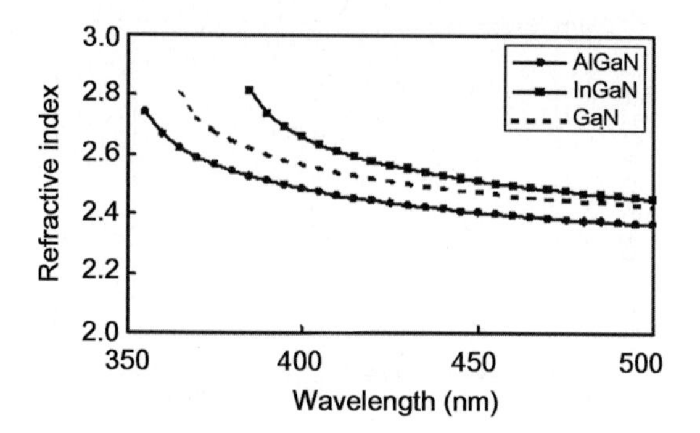

Fig. 3.100. Wavelength dependence of the refractive index of $In_{0.05}Ga_{0.95}N$, GaN and $Al_{0.05}Ga_{0.95}N$

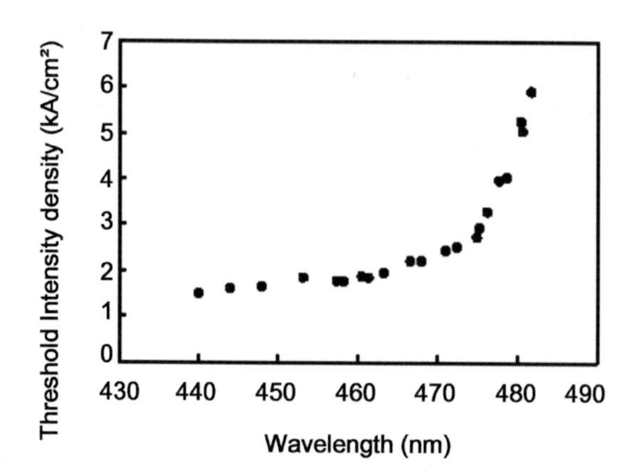

Fig. 3.101. Variation of the emission wavelength with threshold current density of LDs

temperature of the active layer and increasing the In composition of the quantum well. Optical confinement in the perpendicular direction was enhanced by use of an InGaN/GaN guide layer and an AlGaN/AlGaN superlattice. This structure also prevents increases in the threshold current (Fig. 3.100). The LD structure employs a ridge stripe where the stripe has a width of 2.5 μm and a ZrO_2 layer was used for horizontal optical confinement. A high reflection coating was used on both edges.

As shown in Fig. 3.101, the threshold current density of LDs increases with emission wavelength. In particular, the threshold current density increased dramatically for wavelengths greater than 475 nm. These observations are due to the degradation of the crystalline quality of the quantum well layer [236] and as a result the leakage current increases and the emission efficiency falls. Also, the main reason for the observed increases of the threshold current with increasing oscillation wavelength is thought to be the spatial fluctuations of the In composition of the InGaN active layer [237, 238]. These fluctuations

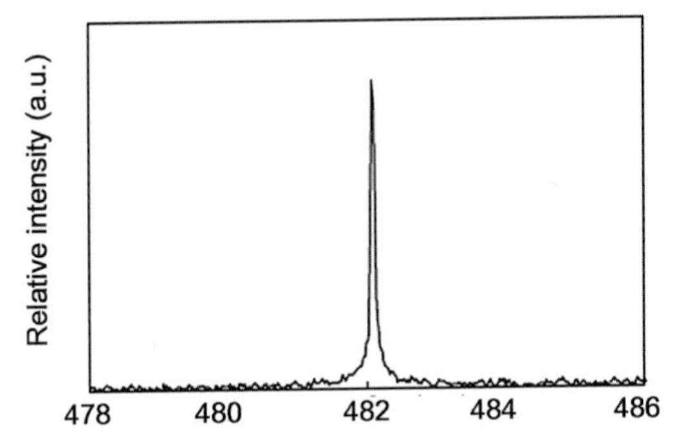

Fig. 3.102. Room temperature, long wavelength continuous wave emission observed from a laser diode fabricated by Nishia Corp

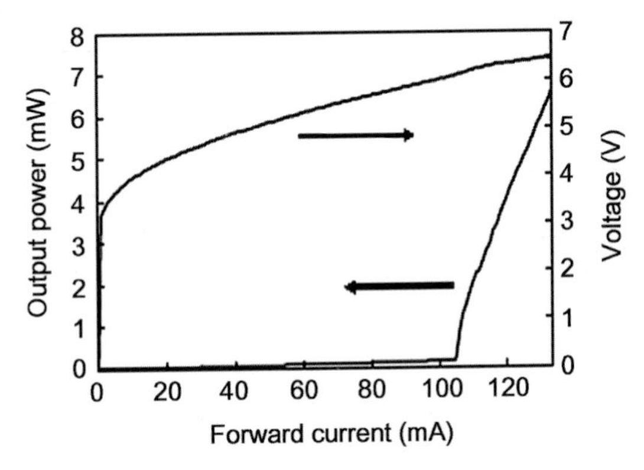

Fig. 3.103. I–L and I–V characteristics of the LD with characteristics shown in Fig. 3.102 with peak emission at 482 nm

contributed to improvements of the emission efficiency of LEDs, but, in the case of LDs, they do not lead to optical gain and in fact act to produce internal losses. Thus, for long wavelength LDs, that is devices with large In content, the internal loss and threshold current densities are larger than for shorter wavelength structures.

Figure 3.102 shows the room temperature, long wavelength continuous wave emission observed from a laser diode fabricated by Nishia Corp. The output power is 5 mW, the case temperature 25°C and the peak wavelength 482 nm. Figure 3.103 shows the I–L and I–V characteristics of this LD, where the threshold current is 105 mA (equivalent to a density of 6.2 kA cm^{-2}), which is larger than for LDs emitting in the 400 nm wavelength region. The emission at 482 nm was achieved by reducing the defect density of the substrate thereby improving the crystal quality of the InGaN active layer grown subsequently.

Extending the optical emission to even longer wavelengths will require further reductions of the dislocation density of substrates and suppression of the fluctuations of In composition in the InGaN active layers.

High-Power Blue Semiconductor Lasers

Color displays require blue LDs with emission wavelengths ranging from 440 to 460 nm. Laser diodes manufactured by Nichia Corp are designed for emission at 445 nm and have both single and multi transverse modes.

Current Status of Single Transverse Mode Laser Diodes

Figure 3.104 shows the I–L and I–V characteristics of a 445 nm LD. The threshold current is 24 mA (current density of 1.8 kA cm^{-2}), and the current and voltage at 50 mW power are 105 mA and 5.4 V, respectively. The maximum power output achieved to-date at these wavelengths was 20 mW. However, reductions of the dislocation density of GaN substrates and improvements of the crystalline quality of the InGaN active layer have enabled fabrication of 50 mW, kink-free LDs.

Figure 3.105 shows the far-field pattern (FFP) of a 50 mW LD with a case temperature of 25°C. It can be seen that the horizontal transverse mode emission is a single mode oscillation. Figure 3.106 shows the results of APC reliability testing at an output power of 50 mW and a case temperature of 60°C. The lifetime after 1,000 h was estimated to be 10,000 h, which indicates the high reliability of the LDs.

Current Status of Transverse Multimode Laser Diodes

The output power of LDs can be increased even further by the use of single wide stripe structures. Figure 3.107 shows the I–L and I–V characteristics of a 445 nm LD with a ridge width of 7 μm and a threshold current density of 1.8 kA cm^{-2}. At an output of 200 mW, the current and voltage are 250 mA and 4.9 V, respectively. Figure 3.108 shows the I–L output characteristics during pulsed and CW operation. Under pulsed conditions, the output reached 1.4 W without optical damage. In contrast, under CW conditions, the output

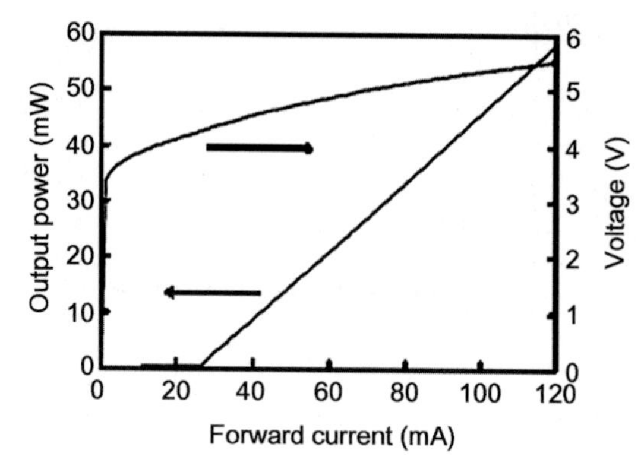

Fig. 3.104. I–L and I–V characteristics of an LD with peak emission at 445 nm

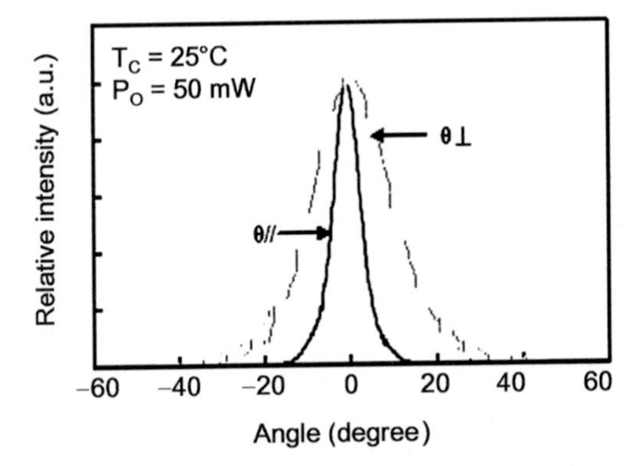

Fig. 3.105. Far-field pattern (FFP) of a single mode 50 mW LD with a case temperature of 25°C

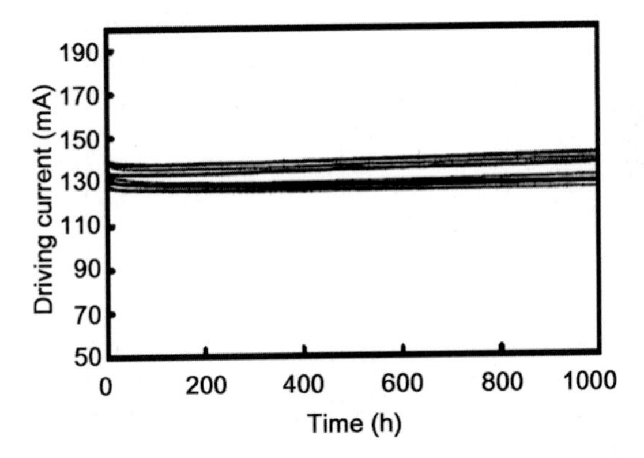

Fig. 3.106. Lifetime characteristics of a single mode blue LD

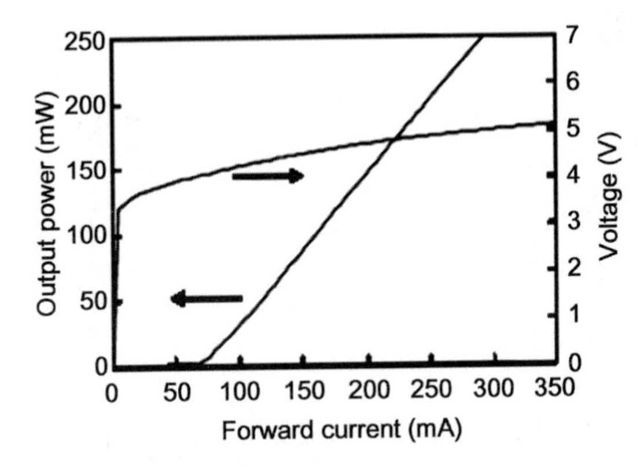

Fig. 3.107. I–L and I–V characteristics of a multimode blue LD

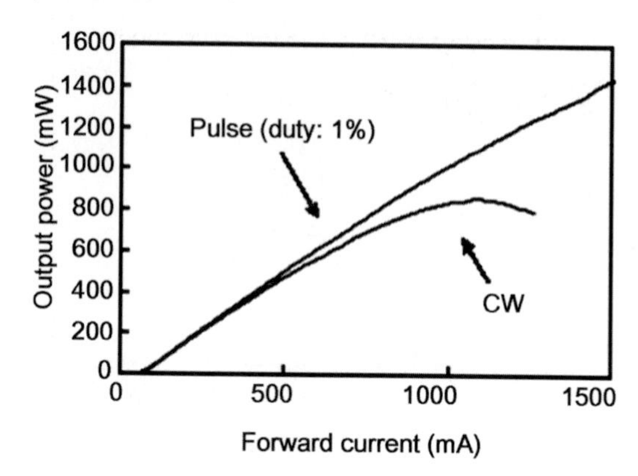

Fig. 3.108. Comparison of the I–L characteristics under pulsed and dc injection

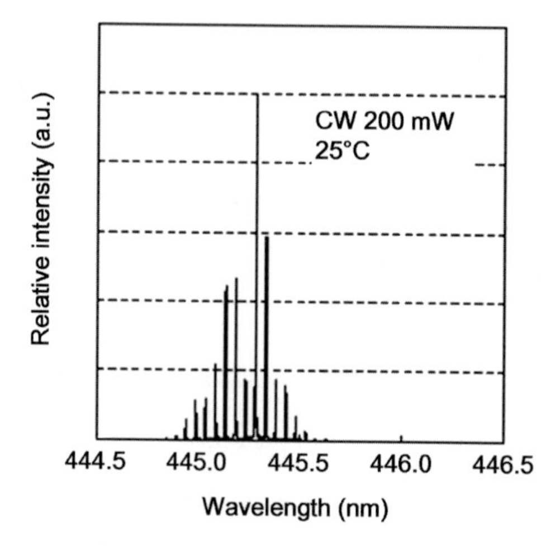

Fig. 3.109. Oscillation spectrum of a multimode blue LD

saturates at ∼800 mW. In these experiments, a commercially available 5.6 mm diameter package with a high thermal resistance was used. It is thought that above 800 mW the temperature of the active layer increased due to an increase in the injection current, which ultimately led to thermally limited saturation of the output.

Figure 3.109 shows the optical spectrum of this LD at an output of 200 mW where the oscillation peak is at 445.3 nm and the emission is longitudinal multimode oscillation. Further, the horizontal transverse mode is a multimode oscillation (Fig. 3.110). In particular, the FFP of the horizontal direction has multiple peaks (Fig. 3.111) due to the use of a 7 μm ridge width which leads to multimode oscillations inside the waveguide.

Figure 3.112 shows reliability results during APC operation, where the measurements were made at an output of 200 mW and a case temperature

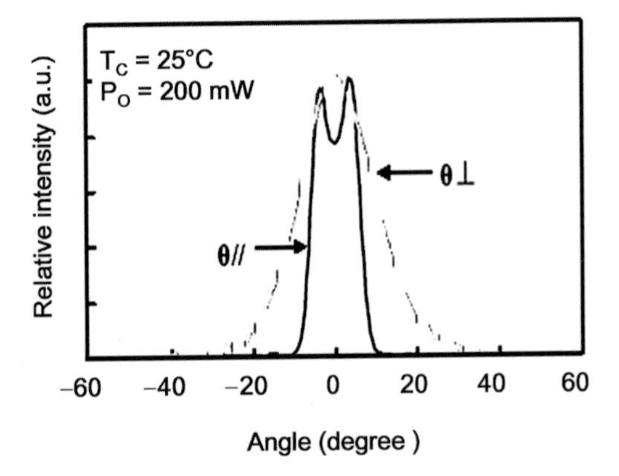

Fig. 3.110. FFP of multimode blue LD

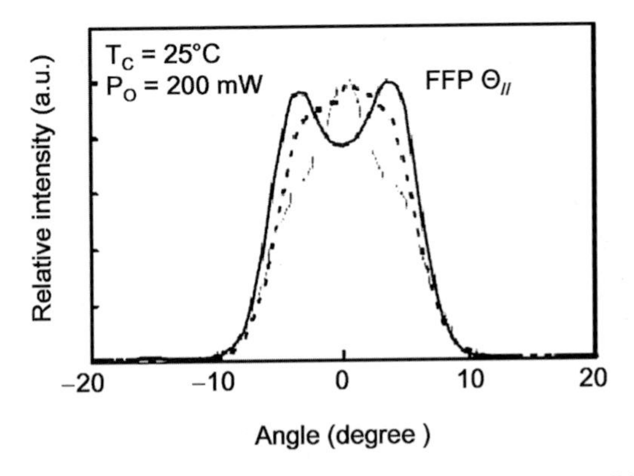

Fig. 3.111. Horizontal FFP of a multimode blue LD

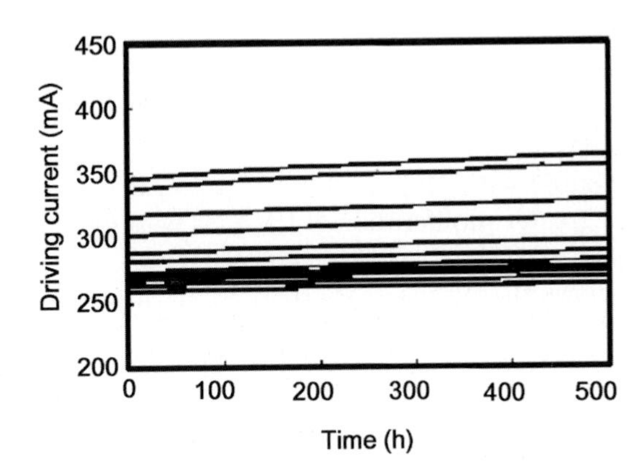

Fig. 3.112. Lifetime characteristics of a multimode blue LD

of 60°C. The lifetime after 500 h was estimated at 10,000 h, which shows the high reliability of these LDs.

Further improvements in the output power can be achieved by optimizing the chip structure [239, 240] and the use of packages with low thermal resistance.

3.5.3 Advances in UV Laser Diodes (H. Amano)

Ultraviolet (UV) light has many important applications. For example, the sterilization effects of UV light are well known. Also, UV light is used for treating psora and atopy [241]. Another application is for the early detection of cancer in the digestive system by encapsulation of a compact UV laser diode together and a white LED into a remotely controlled compact capsule that emulates the action of an endoscope as it passes through the body. High-power nitride semiconductor lasers are also expected to be used in the electronics industry as light sources for optical lithography and fabrication of devices. Compact, short wavelength light sources can be produced by (1) integration of a long wavelength LD with an SHG crystal for up conversion, and (2) fabrication of a semiconductor laser emitting at short wavelengths.

In case (1), commercial systems operating in the visible spectrum are available consisting of an infrared LD and KTP crystals (niobium oxide, tantalum oxide lithium, or potassium tantalum phosphor) for generating blue or green light. On the other hand, UV light is generated using solid state or large gas laser systems combined with crystals for modulating the light's frequency. However, there are very few reports on compact, UV systems. A UV system composed of a blue GaN-based LD (420 nm) and a BBO (beta-barium borate) crystal was used to produce laser emission of tens of microwatts output at 210 nm [242]. The wavelength modulation efficiency of SHG crystals increases with increasing power of the primary light source. The output power of a 420 nm blue LD has been reported to be ~1 W per chip [243] and, in the future, particularly with improvements in the quality of SHG crystals, it will be possible to produce high-power UV light sources (~200 nm emission) using high-power blue LDs and SHG crystals.

However, compared with (1), method (2) is cheaper and easier to implement but requires greater proliferation of UV LDs before practical UV sources can be produced.

Figure 3.113 shows the trends of short-wavelength group-III nitride semiconductor LDs. As of August 2005, there were reports of 343 nm emission from laser structures grown on SiC substrates with AlGaN buffer layers [243–246]. The use of AlN pn junctions would enable emission at 205 nm. However, current research is centered on AlGaN-based double heterostructures with gain reported from optically excited AlGaN crystals emitting at 241 nm [247, 248]. The requirements for the development of deep UV LDs include improvement of the internal quantum efficiency, increase of the electron and hole density of the clad layer and improvement of carrier confinement.

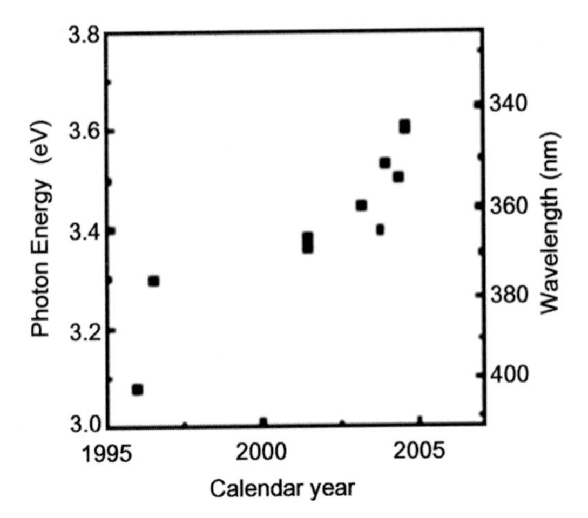

Fig. 3.113. Trends of short-wavelength group-III nitride semiconductor LDs

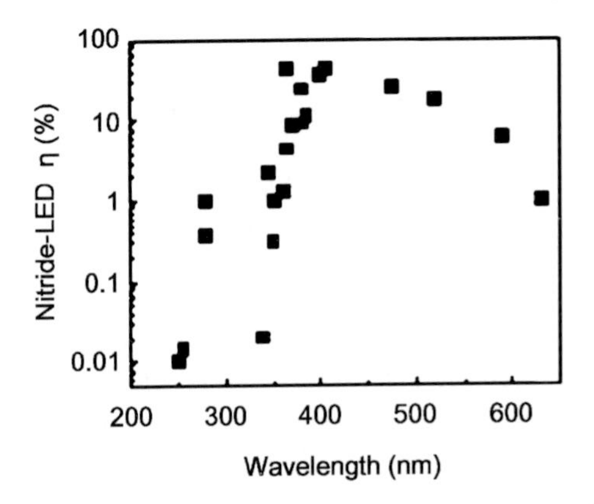

Fig. 3.114. Maximum external efficiencies of group-III nitride LEDs at the peak emission wavelengths up to August 2005

Figure 3.114 shows reports of the maximum external efficiencies of group-III nitride LEDs at the peak emission wavelengths up to August 2005. The light extraction efficiency is only 10–15%, which implies that the internal quantum efficiencies of the LEDs are approximately six times that of the external quantum efficiencies. It is difficult to estimate the figures of merit of LDs using data for LEDs but it can be said that sub 300 nm LDs will become possible in the not too distant future.

3.5.4 ZnSe Laser Diodes (K. Akimoto)

In 1991, 3M reported the successful pulsed operation of ZnSe-based II–VI semiconductor lasers at 77 K. In 2000, ZnSe lasers emitting at 514 nm were operated continuously for 500 h. However, the development of ZnSe-based laser diodes was suddenly terminated due to reliability problems and the superior performance of GaN lasers. This section is a summary of the knowledge assimilated during the development of ZnSe laser technology.

Events Leading to the Observation of Oscillations in ZnSe Lasers

It is well known that strongly ionic II–VI semiconductors can be doped either n- or p-type, that is, unipolar materials. Based on thermal equilibrium in the vapor and solid phases, Kroger [249,250] and Brebrick [251,252] proposed the origin to be due to the formation of defects that caused self-compensation and, later, these theoretical results were experimentally confirmed by Mandel et al. [253,254].

If carriers are generated by n-type doping, then the electron energy of the semiconductor system would be raised to a higher level, which could be considered as being an excited state. Furthermore, if acceptor-like defects were to be formed then the excited state could be relaxed by electron recombination into the acceptor levels. When the formation energy of acceptor-like defects is less than the recombination energy, the acceptor-like defects will be generated. If the energy for defect formation is represented by cohesive energy, E_h, a recombination energy by the bandgap energy, E_g, and considering the divalent defects in II–VI semiconductors, then the conditions for the occurrence of self compensation can be approximated as $2E_h = E_g$.

Further, it has been shown that the ease of the defects formation increases with the bonding ionicity and the ease with which cation and anion vacancies are formed is determined by the ratio of the covalent bond radii. Optical and ESR measurements show the existence of vacancies near dopants which are displaced from the lattice sites in ZnSe and ZnS [255–257].

The difficulties in the development of ZnSe LEDs could have been forecast based on its physical properties, particularly the $2E_h = E_g$ boundary condition. Nonequilibrium growth and impurity doping were recognized as important issue for suppressing defect formation. Thus in the 1980s ZnSe were grown mainly by molecular beam epitaxy (MBE) and metalorganic chemical vapor deposition (MOCVD), and ion doping technology proposed by Mitsuyu et al. is one of the nonequilibrium doping [258]. Also, the ZnSe laser successfully fabricated by 3M employed nonequilibrium nitrogen plasma doping.

Difficulties in ZnSe growth were not only limited to point defects in the epilayers. The extremely small lattice mismatch of only 0.2% between GaAs substrates and ZnSe epilayers was sufficient to cause degradation of the optical and electrical properties of ZnSe and poor surface morphology [259,260]. Figure 3.115 shows the variation of full-width at half maximum of X-ray diffraction peaks with lattice mismatch where it can be seen that high quality crystals can be grown under lattice matched conditions ($x = 0.06$). Also, Fig. 3.116 shows a TEM cross-section image reported by Ponce et al. [261].

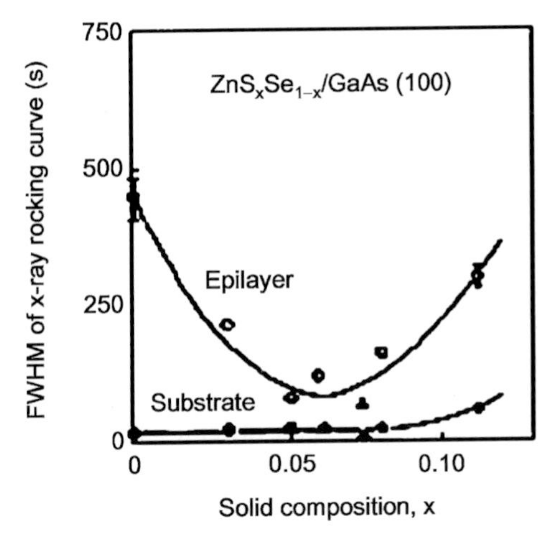

Fig. 3.115. Variation of full-width at half maximum of X-ray diffraction peaks with lattice mismatch of ZnSe grown on GaAs substrates

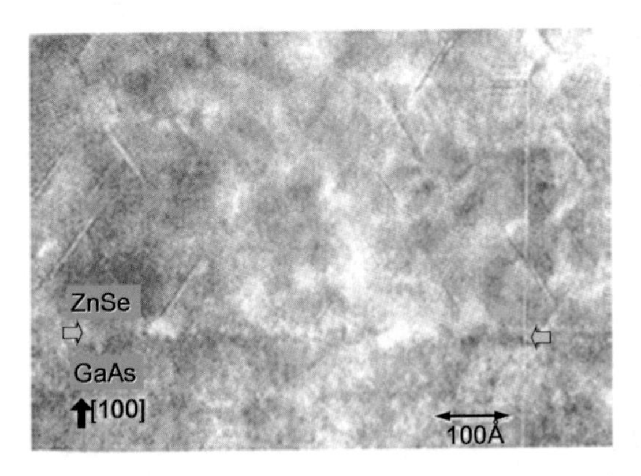

Fig. 3.116. TEM image of ZnSe grown on a GaAs substrate

Misfit dislocations are not seen at the interface itself but stacking faults are seen at a short distance from the interface. This result shows the difficulty in growing II–VI compound semiconductors on III–V substrates.

Successful Growth of p-Type ZnSe and Laser Oscillations

Until the middle of the 1980s, n-type dopant for ZnSe used was Ga and a maximum carrier concentration of $5 \times 10^{17}\,\mathrm{cm}^{-3}$ was obtained. Ohkawa et al. reported a carrier concentration of $\sim 10^{19}\,\mathrm{cm}^{-3}$ using chlorine (Cl) [262]. There were reservations about the use of Cl especially because of memory effects in the growth equipment and of defect formation due to high ionicity, however, such reservation was found to be negligible. Then the Cl doping was widely used.

Reports on the p-type doping of ZnSe were centered on issues related to reproducibility and carrier concentration. In 1990, the potential of p-type doping was demonstrated by a report based on collaboration between Florida University and 3M where plasma-nitrogen was used to dope ZnSe for fabricating LEDs [263]. Independently, Ohkawa et al. also reported on the use of nitrogen plasma doping [264]. These were major breakthroughs in the development of ZnSe-based compound semiconductor optical devices. Due to the necessity for a vacuum, MBE became the main growth technique following these reports.

In 1991, 3M reported the 77 K pulsed operation of a ZnSe-based laser that was fabricated using plasma doping [265] with an active layer of ZnCdSe and a cladding layer of ZnSSe, resulting in an emission wavelength of 490 nm, a threshold current density of 320 A cm^{-2} and an operating voltage of 20 V. Figure 3.117 shows the device structure and Fig. 3.118 its emission spectrum.

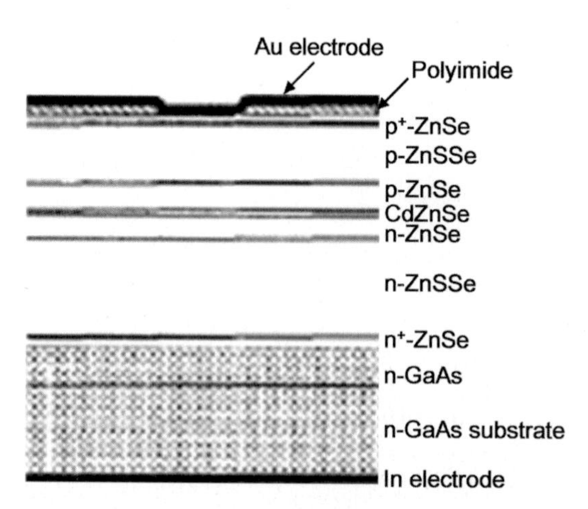

Fig. 3.117. Structure of the first ZnSe-based device exhibiting laser oscillation

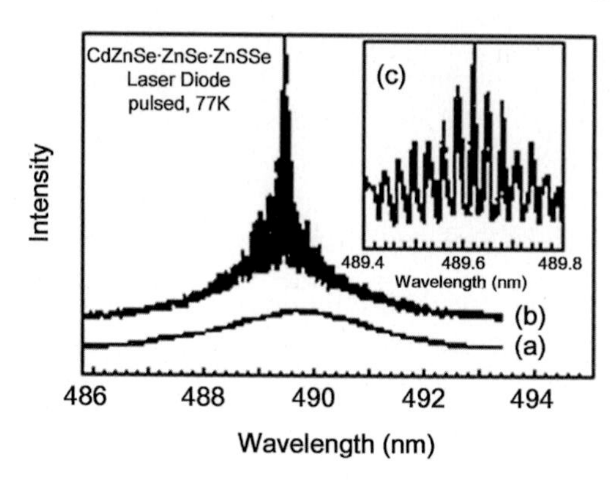

Fig. 3.118. Emission spectrum of the device shown in Fig. 3.117

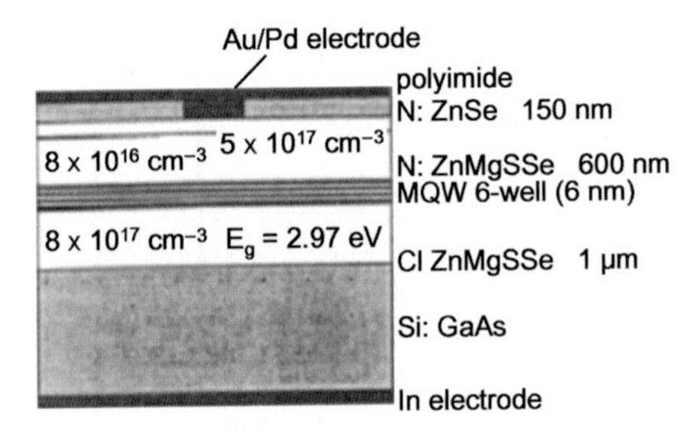

Fig. 3.119. First Mg alloy structure used for fabricating lasers

This was an important development in the field of II–VI optical devices, but, unfortunately, the small 0.2 eV difference in the bandgaps of the active and cladding layers was not enough for CW operation at room temperature.

Development of the Cladding Layer Material for Room Temperature Laser Oscillation

The group from Sony Corp. reported on the use of quaternary ZnMgSSe compound semiconductors as a cladding layer for II–VI semiconductor lasers [266]. Materials for cladding layers should have the following properties (1) a larger bandgap than the active layer; (2) formation of a type I structure with the active layer; (3) a smaller refractive index than the active layer; and (4) be lattice matched to the substrate and active layer. The quaternary ZnMgSSe was well designed so as to satisfy the above properties required for cladding layer. To increase the bandgap energy, lighter element in the periodic table should be added. To fulfill the formation of type I structure, the lighter element of both cation and anion should be added, since the bottom of the conduction band and the top of the valence band mainly composed of s-orbital character of cation and p-orbital character of anion, respectively. It is generally difficult, however, to increase the bandgap energy keeping the lattice constant, because the lighter element has shorter bond length.

Sony researchers appreciated the Mg which has longer bond length than Zn despite the Mg is lighter element than Zn. Figure 3.119 shows the structure of the Mg alloy laser. It is easy to find that the bandgap energy of MgZnSSe can vary from 2.8 eV to around 4 eV with keeping lattice constant as shown in Fig. 1.1.

In 1992, this quaternary compound was used for the fabrication of a laser that showed CW operation at 77 K [267]. Figure 3.120 shows the output-current dependence of the laser structure, which has an emission wavelength at 447 nm and a threshold current of 225 A cm^{-2}.

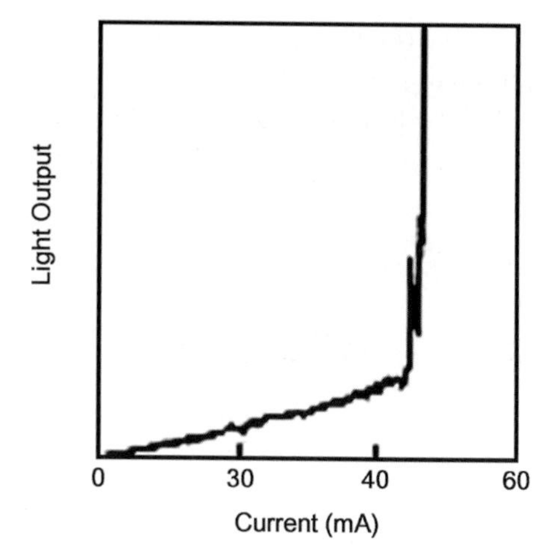

Fig. 3.120. L–I characteristics of the laser structure of Fig. 3.119

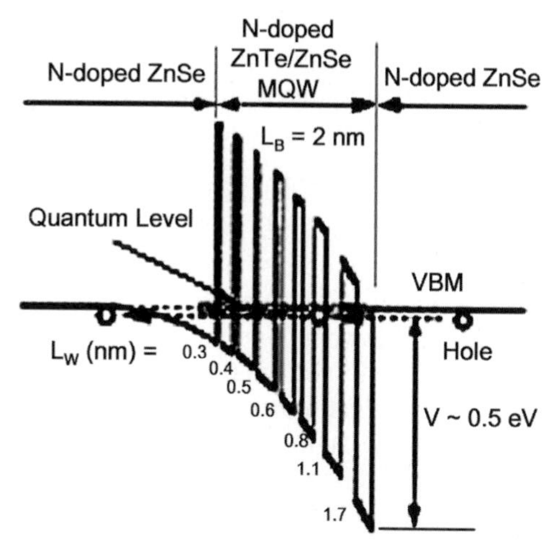

Fig. 3.121. Modulated quantum wells used for forming resonant tunneling ohmic electrodes composed of ZnSe and ZnTe proposed by Sony Corp

Development of Electrodes Based on Quantum Wells

The operating voltages of the lasers developed by 3M and Sony Corp. were $\sim$20 V due to the high contact resistance at p-electrodes. Formation of low resistance ohmic contacts to p-type ZnSe requires the use of metals with large work functions of approximately 7 eV and/or heavy doping. However, the work function of metals is typically 5 eV at largest, and the carrier concentration is $\sim 10^{18}\,\mathrm{cm}^{-3}$ at most. This problem was solved by the use of superlattices.

Figure 3.121 shows the use of modulated quantum wells for forming resonant tunneling ohmic electrodes composed of ZnSe and ZnTe proposed by Sony Corp. [268]. Almost at the same time, Gunshor et al. of Purdue

University reported the successful formation of ohmic contacts using ZnSe, ZnTe quasialloy structures [269]. Basic concept for the use of superlattice to get Ohmic contact is different each other, however, it is interesting to find the structural similarity between them.

In 1993, based on this technology, Sony, 3M, Philips, and Purdue University achieved pulsed operation at room temperature [270–272]. Further, in the same year Sony and Purdue also reported CW operation at room temperature for wavelengths between 490 and 523 nm [273–275]. The threshold current density, operating voltage, and lifetime were $1.5\,\mathrm{kA\,cm^{-2}}$, 6 V, and 20 s, respectively. The device structure was consisted of a ZnCdSe single quantum well active layer, a ZnSe guiding layer, a ZnMgSSe cladding layer, and a ZnSe/ZnTe superlattice p-type contact layer.

Origin and Suppression of Rapid Degradation

Guha et al. of 3M and Hua et al. of Purdue studied the TEM images after degradation and found the dark spots in the active layer to evolve into dark lines [276–278]. Figure 3.122 illustrates these observations.

Reducing the stacking defects at the substrate interface is the only way of resolving these problems by careful control of the initial growth parameters.

Prior to growth, GaAs substrates are annealed in a vacuum to remove the thin oxide layer on their surface. At this time, the GaAs surface becomes Ga rich and the growth of ZnSe on these surface results in the RHEED exhibiting a spot pattern indicating 3D growth. However, As rich (2×4) surfaces have been found to produce flat ZnSe films suitable for device applications. Further, the use of a buffer layer has been reported to reduce the density of stacking defects from 10^7 to $10^5\,\mathrm{cm^{-2}}$ [279–281].

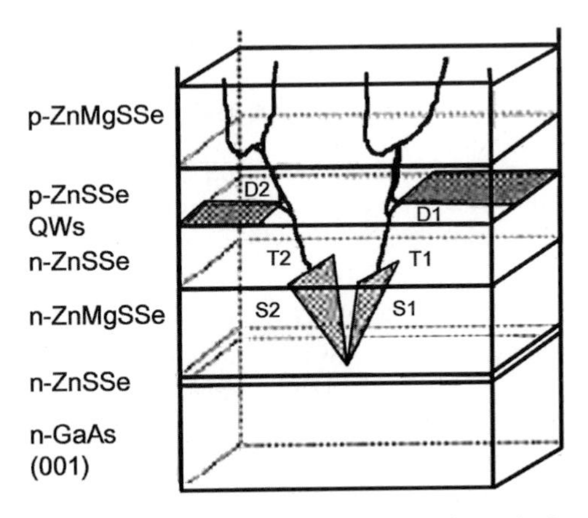

Fig. 3.122. Illustration of sudden degradation

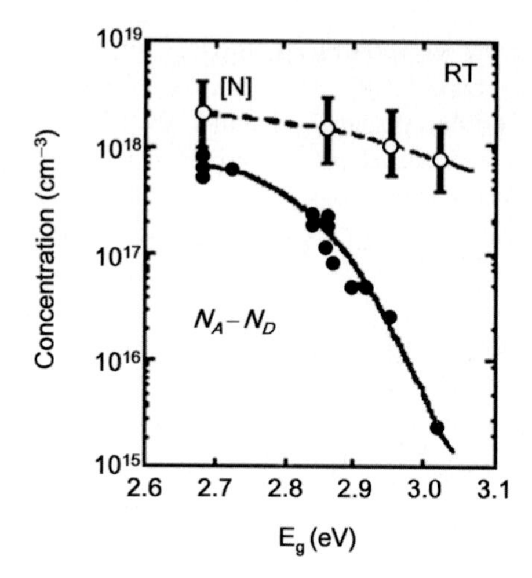

Fig. 3.123. Hole concentration dependence on the ZnMgSSe bandgap

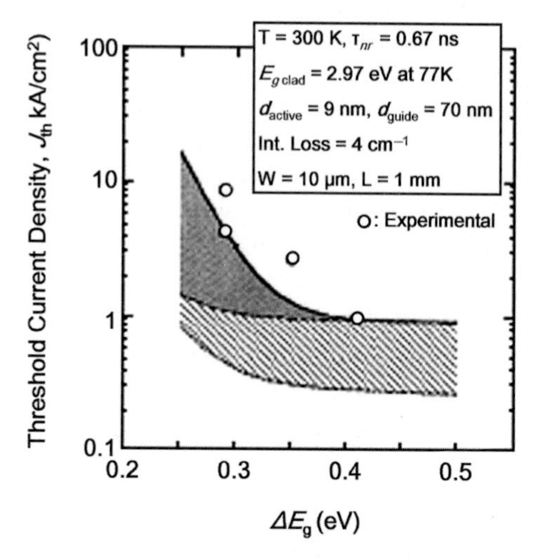

Fig. 3.124. Simulation results of the variation of threshold current density with difference in the bandgap energies of the active and clad layers

Exposure of GaAs substrates to Zn vapor prior to initiating ZnSe growth was found to improve the flatness of the epilayers [282] due to the prevention of a Ga_2Se_3-like layer formation at the substrate interface [283]. The stacking fault density was found to be 10^3–10^4 cm^{-2} using this technique.

As shown in Fig. 3.123, the hole concentration depends on the ZnMgSSe bandgap [284]. Figure 3.124 shows simulation results of the variation of threshold current density with difference in the bandgap energies of the active and cladding layers [285].

Such improvements in growth technology enabled fabrication of laser diodes having lifetimes as long as 100 h [286]. The wavelength of the laser oscillation, threshold current densities, and the dislocation densities of the laser diodes were 514 nm, 533 A cm^{-2}, and $\sim 10^3\,\mathrm{cm}^{-2}$, respectively.

Degradation Mechanisms and Lifetime

Figure 3.125 shows the relationship between laser lifetime and operating temperature [287]. The data show the trends for large and small densities of stacking faults. The activation energy associated with degradation in the case of a large density of stacking faults is smaller than for a small density of defects. Figure 3.126 shows the cathode luminescence of the active layer after degradation [288] where a dark line was formed along the <100> direction in the strip (Fig. 3.126b). Figure 3.127 shows the emission spectra corresponding to Fig. 3.126a–c. Degradation leads to the emergence of emission peaks in regions at shorter and longer wavelengths due to mutual diffusion of Cd and Zn. TEM results after degradation show the presence of dislocation loops and clusters as is observed in GaAs lasers [289]. The dislocation loops and clusters are considered to be formed by gathering point defects, therefore, decreasing point defect density was focused to extend the device life time.

Optimal conditions for the growth of group II–VI compound semiconductors were reported to be those of 1:1 stoichiometry corresponding the co-existence of c(2X2) and (2X1) RHEED patterns, although Se rich growth is known to yield flat, compositionally uniform films [290]. Figure 3.128 shows

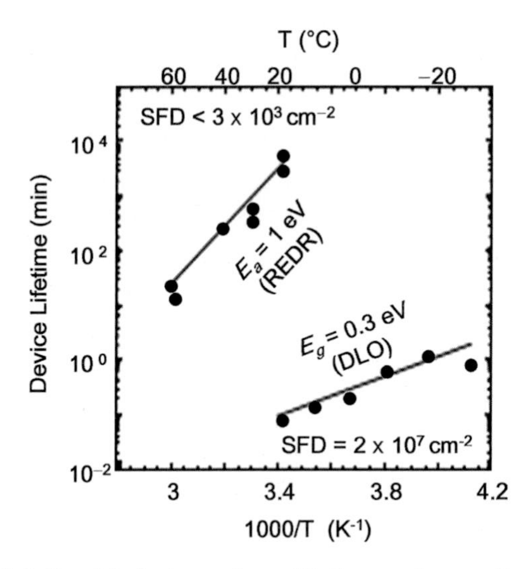

Fig. 3.125. Relationship between laser lifetime and operating temperature

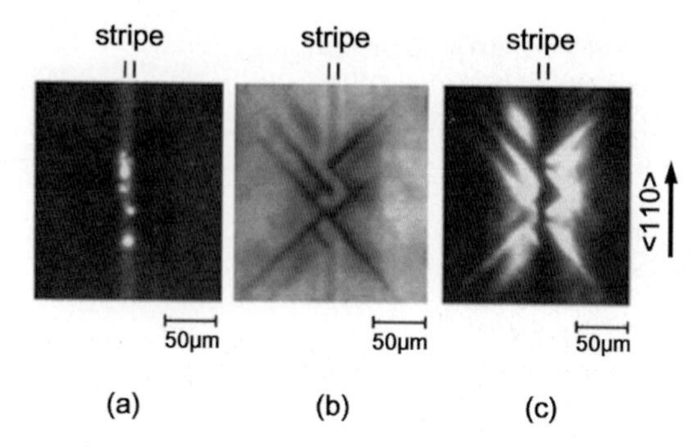

Fig. 3.126. Cathode luminescence of the active layer after degradation. (**a**) short wavelength; (**b**) just; (**c**) long wavelength

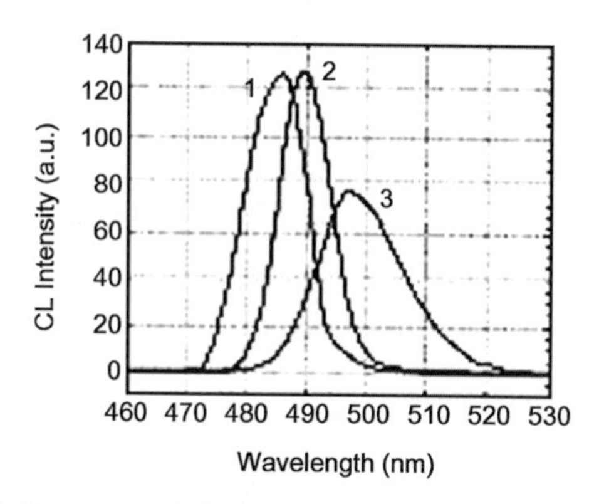

Fig. 3.127. Spectrum of the bright areas in the images of Fig. 3.126

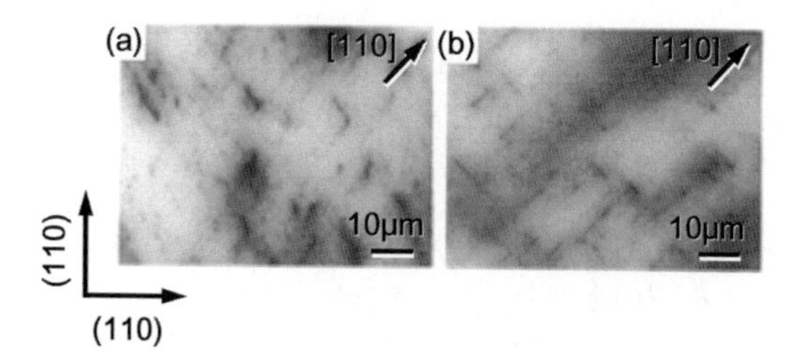

Fig. 3.128. Electroluminescence images after degradation for devices grown under small (**a**) and large (**b**) VI/II ratios

electroluminescence images after degradation for the devices grown under small and large VI/II ratios where distinct differences in the degree of degradation are visible.

In 2000, Sony Corp. reported the fabrication of ZnSe laser structures grown under Se rich conditions that showed room temperature lifetimes as long as 500 h [287]. Figure 3.129 shows the structure of such a laser which

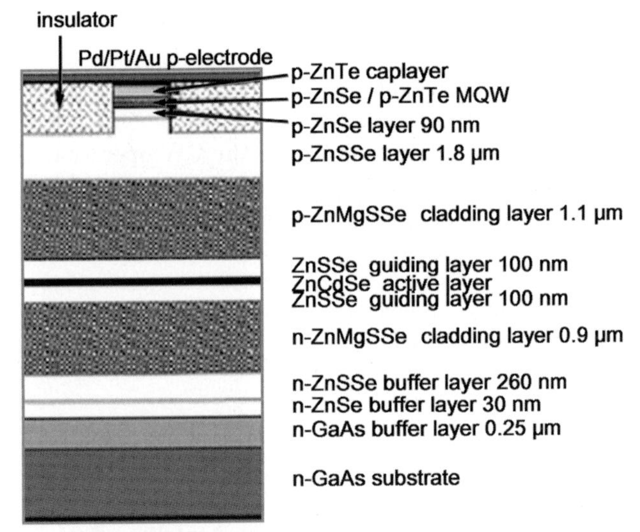

Fig. 3.129. Device structure of a laser with a lifetime of 500 h

had a threshold current of 431 A cm^{-2}, operating voltage of 5.3 V, a ZnMgSSe layer with a carrier concentration of 10^{17} cm^{-3} and a bandgap difference with the active layer of 0.4 eV.

Experiments on the Use of Be-Based Alloys as Cladding Layers

As described above, defects are easily formed in ZnSe materials due to self-compensation effects. A group from the University of Wurzburg proposed the use of Be alloys to suppress such effects [291]. The addition of Be leads to an increase in the bandgap and was also expected make it easier to produce p-type material due to high covalency. Figure 1.1 shows the relationship between the bandgap energy and lattice constants of Be-based alloys. The use of a BeTe interface layer on the GaAs substrate (the lattice constant of BeTe is almost the same as GaAs) enabled the growth of layers with stacking faults of $\sim 10^3$ cm^{-3} [292]. In 1998, a laser structure with a BeMgZnSe cladding layer was reported to have a device life time of 57 h [293]. The oscillation wavelength was 514 nm and the threshold current density was 700 A cm^{-2}. The toxicity of Be curtailed further development of Be-based laser structures.

Green Lasers

SHG green lasers are commercially available but they are much larger than semiconductor laser systems. The University of Sophia group succeeded in the pulsed operation of semiconductor laser at 77 K with the emission wavelength of 560 nm. The laser structure is consisted of InP substrate, p-MgSe/BeZnTe superlattice cladding layer, n-MgSe/ZnCdSe superlattice cladding layer, and ZnCdSe active layer. The threshold current density and operation voltage were 2.5 kA cm^{-2} and 30 V, respectively [294].

The Bremen University group succeeded the CW operation at room temperature with the wavelength of 560 nm from a structure consisting of: ZnSe or GaAs substrate, ZnCdSSe active layer, and ZnMgSSe clad layer. The threshold current density, operating voltage and lifetime were 560 A cm^{-2}, 6 V, and 1 min, respectively [295]. The use of ZnSe substrates eliminates problems due to defects at the substrate/epilayer interface, but methods and technology for cleaning the ZnSe substrates then become important. These two groups are still continuing this research.

References

1. H. Sugawara, M. Ishikawa, and G. Hatakoshi; Appl. Phys. Lett. 58 (1991) 1010.
2. H. Sugawara, K. Itaya, and G. Hatakoshi; Jpn. J. Appl. Phys. P.1 33 (1994) 6195.
3. I. Schnitzer, E. Yablonovitch, C. Caneau, T.J. Gmitter, and A. Scherer; Appl. Phys. Lett. 63 (1993) 2174.
4. M.R. Krames, M. Ochiai-Holcomb, G.E. Hofler, C. Carter-Coman, E.I. Chen, I.H. Tan, P. Grillot, N.F. Gardner, H.C. Chui, J.W. Huang, S.A. Stockman, F.A. Kish, M.G. Craford, T.S. Tan, C.P. Kocot, M. Hueschen, J. Posselt, B. Loh, G. Sasser, and D. Collins; Appl. Phys. Lett. 75 (1999) 2365.
5. T. Baba, K. Inoshita, H. Tanaka, J. Yonekura, M. Ariga, A. Matsutani, T. Miyamoto, F. Koyama, and K. Iga; J. Lightwave Technol. 17 (1999) 2113.
6. M. Fujita, S. Takahashi, Y. Tanaka, T. Asano, and S. Noda; Science 308 (2005) 1296.
7. W.L. Barnes; J. Lightwave Technol. 17 (1999) 2170.
8. K. Okamoto, I. Niki, A. Shvartser, Y. Narukawa, T. Mukai, and A. Scherer; Nat. Mater. 3 (2004) 601.
9. R.C. Miller, D.A. Kleinman, W.A. Nordland Jr., and A.C. Gossard; Phys. Rev. B 22 (1980) 863.
10. I.H. Ho and G.B. Stringfellow; Appl. Phys. Lett. 69 (1996) 2701.
11. R. Zimmermann; J. Cryst. Growth 101 (1990) 346.
12. S. Chichibu, T. Azuhata, T. Sota, and S. Nakamura; Appl. Phys. Lett. 69 (1996) 4188.
13. Y. Narukawa, Y. Kawakami, M. Funato, S. Fujita, S. Fujita, and S. Nakamura; Appl. Phys. Lett. 70 (1997) 981.
14. C. Gourdon and P. Lavallard; Phys. Status Solidi B 153 (1989) 641.
15. Y. Kawakami, in "Low Dimensional Nitride Semiconductors" (B. Gil, ed.), Series on Semiconductor Science and Technology, Oxford University Press, New York City, 2002, p. 233–256.
16. A. Kaneta, K. Okamoto, Y. Kawakami, S. Fujita, G. Marutsuki, Y. Narukawa, and T. Mukai; Appl. Phys. Lett. 81 (2002) 4353.
17. A. Kaneta, T. Mutoh, Y. Kawakami, S. Fujita, G. Marutsuki, Y. Narukawa, and T. Mukai; Appl. Phys. Lett. 83 (2003) 3462.

18. R.G. Palmer, D.L. Stein, E. Abrahams, and P.W. Anderson; Phys. Rev. Lett. 53 (1984) 958.
19. D.L. Huber; Phys. Rev. B 31 (1985) 6070.
20. A. Hangleiter, J.S. Im, H. Kollmer, S. Heppel, J. Off, and F. Scholz; MRS Internet J. Nitride Semicond. Res. 3 (1998) 15. (http://nsr.mij.mrs.org/3/15/).
21. M. Sugawara; Phys. Rev. B 51 (1995) 10743.
22. J. Feldmann, G. Peter, E.O. Göbel, P. Dawson, K. Moore, C. Foxon, and R.J. Elliott; Phys. Rev. Lett. 59 (1987) 2337.
23. H. Gotoh, H. Ando, T. Takagahara, H. Kamada, A. Chavez-Pirson, and J. Temmyo; Jpn. J. Appl. Phys. P.1 36 (1997) 4204.
24. Y. Narukawa, Y. Kawakami, S. Fujita, and S. Nakamura; Phys. Rev. B 59 (1999) 10283.
25. Y. Kawakami, Y. Narukawa, K. Omae, S. Fujita, and S. Nakamura; Phys. Status Solidi A - Appl. Res. 178 (2000) 331.
26. S. Kamiyama, M. Suzuki, T. Uenoyama, and Y. Ban, in Proc. "Topical Workshop on III-V Nitrides (TWN'95)", Nagoya, 21–23 September, 1995, p. 205.
27. M. Suzuki and T. Uenoyama; Jpn. J. Appl. Phys. P.1 - Reg. Pap. 35 (1996) 1420.
28. M. Suzuki, T. Uenoyama, and A. Yanase; Phys. Rev. B 52 (1995) 8132.
29. I. Akasaki, H. Amano, and I. Suemune, in Proc. "The Sixth International Conference on Silicon Carbide and Related Materials" (S. Nakashima, H. Matsunami, S. Yoshida, and H. Harima, eds.), Institute of Physics Conference Series, Vol. 142, Institute of Physics Publishing, Kyoto, 18–21 September, 1995, p. MoA-I-2.
30. S. Kamiyama, M. Iwaya, H. Amano, and I. Akasaki; Jpn. J. Appl. Phys. P.1 - Reg. Pap. 39 (2000) 390.
31. S. Nakamura, M. Senoh, S. Nagahama, N. Iwasa, T. Yamada, T. Matsushita, H. Kiyoku, Y. Sugimoto, T. Kozaki, H. Umemoto, M. Sano, and K. Chocho; Jpn. J. Appl. Phys. P.2 - Lett. 37 (1998) L309.
32. O. Matsumoto, S. Goto, T. Sasaki, Y. Yabuki, T. Tojyo, S. Tomiya, K. Naganuma, T. Asatsuma, K. Tamamura, S. Uchida, and M. Ikeda, in "Extended Abstracts of the 2002 International Conference on Solid State Devices and Materials (SSDM)", Business Center for Academic Societies, Japan, Nagoya, 17–19 September 2002, p. 832–833.
33. K. Motoki, T. Okahisa, N. Matsumoto, M. Matsushima, H. Kimura, H. Kasai, K. Takemoto, K. Uematsu, T. Hirano, M. Nakayama, S. Nakahata, M. Ueno, D. Hara, Y. Kumagai, A. Koukitu, and H. Seki; Jpn. J. Appl. Phys. P.2 - Lett. 40 (2001) L140.
34. H. Sugawara, K. Itaya, M. Ishikawa, and G. Hatakoshi; Jpn. J. Appl. Phys. P.1 31 (1992) 2446.
35. M. Suzuki, Y. Nishikawa, M. Ishikawa, and Y. Kokubun; J. Cryst. Growth 113 (1991) 127.
36. M. Suzuki, K. Itaya, Y. Nishikawa, H. Sugawara, and M. Okajima; J. Cryst. Growth 133 (1993) 303.
37. M. Suzuki, K. Itaya, and M. Okajima; Jpn. J. Appl. Phys. P.1 33 (1994) 749.
38. H. Sugawara, M. Ishikawa, and G. Hatakoshi; Appl. Phys. Lett. 58 (1991) 1010.
39. G. Hatakoshi, K. Nitta, and M. Ishikawa, in Proc. "The Fifth Microoptics Conference (MOC'95)", Technical Digest, Hiroshima, 1995, p. 212–215.
40. G. Hatakoshi and H. Sugawara; Display and Imaging [Jpn. Ed.] 5 (1997) 101.

41. H. Sugawara, K. Itaya, H. Nozaki, and G. Hatakoshi; Appl. Phys. Lett. 61 (1992) 1775.
42. H. Sugawara, K. Itaya, and G. Hatakoshi; J. Appl. Phys. 74 (1993) 3189.
43. H. Sugawara, K. Itaya, and G. Hatakoshi; Jpn. J. Appl. Phys. P.1 33 (1994) 6195.
44. F.A. Kish, F.M. Steranka, D.C. DeFevere, D.A. Vanderwater, K.G. Park, C.P. Kuo, T.D. Osentowski, M.J. Peanasky, J.G. Yu, R.M. Fletcher, D.A. Steigerwald, M.G. Craford, and V.M. Robbins; Appl. Phys. Lett. 64 (1994) 2839.
45. K. Streubel, U. Helin, V. Oskarsson, E. Backlin, and A. Johansson; IEEE Photon. Technol. Lett. 10 (1998) 1685.
46. K. Takaoka and G. Hatakoshi, in The IEICE Technical Report, LQE2000-128 (The Institute of Electronics Information and Communication Engineers of Japan), Tokyo, 2001, p. 51–56.
47. M.R. Krames, M. Ochiai-Holcomb, G.E. Hofler, C. Carter-Coman, E.I. Chen, I.H. Tan, P. Grillot, N.F. Gardner, H.C. Chui, J.W. Huang, S.A. Stockman, F.A. Kish, M.G. Craford, T.S. Tan, C.P. Kocot, M. Hueschen, J. Posselt, B. Loh, G. Sasser, and D. Collins; Appl. Phys. Lett. 75 (1999) 2365.
48. M.R. Krames, H. Amano, J.J. Brown, and P.L. Heremans, eds., "The issue on high-efficiency light-emitting diodes", IEEE Journal of Selected Topics in Quantum Electronics, Vol. 8(2), IEEE, 2002.
49. G. Hatakoshi, in Proc. "The 10th International Display Workshop (IDW'03)", Fukuoka, December 3–5, 2003, p. PH4 - 1.
50. G. Hatakoshi; O plus E 26(292) No. 3 (2004) 263.
51. G. Hatakoshi, in Proc. "IEEE/LEOS 3rd International Conference on Numerical Simulation of Semiconductor Optoelectronic Devices", Vol. 03EX726, IEEE, Tokyo, October 14 -16, 2003, p. 21–24.
52. G. Hatakoshi and M. Yamamoto, in Proc. "The 10th Microoptics Conference (MOC'04)", Jena, 1–3 September, 2004, p. J-6.
53. H. Amano, N. Sawaki, I. Akasaki and Y. Toyoda: Appl. Phys. Lett. 48. 353 (1986)
54. H. Amano, T. Asahi, M. Kito, I. Akasaki: Stimulated emission in MOVPE-grown GaN film Journal of Luminescence 48/49, 889 (1991).
55. H. Amano, M. Kito, K.Hiramatsu and I. Akasaki: J. Electrochem. Soc. 137, 1639 (1990).
56. H. Amano, H. Kito, K. Hiramatsu and I. Akasaki: Jpn. J. Appl. Phys. 28 L2112 [IPAP] (1989)
57. N. Yoshimoto, T. Matsuoka, T. Sasaki, A. Katsui: Applied Physics Letters 59 (18), 2251 (1991).
58. S. Nakamura, M. Senoh and T. Mukai, Appl. Phys. Lett. 62, 2390 (1993)
59. S. Nakamura, T. Mukai, and M. Senoh, J. Appl. Phys. 76, 8189 (1994).
60. S. Nakamura, M. Senoh, N. Iwasa and S. Nagaham: Jpn. J. Appl. Phys. 34 L797 [IPAP] (1995).
61. M. Yamada, T. Mitani, Y. Narukawa, S. Shioji, I. Niki, S. Sonobe, K. Deguchi, M. Sano and T. Mukai "InGaN-Based Near-Ultraviolet and Blue-Light-Emitting Diodes with High External Quantum Efficiency Using a Patterned Sapphire Substrate and a Mesh Electrode" Jpn. J. Appl. Phys. Vol. 41 L1431-L1433 (2002).
62. T. Mukai, M. Yamada and S. Nakamura: Jpn. J. Appl. Phys. 37 L1358 [IPAP] (1998).

63. T. Mukai and S. Nakamura: Jpn. J. Appl. Phys. 38, 5735 (1999).
64. M. Yamada, Y. Narukawa and T. Mukai: Jpn. J. Appl. Phys. 41 L246[IPAP] (2002)
65. N. Nakayama, S. Itoh, T. Ohata, K. Nakano, H. Okuyama, M. Ozawa, A. Ishibashi, M. Ikeda, and Y. Mori; Electron. Lett. **29** (1993) 1488.
66. K. Kishino and I. Nomura; IEEE J. Sel. Top. Quantum Electron. **8** (2002) 773.
67. T. Morita, K. Kikuchi, I. Nomura, and K. Kishino; J. Electron. Mater. **25** (1996) 425.
68. S.B. Che, I. Nomura, W. Shinozaki, A. Kikuchi, K. Shimomura, and K. Kishino; J. Cryst. Growth **214** (2000) 321.
69. S.B. Che, I. Nomura, A. Kikuchi, K. Shimomura, and K. Kishino; Phys. Status Solidi B - Basic Res. **229** (2002) 1001.
70. Y. Takashima, I. Nomura, Y. Nakai, A. Kikuchi, and K. Kishino; Phys. Status Solidi B - Basic Res. **241** (2004) 747.
71. A. Žukauskas, M.S. Shur, and R. Gaska, "Introduction to Solid State Lighting", John Wiley & Sons, New York City, 2002.
72. H. Hirayama; J. Appl. Phys. **97** (2005) Art. No. 091101.
73. H. Hirayama, K. Akita, T. Kyono, T. Nakamura, and Y. Aoyagi; Rev. Laser Eng. **32** (2004) 402.
74. D. Morita, M. Sano, M. Yamamoto, T. Murayama, S.-i. Nagahama, and T. Mukai; Jpn. J. Appl. Phys. P.2 **41** (2002) L1434.
75. T. Nishida, H. Saito, and N. Kobayashi; Appl. Phys. Lett. **79** (2001) 711.
76. V. Adivarahan, W.H. Sun, A. Chitnis, M. Shatalov, S. Wu, H.P. Maruska, and M.A. Khan; Appl. Phys. Lett. **85** (2004) 2175.
77. W.H. Sun, V. Adivarahan, M. Shatalov, Y.B. Lee, S. Wu, J.W. Yang, J.P. Zhang, and M.A. Khan; Jpn. J. Appl. Phys. P.2 **43** (2004) L1419.
78. H. Hirayama, K. Akita, T. Kyono, T. Nakamura, and K. Ishibashi; Jpn. J. Appl. Phys. P.2 **43** (2004) L1241.
79. H. Hirayama, A. Kinoshita, T. Yamabi, Y. Enomoto, A. Hirata, T. Araki, Y. Nanishi, and Y. Aoyagi; Appl. Phys. Lett. **80** (2002) 207.
80. H. Hirayama, Y. Enomoto, A. Kinoshita, A. Hirata, and Y. Aoyagi; Appl. Phys. Lett. **80** (2002) 1589.
81. T. Obata, H. Hirayama, Y. Aoyagi, and K. Ishibashi; Phys. Status Solidi A - Appl. Res. **201** (2004) 2803.
82. I. Akasaki, ed., "Blue-light-emitting devices", Kogyo Chosakai Publishing, Tokyo, 1997 (in Japanese)
83. S. Ohkubo; Nikkei Electronics **898** (2005) p. 79 (in Japanese)
84. P. Yu, Z.K. Tang, G.K.L. Wong, M. Kawasaki, A. Ohtomo, H. Koinuma, and Y. Segawa, in Proc. "23rd International Conference on the Physics of Semiconductors" (M. Scheffler and R. Zimmermann, eds.), World Scientific, Singapore, 1996, p. 1453
85. Y. Segawa, A. Ohtomo, M. Kawasaki, H. Koinuma, Z.K. Tang, P. Yu, and G.K.L. Wong; Phys. Status Solidi B - Basic Res. **202** (1997) 669
86. A. Ohtomo and M. Kawasaki; Parity: Phys. Sci. Mag. **13** (1998) 48 (in Japanese)
87. M. Kawasaki and A. Ohtomo; Kotai Butsuri **33** (1998) 59 (in Japanese)
88. A. Tsukazaki, A. Ohtomo, T. Onuma, M. Ohtani, T. Makino, M. Sumiya, K. Ohtani, S.F. Chichibu, S. Fuke, Y. Segawa, H. Ohno, H. Koinuma, and M. Kawasaki; Nature Mater. **4** (2005) 42

89. K. Maeda, M. Sato, I. Niikura, and T. Fukuda; Semicond. Sci. Technol. **20** (2005) S49
90. T. Makino, Y. Segawa, M. Kawasaki, A. Ohtomo, R. Shiroki, K. Tamura, T. Yasuda, and H. Koinuma; Appl. Phys. Lett. **78** (2001) 1237
91. A. Ohtomo, M. Kawasaki, T. Koida, K. Masubuchi, H. Koinuma, Y. Sakurai, Y. Yoshida, T. Yasuda, and Y. Segawa; Appl. Phys. Lett. **72** (1998) 2466
92. A. Ohtomo and A. Tsukazaki; Semicond. Sci. Technol. **20** (2005) S1
93. M. Kawasaki, T. Makino, Y. Segawa, and H. Koinuma; Oyo Butsuri **70** (2001) 523 (in Japanese)
94. A. Ohtomo, T. Makino, K. Tamura, Y. Matsumoto, Y. Segawa, Z.L. Tang, G.K.L. Wong, H. Koinuma, and M. Kawasaki, in Proc. "Combinatorial and Composition Spread Techniques in Materials and Device Development", Vol. 3941, SPIE, San Jose, CA, USA, 2000, p. 70–81
95. M. Ohtani, M. Lippmaa, T. Ohnishi, and M. Kawasaki; Rev. Sci. Instrum. **76** (2005) 62218
96. T. Makino, Y. Segawa, M. Kawasaki, and H. Koinuma; Kotai Butsuri **36** (2001) 297 (in Japanese)
97. T. Makino, Y. Segawa, M. Kawasaki, and H. Koinuma; Semicond. Sci. Technol. **20** (2005) S78
98. T. Ohnishi, A. Ohtomo, M. Kawasaki, K. Takahashi, M. Yoshimoto, and H. Koinuma; Appl. Phys. Lett. **72** (1998) 824
99. A. Ohtomo, K. Tamura, K. Saikusa, K. Takahashi, T. Makino, Y. Segawa, H. Koinuma, and M. Kawasaki; Appl. Phys. Lett. **75** (1999) 2635
100. E.S. Hellman, in Proc. "Symposium N "III-V Nitrides"" (T. Moustakas, I. Akasaki, B. Monemar, and F. Ponce, eds.), Materials Research Society Symposium Proceedings, Vol. 395, Materials Research Society, Boston, MA, December 2–6, 1996, p. 51
101. B. Wessler, A. Steinecker, and W. Mader; J. Cryst. Growth **242** (2002) 283
102. A. Ohtomo, H. Kimura, K. Saito, T. Makino, Y. Segawa, H. Koinuma, and M. Kawasaki; J. Cryst. Growth **214** (2000) 284
103. K. Tamura, A. Ohtomo, K. Saikusa, Y. Osaka, T. Makino, Y. Segawa, M. Sumiya, S. Fuke, H. Koinuma, and M. Kawasaki; J. Cryst. Growth **214** (2000) 59
104. A. Tsukazaki, H. Saito, K. Tamura, M. Ohtani, H. Koinuma, M. Sumiya, S. Fuke, T. Fukumura, and M. Kawasaki; Appl. Phys. Lett. **81** (2002) 235
105. A. Tsukazaki, A. Ohtomo, and M. Kawasaki; Appl. Phys. Lett. **88** (2006) 152106
106. A. Tsukazaki, A. Ohtomo, S. Yoshida, M. Kawasaki, C.H. Chia, T. Makino, Y. Segawa, T. Koida, S.F. Chichibu, and H. Koinuma; Appl. Phys. Lett. **83** (2003) 2784
107. M. Kawasaki, K. Terai, and A. Shikazeki; Nihon Kessho Seicho Gakkaishi **32** (2005) 74 (in Japanese)
108. A. Tsukazaki, M. Kubota, A. Ohtomo, T. Onuma, K. Ohtani, H. Ohno, S.F. Chichibu, and M. Kawasaki; Jpn. J. Appl. Phys. P.2 **44** (2005) L643
109. V.Z. Mordkovich, H. Hayashi, M. Haemori, T. Fukumura, and M. Kawasaki; Adv. Funct. Mater. **13** (2003) 519
110. H. Watanabe, K. Hayashi, D. Takeuchi, S. Yamanaka, H. Okushi and K. Kajimura, Appl. Phys. Lett. 73 (1998) 981.
111. P.J. Dean, E.C. Lightowlers and D. R. Wight: Phys. Rev. 140 (1965) A352.

112. S. Koizumi, K. Watanabe, M. Hasegawa and H. Kanda: Science **292**, (2001) 1899.
113. S. Koizumi, M. Kamo, Y. Sato, H. Ozeki and T. Inuzuka: Appl. Phys. Lett. **71**, (1997) 1065.
114. K. Horiuchi, A. Kawamura, T. Ida, T. Ishikura, K. Nakamura and S. Yamashita: Jpn J. Appl. Phys., **40**, (2001) L275.
115. K. Horiuchi, A. Kawamura, Y. Okajima and T. Ida: New Diamond (in Japanese), **20** (2004) 6.
116. H. Okushi: Diamond Relat. Mater. **10**, (2001) 281.
117. H. Kato, S. Yamasaki and H. Okushi: Appl. Phys. Lett., **86**, 222111 (2005).
118. T. Makino, H. Kato, H. Ogura, H. Watanabe, S-Gi. Ri, S. Yamasaki and H. Okushi: Jpn. J. Appl. Phys. **49**, (2005) L1190.
119. H. Okushi, H. Watanabe and S. Kanno: Phys. Stat. Sol. (a), **202**, (2005) 2051.
120. Private communication.
121. H. Watanabe, D. Takeuchi, S. Yamanaka, H. Okushi, K. Kajimura and T. Sekiguchi: Diamond Relat. Mater. **8**, (1999) 1272.
122. Y.G. Chen, H. Ogura, S. Yamasaki and H. Okushi: Semicond. Sci. & Technol. 20, (2005) 860.
123. T. Teraji, S. Koizumi, H. Kannda: Appl. Phys. Lett. **76**, (2000) 1303.
124. F. Fichter; Z. Anorg. Chem. **54** 322 (1907).
125. Y. Taniyasu, M. Kasu, and T. Makimoto; Nature **441**, 325 (2006).
126. M.A. Khan, M. Shatalov, H.P. Maruska, H.M. Wang and E. Kuokstis; Jpn. J. Appl. Phys. P.1 **44**, 7191 (2005).
127. T.G. Mihopoulos, V. Gupta, and K.F. Jensen; J. Cryst. Growth **195**, 7333 (1998).
128. H. Amano, A. Miyazaki, K. Iida, T. Kawashima, M. Iwaya, S. Kamiyama, I. Akasaki, R. Liu, A. Bell, F.A. Ponce, S. Sahonta, and D. Cherns; Phys. Status Solidi A - Appl. Res. **201** (2004) 2679.
129. Y. Taniyasu, M. Kasu, and T. Makimoto; Appl. Phys. Lett. **89**, (2006) 182112
130. Y. Taniyasu, M. Kasu, and T. Makimoto; J. Cryst. Growth, (2006) in press, doi:10.1016/j.jcrysgro.2006.10.032.
131. Y. Taniyasu, M. Kasu, and N. Kobayashi; Appl. Phys. Lett. **81**, 1255 (2002).
132. Y. Taniyasu, M. Kasu, and T. Makimoto; Appl. Phys. Lett. **85**, 4672 (2004).
133. C. Stampfl and C.G. Van der Walle; Appl. Phys. Lett. **72**, 459 (1998).
134. S. Nakamura, T. Mukai, M. Senoh, and N. Iwasa; Jpn. J. Appl. Phys. P.2 **31**, L139 (1992).
135. K.B. Nam, M.L. Nakarmi, J. Li, J.Y. Lin, and H.X. Jiang; Appl. Phys. Lett. **83**, 878 (2003).
136. Y. Taniyasu, M. Kasu, K. Kumakura, T. Makimoto, and N. Kobayashi; Phys. Stat. Solidi A **200**, 40 (2003).
137. S. Nakamura, M. Senoh, S.-I. Nagahama, N. Iwasa, T. Yamada, T. Matsushita, Y. Sugimoto, and H. Kiyoku; Appl. Phys. Lett. **69** (1996) 4056.
138. H. Amano, M. Iwaya, N. Hayashi, T. Kashima, M. Katsuragawa, T. Takeuchi, C. Wetzel, and I. Akasaki; MRS Internet J. Nitride Semicond. Res. **4S1** (1999) G10.1.
139. C. Pernot, A. Hirano, M. Iwaya, T. Detchprohm, H. Amano, and I. Akasaki; Jpn. J. Appl. Phys. P.2 **39** (2000) L387.
140. R. McClintock, A. Yasan, K. Mayes, D. Shiell, S.R. Darvish, P. Kung, and M. Razeghi; Appl. Phys. Lett. **84** (2004) 1248.

141. R. Mouillet, A. Hirano, M. Iwaya, T. Detchprohm, H. Amano, and I. Akasaki; Jpn. J. Appl. Phys. P.2 **40** (2001) L498.
142. E. Monroy, F. Calle, J.L. Pau, E. Munoz, M. Verdu, F.J. Sanchez, M.T. Montojo, F. Omnes, Z. Bougrioua, I. Moerman, and E. San Andres; Phys. Status Solidi A - Appl. Res. **188** (2001) 307.
143. M.A. Khan, M.S. Shur, Q. Chen, J.N. Kuznia, and C.J. Sun; Electron. Lett. **31** (1995) 398.
144. C. Pernot, A. Hirano, N. Amano, and I. Akasaki; Jpn. J. Appl. Phys. P.2 **37** (1998) L1202.
145. A. Hirano, C. Pernot, M. Iwaya, T. Detchprohm, H. Amano, and I. Akasaki; Phys. Status Solidi A - Appl. Res. **188** (2001) 293.
146. K. Bando, K. Sakano, Y. Noguchi, and Y. Shimizu; J. Light Vis. Env. **22** (1998) 2.
147. K. Sakuma, Y. Yamamoto, K. Omichi, N. Hirosaki, R.-J. Xie, M. Ohashi, N. Kimura, T. Suehiro, and D. Tanaka, in Proc. "The 11th International Display Workshop (IDW'04)", Niigata, December 8–10, 2004, p. 1115 (PHp -1).
148. R.J. Xie, N. Hirosaki, M. Mitomo, Y. Yamamoto, T. Suehiro, and K. Sakuma, in Proc. "The 11th International Display Workshop (IDW'04)", Niigata, December 8–10, 2004, p. 1105 (PH4 -1).
149. K. Ueda and et al., Proc. the 182nd Meeting, (The 125th JSPS Research Committee on Mutual Conversion between Light and Electricity), Tokyo, 2003, p. 42.
150. Y. Shimomura and N. Kojima, "65th Fall Meeting of the Japanese Society of the Applied Physics", Fukuoka, September 7, 2004, Ext. Abstracts, 3, 1285 (2p-ZL-19).
151. K. Ueda, N. Hirosaki, Y. Ogawa, A. Yamamoto, and Y. Yamamoto, "65th Fall Meeting of the Japanese Society of the Applied Physics", Fukuoka, September 7, 2004, Ext. Abstracts, 3, 1283 (2p-ZL-13).
152. H. Amano, N. Sawaki, I. Akasaki, and Y. Toyoda; Appl. Phys. Lett. **48** (1986) 353.
153. H. Amano and I. Akasaki, in Proc. "1989 MRS Fall Meeting Symposium - Characterization of Plasma-Enhanced CVD Processes" (G. Lucovsky, D.E. Ibbotson, and D.W. Hess, eds.), MRS Symposium Proceedings, Vol. 165, Materials Research Society, 1990, Boston, Massachusetts, Nov. 27 - Dec. 2, 1989, p. EA - 21.
154. S. Michinari, M. Katsuhide, M. Akira, K. Hisayoshi, H. Masafumi, and A. Isamu, Japan Pat. No. Hei3-252175 (JP3252175), 1991-11-11, 1991
155. H. Amano, M. Kito, K. Hiramatsu, and I. Akasaki; Jpn. J. Appl. Phys. P.2 28 (1989) L2112.
156. N. Koide, H. Kato, M. Sassa, S. Yamasaki, K. Manabe, M. Hashimoto, H. Amano, K. Hiramatsu, and I. Akasaki; J. Cryst. Growth 115 (1991) 639.
157. M. Koike, N. Shibata, H. Kato, S. Yamasaki, N. Koide, H. Amano, and I. Akasaki, in Proc. "Topical Workshop on III-V Nitrides (TWN'95)", Nagoya, 21–23 September, 1995, p. 45 (C2).
158. M. Koike, N. Shibata, S. Yamasaki, S. Nagai, S. Asami, H. Kato, N. Koide, H. Amano, and I. Akasaki, in Proc. "MRS Symposium: Gallium Nitride and Related Materials" (F.A. Ponce, R.D. Dupuis, S. Nakamura, and J.A. Edmond, eds.), Material Research Society Symposium Proceedings, Vol. 395, Materials Research Society, Pittsburgh, PA, 1996, p. 889–895.

159. S. Nakamura, M. Senoh, and T. Mukai; Appl. Phys. Lett. **62** (1993) 2390.
160. S. Nakamura, M. Senoh, N. Iwasa, and S. Nagahama; Appl. Phys. Lett. **67** (1995) 1868.
161. K. Bando, K. Sakano, Y. Noguchi, and Y. Shimizu; J. Light Vis. Env. **22** (1998) 2.
162. M. Yamada, T. Mitani, Y. Narukawa, S. Shioji, I. Niki, S. Sonobe, K. Deguchi, M. Sano, and T. Mukai; Jpn. J. Appl. Phys. P.2 **41** (2002) L1431.
163. Y. Narukawa, I. Niki, K. Izuno, M. Yamada, Y. Murazaki, and T. Mukai; Jpn. J. Appl. Phys. P.2 **41** (2002) L371.
164. K. Tadatomo, H. Okagawa, Y. Ohuchi, T. Tsunekawa, Y. Imada, M. Kato, and T. Taguchi; Jpn. J. Appl. Phys. P.2 **40** (2001) L583.
165. Y. Narukawa, J. Narita, T. Sakamoto, T. Yamada, H. Narimatsu, M. Sano, T. Mukai, "International Workshop on Nitride Semiconductors (IWN2006)", Kyoto, October 22–27, 2006, submitted.
166. K. Katayama, H. Matsubara, F. Nakanishi, T. Nakamura, H. Doi, A. Saegusa, T. Mitsui, T. Matsuoka, M. Irikura, and T. Takebe; J. Cryst. Growth **214–215** (2000) 1064.
167. T. Shirakawa; Mater. Sci. Eng. B - Solid State Mater. Adv. Technol. **91** (2002) 470.
168. S.L. Chuang, A. Ishibashi, S. Kijima, N. Nakayama, M. Ukita, and S. Taniguchi; IEEE J. Quantum Electron. **33** (1997) 970.
169. M. Adachi, Z.M. Aung, K. Minami, K. Koizumi, M. Watanabe, S. Kawamoto, T. Yamaguchi, H. Kasada, T. Abe, K. Ando, K. Nakano, A. Ishibashi, and S. Itoh; J. Cryst. Growth **214** (2000) 1035.
170. T. Fujita; The TRC News **84** (2003) 31 (in Japanese).
171. K. Katayama and T. Nakamura; J. Appl. Phys. **95** (2004) 3576.
172. T. Nakamura, S. Fujiwara, H. Mori, and K. Katayama; Jpn. J. Appl. Phys. P.1 **43** (2004) 1287.
173. K. Tadatomo and T. Taguchi; Kino Zairyo **24** (2004) 7.
174. in "Nikkei Electronics," Vol. 844, 2003, p. 105.
175. White Light LEDs Report (Illuminating Engineering Institute of Japan), Tokyo, 2004.
176. J. Shimada, Y. Kawakami, and S. Fujita, in Proc. "Solid State Lighting and Displays", Vol. 4445, SPIE, San Diego, CA, USA, 2001, p. 13–22.
177. Y. Kawakami, J. Shimada, and S. Fujita; Oyo Butsuri **71** (2002) 1381.
178. Innovative Cluster Program Report (Yamaguchi Prefecture Industrial Development Agency), 2005.
179. K. Tadatomo, H. Okagawa, T. Jyoichi, M. Kato, M. Harada, and T. Taguchi; Mitsubishi Cable Industries Review **99** (2002) 35.
180. D. Yamaguchi, K. Tadatomo, N. Akita, and K. Hoshino, 66th Fall Meeting of the Japanese Society of the Applied Physics,Tokushima, 7a-K-5, 2005
181. H. Yanai, K. Okita, K. Kikuchi, K. Tadatomo, H. Nakamura, J. Nishikawa, T. Yoshida, and T. Taguchi; Endoscopy (2006) 290.
182. G. Iddan, G. Meron, A. Glukhovsky, and P. Swain; Nature **405** (2000) 417.
183. URL: http://www.givenimaging.com/Cultures/en-US/given/english, July 25 (2006) Given Imaging Ltd, Israel
184. URL: http://www.rfnorika.com, July 25 (2006) RF Norika,
185. URL: http://www.sony.jp/CorporateCruise/Press/200408/08–0819B/, July 25 (2006) SONY Marketing (Japan) Inc.,

186. URL: http://www.lumileds.com/solutions/solution.cfm?id=10, July 25 (2006) Philips Lumileds Lighting Company (LLC) San Jose, CA
187. S. Nakamura, M. Senoh, S.-i. Nagahama, N. Iwasa, T. Yamada, T. Matsushita, H. Kiyoku, and Y. Sugimoto; Jpn. J. Appl. Phys. P.2 **35** (1996) L74.
188. I. Akasaki; Electron. Lett. **32** (1996) 1105.
189. H. Amano, N. Sawaki, I. Akasaki, and Y. Toyoda; Appl. Phys. Lett. **48** (1986) 353.
190. H. Amano, M. Kito, K. Hiramatsu, and I. Akasaki; Jpn. J. Appl. Phys. P.2 **28** (1989) L2112.
191. S. Nakamura; Jpn. J. Appl. Phys. P.2 **30** (1991) L1705.
192. T. Matsuoka, H. Tanaka, T. Sasaki, and A. Katsui, in Proc. "16th International Symposium on GaAs and Related Compounds", Institute of Physics Conference Series, Vol. 106, Insitute of Physics, Bristol, 1990, Karuizawa, September, 1989, p. 141–141.
193. S. Nakamura, M. Senoh, S.-i. Nagahama, N. Iwasa, T. Yamada, T. Matsushita, Y. Sugimoto, and H. Kiyoku; Appl. Phys. Lett. **69** (1996) 4056.
194. S.D. Lester, F.A. Ponce, M.G. Craford, and D.A. Steigerwald; Appl. Phys. Lett. **66** (1995) 1249.
195. W. Qian, M. Skowronski, M.D. Graef, K. Doverspike, L.B. Rowland, and D.K. Gaskill; Appl. Phys. Lett. **66** (1995) 1252.
196. S. Nakamura, M. Senoh, S. Nagahama, N. Iwasa, T. Yamada, T. Matsushita, Y. Sugimoto, and H. Kiyoku; Appl. Phys. Lett. **70** (1997) 868.
197. T. Nishinaga, T. Nakano, and S. Zhang; Jpn. J. Appl. Phys. P.2 **27** (1988) L964.
198. O.H. Nam, M.D. Bremser, T.S. Zheleva, and R.F. Davis; Appl. Phys. Lett. **71** (1997) 2638.
199. A. Usui, H. Sunakawa, A. Sakai, and A.A. Yamaguchi; Jpn. J. Appl. Phys. P.2 **36** (1997) L899.
200. S. Nakamura, M. Senoh, S. Nagahama, N. Iwasa, T. Yamada, T. Matsushita, H. Kiyoku, Y. Sugimoto, T. Kozaki, H. Umemoto, M. Sano, and K. Chocho; Jpn. J. Appl. Phys. P.2 **37** (1998) L309.
201. M. Takeya, K. Yanashima, T. Asano, T. Hino, S. Ikeda, K. Shibuya, S. Kijima, T. Tojyo, S. Ansai, S. Uchida, Y. Yabuki, T. Aoki, T. Asatsuma, M. Ozawa, T. Kobayashi, E. Morita, and M. Ikeda; J. Cryst. Growth **221** (2000) 646.
202. S. Nagahama, N. Iwasa, M. Senoh, T. Matsushita, Y. Sugimoto, H. Kiyoku, T. Kozaki, M. Sano, H. Matsumura, H. Umemoto, K. Chocho, and T. Mukai; Jpn. J. Appl. Phys. P.2 **39** (2000) L647.
203. M. Ikeda and S. Uchida; Phys. Status Solidi A - Appl. Res. **194** (2002) 407.
204. S. Nakamura, M. Senoh, S.-i. Nagahama, N. Iwasa, T. Yamada, T. Matsushita, H. Kiyoku, Y. Sugimoto, T. Kozaki, H. Umemoto, M. Sano, and K. Chocho; Jpn. J. Appl. Phys. P.2 **36** (1997) L1568.
205. K. Domen, R. Soejima, A. Kuramata, K. Horino, S. Kubota, and T. Tanahashi; Appl. Phys. Lett. **73** (1998) 2775.
206. K. Sasanuma and G. Hatakoshi, in Tech. Rep. IEICE Technical Group of Lasers and Quantum Electronics LQE97–152, Tokyo, 1998, p. 49.
207. S. Nakamura, M. Senoh, S. Nagahama, N. Iwasa, T. Yamada, T. Matsushita, H. Kiyoku, Y. Sugimoto, T. Kozaki, H. Umemoto, M. Sano, and K. Chocho; Jpn. J. Appl. Phys. P.2 **37** (1998) L1020.
208. M. Kuramoto, A.A. Yamaguchi, A. Usui, and M. Mizuta; IEICE Trans. Electron. **E83C** (2000) 552.

209. S. Nakamura; IEEE J. Sel. Top. Quantum Electron. **3** (1997) 712.
210. M. Kuramoto, C. Sasaoka, N. Futagawa, M. Nido, and A.A. Yamaguchi; Phys. Status Solidi A - Appl. Res. **192** (2002) 329.
211. M. Takeya, T. Tojyo, T. Asano, S. Ikeda, T. Mizuno, O. Matsumoto, S. Goto, Y. Yabuki, S. Uchida, and M. Ikeda; Phys. Status Solidi A - Appl. Res. **192** (2002) 269.
212. T. Mizuno and et al., in Tech. Rep. IEICE Technical Group of Lasers and Quantum Electronics LQE2002–87, Tokyo, 2002, p. 49.
213. S. Nunoue, M. Yamamoto, M. Suzuki, C. Nozaki, J. Nishio, L. Sugiura, M. Onomura, K. Itaya, and M. Ishikawa; Jpn. J. Appl. Phys. P.1 **37** (1998) 1470.
214. M.F.C. Schemmann, C.J. Vanderpoel, B.A.H. Vanbakel, H. Ambrosius, A. Valster, J.A.M. Vandenheijkant, and G.A. Acket; Appl. Phys. Lett. **66** (1995) 920.
215. T. Tojyo, S. Uchida, T. Mizuno, T. Asano, M. Takeya, T. Hino, S. Kijima, S. Goto, Y. Yabuki, and M. Ikeda; Jpn. J. Appl. Phys. P.1 **41** (2002) 1829.
216. G.E. Bulman, K. Doverspike, S.T. Sheppard, T.W. Weeks, H.S. Kong, H.M. Dieringer, J.A. Edmond, J.D. Brown, J.T. Swindell, and J.F. Schetzina; Electron. Lett. **33** (1997) 1556.
217. A. Kuramata, K. Domen, R. Soejima, K. Horino, S. Kubota, and T. Tanahashi; Jpn. J. Appl. Phys. P.2 **36** (1997) L1130.
218. S. Nakamura, M. Senoh, S. Nagahama, N. Iwasa, T. Yamada, T. Matsushita, H. Kiyoku, and Y. Sugimoto; Appl. Phys. Lett. **68** (1996) 2105.
219. M. Kuramoto, C. Sasaoka, Y. Hisanaga, A. Kimura, A.A. Yamaguchi, H. Sunakawa, N. Kuroda, M. Nido, A. Usui, and M. Mizuta; Jpn. J. Appl. Phys. P.2 **38** (1999) L184.
220. K. Motoki, T. Okahisa, N. Matsumoto, M. Matsushima, H. Kimura, H. Kasai, K. Takemoto, K. Uematsu, T. Hirano, M. Nakayama, S. Nakahata, M. Ueno, D. Hara, Y. Kumagai, A. Koukitu, and H. Seki; Jpn. J. Appl. Phys. P.2 **40** (2001) L140.
221. O. Matsumoto, S. Goto, T. Sasaki, Y. Yabuki, T. Tojyo, S. Tomiya, K. Naganuma, T. Asatsuma, K. Tamamura, S. Uchida, and M. Ikeda, in Proc. "Extended Abstracts of the 2002 International Conference on Solid State Devices and Materials (SSDM)", Business Center for Academic Societies, Japan, Nagoya, 17–19 September 2002, p. 832–833.
222. H. Fujii, Y. Ueno, and K. Endo; Appl. Phys. Lett. **62** (1993) 2114.
223. K. Shinozaki, A. Watanabe, R. Furukawa, and N. Watanabe; J. Appl. Phys. **65** (1989) 2907.
224. S. Tomiya, S. Goto, M. Takeya, and M. Ikeda; Phys. Status Solidi A - Appl. Res. **200** (2003) 139.
225. M. Takeya, T. Mizuno, T. Sasaki, S. Ikeda, T. Fujimoto, Y. Ohfuji, K. Oikawa, Y. Yabuki, S. Uchida, and M. Ikeda; Phys. Status Solidi C **0** (2003) 2292.
226. M. Ikeda, T. Mizuno, M. Takeya, S. Goto, S. Ikeda, T. Fujimoto, Y. Ohfuji, and T. Hashizu; Phys. Status Solidi C **1** (2004) 1461.
227. M. Kneissl, D.P. Bour, L. Romano, C.G. Van de Walle, J.E. Northrup, W.S. Wong, D.W. Treat, M. Teepe, T. Schmidt, and N.M. Johnson; Appl. Phys. Lett. **77** (2000) 1931.
228. R.L. Hartman and R.W. Dixon; Appl. Phys. Lett. **26** (1975) 239.
229. K. Endo, K. Kobayashi, H. Fujii, and Y. Ueno; Appl. Phys. Lett. **64** (1994) 146.

230. S. Nakamura, M. Senoh, N. Iwasa, and S. Nagahama; Jpn. J. Appl. Phys. P.2 **34** (1995) L797.
231. T. Mukai, H. Narimatsu, and S. Nakamura; Jpn. J. Appl. Phys. P.2 **37** (1998) L479.
232. S. Nakamura, M. Senoh, S. Nagahama, N. Iwasa, T. Yamada, T. Matsushita, H. Kiyoku, and Y. Sugimoto; Jpn. J. Appl. Phys. P.2 **35** (1996) L74.
233. S. Masui, Y. Matsuyama, T. Yanamoto, T. Kozaki, S. Nagahama, and T. Mukai; Jpn. J. Appl. Phys. P.2 **42** (2003) L1318.
234. T. Kozaki, T. Yanamoto, T. Miyoshi, Y. Fujimura, S. Nagahama, and T. Mukai, in Proc. "Society for Information Display International Symposium, Seminar & Exhibition (SID'05)", SID, Boston, Massachusetts, May 22–27, 2005, p. 1605–1607.
235. S. Nagahama, M. Sano, T. Yanamoto, D. Morita, O. Miki, K. Sakamoto, M. Yamamoto, Y. Matsuyama, Y. Kawata, T. Murayama, and T. Mukai, in Proc. "Novel In-Plane Semiconductor Lasers II", Vol. 4995, SPIE, San Jose, CA, USA, 2003, p. 108–116.
236. S. Nagahama, T. Yanamoto, M. Sano, and T. Mukai; Jpn. J. Appl. Phys. P.1 **40** (2001) 3075.
237. Y. Narukawa, Y. Kawakami, M. Funato, S. Fujita, S. Fujita, and S. Nakamura; Appl. Phys. Lett. **70** (1997) 981.
238. S. Chichibu, K. Wada, and S. Nakamura; Appl. Phys. Lett. **71** (1997) 2346.
239. M. Takeya, T. Tojyo, T. Asano, S. Ikeda, T. Mizuno, O. Matsumoto, S. Goto, Y. Yabuki, S. Uchida, and M. Ikeda; Phys. Status Solidi A - Appl. Res. **192** (2002) 269.
240. S. Goto, M. Ohta, Y. Yabuki, Y. Hoshina, K. Naganuma, K. Tamamura, T. Hashizu, and M. Ikeda; Phys. Status Solidi A - Appl. Res. **200** (2003) 122.
241. J. Krutmann, H. Stege, and A. Morita, in "Dermatological phototherapy and photodiagnostic methods" (J. Krutmann and et al., eds.), Springer, Berlin-Heidelberg-New York-Tokio, 2001.
242. T. Nishimura, K. Toyoda, M. Watanabe, and S. Urabe; Jpn. J. Appl. Phys. P.1 **42** (2003) 5079.
243. S. Nakamura, M. Senoh, S. Nagahama, N. Iwasa, T. Yamada, T. Matsushita, H. Kiyoku, and Y. Sugimoto; Jpn. J. Appl. Phys. P.2 **35** (1996) L74.
244. M. Kneissl, D.W. Treat, M. Teepe, N. Miyashita, and N.M. Johnson, in Proc. "Novel In-Plane Semiconductor Lasers II", Vol. 4995, SPIE, San Jose, CA, USA, 2003, p. 103–107.
245. K. Iida, T. Kawashima, A. Miyazaki, H. Kasugai, S. Mishima, A. Honshio, Y. Miyake, M. Iwaya, S. Kamiyama, H. Amano, and I. Akasaki; Jpn. J. Appl. Phys. P.2 **43** (2004) L499.
246. J. Edmond, A. Abare, M. Bergman, J. Bharathan, K.L. Bunker, D. Emerson, K. Haberern, J. Ibbetson, M. Leung, P. Russel, and D. Slater; J. Cryst. Growth **272** (2004) 242.
247. J. Holst, L. Eckey, A. Hoffmann, O. Ambacher, and M. Stutzmann; J. Cryst. Growth **190** (1998) 692.
248. T. Takano, Y. Narita, A. Horiuchi, and H. Kawanishi; Appl. Phys. Lett. **84** (2004) 3567.
249. F.A. Kroger; J. Phys. Chem. Solids **26** (1965) 1707.
250. F.A. Kroger and H.J. Vink; J. Phys. Chem. Solids **5** (1958) 208.

251. R.F. Brebrick; J. Phys. Chem. Solids **18** (1961) 116.
252. R.F. Brebrick; J. Phys. Chem. Solids **4** (1958) 190.
253. G. Mandel, F.F. Morehead, and P.R. Wagner; Phys. Rev. **136** (1964) A826.
254. G. Mandel; Phys. Rev. **134** (1964) A1073.
255. R.S. Title, G. Mandel, and F.F. Morehead; Phys. Rev. **136** (1964) A300.
256. A.R. Reinberg, W.C. Holton, M. de Wit, and R.K. Watts; Phys. Rev. B **3** (1971) 410.
257. R.K. Watts, W.C. Holton, and M. de Wit; Phys. Rev. B **3** (1971) 404.
258. T. Mitsuyu, K. Ohkawa, and O. Yamazaki; Appl. Phys. Lett. **49** (1986) 1348.
259. T. Yao, Y. Makita, and S. Maekawa; Jpn. J. Appl. Phys. P.2 **20** (1981) L741.
260. H. Mitsuhashi, I. Mitsuishi, and H. Kukimoto; Jpn. J. Appl. Phys. P.2 **24** (1985) L864.
261. F.A. Ponce, W. Stutius, and J.G. Werthen; Thin Solid Films **104** (1983) 133.
262. K. Ohkawa, T. Mitsuyu, and O. Yamazaki; J. Appl. Phys. **62** (1987) 3216.
263. R.M. Park, M.B. Troffer, C.M. Rouleau, J.M. DePuydt, and M.A. Haase; Appl. Phys. Lett. **57** (1990) 2127.
264. K. Ohkawa, T. Karasawa, and T. Mitsuyu; Jpn. J. Appl. Phys. P.2 **30** (1991) L152.
265. M.A. Haase, J. Qiu, J.M. DePuydt, and H. Cheng; Appl. Phys. Lett. **59** (1991) 1272.
266. H. Okuyama, K. Nakano, T. Miyajima, and K. Akimoto; Jpn. J. Appl. Phys. P.2 **30** (1991) L1620.
267. H. Okuyama, T. Miyajima, Y. Morinaga, F. Hiei, M. Ozawa, and K. Akimoto; Electron. Lett. **28** (1992) 1798.
268. F. Hiei, M. Ikeda, M. Ozawa, T. Miyajima, A. Ishibashi, and K. Akimoto; Electron. Lett. **29** (1993) 878.
269. Y. Fan, J. Han, L. He, J. Saraie, R.L. Gunshor, M. Hagerott, H. Jeon, A.V. Nurmikko, G.C. Hua, and N. Otsuka; Appl. Phys. Lett. **61** (1992) 3160.
270. S. Itoh, N. Nakayama, T. Ohata, M. Ozawa, H. Okuyama, K. Nakano, A. Ishibashi, M. Ikeda, and Y. Mori; Jpn. J. Appl. Phys. P.2 **32** (1993) L1530.
271. M.A. Haase, P.F. Baude, M.S. Hagedorn, J. Qiu, J.M. DePuydt, H. Cheng, S. Guha, G.E. Hofler, and B.J. Wu; Appl. Phys. Lett. **63** (1993) 2315.
272. J.M. Gaines, R.R. Drenten, K.W. Haberern, T. Marshall, P. Mensz, and J. Petruzzello; Appl. Phys. Lett. **62** (1993) 2462.
273. N. Nakayama, S. Itoh, T. Ohata, K. Nakano, H. Okuyama, M. Ozawa, A. Ishibashi, M. Ikeda, and Y. Mori; Electron. Lett. **29** (1993) 1488.
274. N. Nakayama, S. Itoh, H. Okuyama, M. Ozawa, T. Ohata, K. Nakano, M. Ikeda, A. Ishibashi, and Y. Mori; Electron. Lett. **29** (1993) 2194.
275. A. Salokatve, H. Jeon, J. Ding, M. Hovinen, A.V. Nurmikko, D.C. Grillo, L. He, J. Han, Y. Fan, M. Ringle, R.L. Gunshor, G.C. Hua, and N. Otsuka; Electron. Lett. **29** (1993) 2192.
276. G.C. Hua, N. Otsuka, D.C. Grillo, Y. Fan, J. Han, M.D. Ringle, R.L. Gunshor, M. Hovinen, and A.V. Nurmikko; Appl. Phys. Lett. **65** (1994) 1331.
277. S. Guha, J.M. DePuydt, M.A. Haase, J. Qiu, and H. Cheng; Appl. Phys. Lett. **63** (1993) 3107.
278. S. Guha, J.M. DePuydt, J. Qiu, G.E. Hofler, M.A. Haase, B.J. Wu, and H. Cheng; Appl. Phys. Lett. **63** (1993) 3023.
279. J.M. DePuydt, H. Cheng, J.E. Potts, T.L. Smith, and S.K. Mohapatra; J. Appl. Phys. **62** (1987) 4756.

280. M.C. Tamargo, J.L. de Miguel, D.M. Hwang, and H.H. Farrell; J. Vac. Sci.Technol. B: Microelectr. Nanomet. Struct. **6** (1988) 784.
281. W. Xie, D.C. Grillo, R.L. Gunshor, M. Kobayashi, H. Jeon, J. Ding, A.V. Nurmikko, G.C. Hua, and N. Otsuka; Appl. Phys. Lett. **60** (1992) 1999.
282. S. Guha, H. Munekata, and L.L. Chang; J. Appl. Phys. **73** (1993) 2294.
283. L.H. Kuo, L. Salamanca-Riba, B.J. Wu, G. Hofler, J.M. DePuydt, and H. Cheng; Appl. Phys. Lett. **67** (1995) 3298.
284. H. Okuyama, Y. Kishita, T. Miyajima, A. Ishibashi, and K. Akimoto; Appl. Phys. Lett. **64** (1994) 904.
285. S. Itoh, N. Nakayama, T. Ohata, M. Ozawa, H. Okuyama, K. Nakano, M. Ikeda, A. Ishibashi, and Y. Mori; Jpn. J. Appl. Phys. P.2 **33** (1994) L639.
286. S. Taniguchi, T. Hino, S. Itoh, K. Nakano, N. Nakayama, A. Ishibashi, and M. Ikeda; Electron. Lett. **32** (1996) 552.
287. S. Itoh, K. Nakano, and A. Ishibashi; J. Cryst. Growth **214** (2000) 1029.
288. A. Toda, K. Nakano, and A. Ishibashi; Appl. Phys. Lett. **73** (1998) 1523.
289. S. Tomiya, H. Okuyama, and A. Ishibashi; Appl. Surf. Sci. **159** (2000) 243.
290. S. Tomiya, R. Minatoya, H. Tsukamoto, S. Itoh, K. Nakano, E. Morita, and A. Ishibashi; J. Appl. Phys. **82** (1997) 2938.
291. F. Fischer, G. Landwehr, T. Litz, H.J. Lugauer, U. Zehnder, T. Gerhard, W. Ossau, and A. Waag; J. Cryst. Growth **175** (1997) 532.
292. F. Fischer, M. Keller, T. Gerhard, T. Behr, T. Litz, H.J. Lugauer, M. Keim, G. Reuscher, T. Baron, A. Waag, and G. Landwehr; J. Appl. Phys. **84** (1998) 1650.
293. G. Landwehr, A. Waag, F. Fischer, H.J. Lugauer, and K. Schull; Physica E **3** (1998) 158.
294. S.B. Che, I. Nomura, A. Kikuchi, and K. Kishino; Appl. Phys. Lett. **81** (2002) 972.
295. M. Klude and D. Hommel; Appl. Phys. Lett. **79** (2001) 2523.

4

Electronic Devices

4.1 High Frequency Power Devices

4.1.1 Nitride Semiconductor High Frequency Power Devices (H. Miyamoto)

Applications

There are many potential applications of nitride-semiconductor (GaN-based FETs) electron devices because of their extremely high performance at high voltages and high frequencies [1–3]. This operation is possible because of the high breakdown electric field ($3 \times 10^6\,\mathrm{V\,cm^{-1}}$, ten times larger than Si and GaAs) and electron drift velocity ($2 \times 10^7\,\mathrm{cm\,s^{-1}}$) at high electric fields of GaN. These properties of GaN offer major advantages in the trade-off between operating voltage and cut-off frequency that limit the operation of conventional field effect transistors. Further, the polarization effect at the AlGaN/GaN heterojunction interface results in the creation of an electron carrier density of approximately $10^{13}\,\mathrm{cm^{-2}}$; a value that is five times larger than that of AlGaAs/GaAs heterostructures. These characteristics of high frequency, high voltage and high carrier density are particularly effective in enabling the fabrication of smaller devices delivering high output power at high frequencies.

A particularly important application of nitride semiconducting high frequency devices is for microwave power transmission amplifiers for third generation mobile telephone base stations [4–7]. These applications necessitate highly linear device characteristics for amplification of digitally modulated signals. Further, the use of nitride semiconductors enables high power output amplifier modules that incorporate simplified voltage–voltage converter and output matching circuits with high efficiency.

There is also a growing demand for wireless technology for use in broadband internet communication such as fixed wireless access (FWA) and satellite internet access networks. These applications require high efficiency amplifiers with power ratings of $\sim$10 W, operating in the 22–38 GHz frequency range [8, 9]. The traveling wave tube amplifiers used at present in this region are expected to be replaced by nitride semiconductor devices within the next decade.

Device Structure

The main feature of nitride high-frequency power devices is the two-dimensional electron gas at the AlGaN/GaN heterojunction. The first report on the fabrication and operation of AlGaN/GaN heterojunction FETs was by Khan et al. in 1994 [10].

Figures 4.1 and 4.2 shows the cross-sectional diagram of an AlGaN/GaN heterojunction FET and the energy band diagram under the gate electrode, respectively. A high carrier density of $\sim 10^{13}\,\mathrm{cm}^{-2}$ is generated at the AlGaN/GaN heterointerface due to spontaneous and piezoelectric polarization effects [11]. The combination of the large carrier density and high breakdown voltage enables high power output operation. Further, it is possible to realize devices operating in the microwave and millimeter region with this material system because the distance between the gate electrode and the two-dimensional electron gas layer can be reduced to less than 20 nm by varying the thickness of the AlGaN Schottky layer, which enables fabrication channel sizes of a few hundred nanometers.

During the mid-1990s, high resistivity, low cost, and large area sapphire substrates were used for growth of nitride semiconductor devices. Recently, in order to reduce device heating, silicon carbide (SiC) has become the material of choice for substrates because of its high thermal conductivity. However, in spite of problems related to the high cost, low controllability on resistivity,

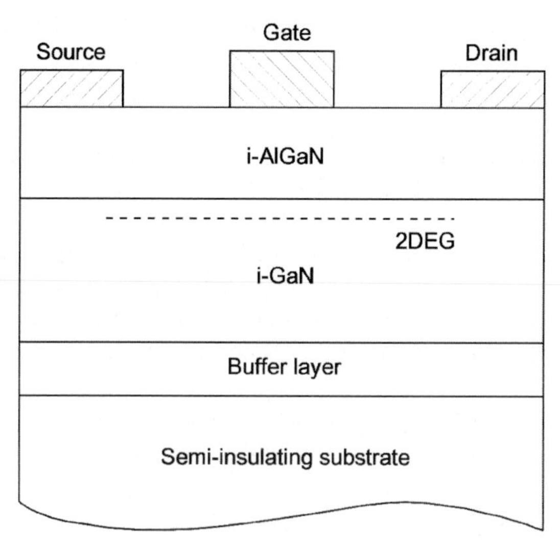

Fig. 4.1. Cross-sectional diagram of an AlGaN/GaN heterojunction FET

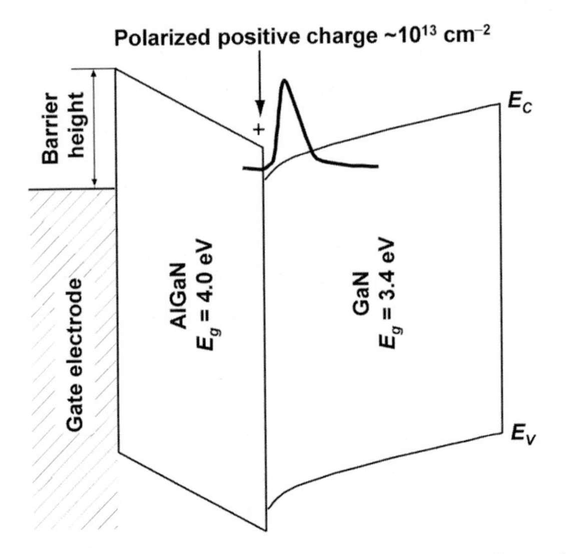

Fig. 4.2. Energy band diagram under the gate electrode

and high dislocation density of SiC substrates, the growth of AlGaN/GaN heterojunction FETs on SiC substrates shows tremendous potential for high frequency devices [4–6]. In 2005, the use of low cost, 4-in. Si substrates [7], as well as high quality GaN and AlN single crystal substrates [12], were reported.

Device Processing

Figure 4.3 shows a typical process used for fabricating AlGaN/GaN heterojunction FETs:

1. Ti/Al-based electrodes are evaporated followed by formation of ohmic contacts by a short period annealling at 800–900°C.
2. The device active layers are covered with a photoresist mask and nitrogen (N) ion implantation is used for producing device isolation.
3. A passivation layer of silicon nitride (SiN) is deposited by plasma CVD.
4. A window is opened in the SiN for the gate region. For recess structures, the procedure is modified by first using the SiN as a mask and removing a part of the AlGaN layer by BCl_3 plasma-dry etching [13].
5. Ni/Au gate electrode is deposited by electron beam evaporation. For base station applications, gate lengths of 0.7–10 μm are used and for millimeter wave applications electron beam lithography is used to form 0.25 μm gate lengths [14].
6. Finally, SiN is deposited as a passivation layer and gold electrodeposition used to form air-bridges for multifinger device structures of the high output power FETs.

Device processing still has several issues to resolve including reduction of gate leakage currents and minimization of ohmic contact resistance.

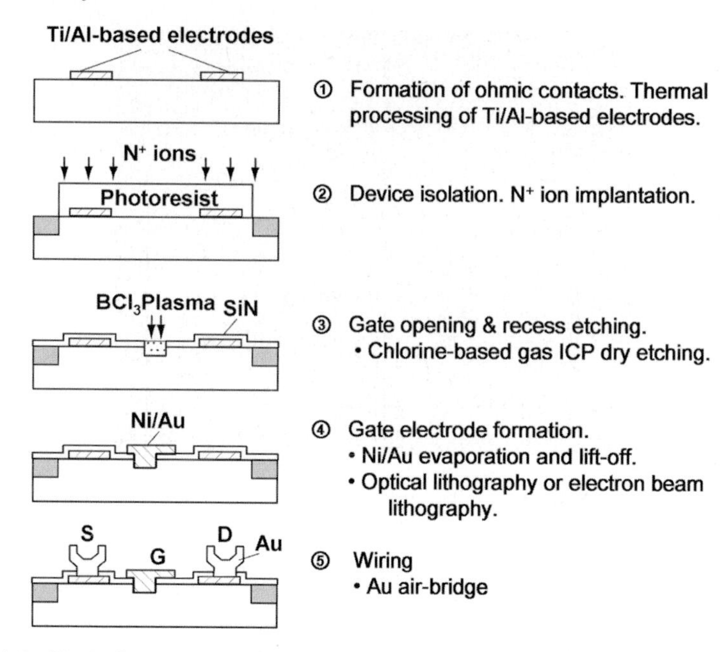

Fig. 4.3. Typical process used for fabricating AlGaN/GaN heterojunction FETs

Crystalline defects in the AlGaN layer have been reported to be a source of leakage currents in AlGaN/GaN devices [15]. In addition to the demand for the improvements in AlGaN layers, there have also been reports on the use Pt and Pd, which have higher Schottky barriers than Ni/Au [16], to reduce the gate leakage current.

The contact resistance can be reduced by use of multilayered metal ohmic structures. For example, compared with a contact resistance of $10^{-5}\ \Omega\,\text{cm}^2$ for Ti/Al electrodes that were annealed at 600°C, values of $\sim 10^{-6}\ \Omega\,\text{cm}^2$ were reported for Ti/Al/Ni/Au, Ti/Al/Mo/Au and Ti/Al/Nb/Au layers annealed at 800–900°C [17].

Improving High Power Performance

Power density of a transistor is estimated to $1/8 \times I_{max} \times BV_{gd}$ under class A operation (where I_{max} is the maximum drain current density, BV_{gd} the Break down voltage).

At a typical I_{max} of $1\,\text{A}\,\text{mm}^{-1}$ and a BV_{gd} of 100 V, the RF output power is expected to be $12.5\,\text{W}\,\text{mm}^{-1}$; a figure of merit that is 10 times that of Si and GaAs high frequency power devices. Problems related to self-heating and current collapse must be resolved for stable high voltage/high power output operation of high frequency power nitride semiconductor devices. Current collapse is the term used to describe the observed large decrease in current at high voltage operation (Fig. 4.4). RF device performance is less than expected from DC characteristics. The origins of

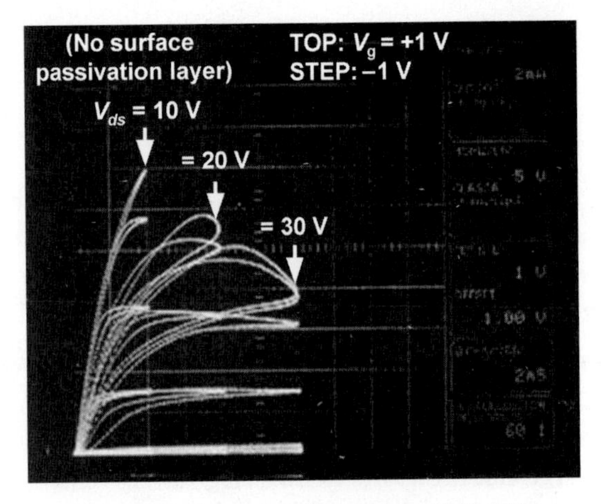

Fig. 4.4. Current collapse of a GaN FET

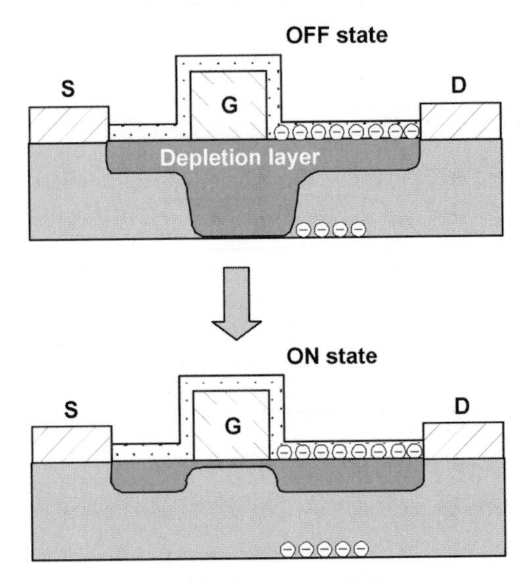

Fig. 4.5. FET under RF operating conditions

current collapse have been reported as being due to deep levels in the AlGaN and GaN layers and at the surface of AlGaN [18], where, under high voltage operating conditions, electrons are injected from the gate electrode into the AlGaN surface or GaN buffer layer and trapped in deep levels, leading to the creation of a depletion layer and subsequent current collapse (Fig. 4.5).

Current collapse can be reduced by depositing a SiN passivation layer on the AlGaN surface [19]. However, when SiN passivation layers are used, a tradeoff arises between current collapse and the maximum operating bias voltage. This is because although the SiN layer is a solution to problems related to current collapse, there is a significant decrease in the breakdown

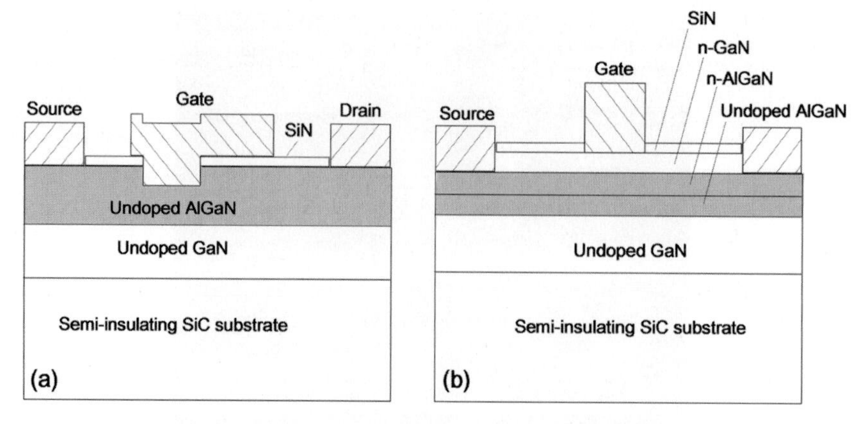

Fig. 4.6. Structures for reducing current collapse

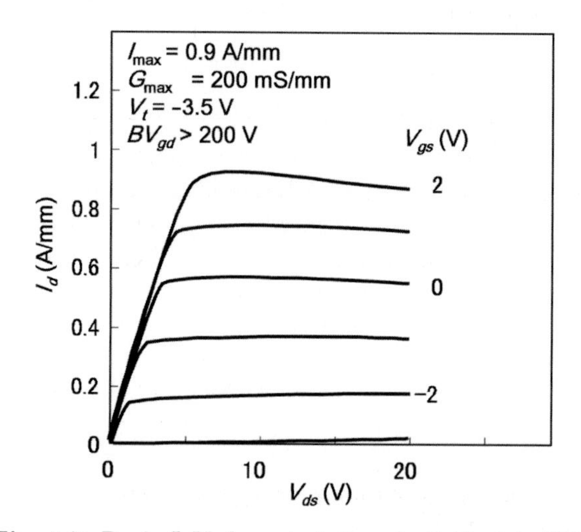

Fig. 4.7. Drain I–V characteristics of a field plate FET

voltage. Figure 4.6 shows a means of resolving the trade-off problem between current collapse and applied voltage where a "field plate" (FP) electrode or an n-GaN gap structure [5, 6] is used to prevent the electric field becoming excessively high near the gate electrode. This method reduces the injection of electrons from the gate electrode into the AlGaN surface.

DC Characteristics of AlGaN/GaN Heterojunction FETs

Figure 4.7 shows the DC characteristics of a recessed FP electrode structure in a AlGaN/GaN FET with a gate length of 1.0 μm. The maximum drain current density is 900 mA mm^{-1}, the threshold voltage −3.5 V, and the transconductance 200 mS mm^{-1}. The gate–drain breakdown voltage is 200 V, a figure of merit that is four times larger than devices without an FP electrode [6].

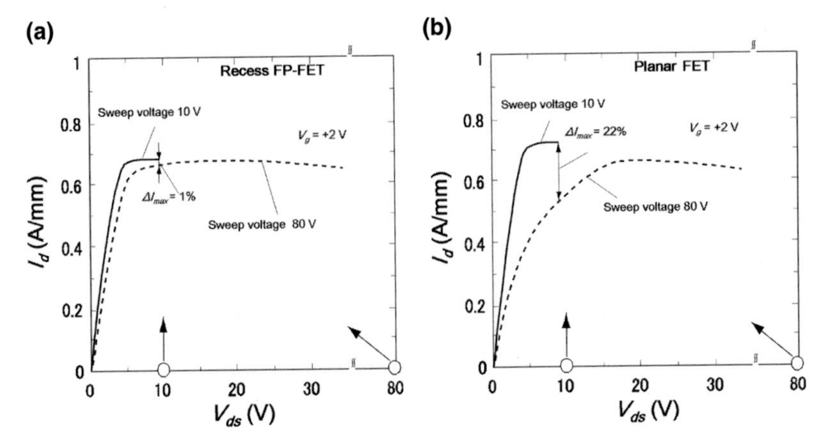

Fig. 4.8. (**a**) Recess structure FP–FET; (**b**) Planar FET without FP

Next, the electrical characteristics of a recessed structure FP–FET and planar structure FET without a FP electrode are described [20]. Figure 4.8, shows the variation of the maximum drain current density (I_{max}) with sweep voltage for the two FET structures. In these measurements, the drain voltage was swept between 0–10 and 0–80 V and the different values of I_{max} were measured at a drain voltage of 10 V. Current collapse was almost negligible for the recessed structure FP–FET (Fig. 4.8a), but the planar structure without FP showed a 20% decrease in current over the 0–80 V sweep at a drain voltage of 10 V (Fig. 4.8b). These results show that the FP structure significantly reduces current collapse together with even better results for recess structures [20].

Output Power Characteristics of AlGaN/GaN Heterojunction FETs

Figure 1.10 shows the increases of the output power of single-chip nitride semiconductor high frequency devices, over the last eight years. The first AlGaN/GaN FET reported by Khan et al. in 1994 was developed for 4–10 GHz bandwidth radar applications. In Japan, there was a dramatic increase in publications from 2001 due to demand for 2 GHz power devices for third generation mobile telephone base stations. Recently, there have also been reports on the power characteristics of devices operating in the 5 GHz bandwidth for use in next generation mobile telephone technology [21]. From 2001 to 2003, the increases in the maximum power output of devices can be attributed to improvements of the crystalline quality of GaN buffer and AlGaN Schottky layers as well as the suppression of surface deep levels by use of SiN passivation layers. Then, in the 2003–2004 period, device performance was further improved by reduction of electric field concentration about the gate using field plate electrode structures [6] and n-GaN cap layers [5]. Development of

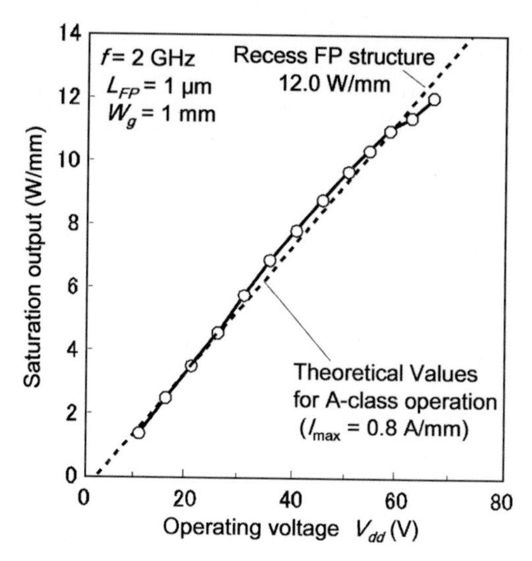

Fig. 4.9. Variation of saturation power with operating voltage for a device with a gate width of 4 mm. The *broken line* shows the theoretical operating characteristics of a Class-A FET

device pattern layouts for better heat dissipation also enabled the implementation of large area device structures. Recently, devices with power outputs of 200 W have been reported [5, 6]. Figure 4.9 shows the variation of saturation power (P_{sat}) with operating voltage (V_{dd}) for a device with a gate width of 4 mm, where the broken line shows the theoretical calculation under Class-A FET operation. The output power increases up to 66 V because the recessed structure field plate suppresses current collapse [13].

Enhancement of Operating Frequency

Further increases in the operating frequency require reduction of the gate length while maintaining a satisfactory aspect ratio of the gate length to gate-to-channel distance. There have been reports on the use of electron beam lithography to fabricate 50 nm devices exhibiting current gain cutoff frequencies (f_T) of 110 GHz and maximum oscillation frequencies (f_{max}) of ~140 GHz [22]. Typical nitride semiconductor millimeter-wave power devices for the 30 GHz band have T-shaped gates, with lengths of 0.25 μm and widths of 1 mm [14]. At a drain bias of 30 V, the saturated power output is 5.8 W mm^{-1}, power efficiency is 43% and linear gain 9.2 dB. A single gallium nitride FET can provide several watts of power; such performance is not possible with conventional GaAs-FETs.

Higher frequency operation will require still further reduction in the thickness of the AlGaN layer and greater carrier confinement at the AlGaN/GaN heterointerface.

Future Applications

Nitride semiconductor power devices are developing for use in power amplifiers of W-CDMA mobile telephone base stations. A push–pull amplifier with two 36 mm devices and peak saturation power of 250 W was reported [5].

AlGaN/GaN FETs, when combined with compensation circuits, satisfy W-CDMA specifications (adjacent channel leakage power of −50 dBc) exhibiting drain efficiencies of 35% [5, 7]. Studies are in progress to reduce the cost of devices by combining high resistance AlN buffer layers with low cost n-SiC substrates [23] and high resistance Si substrates [5, 7].

Summary

The present status and future prospects of GaN high frequency power devices was presented. Suppression of current collapse by use of field plates has enabled the fabrication of high frequency, high power AlGaN/GaN FETs operating at 2 GHz, with an output power of 200 W. Further, electron beam lithography has enabled fabrication of devices with gates lengths of 0.25 µm operating at 30 GHz with an output power of 5.8 W. Such figures of merit are not possible using single-chip GaAs technology. Nitride semiconductor AlGaN/GaN devices are being aggressively developing for use in third generation (3G), 2 GHz mobile telephone base stations.

4.1.2 SiC (M. Arai)

Efficient, high-voltage electronic devices are required for the further proliferation of mobile communications and as alternative, solid state microwave sources to replace vacuum tubes for radar and heating equipment. Use of wide bandgap semiconductors to replace Si and GaAs electronics is increasing for high frequency devices to go beyond current limits of performance.

Silicon carbide (SiC) is a promising wide bandgap semiconductor for microwave applications because of its unique physical properties (1) its thermal conductivity is three times that of Si and ten times larger than GaAs, (2) the electron saturation velocity is two times larger than Si, and (3) and the breakdown electric field is ten times larger than Si. The characteristics of high-power, high-frequency SiC devices are described in this section.

Metal-Semiconductor FETs (MESFETs)

Figure 4.10 shows the cross-section of a MESFET structure that consists of a buffer, n^--channel (300 nm, 3.0×10^{17} cm^{-3}) and n^+ (200 nm, 10^{20} cm^{-3}) layers, grown epitaxially on a semi-insulating SiC substrate.

A typical SiC MESFET structure is fabricated by the following process (1) a dry etch to produce a mesa for device isolation and a recess for forming the gate; (2) electron beam deposition of Ni for forming the source and drain

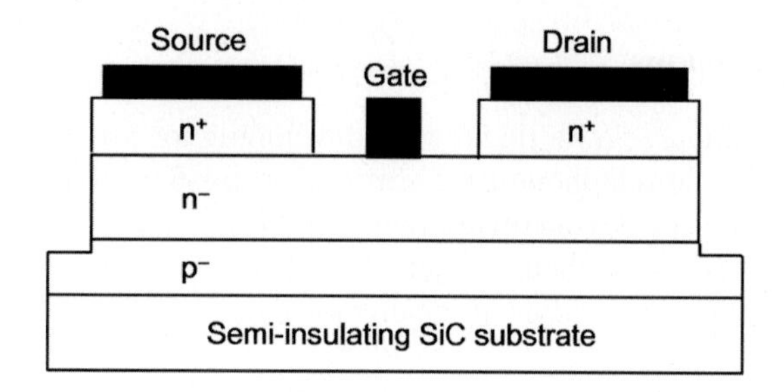

Fig. 4.10. Cross-sectional structure of a SiC-MESFET

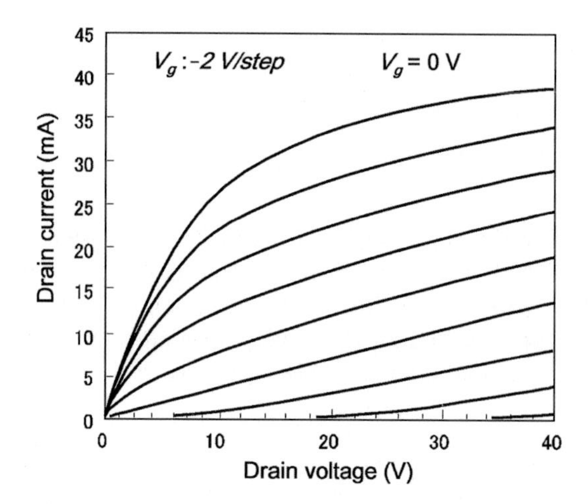

Fig. 4.11. DC characteristics of a SiC-MESFET

electrodes followed by an anneal at 1,000°C to alloy the Ni with the SiC and form ohmic contacts; and finally, (3) deposition of gate electrodes by electron beam evaporation and lift-off. Thus MESFET structures are simple to produce because they do not require complicated processes such as ion beam implantation and high temperature annealing.

The principle of operation is as follows. Current flows along the channel layer from the drain to the source electrode. The channel current is controlled by the gate voltage which changes the width of the depletion layer formed by the Schottky barrier at the gate/SiC interface. Application of a negative gate voltage causes an increase of the depletion layer which results in a decrease of the drain current. The operating speed of MESFETs is limited by the time taken for channel electrons to travel under the gate electrode. Thus it is possible to achieve extremely high frequency operation by reducing the gate length.

Figure 4.11 shows the dc characteristics of a SiC MESFET with a gate length of 0.5 μm, a gate width of 100 μm and a gate–drain separation of 1.5 μm.

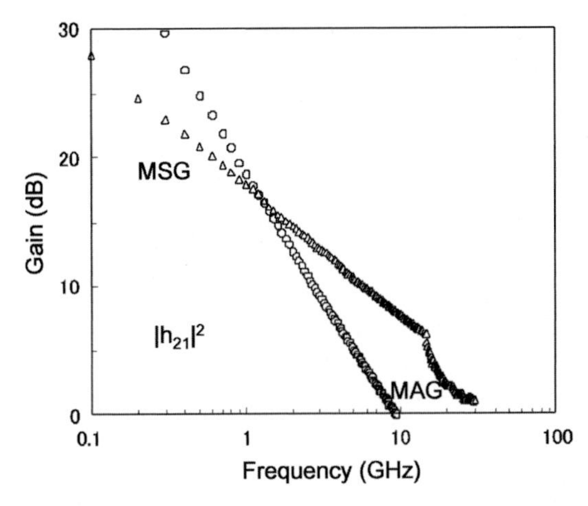

Fig. 4.12. Small signal characteristics of a SiC-MESFET

At a drain voltage of 40 V and drain saturation current of 38.7 mA, the transconductance is 23.5 mS mm^{-1}. The off-breakdown-voltage of this device exceeded 100 V.

Figure 4.12 shows the small signal characteristics of a MESFET with a gate length of 0.5 μm and total gate width of 100 μm. The maximum oscillation frequency (f_{max}) and cut-off frequency (f_t) are used as figures of merit to compare the performance of MESFETs. The f_{max} is the frequency where the maximum available gain (MAG) equals one (0 dB) and f_t is the frequency where the current gain |h21| equals one (0 dB). From the small signal characteristics shown in Fig. 4.12, this MESFET has an f_t of 9.3 GHz and f_{max} of 34.0 GHz. The maximum values of f_t and f_{max} reported to date are 22 GHz and 50 GHz, respectively [24]. MESFETs with following output power characteristics have been reported (1) 120 W at 3.1 GHz under pulsed conditions and 80 W for continuous (CW) operation; and (2) 1.1 W at 10 GHz CW with power added efficiency of 20% and power gain of 9 dB [25,26]. Further, a device with normalized output power of 7.2 W mm^{-1} at a drain voltage of 70 V and frequency of 3.0 GHz has also been reported [26]. Figure 4.13 shows a comparison of the power output of SiC-MESFETs with Si and GaAs devices. The SiC-MESFETs are seen to exhibit a large output power density at high voltages. The main features of transmitters incorporating SiC-MESFETs would be (1) the possibility of reducing the surface area of devices; (2) matching circuits being easier to produce due the large impedance of the input/output circuits; and (3) simplification of power conversion circuits for power supplies. Also, lifetime tests have shown that the drain current of a 10 W class SiC-MESFET only decreased by 2.5% after 1,150 h, with the junction temperature held at 165°C [26]. However, the drain current has been found to show hysteresis due to applied voltage stress. DLTS measurements have shown this hysteresis to be strongly linked with the formation of deep levels due to crystalline defects in the SiC substrate and the channel layer.

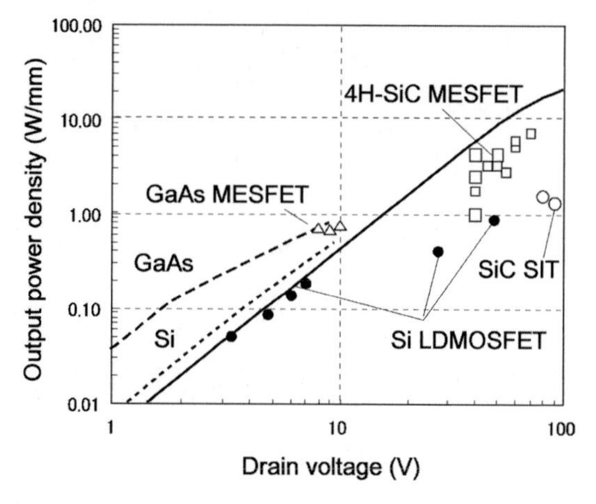

Fig. 4.13. Output power density of high frequency semiconductor devices

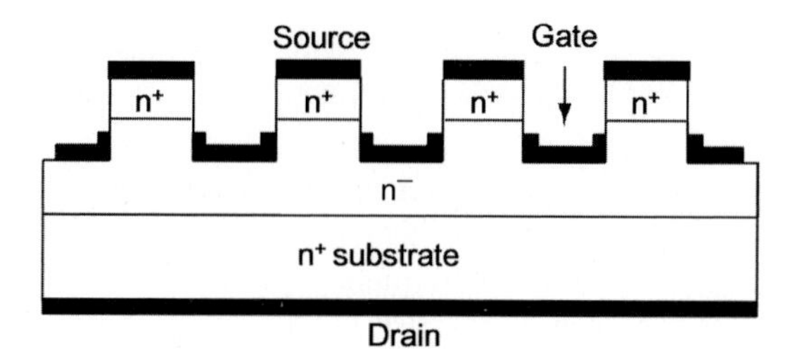

Fig. 4.14. Cross-section of a SiC-SIT

Static Induction Transistor (SIT)

Figure 4.14 shows the cross-section of a static induction transistor (SIT), where, since the drain current flows in the vertical direction toward the source, the presence of defects on the device surface or at the epilayer/substrate interface is less likely to have adverse affects on device performance. Other features of SIT structures include the possibility of attaining a larger current density compared with MESFET devices and the use of conducting substrates will enable reductions of manufacturing costs.

However, certain problems must be overcome for practical applications of SIT structures (1) it is necessary to reduce the size of the source and gate electrodes in order improve high frequency performance, which will require extremely precise alignment; (2) formation of gate electrodes in trenches with large aspect ratios; and (3) difficulties in reproducibility and yield of the process. Thus, at this stage, the main applications are thought to be the relatively low frequency, high power output devices for UHF and L-band components. So far, SIT devices with source mesa widths of 0.5 μm and gate trench widths of 1.0 μm have shown f_t values of 7.0 GHz. SIT modules with an average output of 200 W and a pulse peak of 1 kW are being developed for high

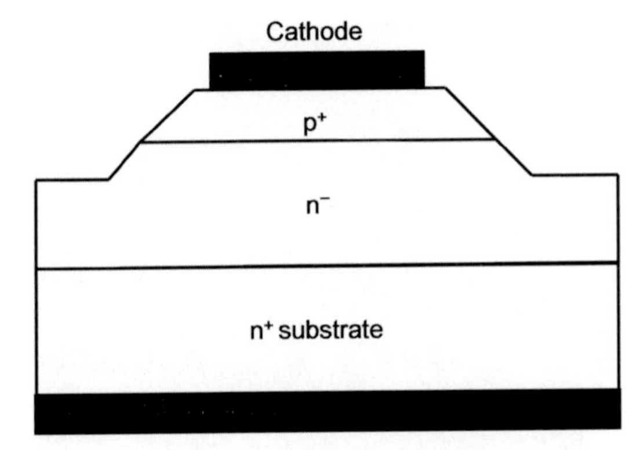

Fig. 4.15. Cross-section of a SiC-IMPATT diode

definition TV (HDTV) and a HDTV broadcast demonstration was carried out in 1996 [27].

IMPATT Diodes

Impact ionization and time transit diodes are well known because they exhibit negative resistance at microwave frequencies, avalanche breakdown and transit time effects. IMPATT diodes have simple structures and show large continuous oscillation output above the millimeter bandwidth but are known to have a large noise figure due to fluctuations in the avalanche breakdown. These devices are currently used in airborne radar systems.

Figure 4.15 shows the structure of a $p^+/n^-/n$ type SiC IMPATT diode. The epitaxial n^- layer is 2.2 μm thick with an impurity density of $10^{17}\,cm^{-3}$. The IMPATT diode was put into a cavity oscillator and the oscillation spectrum measured (Fig. 4.16). A peak output of 1.8 W was obtained at an oscillation frequency of 11.93 GHz [28]. These results are still inferior to those of Si and GaAs IMPATT devices but improvements are expected through reducing the relative resistance and contact resistance of the p-type layer.

4.1.3 Diamond (N. Fujimori, H. Kawarada)

High-Power, High-Frequency Diamond Field Effect Transistors

Diamond has a breakdown electric field of $10^7\,V\,cm^{-1}$ (three times that of SiC and GaN) and a thermal conductivity of $22\,W\,(cm\,K)^{-1}$ (four times SiC and five times GaN).

Figure 4.17 shows a comparison for the cut-off frequency (f_T) dependence of the maximum operating voltage (V_{max}) of diamond with other semiconductors. The product $f_T V_{max}$ offers a practical evaluation of the performance of

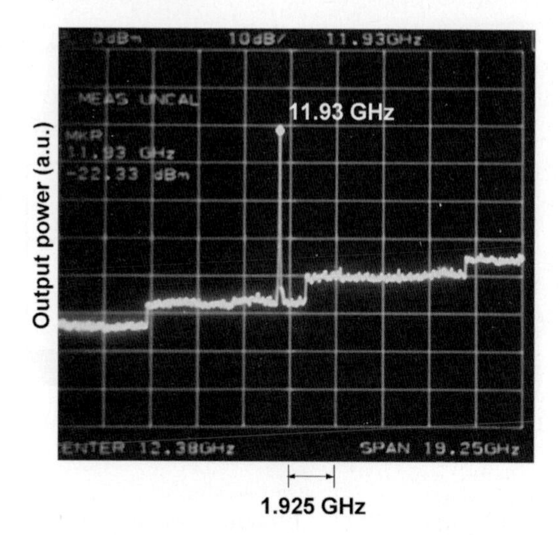

Fig. 4.16. Oscillation spectrum of SiC-IMPATT diode

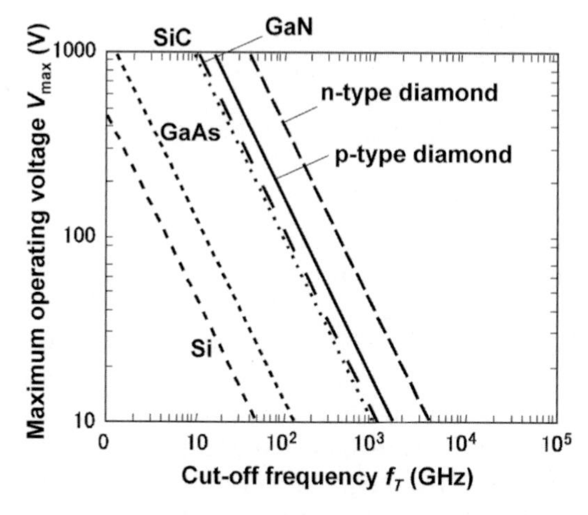

Fig. 4.17. Comparison of the cut-off frequency (f_T) dependence of the maximum operating voltage ($V_{\max}$) of diamond with other semiconductors

a transistor operated under conditions of saturated carrier velocity and maximum gate voltage. The following equations show the relationship between the device operation parameters:

$$f_T = \frac{\nu_s}{2\pi L_g} \tag{4.1}$$

$$V_{\max} = E_b L_g \tag{4.2}$$

Here, ν_s is the saturation carrier concentration, E_b the breakdown electric field and L_g the gate length. Since the magnitudes of ν_s and E_b are material dependent, the product $f_T V_{\max}$ has a constant value for each material.

Diamond has the largest E_b of all the materials and v_s is comparable to the other semiconductors. Thus the $f_T V_{max}$ product is extremely high. Equation (4.2) is the maximum voltage in the case of a uniform electric field distribution in the channel but in real devices the electric field has a maximum on the drain side and the actual maximum voltage is less than that given by V_{max}. However, diamond still has tremendous potential for high-power, high-frequency device applications compared with other semiconductors.

Generally, the cut-off frequency, f_T is written in terms of the transconductance, g_m, and the gate capacitance, C_g:

$$f_T = \frac{g_m}{2\pi C_g} \tag{4.3}$$

This equation shows that low C_g and large g_m are necessary for high frequency operation. Reduction of the gate length is effective in achieving these requirements. Large magnitudes of g_m are achieved in Si-MOSFETs and modulation doped FETs with two-dimensional electron gas structures by the use of thin (~10 nm) conducting channels.

Two-dimensional conducting structures in diamond have been reported as a result of p-type accumulation layers resulting from hydrogen terminated surfaces [29]. The surface carrier concentration in such structures was reported to be $\sim 10^{13}\,\mathrm{cm}^{-2}$ [30] with 90% of the carriers existing within 10 nm of the surface [31]. The sheet resistance was $\sim 10\,\mathrm{k\Omega/\square}$. If MESFET or MISFET structures are formed on diamond, as shown in Fig. 4.18, a high g_m can be attained by controlling the carrier density through varying the surface potential. It is possible to use SiO_2, CaF_2 [32], and Al_2O_3 [33] as gate insulation materials. Boron delta doping has also been used for producing two-dimensional conducting structures [34].

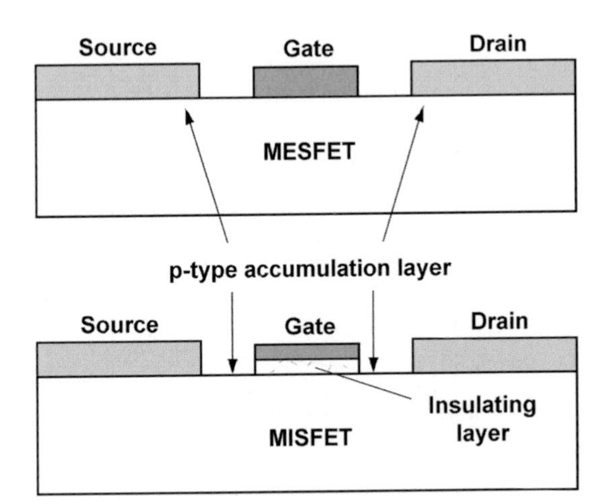

Fig. 4.18. Hydrogen terminated MESFET and MISFET structures

In the p-type accumulation layer of the hydrogen terminated surface, ohmic contacts with resistances of less than 10^{-5} $\Omega\,\mathrm{cm}^2$ have been obtained using metals with large work functions. Since the sheet resistance, ρ_s, is $\sim 10\,\mathrm{k}\Omega/\square$, it would be possible to pass drain currents greater than $100\,\mathrm{mA\,mm}^{-1}$ by reducing the source– and gate–drain separations to less than $1\,\mu\mathrm{m}$ to reduce parasitic resistances. Also, insulating layers formed at low temperature do not damage the surface, leading to a reduction of the surface state density ($<10^{11}\,\mathrm{cm}^{-2}$) [32] of the bandgap. The high frequency characteristics of MESFET and MISFET device structures incorporating hydrogen terminated, p-type accumulation layers and exhibiting $g_m \sim 100\,\mathrm{mS\,mm}^{-1}$ [35] are being studied for practical applications [36, 37].

Carrier Transport in Diamond MISFETs

Transconductance

Figure 4.19 shows the DC characteristics of MISFET structures with a $0.5\,\mu\mathrm{m}$ gate length. The pinch off and saturation regions are clearly visible. The g_m is $165\,\mathrm{mS\,mm}^{-1}$, the largest value reported for a diamond FET [38]. The channel mobility is $\sim 200\,\mathrm{cm}^2\,\mathrm{Vs}^{-1}$. Further improvements in g_m will require fabrication of devices with accumulation layers with even lower series resistances.

Electric Field Dependence of the Carrier Velocity

Figure 4.20 shows is a comparison of the electric field dependence of carrier velocity for various semiconductors. The results were obtained mainly by time of flight measurements. The maximum velocity is reached at low electric fields for semiconductors with large carrier mobilities, such as GaAs and InGaAs.

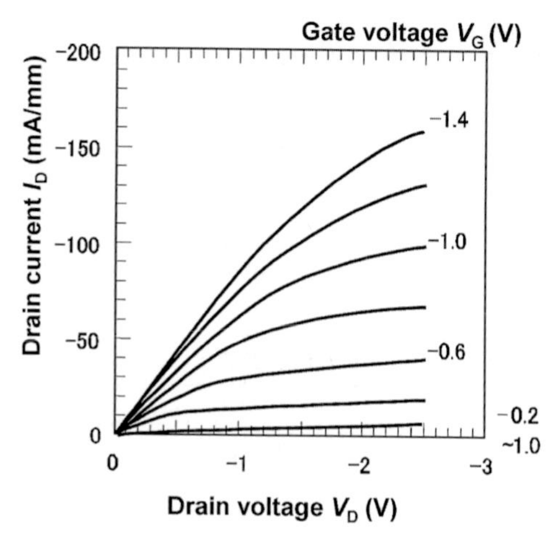

Fig. 4.19. DC characteristics of MISFET structures with a $0.5\,\mu\mathrm{m}$ gate length

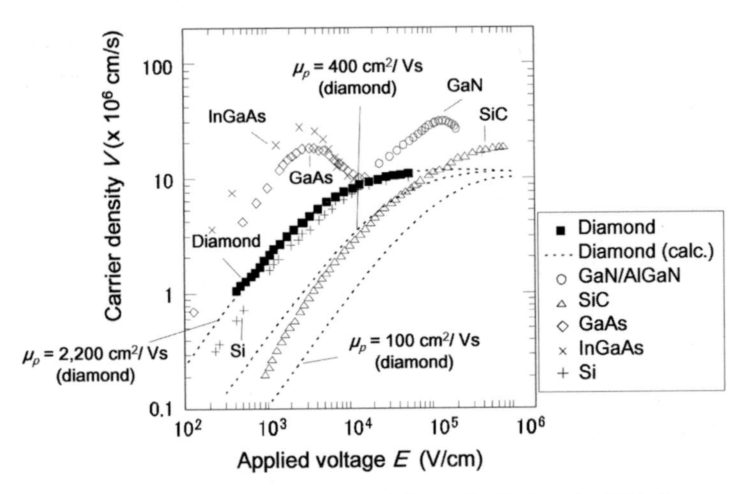

Fig. 4.20. Variation of carrier density with applied electric field for a range of semiconductors

The mobility of carriers in wide bandgap semiconductors (diamond, GaN, and SiC) is not as high and thus a larger electric field is required to achieve the maximum carrier velocity. The relationship between the carrier velocity, v, and carrier mobility, μ can be approximated by the following equation [39]:

$$\nu(E) = \frac{\mu E}{\left[1 + \left(\frac{\mu E}{\nu_s}\right)^a\right]^{\frac{1}{a}}} \tag{4.4}$$

Here, v_s is the saturation carrier velocity and a is the coefficient. Fitting the experimental results [40] to this equation gives a hole mobility of 2,200 cm^2 Vs^{-1}, a hole saturation velocity v_s of 1.2×10^7 cm s^{-1} and a value of a of 0.998. Figure 4.20 also shows the v–E relationship, assuming $\mu = 400$ and 100 cm^2 Vs^{-1}, for the same values of v_s and a. It can be seen that even if the mobility is small, increasing the operating voltage enables high frequency operation near the region of saturation velocity.

The channel mobility determined from an actual MISFET diamond device with a 0.2 μm gate length yields values of μ in the range 200–300 cm^2 Vs^{-1} [38].

Analysis of Carrier Transport in Diamond Transistors Based on Cut-off Frequency

Figure 4.21 shows the relationship between the gate length and cut-off frequency [41], where the results for AlGaN/GaN are also shown for comparison with the diamond FETs. The dotted line shows $f_T - L_g^{-1}$ results calculated using (4.1) for values of v_s between 10^6 and 2×10^7 cm s^{-1}. For a diamond MISFET incorporating a CaF_2 insulating gate the cut-off frequency exhibits an $f_T \propto L_g^{-1}$ dependence on the gate length for L_g between 1 and 0.5 μm. The v_s is 5×10^6 cm s^{-1}. Further, when Al_2O_3 gate insulator is used,

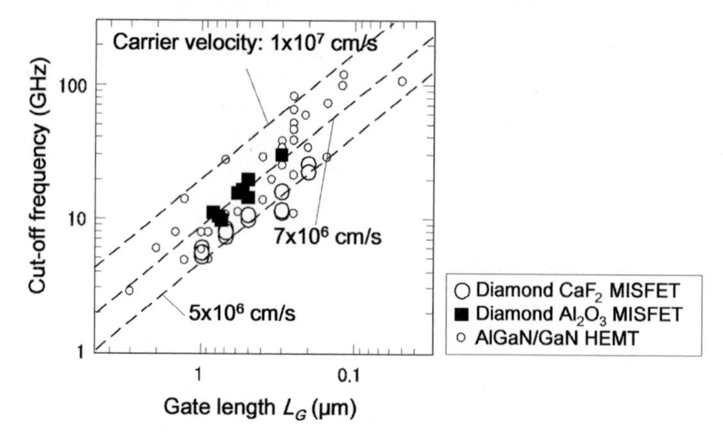

Fig. 4.21. Comparison of the gate length dependence of the cut-off frequency for AlGaN/GaN the diamond FETs

the results lie on a $7 \times 10^6\,\mathrm{cm\,s^{-1}}$ line because it is possible to produce a much thinner layer of Al_2O_3, yielding a much higher g_m. The maximum f_T reported for a diamond MISFET with a gate length of 0.3 μm was 30 GHz [42] and that for a MESFET with a gate length of 0.2 μm was 25 GHz [43]. Further increases in f_T will require lower parasitic components as well as shorter gate lengths. If a saturation velocity of $7 \times 10^6\,\mathrm{cm\,s^{-1}}$ were possible, then it would be possible to realize devices operating at 40 GHz at an L_g of 0.2 μm and 80 GHz at 0.1 μm.

The f_T of diamond FETs with p-type accumulation layers is comparable to SiC-MESFETs, but still about one third that of AlGaN/GaN-based devices. The reason for the poor results compared with nitrides is that the carrier mobility and the hole saturation velocity for diamond are lower than for the electrons in the channels of AlGaN/GaN structures.

Present Status and Future Prospects of High Frequency Diamond Devices

Diamond FET devices (metal/insulator/semiconductor) using hydrogen terminated p-type accumulation layer channels exhibit transconductance of $\sim 100\,\mathrm{mS\,mm^{-1}}$ and cut-off frequencies of 20 GHz. Analysis of the cut-off frequency and applied gate fields show the channel mobility to be $200\,\mathrm{cm^2\,Vs^{-1}}$ and the average velocity to be $7 \times 10^6\,\mathrm{cm\,s^{-1}}$ (~70% of the saturation velocity). At a gate length of 0.2 μm, devices show an f_T of 40 GHz with expectations of a maximum frequency, f_{max} of 100 GHz. Power outputs of $2.1\,\mathrm{W\,mm^{-1}}$ have been reported [43].

Issues remaining to be resolved include the formation of shallow dopant levels and the development of technology for the production of large sized single crystal wafers and heterostructures.

4.2 High Breakdown Voltage, High Current Density Power Devices

4.2.1 GaN (S. Yoshida)

Introduction

The physical properties of GaN-based nitride semiconductors show excellent potential for the fabrication of low power loss, high power electronic devices [44–51]. The figures of merit for high power, high speed, high frequency, and low loss devices are widely accepted as being the Johnson index and Baliga's high frequency index, which are larger for GaN than Si, GaAs, and SiC [44,45].

Figure 4.22 shows theoretical limits for the on-resistance and reverse voltage of a lateral FET calculated by Saito et al. [46]. The results imply that it should be possible to produce lateral GaN devices with on-resistances comparable with vertical Si and SiC devices. Figure 4.23 shows the relationship

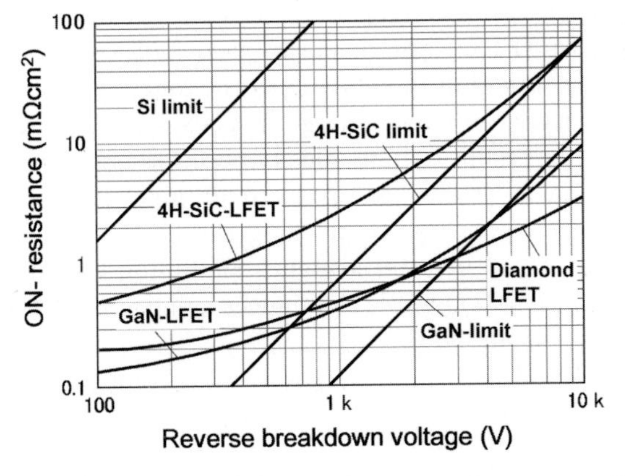

Fig. 4.22. Theoretical limits for the on-resistance and reverse voltage of a lateral FET calculated by Saito et al. [46]

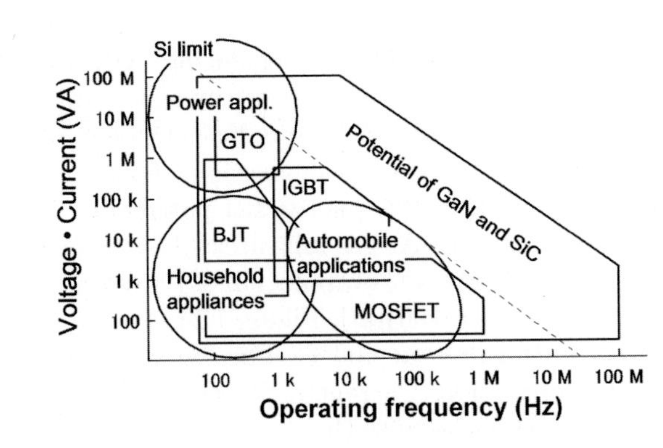

Fig. 4.23. Frequency dependence of the output power GaN power devices

between the output power and frequency of GaN power devices. It can be seen that GaN shows potential for high power, high frequency devices. There have been reports on the fabrication of 1,000 V lateral FETs on large area Si substrates, which will enable lower manufacturing costs [47].

Compact Low Loss Switching Power Supplies Using GaN

Compact low loss inverters require highly efficient, high frequency devices with low on-resistances [45]. For these reasons, the on-resistance, frequency, and high speed characteristics of unipolar FETs are better than those of insulated gate bipolar transistor (IGBT) structures. The switching loss of conventional Si MOSFETs is low but the steady state loss is large. In the case of Si IGBTs, the steady state loss can be reduced but the problem of a large switching loss remains. The use of GaN FETs will enable the possibility of simultaneously reducing the steady state and switching losses.

However, even if the FET performance is improved, the large switching loss of peripheral free wheel diodes (FWD) must be reduced for practical applications and increasing the speed of FWD is an important issue. Thus, fast recovery diodes (FRD) are generally used as FWDs. A large switching loss occurs in Si-pn junction diodes due to the large reverse recovery charge. In contrast, the reverse recovery charge in GaN and SiC Schottky diodes is extremely small thus enabling high speed operation and low switching losses. In particular, when inverters are operated at high speed, it is possible to instantly pass a regenerative current using the diodes, which greatly affects the switching loss. Thus, inverter systems combining GaN FETs and diodes show potential for use in household appliances and compact, low loss switching power supplies for automobiles.

Examples of GaN Electron Devices [49–52]

GaN MESFET

A typical GaN MESFET structure consists of a sapphire substrate/GaN buffer layer (50 nm)/undoped GaN (1,000 nm)/insulating AlN layer (50 nm)/Si-doped GaN active layer (200 nm). The carrier concentration and mobility of such a layer are typically $\sim 3 \times 10^{17}\,\mathrm{cm}^{-3}$ and $200\,\mathrm{cm}^2\,\mathrm{Vs}^{-1}$, respectively. The device fabrication process involves (1) deposition of 200 nm silicon dioxide layer on the GaN epitaxial layer; (2) photoresist patterning; (3) ECR plasma etching of GaN using a $CH_4/H_2/Ar$ gas mixture; (4) evaporation of Al/Ti for source and drain ohmic contacts; (5) evaporation of Pt Schottky gate electrode (length = 2.5 μm and width = 100 μm). Figure 4.24 shows the stable current voltage characteristics measured at 400°C. The I_{ds} of these devices was stable even after 1,800 h operation. Currently, the maximum reverse breakdown voltage of GaN-MESFETs is 550 V with a maximum current of 15 A [51].

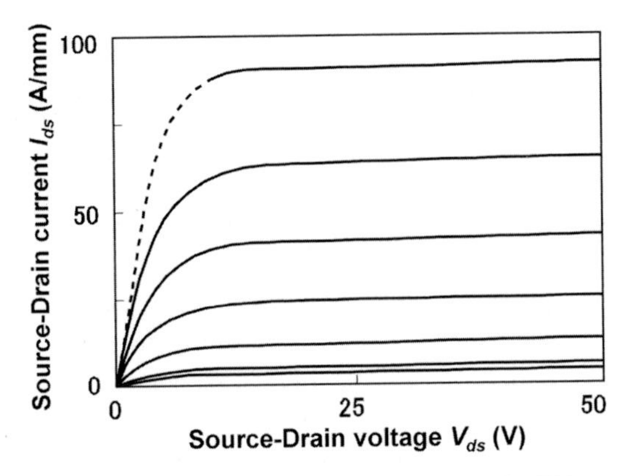

Fig. 4.24. Stable current voltage characteristics of MESFET measured at 400°C

GaN Bipolar Junction Transistor

GaN-bipolar junction transistors (BJT) with p–n–p structures are difficult to produce due to problems related to the growth of p-type GaN epilayers. GaN multilayer BJT devices with n–p–n structures have been studied [49–52]. Typical n- and p-type carrier concentrations are $3–5 \times 10^{17}$ and $2 \times 10^{17}\,\text{cm}^{-3}$, respectively. The BJT structures are fabricated according to the following procedure (1) ECR plasma etching ($\sim$6 nm min^{-1} for p-type GaN) to produce device isolation; (2) evaporation of Al/Ti for n-type ohmic electrodes and Pt/Au for p-type electrodes. These structures showed stable operation at 300°C with $h_{fe} = dI_c/dI_b$ 10–30, which is the largest value reported for GaN-BJTs. The carrier lifetime for a GaN-BJT is 340 ps, which is one order of magnitude larger than GaAs-based devices. Optimization of the growth parameters and carrier concentrations will produce transistors with even higher figures of merit. Further, high current gain has been reported using heterobipolar transistor structures with large bandgap materials for the emitter. For example, GaN/SiC, AlGaN/GaN, and GaN/InGaN structures have been reported [52].

Inverters Using Normally on-AlGaN/GaN HFETs

AlGaN/GaN heterojunction FETs (HFET) are promising as devices for large current power devices. The structures exploit the formation of a two-dimensional electron gas (2DEG) with a sheet carrier concentration of $\sim 1 \times 10^{13}\,\text{cm}^{-2}$ and an electron mobility of $1{,}500\,\text{cm}^2\,\text{Vs}^{-1}$ at the hetero-interface for device operation. Compared with MESFET and BJT structures, HFET offer the potential advantage of attaining high power output at low ON-resistances.

Figure 4.25 shows a typical AlGaN/GaN HFET structure that has been operated at 20 A, 370 V and has an on-resistance of $5\,\text{m}\Omega\,\text{cm}^2$ [53–55] as shown

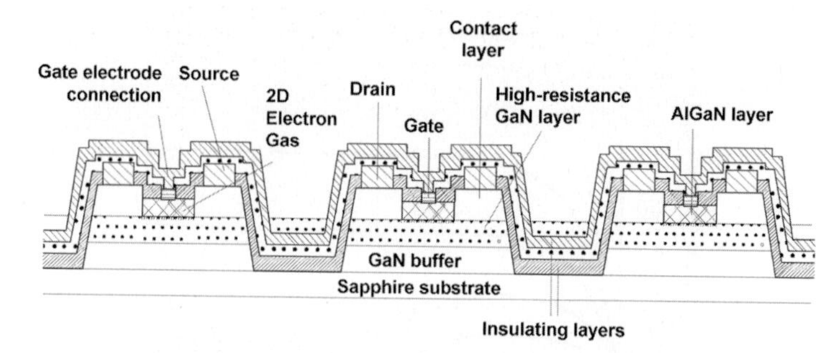

Fig. 4.25. Typical large current operation AlGaN/GaN HFET structure

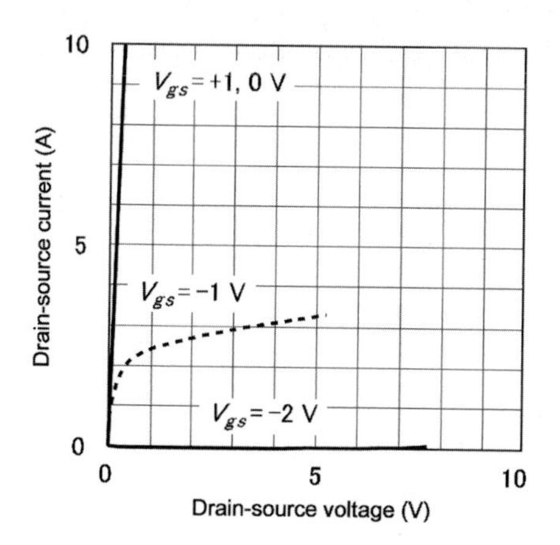

Fig. 4.26. Current–voltage characteristics of an AlGaN/GaN HFET

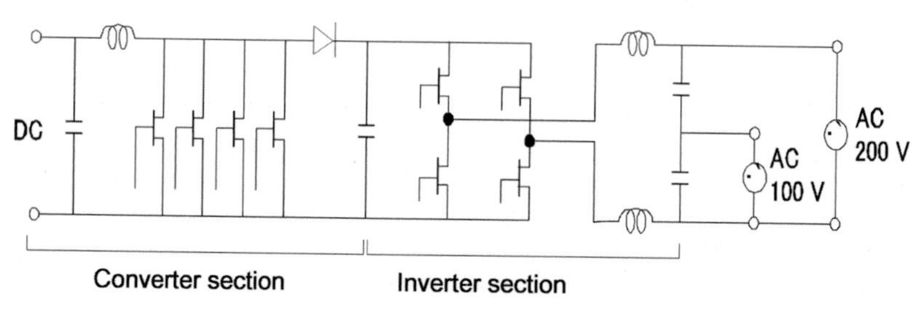

Fig. 4.27. Schematic diagram of an inverter incorporating an AlGaN/GaN HFET

in Fig. 4.26, which is higher than theoretical values for Si devices. Figure 4.27 shows an inverter with 4-HFET in the up-converter section and 4-HFETs in the DC converter. This circuit produced an AC 100 V at DC 30 V input, as shown in Fig. 4.28. The peak value is ±140 V. Although still in the experimental stages, these are the first results for the use of AlGaN/GaN HFETs in inverter circuits [55]. There have also been reported on the use of HFETs for DC–DC conversion in the range 12–42 V. Recently, there have been increasing

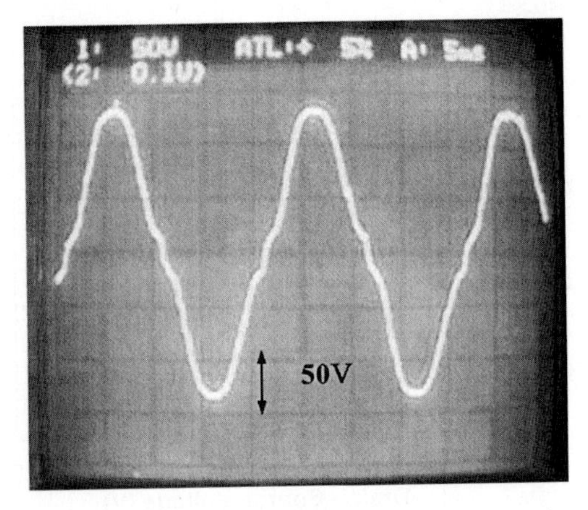

Fig. 4.28. AC output after inverter conversion

numbers of reports on AlGaN/GaN HFETs grown on Si substrates operating at 150 A [56].

Normally off-AlGaN/GaN HEMT Structures

The use of GaN devices in power sources will require the development of normally OFF FET structures [57,58]. Here, control of the threshold voltage of AlGaN/GaN for producing normally off HFET structures is described. There are two methods for producing normally off AlGaN/GaN HEMT structures. The first involves minimizing the thickness of the AlGaN layer and operating at a threshold voltage (V_{th}) close to 0 V. In the second method, only the gate region is recess etched to reduce the thickness of the AlGaN layer. The main problem with the first method is that the sheet carrier density decreases because of a weaker piezoelectric effect at the AlGaN/GaN heterointerface and a thin AlN layer is inserted between the AlGaN and GaN layers to compensate for these adverse effects. Insertion of the AlN layer enables the realization of a large piezoelectric effect and thus a large sheet carrier density and large carrier mobility. That is, by reducing the thickness of the AlGaN layer of a conventional normally-on structure it is possible to reduce the size of the depletion region and thereby reduce V_{th}, to between 0 and +1 V. In the second method, high precision technology is required to produce the recess region. Here, only the first method will be considered in more detail.

The procedure for fabricating an AlGaN/GaN HFET by MOCVD on low cost Si substrates involves (1) deposition of an AlGaN buffer layer at 850°C using ammonia and trimethylgallium sources; (2) at a substrate temperature of 1,050°C, growth of 2 μm layer of high resistivity GaN, 1 nm layer of undoped AlN, and 5 nm of $Al_{0.2}Ga_{0.8}N$. An HFET device structure is then fabricated using this epiwafer [57–59]. For AlGaN layers less than 5 nm, V_{th} exceeds 0 V thus confirming experimental conditions for normally OFF operation.

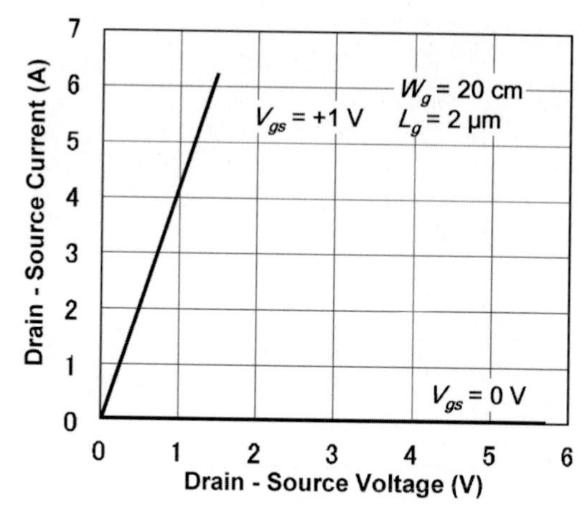

Fig. 4.29. I_{ds}–V_{ds} characteristics of a large device with a gate width of 20 cm

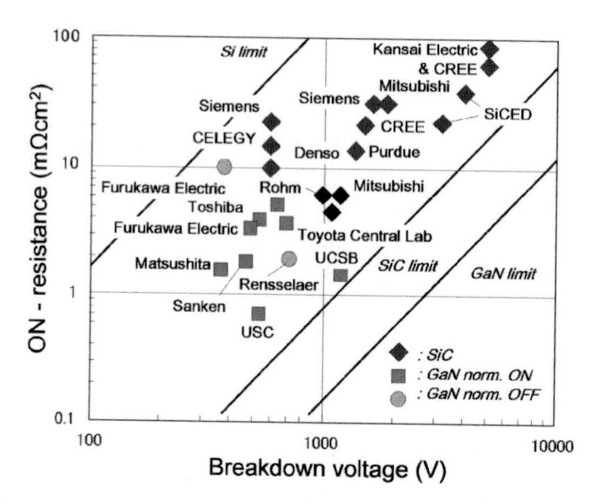

Fig. 4.30. Comparison of recent reports for SiC and GaN-FET on-resistances

Figure 4.29 shows the I_{ds}-V_{ds} characteristics of a large device with a gate width of 20 cm. The device is OFF at $V_{gs} = 0$ V. The maximum current and reverse breakdown voltages are 10 A and 380 V, respectively. Figure 4.30 is a comparison of recent reports for SiC and GaN-FET on-resistances. A group from the Rensselaer Polytechnic Institute in the USA reported on the fabrication of normally off GaN-MOSFETs with reverse breakdown voltages of 700 V, on-resistances of 1.9 mΩ cm^2 and a channel mobility of 45 cm^{2}Vs^{-1} [60].

Low Loss Diodes (Field Effect Schottky Barrier Diode, FESBD)

The construction of low loss, high efficiency power supplies requires not only FET devices but also low loss diodes. The reverse recovery time of Schottky

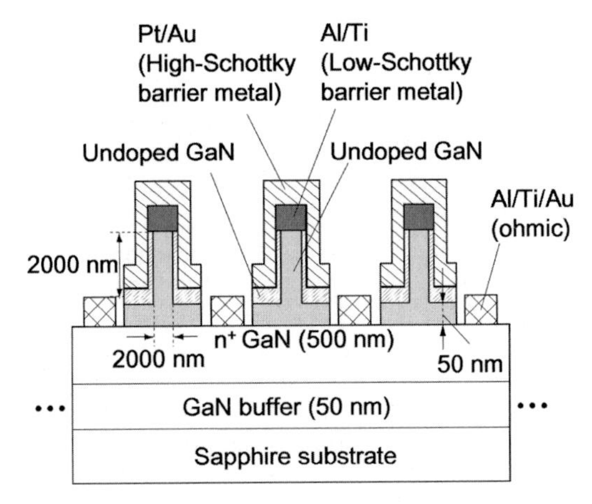

Fig. 4.31. Vertical FESBD structure consisting of a sapphire substrate/50 nm GaN/500 nm n-type GaN/25 nm $Al_{0.25}Ga_{0.75}N$

diodes made using nitride semiconductors is extremely small with the possibility of high speed operation at high reverse breakdown voltages. Zhang et al. reported on a GaN-based Schottky diode with a breakdown voltage of 9,700 V [61], but this device did not show low loss characteristics. Losses in Schottky diodes arise due to the existence of the on-voltage (Von) corresponding to the Schottky barrier height. Thus the magnitude of V_{on} must be reduced to zero in order to reduce such losses. New diode structures have been developed with zero on-voltages for low loss, compact power sources. These field effect Schottky barrier diode (FESBD) structures incorporate a double electrode structure, and their operation is governed by application of electric fields and pinch-off effects [62, 63]. Figure 4.31 shows a vertical FESBD structure consisting of a sapphire substrate/50 nm GaN/500 nm n-type GaN/25 nm $Al_{0.25}Ga_{0.75}N$. Devices are fabricated by dry etching followed by SiO_2 deposition and patterning to form vertical AlGaN structures. Ohmic contacts on AlGaN are made using Al/Ti and Schottky barriers using Pt/Au. Ohmic contacts for n-GaN are formed using Al/Ti/Au. The V_{on} of this vertical structure is almost 0 V. Similar results have been confirmed in a planar FESBD with a normally-off AlGaN/AlN HEMT structure. Figure 4.32 shows the forward and reverse characteristics of this planar diode structure where the V_{on} is almost 0 V, the reverse bias breakdown voltage is ~400 V and rise time is ~3.9 ns, thus confirming high speed operation.

4.2.2 SiC (T. Shinohe)

For the last forty years the power electronics industry has relied on Si semiconductor devices. However, the physical limitations of Si have triggered the development of alternative devices such as super junction FETs incorporating

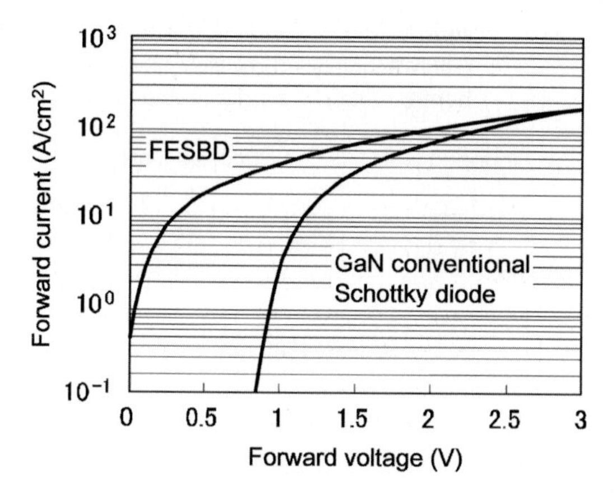

Fig. 4.32. Forward and reverse characteristics of planar diode structure

three-dimensional structures [64], floating junction FETs [65], as well as the use of wide bandgap semiconductors, including SiC [66–68].

Currently there is a shift in the production of SiC wafers from two to three and four inches. Micropipe defect densities have been dramatically reduced, and the main efforts to improve the crystal quality are being shifted to reduce other microdefects, such as basal plane, screw, and threading edge dislocations. Within the last ten years, a wide range of high performance device structures have been demonstrated: a PiN diode with blocking voltage (V_b) of 19.5 kV [68]; an SiCGT (SiC commutated gate turn-off thyristor) with V_b of 12.7 kV [69]; a Schottky barrier diode with current rating of 100 A [70]; a JFET with specific on-resistance ($R_{on}S$) of 3 mΩ cm^2 (V_b = 600 V) [71]; an SIT with $R_{on}S$ of 1.01 mΩ cm^2 (V_b = 700 V) [72]; and a MOSFET with $R_{on}S$ of 2.7 mΩ cm^2 (V_b = 700 V) [73]. Infineon Technologies AG and Cree Inc. already sell Schottky barrier diodes with values of V_b ranging from 300 to 1,200 V for power factor correction (PFC) circuits of switching mode power supplies. Germany's SiCED GmbH & Co. KG. is shipping samples of 4–50 A hybrid SiC/Si cascode switches consisting of a normally-ON SiC JFET (V_b = 1,200 V) and a Si-MOSFET (V_b = 60 V) connected in series in a package.

Design of SiC Power Devices

The breakdown electric field (E_c) of SiC is about ten times that of Si (E_c(4H) = 2.7 MV cm^{-1}), which offers the following advantages of SiC power devices over Si counterparts (1) the thickness of the drift layer of SiC can be reduced to 1/10 of that of Si, and also higher doping is possible of the drift layer of SiC, and these combined enable reduction of the drift layer resistance to 1/300 that of Si in unipolar devices (Fig. 1.6); and (2) the use of the same drift layer thickness as Si enables a blocking voltage of ten times that of Si. The characteristics (1) would enable further reduction of the on-resistance of unipolar devices as well raise the possibility of using unipolar SiC power devices above 600 V, which would contribute to reducing the size of inductive and capacitive components of power supplies. The characteristics (2) would enable the fabrication of devices operating at ultrahigh blocking voltages (∼ some tens of kV) and it would be possible to reduce the number of series connected devices required for construction of high power converter systems. Further, the wide bandgap of SiC (E_g(4H) = 3.26 eV) enables high-temperature operation, and its high thermal conductivity compatible with copper (4.9 W (cm K)$^{-1}$) is useful for effective cooling, which lead to the construction of compact power supplies.

Among many polytypes (4H, 6H, 3C, ...) of SiC, 4H is used for SiC power devices because of its high bulk mobility of 1,000 cm^2 V^{-1}s^{-1} [67]. The 6H is mainly used as substrates for the growth of GaN epilayers, and 3C shows promise for growing 6-in. wafers by using heteroepitaxial growth on Si substrates. To design 4H-SiC power devices, it is necessary to note the following issues (1) the anisotropic electrical properties (bulk mobility, breakdown voltage) due to hexagonal crystal structure, (2) appropriate selection of physical models suitable for the high electric field region, (3) higher hole density at elevated temperatures and (4) dynamic punch-through effects due to deep acceptor levels.

Diodes

Schottky Barrier Diodes

Schottky barrier diodes are unipolar devices, which exploit the full potential of the physical properties of SiC. Device characteristics near the unipolar limit of 4H-SiC have been attained (1) an SBD with a three-zone junction termination expansion (JTE) structure exhibiting V_b of 4,150 V and $R_{on}S$ of 9.07 mΩ cm^2 (CRIEPI); and (2) a floating junction SBD with buried p-type layers in the drift layer exhibiting V_b of 2,427 V and $R_{on}S$ of 3.03 mΩ cm^2 (Toshiba). As mentioned earlier, 10 A class SBDs are being commercialized, and used in switching mode power supplies. The hybrid-pair of a Si-IGBT/Si-MOSFET and a SiC-SBD is the most reliable method at present to reduce power-loss of power supplies.

Cree reported on the characteristics of a 100 A class SBD with V_b of 600 V fabricated using 3-in. diameter substrates (micropipe density of ∼0.92 cm^{-2}) with a yield of 77% [70]. Such yields would enable mass production for automotive, industrial equipment and other high power applications of power electronics.

PiN Diodes

PiN diodes are being studied as devices with reverse blocking voltages greater than 3 kV, where on-state voltages of greater than 2.5 V are acceptable for

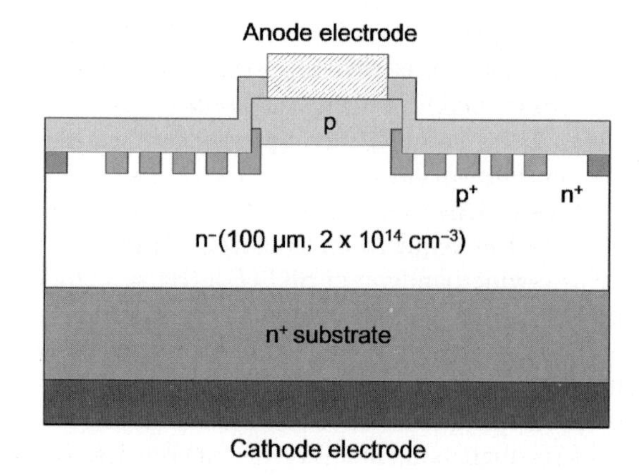

Fig. 4.33. Device structure of a PiN diode (10 kV, 50 A)

device applications. KEPCO reported the highest confirmed V_b at 19.5 kV (epilayer thickness of 200 μm) with a current less than 1 A (electrode diameter of 200 μm–1 mm), an on-state voltage of 6.5 V (100 A cm^{-2}) and a reverse recovery time of 28–100 ns [68]. The same group also successfully fabricated 6 × 6 mm diodes with a large current rating of 100 A (on-state voltage of 4.2 V), although V_b is 5.2 kV. Cree reported an 8.5 × 8.5 mm diode with V_b of 10 kV, current rating of 50 A (on-state voltage of 4.2 V, current density of 100 A cm^{-2}), as shown in Fig. 4.33 [74], where an on-state voltage of 5.9 V was obtained at 330 A (660 A cm^{-2}). PiN diodes also show potential for high-temperature applications because of their superior characteristics, such as lower on-resistance and lower leakage current even at elevated temperatures compared to SBDs. ABB reported on 400 A switching experiments combining 4.5 kV class Si-IGBTs and SiC-PiN diodes, and KEPCO demonstrated the use of 4.5 kV, 1.2 kA Si-IGBT modules and 3 kV, 600 A SiC-PiN diode modules for the construction of a 200 kVA class electric power stabilizing system. Although SiC PiN diodes had suffered from on-state voltage degradation problems due to crystal defects, these issues have almost been resolved.

Switching Devices

Unipolar Devices

MOSFETs are at present the main battlefield of SiC power device research and development. SiC MOSFET devices with $R_{on}S$ of less than 10 mΩ cm^2 (V_b of ∼1,000 V) were reported recently [73,75]. This specific on-resistance is still at a level that is a little less than that of 600 V Si-super junction FETs (commercial devices ∼20 mΩ cm^2 and experimental devices ∼10 mΩ cm^2) and has not enabled exploitation of the full potential of the physical properties of SiC. The reason for the poor performance of SiC MOSFET is because the MOS interface state density is one order of magnitude larger than Si and electrons injected from the source are trapped by the interface traps and their contribution to electrical conduction is reduced, or the interface states become negatively charged and become electron scattering centers, resulting in a decrease of the carrier mobility of the MOS channel. Carbon clusters at the SiC/SiO_2 interface and near-interfacial defects in SiO_2 have been proposed as the origin of these effects.

The mobility of the MOS channel has been improved to $\sim$200 $cm^2\,V^{-1}s^{-1}$ using methods such as modification of the oxidization method, annealing in H_2, NO, or NO_2 atmospheres, use of a buried channel, delta doped channel, use of $(11\bar{2}0)$ or $(000\bar{1})$ surface, and use of stacked layers [66, 67]. However, fabrication of low on-resistance MOSFETs further requires suppression of the adverse effects of annealing used after ion implantation (>1,500 °C) and for forming ohmic contacts (>900°C). Also, reliability of the gate oxide layer is the key factor for commercialization of SiC MOSFETs. There are several issues to consider for improving the reliability (1) the smaller energy gap (2.70 eV for 4H-SiC) between the conduction band edges of SiC and SiO_2 compared with that of Si (3.15 eV); (2) remnant carbon in the SiO_2 and transition layers; (3) adverse effects of crystalline defects of the SiC surface; (4) exposure to high electric fields of ten times that of Si; and (5) requirements for operation at higher temperatures than Si devices. At the present time, the magnitude of Q_{bd} determined from time dependent dielectric breakdown (TDDB) tests is ten times worse than that of Si. Methods for improving the crystal quality and formation of oxides are being aggressively studied to establish technology for producing reliable gate oxide layers. From the device design point of view, incorporation of electric field shield structures using buried p-type layers has been found to be effective in reducing the adverse affects of the high electric fields of issue (4).

A wide range of SiC power MOSFETs have been investigated including the conventional DMOSFET, UMOSFET and ACCUFET, ECFET, SIAFET, SEMOSFET, DACFET, and IEMOSFET structures which employ mobility of accumulation layers [66–68, 73]. In particular, the unique features of the SiC structure include (1) the low mobility of the MOS channel can be overcome by using accumulation layer mobility, epichannel, buried channel, and $(11\bar{2}0)$ or $(000\bar{1})$ surfaces; and, (2) reduction of the electric field at the gate oxide interface by using DMOSFET structures and formation of a p-type layer at the bottom of the trench gate of UMOSFETs.

It is difficult for SiC MOSFETs to compete with Si-super-junction FETs and Si-IGBTs in the 600–1,200 V blocking voltage region. But the performance of SiC-MOSFETs ($R_{on}S$ of 3.1 $m\Omega\,cm^2$ for V_b of 900 V, Rohm Co. Ltd. [75]; $R_{on}S$ of 2.7 $m\Omega\,cm^2$ for V_b of 700 V, AIST [73]) has finally been shown to be superior to that of Si-IGBTs. Figure 4.34 shows the structure of an IEMOSFET [73]. Further improvements are still necessary to meet industrial application requirements ($R_{on}S$ of 1–2 $m\Omega\,cm^2$). It has been reported that the use

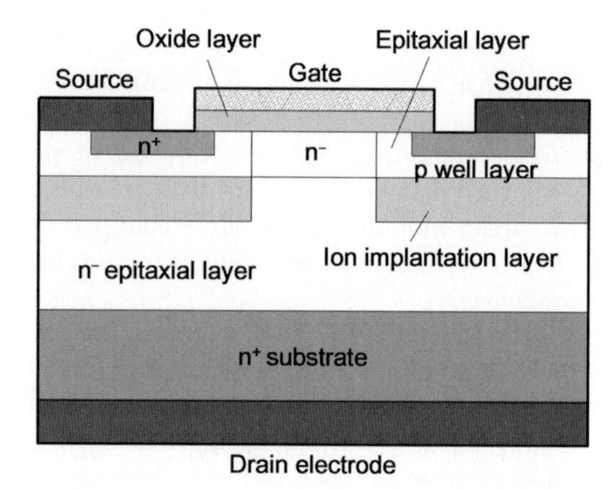

Fig. 4.34. Device structure of IEMOSFET

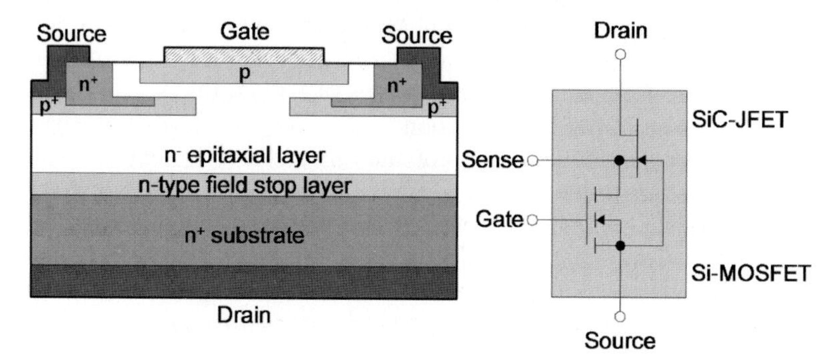

Fig. 4.35. Hybrid SiC/Si cascode switch

of SiC-MOSFETs instead of Si-IGBTs + Si-PiN diodes would lead to a 60% reduction of power losses in inverters for hybrid cars.

SiC MOSFETs show superior performance above 1,200 V blocking voltages as shown by reports of $V_b = 1{,}200\,V$, $R_{on}S = 12\,m\Omega\,cm^2$ (SiCED); $V_b = 1{,}800\,V$, $R_{on}S = 8\,m\Omega\,cm^2$ (Cree); $V_b = 3{,}000\,V$, $R_{on}S = 45\,m\Omega\,cm^2$ (SiCED); and $V_b = 5{,}020\,V$, $R_{on}S = 88\,m\Omega\,cm^2$ (KEPCO).

JFET structures have received attention from the SiC community because they have the potential to avoid problems related to MOS channel mobility and gate oxide layer reliability. However, the focus of research and development is shifting to MOSFETs now. As described, sample shipments of hybrid SiC/Si cascode switches (Fig. 4.35) are already available. Initial commercial applications are likely to be for resonant converters and auxiliary power supplies. A minimum value for the specific on-resistance ($1.01\,m\Omega\,cm^2$) was reported by the AIST group using an SIT structure with V_b of 700 V [72]. Rockwell Scientific Co. reported an MOS-enhanced JFET structure with $R_{on}S$ of $3\,m\Omega\,cm^2$ and V_b of 600 V [71]. There have also been reports on a compact 6.5 kVA class 3-phase PWM inverter (DC bus voltage = 250 V; efficiency = 99.4%) using 600 V JFETs (25 A with four chips) and SiC-SBDs [71]; and a high voltage version (DC bus voltage = 600 V) composed of 1,500 V/10 A modules. Reports on the high blocking voltage region include JFETs with buried p-type layers with $R_{on}S$ of 10 mΩ cm^2 and 25 mΩ cm^2 at V_b of 1,200 V and 3,300 V, respectively, (SiCED); and a SEJFET (static expansion channel JFET) with $R_{on}S$ of 218 mΩ cm^2 at V_b of 5,500 V (KEPCO). The normally-ON feature is the major weakness of this device structure and there are efforts underway to fabricate normally-OFF devices by reducing the size of the channel region. The Rutgers group has reported on a normally-OFF JFET where a trench structure was used to reduce minimum feature sizes to 0.55 μm, resulting in a device with V_b of 1,726 V and $R_{on}S$ of 3.6 mΩ cm^2 at a V_{GS} of 5 V [76].

Figure 4.36 shows the reported specific on-resistance values for SiC switching devices (including bipolar structures). Further improvements in device performance will require increasing the channel density by reducing the size of unit cells to several micrometers, increasing the mobility of the MOS channel itself and reducing the parasitic resistance.

Bipolar Devices

Bipolar transistors exhibit a sharp rise of the forward I–V characteristics from the zero collector voltage, which is most desirable in terms of on-state voltage characteristics. Also, low channel mobility and gate insulation layers do not pose problems in bipolar structures. However, improvement of the

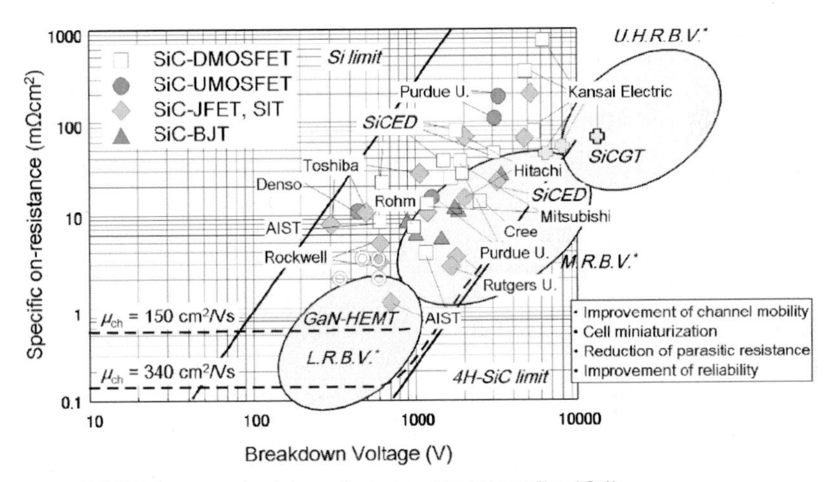

Fig. 4.36. Reported specific on-resistance values for SiC switching devices (including bipolar structures)

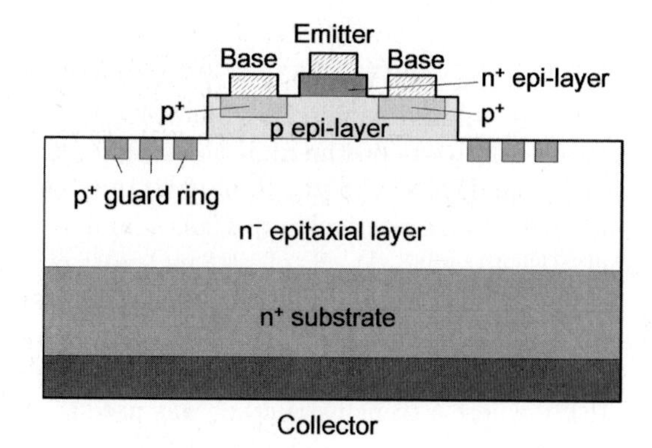

Fig. 4.37. Cross-section of a typical SiC bipolar transistor

current gain, β, is extremely important because a continuous base current must be passed. Figure 4.37 shows a cross-section of a typical SiC bipolar transistor [77]. The emitter layer is grown epitaxially and the thickness and carrier density of the base are optimized for both a high β and large blocking voltage. The current gain can be improved by (1) continuous epitaxial growth of the base and emitter layers; (2) separating the ion implanted base contact region from the emitter layer; and, (3) use of a high quality silicon dioxide passivation layer with a low density of interface levels. These modifications have yielded $\beta = 40$–60 and even values of $\beta = 120$ have been achieved using small area devices. The energy levels of the p-type impurities in SiC are deep and at high temperatures the hole density of the p-type base increases, which results in a reduction of the current gain of npn transistors. Aluminum (Al) is used as a p-type dopant of the base layer. For a base layer with Al $= 2 \times 10^{17}\,\mathrm{cm}^{-3}$, only about 10% activation is achieved at room temperature; 50% at 200°C; and 100% at 275°C. The main reports on SiC bipolar transistors are from Cree, Rutgers and Purdue are given in the below tables.

Cree:

V_b	$R_{on}S$	I	V_F	β
1,000 V	$6.0\,\mathrm{m\Omega\,cm^2}$	30 A	2 V	40
1,450 V	$5.4\,\mathrm{m\Omega\,cm^2}$	20 A	1.2 V	14
1,800 V	$10.8\,\mathrm{m\Omega\,cm^2}$	2.7 A	2 V	20

Rutgers:

V_b	$R_{on}S$	I	V_F	β
858 V	$8.7\,\mathrm{m\Omega\,cm^2}$	7 A	5.5 V	47
1,750 V	$12\,\mathrm{m\Omega\,cm^2}$	4.9 A	6.5 V	24.8

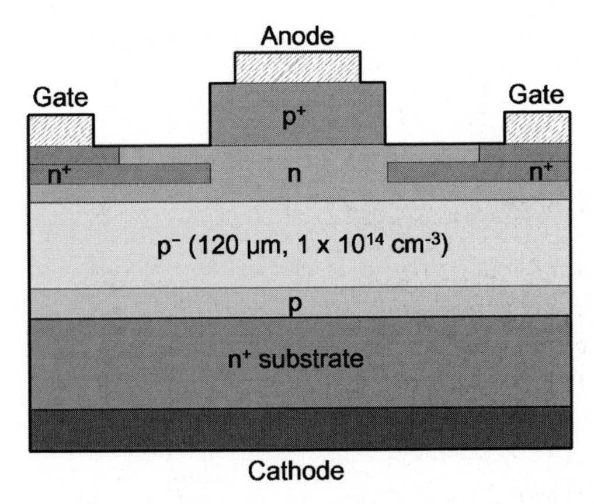

Fig. 4.38. Device structure of a SICGT (reverse voltage of 12.7 kV)

Purdue:

V_b	$R_{on}S$	I	V_F	β
3,200 V	28 mΩ cm^2			20

Rutgers has also reported on the construction of a 20 kHz, 7.4 HP DC-AC PWM inverter for induction motors consisting of 600 V–40 A class SiC bipolar transistors and SiC-MPS diodes.

A one chip, monolithic Darlington transistor was reported to have a $\beta = 2{,}400$ for high-voltage applications. New structures have been proposed by SiCED, the BIFET with a $V_b = 4.5$ kV; by KEPCO, the JGBT with 5 kV and MAGBT with $V_b = 22$ kV. SiC thyristors have also been reported. KEPCO announced a SICGT (Fig. 4.38) with figures of merit surpassing those of 4H-SiC unipolar limit [69]. The characteristics of a 4.2×3.8 mm, 4.5 kV device were a maximum turn-off current = 100 A, a turn-on time = 0.27 μs, and a turn-off time = 0.55 μs. These switching times are 1/40 of those of Si-GTO and 1/6 of those of Si-IGBT devices with the same blocking voltages. Small size devices (1×1 mm) were reported with $V_b = 6.2$ kV, 8.1 kV, 12.7 kV and on-state voltages corresponding to 4.2 V, 5.4 V, and 6.6 V, respectively, at 100 A cm^{-2}. A device with a $V_b = 12.7$ kV showed a turn-on time of 0.22 and 2.68 μs. An 18 kVA three-phase PWM inverter consisting of 30 A modules with 3 kV SICGTs and PiN diodes was also reported [78].

4.3 New Functional Devices

4.3.1 Electron Emitters (A. Hiraki, H. Hiraki)

Introduction

Negative electron affinity (NEA) is found in several wide bandgap semiconductors such as diamond and AlN due to the bottom of the conduction band

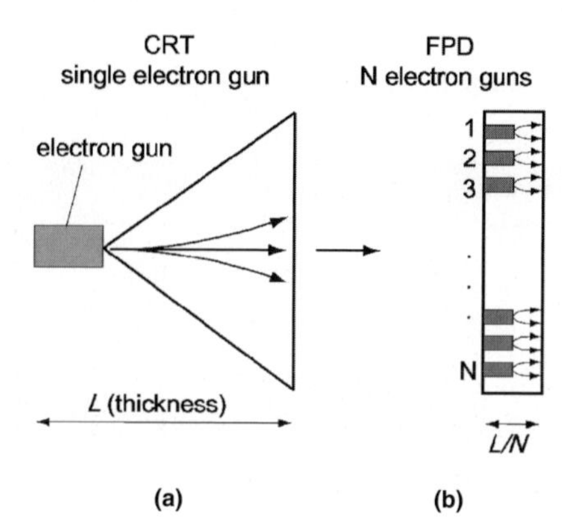

Fig. 4.39. Principle of making display thickness thin with increasing number of electron guns

being at a higher energy than the vacuum level. Further, in the case of diamond, the differences in electronegativity between absorbed hydrogen layers and the underlying the carbon surface are also origins of NEA. There have been many reports on electron emission from CVD diamond thin films. The main motivation for such studies is flat panel display (FPD) applications to replace CRT technology, which has the disadvantages of being too thick and heavy. The solutions to these problems are thin and lightweight displays. As shown in Fig. 4.39a, the reason for the thick CRT screens is that only one electron gun is used. The emitted electrons (a hot cathode is widely used) are electrically accelerated and magnetically scanned over a phosphor that emits light creating the image. The use of N electron guns (cold cathode would be preferable) would enable a thickness reduction to L/N (Fig. 4.39b). Thus an array of electron guns located on a planar surface would enable ultrathin and lightweight displays or pixels. A emission intensity of ~100,000 cd m^{-2} has been attained from carbon emitters (such as carbon nanotubes [79, 80] and diamond [81]). The value for commercially available fluorescent tubes is ~3,000 cd m^{-2}.

Electron Emission from Diamond and Carbon Materials (Carbon Nanotubes and others)

The degree of electron emission observed in a variety of carbon materials (CVD-diamond thin films, carbon nanotubes, and nanowalls) is thought to be governed by the existence of graphite with high electrical conductivity in the materials.

For example, the emission current of polycrystalline CVD-diamond thin films is inversely proportional to the grain size and it has been concluded to originate from the existence of graphite layers at the grain boundaries. Thus selection of deposition conditions that increase the graphite content within the CVD-diamond (that is, conditions that would not be favorable for the

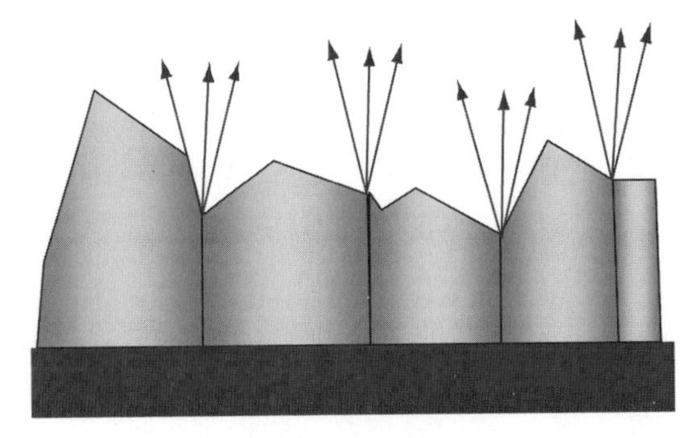

Fig. 4.40. Model for diamond thin film electron emission

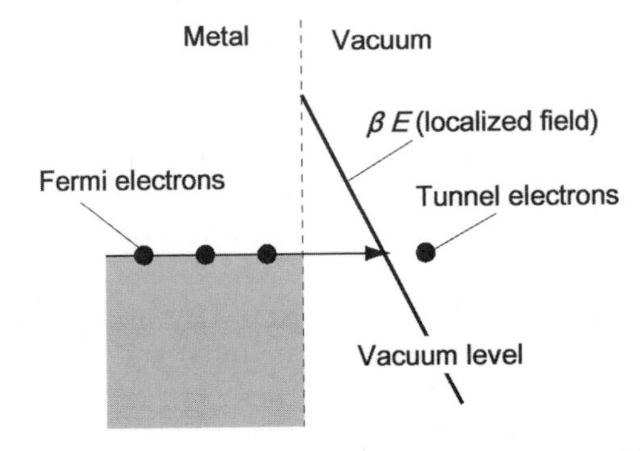

Fig. 4.41. Electron (or electric field) emission phenomena model (tunneling is easier at sharp slope for large β)

growth of high quality diamond films) yield films with high electron emission properties.

Thus, as schematically shown in Fig. 4.40, diamond films with good electron emission properties are composed of high resistivity diamond grains with gaps or grain boundaries filled with high conductivity graphite rods or plates; referred to as "nanographite chips." The device structure is formed by providing an electrode (cathode) below the diamond film; an anode is located at a small distance from the surface of the diamond, as will be described later. Application of an electric field between the cathode (cold cathode) and anode results in the emission of electrons. Thus, if the nanographite chip is thought of as a small raised section on the metal electrode, then, as shown in Fig. 4.41, the vacuum level is inclined when a strong field is applied and electrons are emitted from the protruding area due to a tunneling process similar to field emission from metals. The local electric field (E_{loc}) at the ultrafine protrusion depends on its shape and can be much larger (β-times) than the applied field, E, and can be expressed as, $E_{loc} = \beta E$. Needless to say, the larger the factor β, the easier it becomes for electron emission. If the protrusion is considered as an electrically conducting chip with a length, ℓ, and radius, r, then the factor

$\beta \sim \ell/r$. From this relationship it can be seen that carbon nanotubes, which have high electrical conduction and aspect ratios, should exhibit excellent electron emission characteristics. The current density due to electron emission under external electric fields can be deduced from the Fowler-Nordheim equation:

$$J = aE^2 \exp\left(-b\varphi^{3/2}/\beta E\right) \tag{4.5}$$

Here E is the average applied electric field and φ the height of the potential barrier and a, b are constants. Furthermore, the emission current density, J can be phenomenologically expressed as a

$$J \propto NC_{rod}E_{loc} \tag{4.6}$$

where, N is the density of protrusions ("rods"), E_{loc} ($= \beta E$) is the local electric field and C_{rod} is the conductance of one rod. It can be seen from (4.6) that the following conditions (a)–(c) must be satisfied in order to achieve high current densities (a) high density of rods, N; (b) highly conducting rods; and, (c) strong localized electric field. However, the current density can decrease if the density of rods is too large due to so called "screening effects" described later in "Design Considerations for Fabrication of Carbon Nano-material Emitters."

As shown in Fig. 4.42, the equipotential lines and thus the electric field strength is different for one rod (or isolated rods) and three rods (placed near

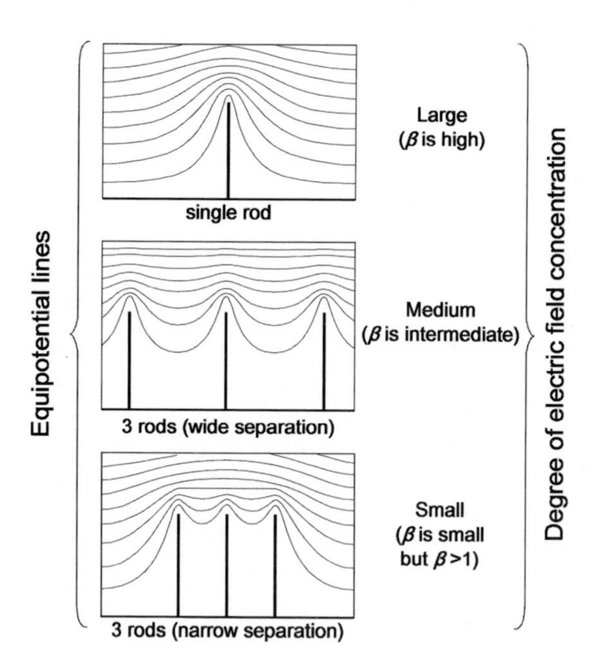

Fig. 4.42. Simulated variation of equipotential distribution with number and separation of the same metal (or highly conductive) rods. Degree of electric field concentration (β) and localization of electric field decreases with increasing number of rods

each other); the electric field strength being weaker in the case of three rods. Thus in order to achieve high electric field strength the separation between the rods must be $\sim 3\ell$, where ℓ is the height of the rod. This result shows that although rods with large aspect ratios would enable high electric fields, the use of such rods would limit the number of rods per unit area due to requirements for a $\sim 3\ell$ separation between the rods.

The following equation represents the main parameters for maximization the current density based on optimization of the rod height and separation:

$$J \propto \alpha_{screening} N C_{rod} E_{loc}, (0 < \alpha_{screening} \leq 1) \tag{4.7}$$

Here, $\alpha_{\text{screening}}$ is a function of ℓ and N.

Based on the aforementioned parameters, Hiraki designed and fabricated electron emitters as described in "Design Considerations for Fabrication of Carbon Nano-material Emitters."

Design Considerations for Fabrication of Carbon Nano-material Emitters

The design rules are based on nanoseeding technology developed by Hiraki [82]. The fabrication of ordered graphite protrusions on electrically conducting substrates is first considered using two types of carbon nanotubes (CNTs) with the same diameters but differing lengths, $\ell_1 \sim \ell_2$. The CNTs are arranged in two-dimensional arrays so that $\alpha_{\text{screening}} \sim 1$ is guaranteed. As shown in Fig. 4.43, for simplicity of analysis, the separation (which should be $\ell_{1(2)} \sim 3\ell_{1(2)}$) is made $\ell_{1(2)}$. If the conductance of the CNTs is approximately the same ($C_{\ell 1} \sim C_{\ell 2}$), then the β of the rods differs by a factor ℓ_1/ℓ_2. Since the total number of protrusions differs by $(\ell_1/\ell_2)^2$, then, from (4.6), the emission current by the shorter rods, ℓ_2, is (ℓ_1/ℓ_2) times larger than that by the longer rods, ℓ_1, assuming $\ell_1 > \ell_2$. If, for example, $\ell_1 \sim 1\,\mu\text{m}$ and $\ell_2 \sim 100\,\text{nm}$, then there is a factor of ten difference in the magnitude of the current. Although this is a rough approximation of the configuration, Hiraki used a protrusion length of $\sim$100 nm for the design and fabrication of electron emitters, as shown in Fig. 4.44, where 100 nm nanodiamond grains were deposited on a metal or ITO electrode (cathode) on a glass substrate by the nanoseeding method. Next, room temperature cathodic arc discharge was used to deposit

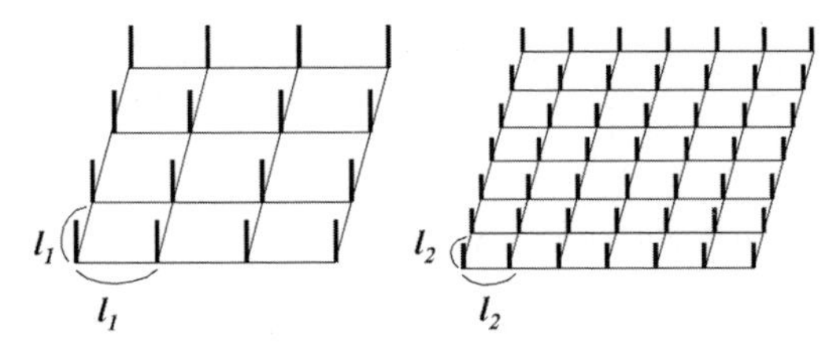

Fig. 4.43. Differences in emission sites of CNT rods with ideal separation but differing lengths for electron emission

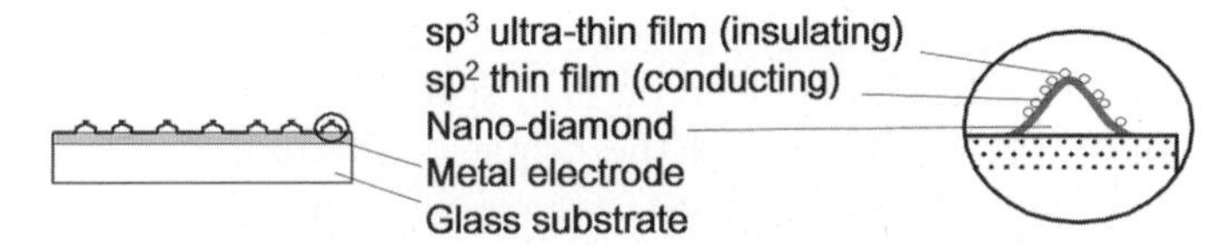

Fig. 4.44. Structure of electron-emitting section of the emission device (emitter)

Fig. 4.45. SEM image of silica spheres

(1) 10 ∼ 100 nm thick graphite-like sp^2 carbon films onto the nanodiamond grains, and (2) extremely thin layers of diamond-like, insulating sp^3 carbon films (nanoparticles of diamond can also be used). The sp^3 films enhance electron tunneling as will be described later, in connection with Fig. 4.44.

The main characteristic of the cathodic arc deposition method is that sp^2 and sp^3 carbon films can be readily deposited at room temperature. That is, it is possible to form electron emitters on glass substrates at room temperature.

Electron Emitters Fabricated by Chemical Vapor Deposition

The characteristics of carbon emitters are (a) sturdiness (they can even be formed in a plasma); (b) high thermal conductivity (the current density in nanostructures is high, thus it is vital to have effective heat dissipation); (c) hydrogen terminated diamond exhibits NEA; and, (d) thin film nanocarbon emitters can be readily fabricated.

Silica Ball Emitter

Figure 4.45 shows an array of commercially available, 100 nm diameter silica balls deposited by (a) dispersion in ethanol; (b) spraying onto the surface of a high conductivity n^{++} Si substrate; (c) annealed at 60°C to desorb the ethanol; (d) finally the heated filament method was used to deposit a graphite layer which serves as the emitter.

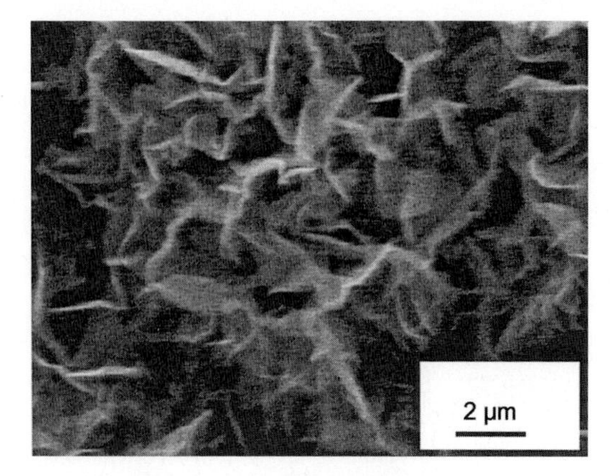

Fig. 4.46. SEM of CNW films

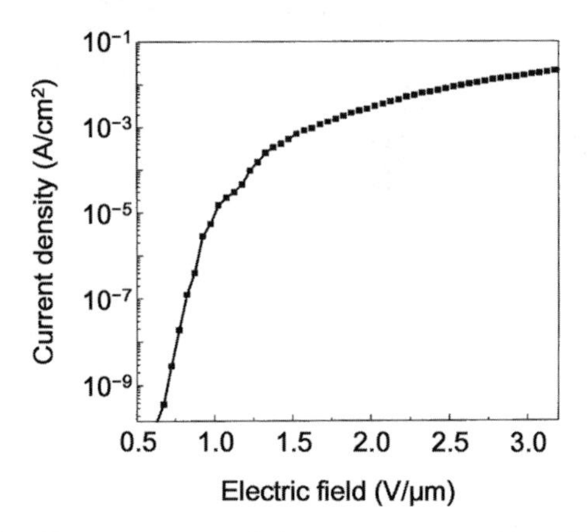

Fig. 4.47. Characteristics of electron emission from CNW films

The emitters are evaluated as follows (a) the threshold field ($\mathrm{V\,\mu m^{-1}}$) to produce a current density of $10\,\mathrm{\mu A\,cm^{-2}}$; and, (b) the electric field ($\mathrm{V\,\mu m^{-1}}$) to achieve a current density of $1\,\mathrm{mA\,cm^{-2}}$, required for field emission displays (FED). Good electron emitters should exhibit the low threshold electric field. In the case of the silica spheres the values for (a) and (b) were $5\,\mathrm{V\,\mu m^{-1}}$ and $11\,\mathrm{V\,\mu m^{-1}}$, respectively. These characteristics are promising for carbon emitters.

Carbon Nanowall Emitters

Figure 4.46 shows thin films known as carbon nanowall (CNW) structures. The CNW are thought to be composed of tens of layers of graphene sheets. From Fig. 4.46, the edge or wall separation is a few micrometers and it is concluded that the aforementioned screening effect conditions are satisfied, to provide sufficient concentration of the electric field.

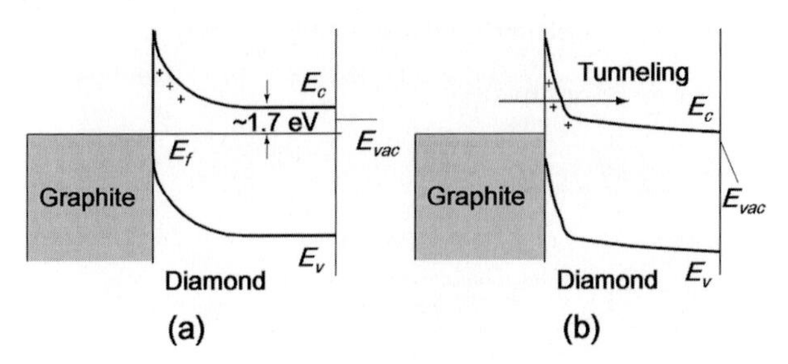

Fig. 4.48. Energy band diagram of graphite/NEA diamond system (**a**) no bias (**b**) voltage applied

As shown in the characteristics of Fig. 4.47, the resulting threshold electric field is $0.5 \sim 1\,\mathrm{V\,\mu m^{-1}}$ and the field for achieving $1\,\mathrm{mA\,cm^{-2}}$ is $3.0\,\mathrm{V\,\mu m^{-1}}$, which implies an exceptional performance with a current density of $100\,\mathrm{mA\,cm^{-2}}$ can be expected at the relatively low electric field. It should be noted that the substrate was a suitable metal and the structures were formed at 900°C by DC or RF-CVD.

NEA Diamond Electron Emitters

Nitrogen (N) forms a donor level at 1.7 eV below the conduction band edge. Figure 4.48 shows the band energy diagram of a graphite (sp^2)/NEA diamond (sp^3) structure with N-doping. The addition of a strong anode voltage to this system (graphite is a cathode) makes the depletion layer at the sp^2/sp^3 interface extremely thin and, as shown in Fig. 4.48b, a current readily flows from the sp^2 to the sp^3 side and, due to NEA, tunnel electrons are easily emitted in a vacuum. The sp^2+sp^3 double layer, shown previously in Fig. 4.44, is also due to this reasoning. Deposition of a 10–100 nm N-doped layer of diamond on the CNW is known to improve the emission characteristics of such devices.

Applications of Carbon Emitters – Optical Sources Rather than Displays

Figure 4.49 shows the operational principle of electron emitters for displays where electrons emitted from a cold cathode are accelerated and used to irradiate a phosphor which emits light (Fig. 4.39b). However, applications other than FPDs are worth considering:

(a) The emission brightness is very high, far higher than required for FPD.
(b) FPD applications do not need such high brightness, but there are many very difficult problems to be resolved.

Thus, Hiraki made a decision to use their emitters as optical sources rather than as FPDs. A company called Dialight Japan [83] was established in the summer of 2004 to commercialize the research results. The characteristics of the field emission (FE) optical sources developed by this company are

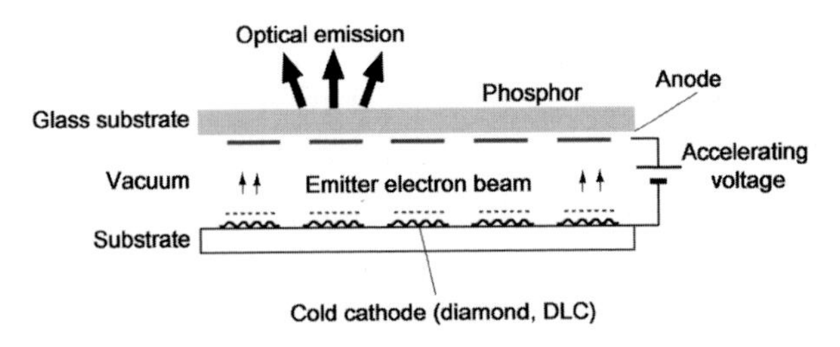

Fig. 4.49. Structure of field-emission display (two-electrode type)

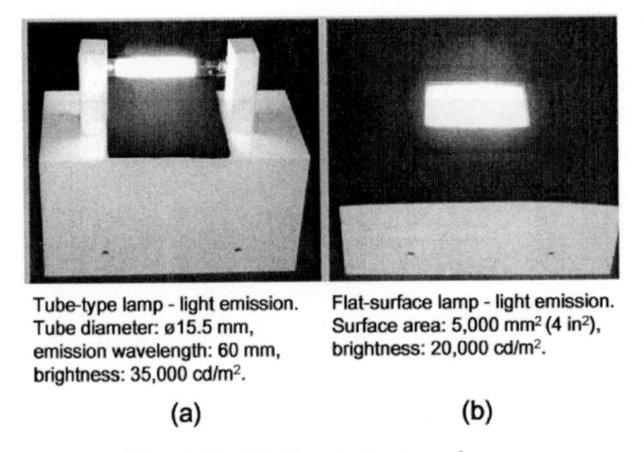

Fig. 4.50. Field-emission type lamps

(a) mercury is not used; (b) low power consumption; (c) low heat dissipation; (d) wide range of colors; (e) long life; (f) high brightness; and (g) continuous emission. Of the many possible applications, Dialight Japan is initially focusing on backlights for LCDs as shown by the compact white light lamp (5T size) shown in Fig. 4.50a. This is probably the first example of field emission carbon emitters being used for high brightness white light sources. This panel's optical characteristics are shown in Fig. 4.50b.

4.3.2 Transparent Devices (H. Koinuma, M. Katayama)

Commercially available smart windows are made of transparent glass whose color is changeable by exposure to light or the application of electric fields. The possibility of incorporating transparent electronic circuits in glass would realize highly functional glass and displays. Once transistors, ferromagnetic materials, and light emission devices could be made transparent, their integration would lead into the fabrication of totally "invisible" devices, which function as switches, memories, and displays. Further, the fabrication of these devices at low temperatures on plastics would contribute to the realization of flexible and wearable devices.

Oxides are promising transparent materials. Transparency of materials has much to do with their wide energy gaps that tend to increase with their high ionicity. Despite the insulating nature of wide-gap transparent materials in general, some oxides exhibit not only semiconducting properties, but also ferromagnetism, light emission, piezoelectricity, and superconductivity [84].

Solid State Properties and Transparency

Definition of Transparency [85]

Transparent materials are colorless with respect to visible radiation. Visible light has a spectrum in the range of 360–830 nm (energy of 3.4–1.5 eV) and if there is no absorption, reflection, and scattering in this spectral range then the material can be considered transparent. The transparency is primarily dependent on the bandgap (E_g) and plasma frequency (ω_p). If E_g is greater than 3.4 eV, band transitions do not occur in the visible radiation spectrum, and such radiation is transmitted through the material. On the other hand, light with energy less than ω_p cannot penetrate inside the plasma and is reflected by the carriers whose plasma frequency, ω_p is given by

$$\omega_p^2 = \frac{nq^2}{\varepsilon m^*} \tag{4.8}$$

where n is the carrier density, q the charge, ε the dielectric constant, and m* the effective mass.

Applications and Functions of Transparent Devices

The transparent devices are demanded to have the same properties as other electronic materials; they must be electrical conductors, semiconductors, or insulators. In addition to these fundamental properties, dielectric and magnetic characteristics would be invaluable for the fabrication of memory and spintronic devices. The properties of light emission, reflection, transmission, and absorption would create extra functionalities that are not just an extension of conventional electronics.

On the other hand, the large bandgap required for transparency also means the existence of a low density of thermally excited carriers. Production of conducting materials requires the incorporation and control of impurities for formation of suitable donor or acceptor levels. Doping is a common problem for both the formation of pn junctions and control of the magnetic properties; thus, proper fabrication technique is required.

Deposition System

Thin Film Fabrication Techniques

Thin films are usually deposited using methods that include sputtering, PLD, CVD, vacuum evaporation, spraying, and dipping. Conducting transparent films are usually produced by sputtering, which enables deposition on large areas [85]. The requirements for fabricating commercial transparent devices are low temperature and deposition on large area substrates. Generally, high substrate temperatures are known to improve the quality of thin films but the physical properties of substrates limits the maximum temperature that can be used. In particular, it is necessary to develop methods for deposition below 150°C on plastic substrates for flexible device applications. Fabrication techniques accompanied by chemical reactions usually need high temperature, and it is also difficult by physical fabrication techniques to fulfill the two requirements simultaneously. Moreover, studies show that fabrication conditions affect postdeposition processing such as a patterning [86].

Combinatorial Method

The combinatorial method has its origins in the organic synthesis of drugs but we recently extended the concept for synthesis of thin films [87]. This method enables the systematic change of the fabrication parameters, such as composition and/or substrate temperature, on a single substrate.

Figure 4.51 illustrates an example of this method. The combination of a movable mask and laser substrate heating enables optimization of temperature and composition simultaneously in one experiment. This method

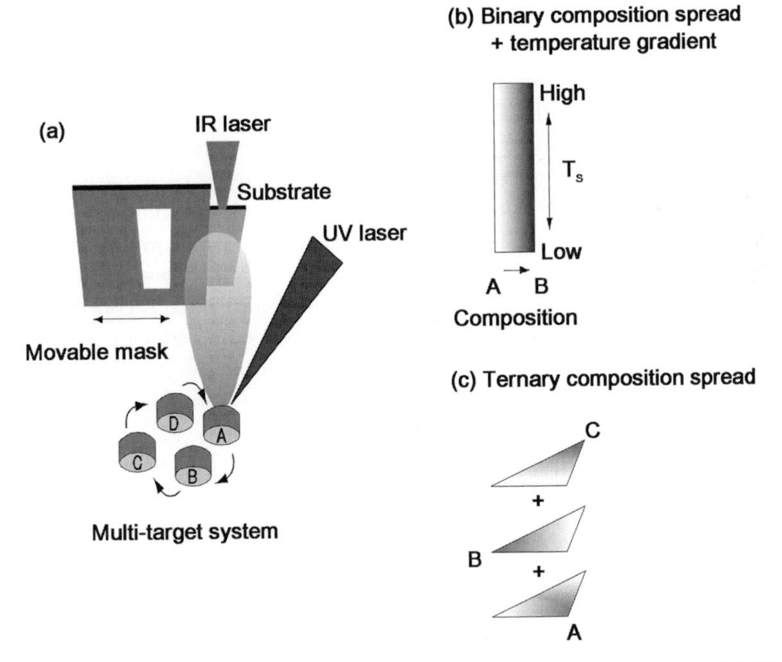

Fig. 4.51. (**a**) Schematic representation of combinatorial PLD system. (**b**) Binary composition spread and temperature gradient method. (**c**) Ternary composition spread film

was pivotal in the development of Co-doped TiO_2, a room temperature ferromagnetic material and p-type ZnO thin films.

Transparent Materials and Devices

Transparent Conducting Films

Transparent conducting films are usually defined as thin films that transmit 80% of incident visible radiation and have specific resistances of less than $10^{-3}\,\Omega$ cm. These films are usable for flat panel displays, touch panels, and electrodes for solar cells. Examples include In_2O_3, SnO_2, ZnO, and CdO with specific resistivity of 10^{-3}–$10^{-4}\,\Omega$ cm and carrier densities of 10^{20}–10^{21} cm^{-3} [85, 86]. Indium tin oxide (Sn: In_2O_3 or "ITO") films exhibit the highest transparency and lowest specific resistivity, and are widely used in industry. However, scarcity of indium have led researchers to investigate alternative materials such as Al or Ga doped ZnO [86]. Furthermore, doping with Mg and Cd would enable control of the bandgap [88] and extend device applications. The aforementioned materials all show n-type conduction with carriers (electrons) generated due to impurities or oxygen deficiency and the conduction bands consist of isotropic s-orbitals. Recently, Nb doped TiO_2, with anisotropic d-bands, was reported as a new transparent conducting electrode [89].

Since the valence band of the majority of oxides is composed of 2p-orbitals of oxygen, even if it was possible to achieve p-type hole doping, the films would be strongly localized and carrier mobility would be low. Thus it has proved difficult to produce p-type transparent oxides. As a solution to this problem, Kawazoe and Hosono et al. reported on the synthesis of transparent conducting films of p-type $CuAlO_2$ based on their original concept that cationic species are required to have a closed shell whose energy is almost comparable to those of the 2p level of oxygen anions [90]. The same group also succeeded in produced homo-pn junctions, light emitting diodes, and have recently extended their studies to the development of p-type conducting oxides including oxysulfides, and Rh^{3+} oxides, which have a quasi-closed shell d-orbital [91]. In particular, amorphous $ZnRh_2O_4$ can be formed at room temperature with potentially wide ranging applications as flexible devices [92].

Transparent Field Effect Transistor

Transparent transistors would extend the applications of glass as well as replacing the amorphous-Si (a-Si) FETs (which absorb in the visible light and require light blocking films for stable operation) currently used in liquid crystal displays. The industrial use of transparent transistors would enable effective use of backlights, reduction of power consumption, and simplification of processing. FETs employing transparent oxide active layers can be

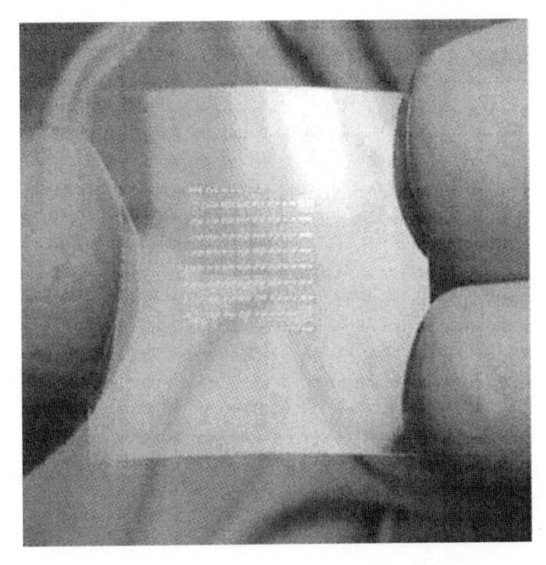

Fig. 4.52. Flexible, transparent a-$InGaO_3(ZnO)_4$ transistor

divided into either thin film or bulk transistors. There is strong interest in commercializing the former and the latter is being studied for controlling the physical properties by electrostatic field carrier doping.

Transparent thin film transistors with ZnO [93, 94] and $InGaO_3(ZnO)_4$ [95,96] active layers were reported to work as good as or even better than a-Si devices (Fig. 4.52). Bulk oxide transistors were fabricated using $SrTiO_3$ [97] and $KTaO_3$ [98] single crystals, exhibiting an on/off ratio and mobility exceeding 10^4 and $0.4\,cm^2\,Vs^{-1}$, respectively. We recently confirmed clear transistor operation in bulk and thin film TiO_2. These results show promise for fabricating homojunctions with Nb-doped transparent conducting films and the Co-doped transparent magnetic films described in the following section.

Transparent Magnetic Materials

TiO_2 is well known as a photocatalyst and for the fabrication of solar cells. It is not known for its magnetic properties. However, since we reported that Co-doped TiO_2 showed ferromagnetism above room temperature [99], it has attracted much attention as a magnetic material (Fig. 4.53). This material is promising for spintronics and an actually TMR device has recently been fabricated [100], which shows the possibility of using this material as a spin injection electrode. There are intriguing aspects of the magnetism in this system and controversy regarding its origin [101].

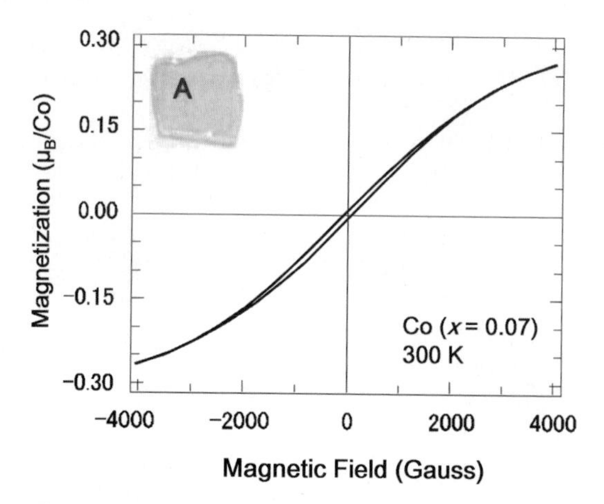

Fig. 4.53. M–H hysteresis curve of Co-doped TiO_2 thin film

References

1. K. Mishra, P. Parikh, and W. Yi-Feng: Proceedings of the IEEE 90 (2002) 1022
2. L.F. Eastman and U.K. Mishra: Spectrum, IEEE 39 (2002) 28
3. Y. Ohno and M. Kuzuhara: IEEE Trans. Electron Devices 48 (2001) 517
4. W.L. Pribble, J.W. Palmour, S.T. Sheppard, R.P. Smith, S.T. Allen, T.J. Smith, Z. Ring, J.J. Sumakeris, A.W. Saxler, and J.W. Milligan: in Proceedings of the 2002 IEEE MTT-S International Microwave Symposium, Vol. 3, 2002, pp. 1819–1822
5. T. Kikkawa, T. Maniwa, H. Hayashi, M. Kanamura, S. Yokokawa, M. Nishi, N. Adachi, M. Yokoyama, Y. Tateno, and K. Joshin: in Proceedings of the 2004 IEEE MTT-S International Microwave Symposium, Vol. 3, 2004, pp. 1347–1350
6. Y. Okamoto, Y. Ando, K. Hataya, T. Nakayama, H. Miyamoto, T. Inoue, M. Senda, K. Hirata, M. Kosaki, N. Shibata, and M. Kuzuhara: IEEE Trans. Microw. Theory Tech. 52 (2004) 2536
7. W. Nagy, S. Singhal, R. Borges, J.W. Johnson, J.D. Brown, R. Therrien, A. Chaudhari, A.W. Hanson, J. Riddle, S. Booth, P. Rajagopal, E.L. Piner, and K.J. Linthicum: in Proceedings of the 2005 IEEE MTT-S International Microwave Symposium, 2005, p. 4
8. Y.F. Wu, M. Moore, A. Saxler, P. Smith, P.M. Chavarkar, and P. Parikh: in Proceedings of the 2003 International Electron Devices Meeting, 2003, pp. 23.5.1–23.5.3
9. M. Kuzuhara: in Proceedings of the Microwave Workshops and Exhibition MWE2003, Yokohama, November 26–28, 2003, p. 85
10. M.A. Khan, J.N. Kuznia, D.T. Olson, W.J. Schaff, J.W. Burm, and M.S. Shur: Appl. Phys. Lett. 65 (1994) 1121
11. O. Ambacher, B. Foutz, J. Smart, J.R. Shealy, N.G. Weimann, K. Chu, M. Murphy, A.J. Sierakowski, W.J. Schaff, L.F. Eastman, R. Dimitrov, A. Mitchell, and M. Stutzmann: J. Appl. Phys. 87 (2000) 334
12. K.K. Chu, P.C. Chao, M.T. Pizzella, R. Actis, D.E. Meharry, K.B. Nichols, R.P. Vaudo, X. Xu, J.S. Flynn, J. Dion, and G.R. Brandes: IEEE Electron Device Lett. 25 (2004) 596

13. Y. Ando, Y. Okamoto, K. Hataya, T. Nakayama, H. Miyamoto, T. Inoue, and M. Kuzuhara: in Proceedings of the 2003 International Electron Devices Meeting, 2003, pp. 23.1.1–23.1.4
14. T. Inoue, Y. Ando, H. Miyamoto, T. Nakayama, Y. Okamoto, K. Hataya, and M. Kuzuhara: in Proceedings of the 2004 IEEE MTT-S International Microwave Symposium, Vol. 3, 2004, pp. 1649–1652
15. T. Hashizume, J. Kotani, and H. Hasegawa: Appl. Phys. Lett. 84 (2004) 4884
16. T. Nanjo, N. Miura, T. Oishi, M. Suita, Y. Abe, T. Ozeki, S. Nakatsuka, A. Inoue, T. Ishikawa, Y. Matsuda, H. Ishikawa, and T. Egawa: Jpn J. Appl. Phys. Part. 1 43 (2004) 1925
17. T. Nakayama, H. Miyamoto, Y. Ando, Y. Okamoto, T. Inoue, K. Hataya, and M. Kuzuhara: Appl. Phys. Lett. 85 (2004) 3775
18. S.C. Binari, P.B. Klein, and T.E. Kazior: in Proceedings of the IEEE 90 (2002) 1048
19. B.M. Green, K.K. Chu, E.M. Chumbes, J.A. Smart, J.R. Shealy, and L.F. Eastman: IEEE Electron Device Lett. 21 (2000) 268
20. Y. Okamoto, Y. Ando, T. Nakayama, K. Hataya, H. Miyamoto, T. Inoue, M. Senda, K. Hirata, M. Kosaki, N. Shibata, and M. Kuzuhara: IEEE Trans. Electron Devices 51 (2004) 2217
21. Y. Okamoto, A. Wakejima, K. Matsunaga, Y. Ando, T. Nakayama, K. Kasahara, K. Ota, Y. Murase, K. Yamanoguchi, T. Inoue, and H. Miyamoto: in Proceedings of the 2005 IEEE MTT-S International Microwave Symposium, 2005, pp. 491–494
22. M. Micovic, N.X. Nguyen, P. Janke, W.S. Wong, P. Hashimoto, L.M. McCray, and C. Nguyen: Electron. Lett. 36 (2000) 358
23. M. Kanamura, T. Kikkawa, and K. Joshin: in Proceedings of the 2004 International Electron Devices Meeting, 2004, pp. 799–802
24. S.T. Allen, R.A. Sadler, T.S. Alcorn, J. Sumakeris, R.C. Glass, C.H. Carter, and J.W. Palmour: Mater. Sci. Forum 264–268 (1998) 953
25. J.W. Palmour, S.T. Allen, S.T. Sheppard, W.L. Pribble, R.A. Sadler, T.S. Alcorn, Z. Ring, and C.H. Carter, Jr.: in Proceedings of the 57th Annual Device Research Conference, IEEE Conference Proceedings, Santa Barbara, 28–30 June, 1999, pp. 38–41
26. J.W. Palmour, S.T. Sheppard, R.P. Smith, S.T. Allen, W.L. Pribble, Z. Ring, A. Ward, and J. Milligan: in Proceedings of the Topical Workshop on Heterostructure Microelectronics (TWHM 2003), Okinawa, January 21–24, 2003, pp. 104–105
27. A.W. Morse, P.M. Esker, S. Sriram, J.J. Hawkins, L.S. Chen, J.A. Ostop, T.J. Smith, C.D. Davis, R.R. Barron, R.C. Clarke, R.R. Siergiej, and C.D. Brandt: in Proceedings of the IEEE MTT-S International Microwave Symposium, IEEE Conference Proceedings, Vol. 1, Denver, 8–13 June, 1997, pp. 53–56
28. M. Arai, S. Ono, and C. Kimura: Electron. Lett. 40 (2004) 1026
29. T. Maki, S. Shikama, M. Komori, Y. Sakaguchi, K. Sakuta, and T. Kobayashi: Jpn J. Appl. Phys. Part. 2 31 (1992) L1446
30. K. Hayashi, S. Yamanaka, H. Okushi, and K. Kajimura: Appl. Phys. Lett. 68 (1996) 376
31. K. Tsugawa, K. Kitatani, H. Noda, A. Hokazono, K. Hirose, M. Tajima, and H. Kawarada: Diam. Relat. Mat. 8 (1999) 927

32. Y. Yun, T. Maki, and T. Kobayashi: J. Appl. Phys. 82 (1997) 3422
33. N. Kawakami, Y. Yokota, K. Hayashi, T. Tachibana, and K. Kobashi: Diam. Relat. Mat. 14 (2005) 509
34. H. Shiomi, Y. Nishibayashi, N. Toda, and S. Shikata: IEEE Electron Device Lett. 16 (1995) 36
35. H. Umezawa, K. Tsugawa, S. Yamanaka, D. Takeuchi, H. Okushi, and H. Kawarda: Jpn J. Appl. Phys. Part. 2 38 (1999) L1222
36. A. Aleksov, A. Denisenko, U. Spitzberg, W. Ebert, and E. Kohn: IEEE Electron Device Lett. 23 (2002) 488
37. H. Taniuchi, H. Umezawa, T. Arima, M. Tachiki, and H. Kawarada: IEEE Electron Device Lett. 22 (2001) 390
38. H. Matsudaira, S. Miyamoto, H. Ishizaka, H. Umezawa, and H. Kawarada: IEEE Electron Device Lett. 25 (2004) 480
39. D.M. Caughey and R.E. Thomas: Proceedings of the IEEE 55 (1967) 2192
40. F. Nava, C. Canali, M. Artuso, E. Gatti, P.F. Manfredi, and S.F. Kozlov: IEEE Trans. Nucl. Sci. 26 (1979) 308
41. H. Umezawa, S. Miyamoto, H. Matsudaira, H. Ishizaka, K.S. Song, M. Tachiki, and H. Kawarada: IEICE Trans. Electron. E86C (2003) 1949
42. K. Hirama et al.: IEEE Electron Device Lett. (2006) submitted.
43. M. Kasu, K. Ueda, H. Ye, Y. Yamauchi, S. Sasaki, S.T. Makimoto, Electron. Lett. 41 (2005) 1249
44. S. Yoshida: Oyo Butsuri 68 (1999) 787 (in Japanese)
45. S. Yoshida: IEICE Transactions on Electronics (Japanese Edition) J86-C (2003) 412 (in Japanese)
46. W. Saito, I. Omura, T. Ogura, and H. Ohashi: Solid State Electron. 48 (2004) 1555
47. H. Gotoh and Y. Otsuka: Research Progress Report, 43 (JSPS Committee on Wide Bandgap Electronic Devices), Tokyo, 2005, pp. 43–49
48. S. Yoshida and J. Suzuki: Jpn J. Appl. Phys. Part. 2 37 (1998) L482
49. S. Yoshida and J. Suzuki: J. Appl. Phys. 85 (1999) 7931
50. S. Yoshida and J. Suzuki: MRS Internet J. Nitride Semicond. Res. 5 (2000) Art. No. W4.8 Suppl. 1
51. S. Yoshida and H. Ishii: Phys. Status Solidi A - Appl. Res. 188 (2001) 243
52. K. Kumakura, Y. Yamauchi, and T. Makimoto: Phys. Status Solidi C 2 (2005) 2589
53. S. Yoshida, H. Ishii, and J. Li: Mater. Sci. Forum 389–393 (2002) 1527
54. S. Yoshida, D.L. Wang, and M. Ichikawa: Jpn J. Appl. Phys. Part. 2 41 (2002) L820
55. S. Yoshida, J. Li, T. Wada, and H. Takehara: in Proceedings of the IEEE 15th International Symposium on Power Semiconductor Devices and ICs (ISPSD'03), IEEE Conference Proceedings, Cambridge, UK, 14–17 April 2003, pp. 58–61
56. M. Hikita, M. Yanagihara, K. Nakazawa, H. Ueno, Y. Hirose, T. Ueda, Y. Uemoto, T. Tanaka, D. Ueda, and T. Egawa: in Proceedings of the IEEE International Electron Devices Meeting (IEDM'04), IEEE Conference Proceedings, San Francisco, CA, 13–15 December 2004, pp. 803–806
57. N. Ikeda, L. Jiang, and S. Yoshida: in Proceedings of the 16th International Symposium on Power Semiconductor Devices and ICs (ISPSD'04), IEEE Conference Proceedings, 24–27 May 2004, pp. 369–372

58. S. Yoshida, J. Li, and N. Ikeda: Phys. Status Solidi C 2 (2005) 2593
59. S. Yoshida, J. Li, T. Wada, and H. Takehara: in Proceedings of the GaN and Related Alloys - Materials Research Society Symposium: Y (H.M. Ng, M. Wraback, K. Hiramatsu, and N. Grandjean, eds.), MRS Symposium Proceedings, Vol. 798, Boston, December 1–5, 2003, p. Y.7.3
60. K. Matocha and R.J. Gutmann: IEEE Trans. Electron Devices 52 (2005) 6
61. A.P. Zhang, J.W. Johnson, F. Ren, J. Han, A.Y. Polyakov, N.B. Smirnov, A.V. Govorkov, J.M. Redwing, K.P. Lee, and S.J. Pearton: Appl. Phys. Lett. 78 (2001) 823
62. S. Yoshida, N. Ikeda, J. Li, T. Wada, and H. Takehara: in Proceedings of the 16th International Symposium on Power Semiconductor Devices and ICs (ISPSD'04), IEEE Conference Proceedings, 24–27 May, 2004, pp. 323–326
63. S. Yoshida, J. Li, N. Ikeda, and K. Hataya: Phys. Status Solidi C 2 (2005) 2602
64. G. Deboy, N. Marz, J.P. Stengl, H. Strack, J. Tihanyi, and H. Weber: in Proceedings of the International Electron Devices Meeting (IEDM'98), IEEE Conference Proceedings, San Francisco, CA, 6–9 December, 1998, pp. 683–685
65. W. Saitoh, I. Omura, K. Tokano, T. Ogura, and H. Ohashi: in Proceedings of the 14th International Symposium on Power Semiconductor Devices and ICs (ISPSD'02), IEEE Conference Proceedings, Santa Fe, NM, 4–7 June 2002, pp. 33–36
66. H. Matsunami, ed.: Technology of Semiconductor SiC and Its Application, The Nikkan Kogyo Shimbun, Tokyo, 2003
67. K. Arai and S. Yoshida: Basics and applications of SiC devices, Ohmsha Ltd., Tokyo, 2003
68. Y. Sugawara: Mater. Sci. Forum 457–460 (2004) 963
69. Y. Sugawara, D. Takayama, K. Asano, A. Agarwal, S. Ryu, J. Palmour, and S. Ogata: in Proceedings of the 16th International Symposium on Power Semiconductor Devices and ICs (ISPSD'04), IEEE Conference Proceedings, Kitakyushu, 24–27 May 2004, pp. 365–368
70. A.R. Powell, R.T. Leonard, M.F. Brady, S.G. Müller, V.F. Tsvetkov, R. Trussell, J.J. Sumakeris, H.M. Hobgood, A.A. Burk, R.C. Glass, and C.H. Carter Jr: Mater. Sci. Forum 457–460 (2004) 41
71. H.R. Chang, E. Hanna, and A.V. Radun: in Proceedings of the IEEE 34th Annual Power Electronics Specialist Conference (PESC'03), IEEE Conference Proceedings, Vol. 1, Acapulco, Mexico, 15–19 June 2003, pp. 211–216
72. Y. Tanaka et al.: The 52nd Spring Meeting of the Japanese Society of Applied Physics, Extended Abstract, Vol. 1, 2005, 459
73. S. Harada et al.: The 52nd Spring Meeting of the Japanese Society of Applied Physics, Extended Abstract, Vol. 1, 2006, 427
74. M.K. Das, J.J. Sumakeris, B.A. Hull, J.T. Richmond, S. Krishnaswami, and A.R. Powell: Mater. Sci. Forum 483–485 (2005) 965
75. http://www.rohm.co.jp/news/sicpower2-j.html, 25 July 2006, Rohm Co. Ltd
76. J.H. Zhao, K. Tone, X. Li, P. Alexandrov, L. Fursin, and M. Weiner: in Proceedings of the IEEE 15th International Symposium on Power Semiconductor Devices and ICs (ISPSD'03), IEEE Conference Proceedings, Cambridge, UK, 14–17 April 2003, pp. 50–53
77. A.K. Agarwal, S.H. Ryu, C. Capell, J.T. Richmond, J.W. Palmour, H. Bartlow, T.P. Chow, S. Scozzie, W. Tipton, S. Baynes, and K.A. Jones: Mater. Sci. Forum 457–460 (2004) 1141

78. Y. Sugawara: Denki-Gakkai Shi 125 (2005) 21
79. L. Nilsson, O. Groening, C. Emmenegger, O. Kuettel, E. Schaller, L. Schlapbach, H. Kind, J.M. Bonard, and K. Kern: Appl. Phys. Lett. 76 (2000) 2071
80. N. Jiang, R. Koie, T. Inaoka, Y. Shintani, K. Nishimura, and A. Hiraki: Appl. Phys. Lett. 81 (2002) 526
81. H. Makita, K. Nishimura, N. Jiang, A. Hatta, T. Ito, and A. Hiraki: Thin Solid Films 282 (1996) 279
82. A. Hiraki: Appl. Surf. Sci. 162 (2000) 326
83. URL: http://www.dialight.com, 25 July 2006, Dialight Ltd
84. H. Koinuma, ed.: Oxide Electronics, Advanced Electronics Series, Vol. I22, Baifukan, Tokyo, 2001
85. Technology of transparent conducting films: Japan Society for the Promotion of Science Committee 166 on Electronic Materials, Ohmsha, Tokyo, 1999
86. T. Minami: Semicond. Sci. Technol. 20 (2005) S35
87. H. Koinuma and M. Kawasaki, eds.: Combinatorial Technology, Maruzen Publishing, Tokyo, 2004
88. K. Matsubara, H. Tampo, H. Shibata, A. Yamada, P. Fons, K. Iwata, and S. Niki: Appl. Phys. Lett. 85 (2004) 1374
89. Y. Furubayashi, T. Hitosugi, Y. Yamamoto, K. Inaba, G. Kinoda, Y. Hirose, T. Shimada, and T. Hasegawa: Appl. Phys. Lett. 86 (2005) Art. No. 252101
90. H. Kawazoe, M. Yasukawa, H. Hyodo, M. Kurita, H. Yanagi, and H. Hosono: Nature 389 (1997) 939
91. H. Ohta, K. Nomura, H. Hiramatsu, K. Ueda, T. Kamiya, M. Hirano, and H. Hosono: Solid State Electron. 47 (2003) 2261
92. S. Narushima, H. Mizoguchi, K. Shimizu, K. Ueda, H. Ohta, M. Hirano, T. Kamiya, and H. Hosono: Adv. Mater. 15 (2003) 1409
93. J. Nishii, F.M. Hossain, S. Takagi, T. Aita, K. Saikusa, Y. Ohmaki, I. Ohkubo, S. Kishimoto, A. Ohtomo, T. Fukumura, F. Matsukura, Y. Ohno, H. Koinuma, H. Ohno, and M. Kawasaki: Jpn J. Appl. Phys. Part. 2 42 (2003) L347
94. T.I. Suzuki, A. Ohtomo, A. Tsukazaki, F. Sato, J. Nishii, H. Ohno, and M. Kawasaki: Adv. Mater. 16 (2004) 1887
95. K. Nomura, H. Ohta, K. Ueda, T. Kamiya, M. Hirano, and H. Hosono: Science 300 (2003) 1269
96. K. Nomura, H. Ohta, A. Takagi, T. Kamiya, M. Hirano, and H. Hosono: Nature 432 (2004) 488
97. K. Shibuya, T. Ohnishi, M. Lippmaa, M. Kawasaki, and H. Koinuma: Appl. Phys. Lett. 85 (2004) 425
98. K. Ueno, I.H. Inoue, T. Yamada, H. Akoh, Y. Tokura, and H. Takagi: Appl. Phys. Lett. 84 (2004) 3726
99. Y. Matsumoto, M. Murakami, T. Shono, T. Hasegawa, T. Fukumura, M. Kawasaki, P. Ahmet, T. Chikyow, S. Koshihara, and H. Koinuma: Science 291 (2001) 854
100. H. Toyosaki, T. Fukumura, K. Ueno, M. Nakano, and M. Kawasaki: Jpn J. Appl. Phys. Part. 2 44 (2005) L896
101. T. Fukumura, Y. Yamada, H. Toyosaki, T. Hasegawa, H. Koinuma, and M. Kawasaki: Appl. Surf. Sci. 223 (2004) 62

5

Novel Nano-Heterostructure Materials and Related Devices

5.1 Nanostructure Devices and Materials

5.1.1 Single Photon Devices (H. Kumano, I. Suemune)

In the field of quantum information science, a single "quantum bit" (qubit), which is a linear superposition of two quantum states forming an orthonormal basis, is treated as a quantum carrier of information. Essential functions required for the devices employed in quantum information science are thus preparation, operation, and detection of single qubits. Among several physical systems to form qubits, photons are one of the most promising systems for robustness against decoherence. After the introduction of first complete protocol by Bennett and Brassard in 1984 (BB84) [1], single photon devices have attracted much attention since these devices can provide absolutely secure communications where eavesdropping will never take place based on the principle of quantum mechanics, so-called quantum key distribution (QKD). In the QKD system, every piece of information will be encoded on single photons. Thus single photon devices that deliver photon qubits on-demand will play a prominent role for the implementation of QKD. In this section, recent advances in the field of semiconductor single-photon devices are overviewed, focusing on the materials that can handle the ultraviolet-near-visible photon wavelength region.

Material Systems for Single Photon Devices and Advantages of Wide Bandgap Semiconductor Quantum Dots

Concerning conventional photon sources for practical QKD systems, weak coherent laser pulses (WCP) [2, 3] have been employed, which provide photons in the coherent state. This inherently leads to inevitable photon number fluctuations originating from the uncertainty principle between the photon numbers and photon phases, and the photon generation events essentially become probabilistic. In the case of BB84, this will lead to low QKD rates since most of the pulses are empty and more importantly the WCP can be never completely robust against a so-called splitter attack due to the residual multiphoton emission probabilities regardless of a selection of average photon number per pulse.

In contrast to coherent light, nonclassical photons in a number state have a pure quantum nature as a consequence of full quantization of the radiation field, in which no photon number fluctuation is involved [4]. Therefore, deterministic single photon sources that will emit photons in the number state are necessary in order to overcome these critical issues. For generating photons in the number state, the preparation of a material system that has an isolated two-level system (for a review, see [5]) is essential. For example, single atoms [6, 7], single molecules [8], nitrogen-vacancy centers in diamond [9], and semiconductor quantum dots (QDs) [10, 11] are among them. From the viewpoint of structural stability, high photon emission rates, current-driving capabilities [12], and well-established processing and device technologies, semiconductor QDs is the most promising system.

So far, in terms of photon wavelengths, most of the researches have been focused on telecommunication wavelengths applicable to QKD in optical-fiber links. However for realizing flexible and expandable secure network systems to satisfy versatile purposes, single photon sources in the ultraviolet-near-visible wavelength region will be of prime importance for constructing complementary links to the optical-fiber networks via free-space links. This will be favored by the availability of highly sensitive and low-noise Si single photon detectors (SPDs). For this purpose, single photon emissions from individual CdSe QDs [10,13,14], GaN QDs [15], impurity states in ZnSe [16], Te clusters in ZnSe [17], and InAlAs QDs [18] have been so far demonstrated. Especially, single photon emissions were reported at room temperature and at 200 K from colloidal [10] and epitaxially-grown CdSe QDs [13], respectively. Among the critical criteria to realize practical high bit rate QKD, wavelength matching of the single photon sources to the SPDs with the highest quantum efficiencies is one of the most important conditions to keep the high total quantum efficiencies of the QKD systems. In this regard, $In_{0.75}Al_{0.25}As$ QDs match quite well to the highest sensitivity wavelength region of the Si-SPD. In the following, $In_{0.75}Al_{0.25}As$ QDs will be mainly focused on and will be dealt with from this viewpoint.

Single Dot Spectroscopy

The QD sample was grown on a (001) GaAs substrate by molecular-beam epitaxy. These QDs were prepared in the Stranski–Krastanow (S–K) growth mode on an $Al_{0.3}Ga_{0.7}As$ layer and were sandwiched with another $Al_{0.3}Ga_{0.7}As$ layer. The topmost surface was terminated with a GaAs capping layer. Figure 5.1 shows a photoluminescence (PL) spectrum measured from a macroscopic region of the $In_{0.75}Al_{0.25}As$ QDs at 22 K. It will be clear that the inhomogeneous, broadened PL spectrum with ∼90 meV full-width at

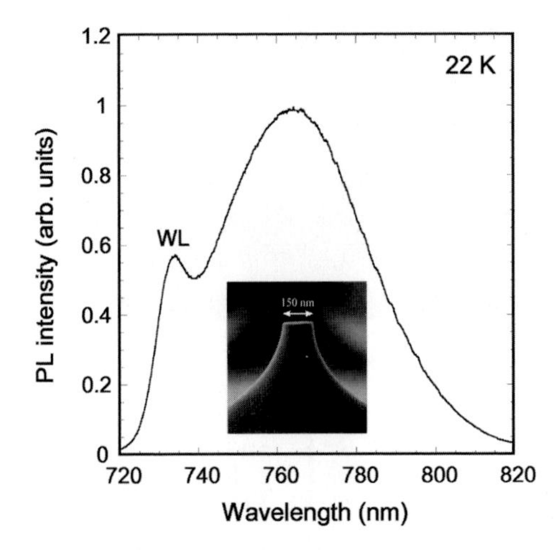

Fig. 5.1. Macroscopic PL spectrum of Stranski–Krastanow InAlAs QDs at 22 K. Inhomogeneous broadening is around 90 meV. Emission from wetting layer is observed at ~730 nm. TEM image of typical mesa structure employed in this study is shown in the inset

half maximum (FWHM) covers the wavelength region with high atmospheric transmittance and the high quantum efficiency of Si-SPDs. The additional structure observed at around 730 nm stems from the wetting layer (WL) adjacent to the S–K QDs. After the growth, the sample was etched into mesa structures with the top lateral size of ~150 nm for isolating single QDs. Scanning electron microscopy of the typical mesa structure is shown in the inset of Fig. 5.1 and the details of this sample preparation are described in [19]. Figure 5.2a shows a streak image obtained from a single InAlAs QD under the WL excitation at the excitation power (P_{ex}) of 2 μW. The corresponding time-integrated PL spectrum is illustrated in Fig. 5.2b. Several sharp emission lines labeled by L1–L5 were well resolved. Although the energy position of these emission lines is different from dot to dot, a series of emission lines in which the energy separation is around 4–5 meV were observed in most of the examined QDs. The L1 and L3 emission lines whose energy separation is 4.5 meV are the typical case. Thus, these two major emission lines L1 and L3 will be mainly discussed here. The physical origin and optical properties of the other emission lines are discussed elsewhere [20]. PL decay curves of the two lines are shown in Fig. 5.2c. Under the weak excitation where the initial mean exciton number in the QD is estimated to be $\mu = 0.4$, both emission lines show single exponential decay after a sharp rise, and the lifetimes of 1.02 ns and 0.55 ns were fitted for the L1 and L3 lines, respectively. The excitation power dependence of the L1 and L3 lines was examined, and the linear and bilinear increases of the PL intensity with higher excitation power were observed as shown in Fig. 5.2d, which are the characteristic features of exciton and biexciton recombinations, respectively. Moreover, higher

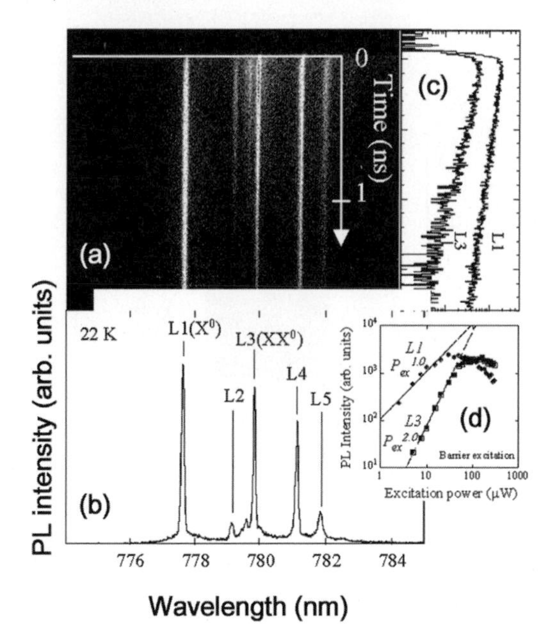

Fig. 5.2. (**a**) Streak image of emission lines from single mesa structure under WL excitation. Full scale of vertical axis is 2 ns. (**b**) Time-integrated PL spectrum corresponds to the streak image. Several emission lines denoted by L1–L5 are observed. PL decay curves (**c**) and excitation power dependence (**d**) of the L1 and L3 lines are also shown

resolution measurements revealed that both the lines exhibit fine-structure splitting. These results reveal that the L1 and L3 peaks are originated from the neutral exciton X^0 and biexciton XX^0 emissions, respectively. Under the weak excitation condition where the mean exciton number in the QD is less than unity, the present QD system composed of the XX^0, X^0, and ground $|0>$ states can be well approximated to be an isolated two-level system. These results suggest that single-photon generation in the number state will be possible based on the Pauli exclusion principle in the thus prepared QD two-level system. This will be examined in "Single Photon Generation".

Single Photon Generation

In the following, nonclassical single-photon generation will be examined employing second-order photon correlation measurements. A direct way to evaluate the statistics of emitted photon numbers is directly "counting" the number of photons per excitation cycle. However, such a photon counter which can distinguish photon numbers with high efficiency is still under development [21]. In this work, therefore, Si-avalanche photodiodes are employed as SPDs and the second-order photon correlation function $g^{(2)}(\tau)$ was measured

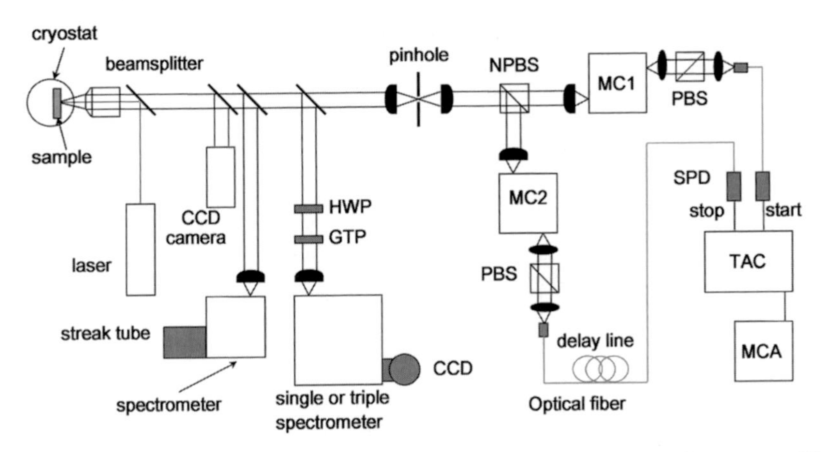

Fig. 5.3. Overall experimental setup. CCD imaging camera, streak camera, CCD detector, and HBT setup can be switched by flipper mirrors. Sample temperatures were 22 K

with the Hanbury-Brown and Twiss (HBT) setup [22] to determine the photon statistics. In Fig. 5.3, our experimental setup is depicted. Photons emitted from a QD are collected by an objective lens with a numerical aperture of 0.42 and are divided into the two optical paths by a 50/50 nonpolarized beamsplitter (NPBS). Each path is then dispersed by the respective monochromator of MC1 and MC2 and a single emission line is filtered with a 1-nm-FWHM transmission bandwidth. The transmitted photons are introduced into each SPD through optical fibers. Each time a photon is detected in one of the two SPDs, an electronic pulse is produced and sent to a time-to-amplitude converter (TAC). The outputs of the SPDs serve as the start and stop signals for the correlation measurements. The temporal difference between the two outputs, which is defined as $\tau = t_{\mathrm{stop}} - t_{\mathrm{start}}$, is translated into a pulse height. Finally, a multichannel analyzer (MCA) accumulates the single event data and produces histograms of the number of counts as a function of the delay time τ. The intuitive explanation of the photon correlation measurement by the HBT setup is as follows: Since a single photon will not be divided into two, the photon never clicks the two SPDs simultaneously, which are located at the same optical distances from the NPBS, when a single photon is generated from the QD.

To confirm the deterministic single-photon emission per pulsed excitation, autocorrelation measurements were performed. The X^0 line in Fig. 5.2b was introduced into the HBT setup. The excitation power was set to satisfy the condition on the average initial exciton number, $\mu < 1$. The measured photon correlation function is illustrated in Fig. 5.4 The observed peak intervals of ~12.2 ns correspond to the excitation laser pulse repetition frequency of 82 MHz. Strong suppression of the second-order correlation function at zero time delay, $g^{(2)}(0) \sim 0$, was observed. This is a clear manifestation of photon

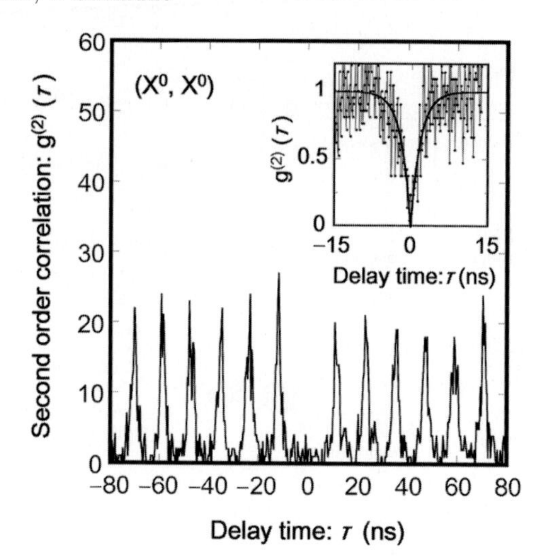

Fig. 5.4. Result of autocorrelation measurement for the X^0 line. The Ti:Sapphire laser with the wavelength of 730 nm was used for the excitation. Observed peak separation of ~12.2 ns corresponds to an excitation laser frequency of 82 MHz. Clear photon antibunching is observed at $\tau = 0$. (Inset) Second-order correlation function $g^{(2)}(\tau)$ under CW excitation

antibunching, i.e., nonclassical photon generation in the number state [4]. This striking feature is completely different from the conventional coherent light in which photon number fluctuation is inevitable and thus photon antibunching never appears. Since the second correlation function at zero time delay is expressed as $g^{(2)}(0) = 1 - 1/n$ for a n-photon number state, the observation of $g^{(2)}(0) \sim 0$ indicates that one and only one photon emission, i.e., single photon emission in the number state ($n = 1$) is realized.

The photon statistics of photons emitted from a single InAlAs QD were also studied under CW excitation. Detailed description of the optical properties on this sample is given in [18] and [23], but an excitonic emission line was used for the CW photon correlation measurement. The inset in Fig. 5.4 shows the measured second-order correlation function $g^{(2)}(\tau)$. In this experiment, the excitation power was set to give a steady-state exciton population in a QD less than 0.1. As was observed in the pulsed excitation, the second-order correlation function at $\tau = 0$ was strongly suppressed. This photon antibunching behavior, described by $g^{(2)}(0) < 1$, is a clear manifestation that the photon emission from the InAlAs QD takes place one by one even under the CW excitation. These results demonstrate the first nonclassical single photon generation from a single InAlAs QD, which will be applicable to single photon sources for free-space links.

Future Prospects

In summary, recent advances in the field of semiconductor single photon sources were overviewed and successful demonstration of the triggered single photon source was briefly introduced. Upgrading these results toward standard and user-friendly devices such as semiconductor laser diodes will require further progress in many aspects. One of the most crucial issues will be the fact that the large fraction of the photons generated from the QD cannot be extracted outside of the device due to the total internal reflection at the air–semiconductor interface. With a simple estimation, the photon extraction efficiency will be as low as $1/4n^2 \sim 0.02$ for a typical semiconductor refractive index of $n = 3.5$. Another issue will be the quantum efficiencies of SPDs. Commercially available Si-SPDs show the high maximum quantum efficiency of ~75% at around the wavelength of 700 nm, but it decreases to below 10% at the wavelength of 400 nm covered by GaN-related wide bandgap materials. Since QKD bit rates are crucially dependent on the above photon extraction efficiencies and the SPD's quantum efficiencies, drastic improvements of the photon extraction efficiencies and the development of SPDs with high sensitivities in the shorter wavelength region will be additional key issues.

5.1.2 GaN Nanocolumn Light Emitting Devices (K. Kishino)

GaN Nanocolumns

GaN nanocolumns self-assembled by rf-plasma-assisted molecular beam epitaxy (rf-MBE) [24–26] are one-dimensional nanocrystals of 50–150 nm in diameter, 0.5–2 μm in height, and $\sim 10^{10}$ cm^2 in density. Self-organized one-dimensional nanocrystals, which are sometimes called nanocolumns, nanorods, nanopillars, nanowhiskers, and nanowires, have been studied for various compound semiconductors such as ZnO, GaAs, InAs, and so on [27–30]. These are expected to be key technologies for realizing the nanometer scale photonic and electronic devices. As shown in Fig. 5.5, by integrating quantum heterosutructure into one-dimensional nanocrystals, various nanodevices such as nanotransistors, nanoresonant-tunnel-diodes (nano-RTDs), nanolaser/LED, and nanodetectors can be developed [27, 29–32]. And periodic arrangement of nanocrystals will produce two-dimensional photonic crystals (PC) [28], which could open the active photonic-band nanodevice field. The research on GaN-based nanocolumns and related nanodevices comes into the accelerated motion.

Let us summarize the history of GaN nanocolumn research. Soon after the first demonstration of GaN nanocolumns in 1996 [24], GaN-based quantum nanostructures, which integrated ten pairs of GaN/AlGaN multiple quantum disks (MQDs) of 6 nm in thickness and 46 nm in diameter, were fabricated [26]. Following to these early reports, during the last decade, several related researches such as growth of the GaN nanocolumns on Si (111) by RF-MBE and characterization of their optical properties [33–35], GaN nanocolumns grown

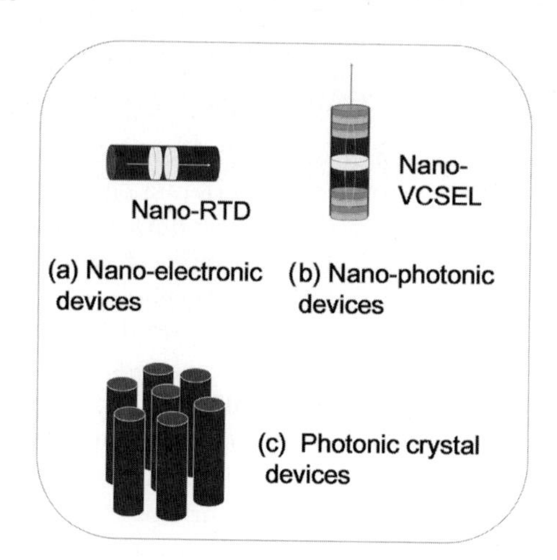

Fig. 5.5. Possible nanophotonic and electronic devices based on GaN nanocolumns

by the other growth techniques such as electron–cyclotron resonance plasma-excited (ECR-) MBE [36] and hydride vapor phase epitaxy (HVPE) [37] were reported. GaN nanocolumns are almost dislocation free [38], thus in principle possesses high emission efficiency. In fact a strong photoluminescence (PL) and a very low threshold excitation density of stimulated emission from GaN nanocolumns [39] were observed. In 2003, the photopump lasing of single GaN nanocolumn [40] and the current injection UV emission from the single GaN nanocolumn pn-junction [41] were reported. Finally the demonstration of InGaN/GaN nanocolumn light emitting diodes (LEDs) were performed by two different technologies of rf-MBE [31] and HVPE [32].

Recently, eight pairs of InGaN/GaN MQD nanocolumns with 3 nm in thickness and around 100 nm in diameter were grown on sapphire substrates, to produce bright blue to red PL emission in naked eye [42,43]. What is interesting is that the PL peak intensity dose not decrease drastically even in red emission of 650 nm in wavelength [44]. At the same time GaN nanocolumns showed high-efficient PL emission, which intensity was 500–600 times stronger than that of standard MOCVD-grown GaN layer (with the etch-pit density of $3–5 \times 10^9\,\mathrm{cm}^{-2}$) [44]. These two characteristics indicate the potential ability to realize highly efficient emitters in the whole visible range from blue to red.

Self-Assembling of GaN Nanocolumns by rf-MBE

The cross-sectional and surface scanning electron microscope (SEM) images of the GaN nanocolumns on (0001) sapphire (Al_2O_3) substrate are shown in Fig. 5.6b and c, respectively. Each nanocolumn is independently prepared in perpendicular to the substrate surface keeping the c-axis. The typical column diameter is $80 \sim 120$ nm, the height is $1 \sim 2\,\mu\mathrm{m}$ and then the column density is $8 \times 10^9 \sim 2 \times 10^{10}\,\mathrm{cm}^{-2}$. Note that the column diameter does not almost change from the bottom to the top in a few-μm range. As schematically shown

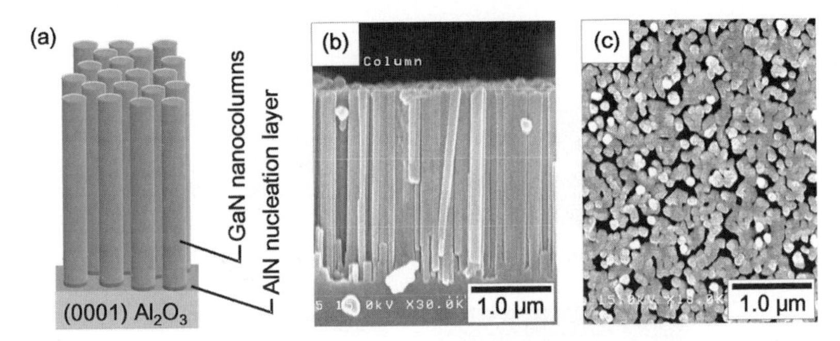

Fig. 5.6. Schematic diagram of GaN nanocolumns (**a**), typical SEM images of GaN nanocolumns grown on (0001) Al_2O_3 substrates observed from cross-section (**b**) and surface (**c**)

in Fig. 5.6a, on sapphire substrates, AlN nucleation layers are utilized for the column formation. Meanwhile, on Si substrates, GaN nucleation islands are adopted to keep a vertical conductivity from nanocolumns to Si, because AlN is highly resistive.

GaN nanocolumns can be self-assembled both on sapphire and Si substrates by rf-MBE, in which an rf-plasma source for nitrogen and conventional effusion cells for Ga, Al, In, Si, and Mg are equipped. The growth is performed either on the 2-in. diameter of (0001) sapphire and low-resistive ($\rho < 0.02\,\Omega$cm) Sb-doped n-type (111) Si substrates. For the growth on Al_2O_3 substrates, the substrate surfaces are thermally cleaned at 900°C for 10 min in the growth chamber and then nitrided under active-nitrogen-beam irradiation for 5 min at 750°C. After depositing AlN nucleation layers, GaN nanocolumns are grown on them at $850 \sim 950$°C at the nitrogen-rich condition. For the Si substrates, the surfaces are thermally cleaned at 830°C for 10 min, followed by deposition of Ga droplets for 25 s at 530°C. The active nitrogen beam is supplied on them typically for 1 min to form dot-like GaN nucleation islands. Then GaN nanocolumns are grown at $870 \sim 922$°C under a nitrogen-rich condition.

The shape of nanocolumn strongly depends on the growth parameters such as growth temperature, V/III beam supply ratio, and nucleation layer thickness. Figure 5.7 shows examples of cross-sectional configuration of GaN nanocolumns grown on (111) Si substrates dependent on the growth temperature. We note that the average column diameter (D) becomes small from 150 to 40 nm with increasing the temperature from 885 to 922°C, but nanocolumns of 80–100 nm in diameters may be suitable for fabrication of nanodevices.

Emission Properties of GaN Nanocolumn Under the Low and Strong Photoexcitation Conditions

In Fig. 5.8, the photoluminescence (PL) spectrum of GaN nanocolumn is compared with those of a GaN substrate prepared by hydride vapor phase

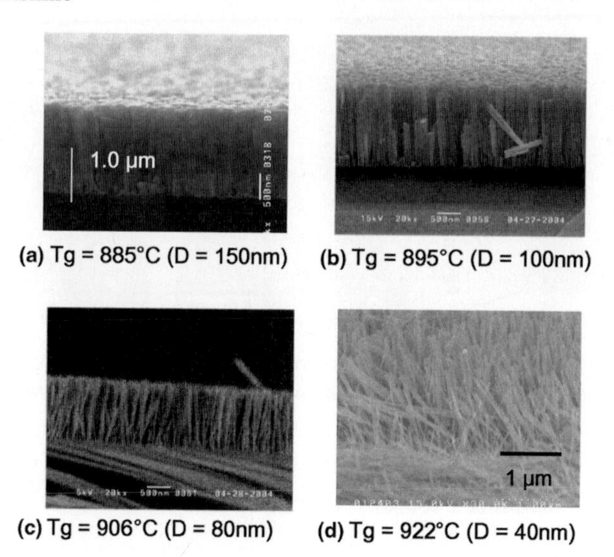

Fig. 5.7. Cross-sectional configurations of GaN nanocolumns grown on (111) Si substrates dependent on the growth temperature (from 885 to 922°C)

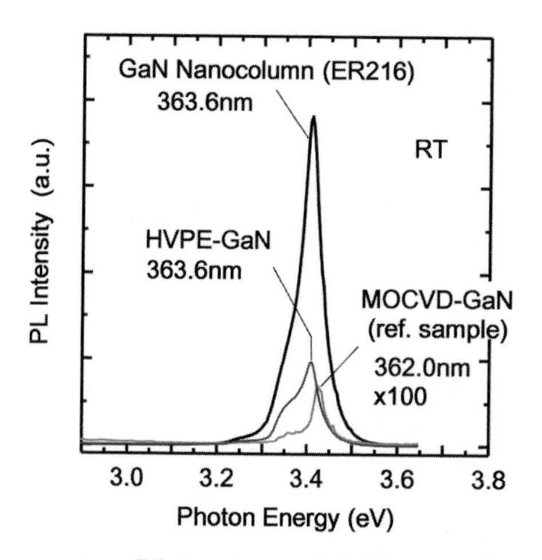

Fig. 5.8. Room temperature PL spectrum of GaN nanocolumns grown on (0001) Al_2O_3 substrate in comparison with those of HVPE-grown GaN substrate and MOCVD-grown GaN film. The crystal specification of HVPE-grown GaN is the threading dislocation (TD) density of $5 \sim 8 \times 10^6\,cm^{-2}$ and the carrier density of $\sim 1 \times 10^{17}\,cm^{-2}$. The MOVPE-grown GaN film shows a standard crystal quality with the TD density of $\sim 5 \times 10^9\,cm^{-2}$ and the thickness of 3.75 μm

epitaxy (HVPE) and a GaN film grown by metal organic vapor phase epitaxy (MOCVD). The excitation light is a 325 nm He–Cd laser with the light power of 9.2 mW. The PL spectrum of MOCVD-grown GaN is plotted magnifying the intensity 100 times from the original one. We notice that the PL peak intensity of GaN nanocolumn is 500–600 times larger than that of MOCVD-grown GaN. When the intensity is compared with that of HVPE-grown GaN substrate, it is still stronger as a factor of 4. The intense PL-emission of GaN nanocolumn can be introduced by excellent nanocolumn character. The dislocation-free character [38] introduces higher internal quantum efficiency, and the one-dimensional nanostructure enhances the light extraction efficiency [45]. As well known, the internal quantum efficiency is a strong function of threading dislocation density [46]. The intensity difference between GaN nanocolumn and HVPE-grown GaN substrate is supposed to be derived from the nanotexture effect of one-dimensional structure. Though more deep investigation is necessary, the surface recombination speed of GaN nanocolumn should be slow, otherwise such a strong PL intensity will not be observed. It is a suitable character for developing nanodevices that the nanocolumn-side-wall is passive against nonradiative recombination. By focusing attention on the PL peak wavelength, we note that the emission wavelength of GaN nanocolumn is 363.6 nm, which is equal to that of HVPE-grown GaN substrate. Meanwhile the MOCVD-grown GaN film emits at 362 nm, where the residual thermal strain in the GaN film brings about the wavelength blue shift. It is, therefore evinced that GaN nanocolumns are free standing from substrates.

When GaN nanocolumns are strongly excited under 355 nm Nd:YAG laser light irradiation (excitation power density: 10 kW cm^{-2} $\sim$ 10 MW cm^{-2}), the spontaneous emission comes into stimulated emission at a certain threshold excitation density as shown in Fig. 5.9 The stimulated emission rises at a longer wavelength side of spontaneous emission peak, here at 370 $\sim$ 366 nm in wavelength. The spontaneous emission intensity in principle increases in proportion to the excitation intensity I_p, while the stimulated emission intensity

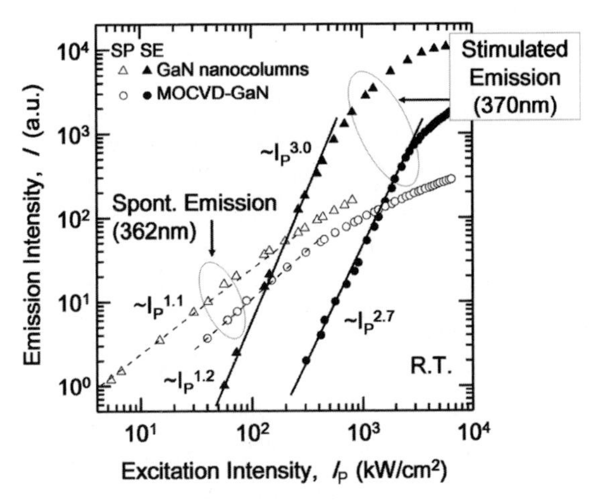

Fig. 5.9. Stimulated and spontaneous emission intensities of GaN nanocolumn and MOCVD-grown GaN film, as a function of optical excitation density

develops with the third power of I_p. The crossing point of two curves in Fig. 5.9 gives the threshold excitation density for stimulated emission. The threshold excitation density of GaN nanocolumn is $198\,\mathrm{kW\,cm^{-2}}$, and it is extremely low compared with the value for MOCVD-GaN ($1.6\,\mathrm{MW\,cm^{-2}}$). This also indicates superior optical quality of GaN nanocolumns.

InGaN/GaN Multiple-Quantum-Well (MQW) Nanocolumns

InGaN/GaN nanocolumn heterostructures can be self-assembled by rf-MBE. To prepare the nanostructures, GaN nanocolumns of ~100 nm in diameter are grown on (0001) sapphire substrates, followed by InGaN/GaN multiple-quantum-wells and then GaN nanocolumn cap layers. InGaN thin layers are embedded into GaN nanocolumn, so that disk-like InGaN nanostructures with a diameter as same as that of nanocolumns, which are sandwiched by GaN in axial direction and surrounded by air in radial direction, can be formed. Thus carriers can be confined effectively into the InGaN quantum disks.

Examples of PL emission spectra from InGaN/GaN MQW nanocolumns with various In-compositions are given in Fig. 5.10 The PL measurement is carried out under the 325 nm He–Cd laser light excitation at the room temperature. The samples consist of AlN nucleation layers, undoped GaN nanocolumns grown at 860°C for 30 min, and two or eight pairs of InGaN (3 nm)/GaN (5 nm) MQW and 30 nm thick GaN cap layers. InGaN/GaN and GaN cap layers are grown at a lower temperature of 650 ~ 700°C. A low temperature growth is necessary for incorporating indium into the crystal. The emission wavelength of InGaN MQWs is controlled from 436 to 614 nm by the growth parameters of growth temperature and In/Ga beam supply ratio,

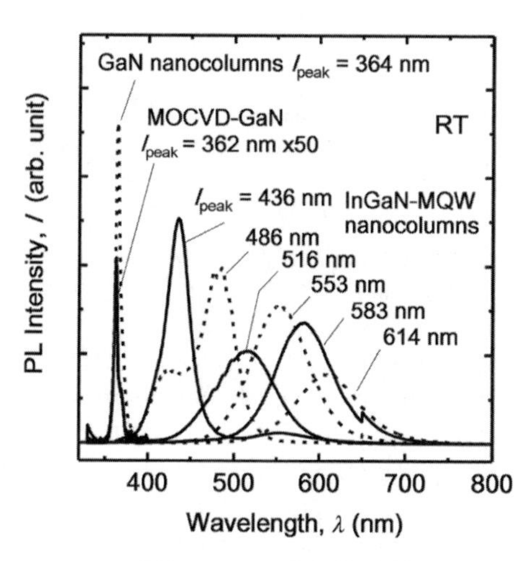

Fig. 5.10. Room temperature PL spectra of InGaN/GaN MQW nanocolumns with the different In composition, GaN nanocolumns, and MOCVD-GaN film

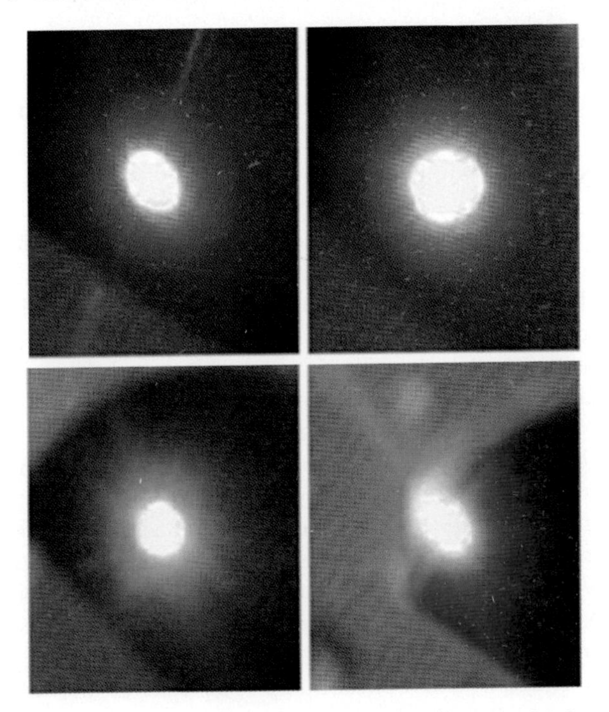

Fig. 5.11. Room temperature PL images of InGaN/GaN multiple quantum well (MQW) nanocolumns

as shown in Fig. 5.10. Bright PL emission with the naked eye from violet to red is observed. Typical emission images are shown in Fig. 5.11.

From Fig. 5.10, we notice that the peak intensity from InGaN-MQW nanocolumns is at least 20 times larger than that of MOCVD-grown GaN film. As the PL-spectra are broad, the integrated PL intensity is employed for the evaluation of emission property. The integrated PL intensities of InGaN-MQW nanocolumns are comparable with that of high-efficient GaN nanocolumns (see Fig. 5.8). The intensity gradually decreases with lengthening the wavelength from blue to red region, but it does not decrease drastically even at the red emission region. In fact, the integrated PL intensity for the red emission (614 nm) is 36% of that of blue emission (410 nm) and is 87% of ref-erence GaN nanocolumns. This tendency is different from the emission property for conventional InGaN LEDs, in which the quantum efficiency decreases steeply with lengthening the wavelength, thus increasing the In-composition. The reason of the drastic decrease can be explained by two factors, increased threading dislocation and increased strain-induced polarization field. The increase of dislocation shortens the nonradiative recombination time (τ_{Nr}). The strong polarization field deforms the potential profile of quantum wells pushing electron and hole wave functions toward the opposite direction, which lowers the oscillation strength and the radiative recombination time (τ_r) becomes long. Thus the internal quantum efficiency (η_{in}) drops as $\eta_{in} = (1+\tau_r/\tau_{Nr})^{-1}$. The experiment earlier, therefore suggests that the different emission mechanism, called here nanocrystal effect acts in InGaN/GaN MQW nanocolumns.

InGaN/GaN MQW Nanocolumn LEDs

The breakthrough in nanocolumn-based current-injection devices is how to fabricate electrodes on the top of nanocrystals of ~100 nm in diameter. The difficulty has been overcome by two schemes and InGaN/GaN MQW nanocolumn LEDs are fabricated; one is to fill the interstice between nanocolumns by a spin on glass (SOG) exposing nanocolumn tops to form electrodes [32] and the other scheme is to use champagne-glass shaped nanocrystals [31] as described later. The champagne-glass nanocolumn LEDs can be prepared through formation of heterojunction-nanocolumns on substrates and subsequent enhanced lateral growth, as shown in Fig. 5.12. An example of growth sequence of champagne-glass nanocolumn LEDs is as follows; Si-doped n-type GaN nanocolumns (~940 nm) are grown on (111) conductive Si substrates, followed by undoped GaN (10 nm), 2 or 8 pairs of InGaN (2 nm)/GaN (3 nm) MQW active region, undoped GaN (10 nm), and Mg-doped p-type GaN (600 nm). Here n-type GaN nanocolumns are grown at 870 ~ 890°C, while the InGaN/GaN MQW active layer and p-type GaN at a lower temperature of 640 ~ 680°C.

For self-assembling of nanocolumns, it is necessary for the growth to be performed at a high temperature under the nitrogen-rich condition. In p-GaN cladding layers, the growth temperature is lowered and then the lateral growth is enhanced. Consequent increase in the nanocolumn diameter coalesces neighboring nanocolumns and finally a continuous film is formed on the

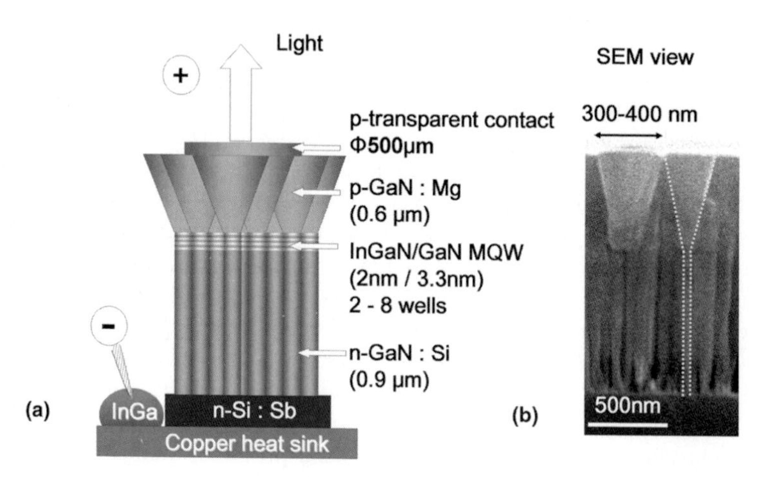

Fig. 5.12. Cross-sectional schematic diagram (**a**) and SEM photograph (**b**) of InGaN/GaN MQW nanocolumn LEDs

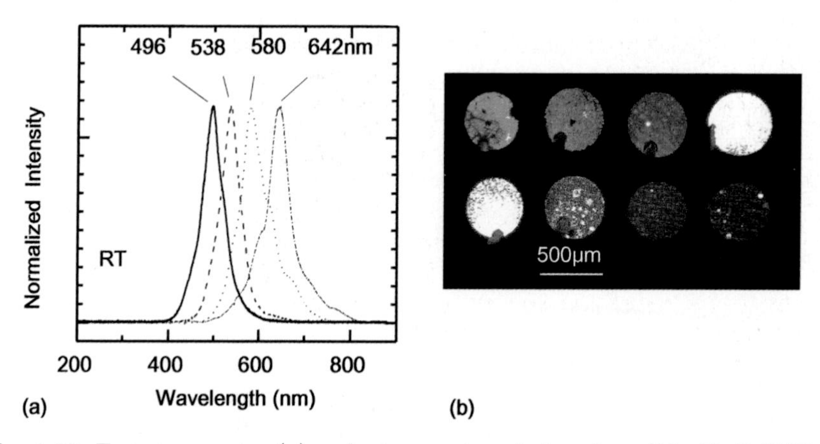

Fig. 5.13. Emission spectra (**a**) and microscopic emission views (**b**) of InGaN/GaN MQW nanocolumn LEDs

surface (see Fig. 5.12). Thus p-type electrodes can be prepared by standard metallization on it. In InGaN/GaN nanocolumn LED, Ni (2 nm)/Au (3 nm) semitransparent electrodes with a diameter of 300 $\sim$ 500 μm are utilized. And the LED samples are mounted on a Cu heat sink in the episide up configuration with the gallium–indium liquid metal as shown in Fig. 5.12. In the nanocolumn LEDs, a columnar structure exists inside from the n-type GaN over the InGaN/GaN MQW active region, so that the superior optical quality of nanocolumns is preserved.

The electroluminescence (EL) of nanocolumn LEDs is observed through the semitransparent electrodes. Figure 5.13a shows typical EL spectra of the InGaN/GaN MQW nanocolumn LEDs with various In-compositions of InGaN active layers. The EL peak wavelengths are 496 nm (blue), 538 nm (green), 580 nm (yellow), and 642 nm (red), respectively. Photographs of near-field-emission-pattern observed by a microscope are shown in Fig. 5.13b. Blue, green, yellow, orange, and red colors of emission are observed in the whole semitransparent electrode area with a diameter of 500 μm. The full-width-at-the-half-maximum (FWHM) of the emission spectrum is changed dependent on the observation area size. The EL spectrum is measured for the whole electrode area and the small observation area with a diameter of about 3 μm (micro-EL observation). In an LED chip, the peak wavelength of EL spectrum for the whole and small area is observed to be 489 and 522 nm, respectively. FWHM becomes narrower from 73.5 to 37.0 nm by decreasing the detection area. The FWHM value for micro-EL observation depends on LED chips and different detection parts in the same emission area. In fact, for a different LED chip it is about 20 nm for the 530 nm emission.

In state of the art nanocolumn LEDs, therefore the emission area includes different character of nanocolumns with various emission wavelengths, but each InGaN nanocolumn has a narrower emission FWHM. Therefore, how to attain homogeneity of nanocolumn is necessary, and the periodic arrangement of nanocolumns could be essential to increase the homogeneity of the emission wavelength and the column diameter in the whole emission area.

5.1.3 Wide Bandgap Semiconductor Nanostructures and Devices (S. Fujita)

Self-assembly of Nanostructures

One Dimensional Structures: Nanorods, Nanobelts, or Nanotubes

Due to the high growth rate along the c-axis orientation for GaN and ZnO, there is a strong tendency for hexagonal columnar structures to be easily formed. This results in various kinds of one-dimensional nanostructure. For ZnO, well-defined nanorods or nanowires (hereafter, we group these structures together as "nanorods") were originally formed by the vapor–liquid–solid (VLS) growth mode with gold catalysis [47]. Later MOCVD allowed similar structures without the catalysts [48]. Nanobelts [49] or nanotubes [50] have also been reported, as well as other modified structures (nanowalls, nanotrees, nanonails, nanobridges, etc.). Not only simple ZnO nanorods but also compositionally-modulated nanorods such as ZnO/ZnMgO MQWs embedded into a nanorod have been synthesized [51]. Diameters of nanorods have been in the range 10–100 nm, while the lengths differ significantly among the reports from 0.3 to 30 μm. Once the rods become longer than 1 μm they no longer grow perpendicularly to the substrate anymore, and may be better to be named "nanowires."

GaN nanorods have also been fabricated by the VLS process [52, 53], but generally the aspect ratio is lower and the geometry is inferior compared to ZnO nanorods. This is probably because the tendency of columnar growth is much stronger in ZnO. With growing GaN on ZnO nanorods, core-shell nanorod structures were successfully demonstrated [54].

Zero-Dimensional Structures: Nanodots

Self-assembled ZnO nanodots have been reported on Si and SiO_2. The formation mechanism on these substrates is supposed to be Volmer-Weber (VW) nucleation due to high interface mismatching and hence accumulation of strain at the interface. Despite the growth on SiO_2, the nanodots exhibit single crystalline properties without significant defects. Quantum effects, as revealed by blue-shift of the luminescence spectrum, have been confirmed for the nanodots with their dimensions about 5–20 nm [55]. Efforts were reported to artificially arrange the nanodots by using SNOM-assisted deposition [56] or prepatterning the substrate surface by a focused ion beam [57].

GaN nanodots have been formed on AlGaN and AlN with Stranski–Krastanov (S–K) mode [58–60]. The selective growth using anodic aluminum oxide films [61] or stacked nanodots [62] is also reported. The details appear in Sect. 5.3.

Examples of Nanodevices

Light Emitting or Detecting Devices

ZnO-based nanorods with ZnO/ZnMgO MQWs exhibit strong emission by excitons confined in the MQWs [63]. However, the difficulty in obtaining p-type ZnO or ZnMgO conduction is against the LED application. The potential of ZnO nanorods for LEDs is demonstrated by UV emission from a ZnO nanorod-embedded n-ZnO/p-GaN heterojunction [64]. On the other hand, with the large surface area of nanorods, a highly sensitive UV photodetector was demonstrated [65].

Light emitters by GaN-based nanorods are described in Sect. 5.4. A UV LED with GaN nanodots embedded between AlGaN confinement layers has been reported [66]. GaN nanowires have also been used for photodetectors [67].

Gas or Biology Sensors

The surface of ZnO nanorods is sensitive to many chemical species. This results in accumulation or depletion of electrodes at the surface with the absorption of certain molecules. Even for small variations of surface charge, the average charge density changes greatly because of the small diameter of nanorods. This gives rise to a remarkable change in the resistivity of nanorods. By placing the nanorod(s) across the electrodes, as shown in Fig. 5.14 [66], and monitoring the current, one can apply this nanorod as a high sensitive gas or biology sensor with low power consumption. Sensing characteristics for oxygen [68], humidity [69], ethanol [70], and so on have been reported.

Transistors

A single nanorod can work as a transistor with placing a gate insulator and electrode. A vertical surround-gate FET structure, as shown in Fig. 5.15 [71],

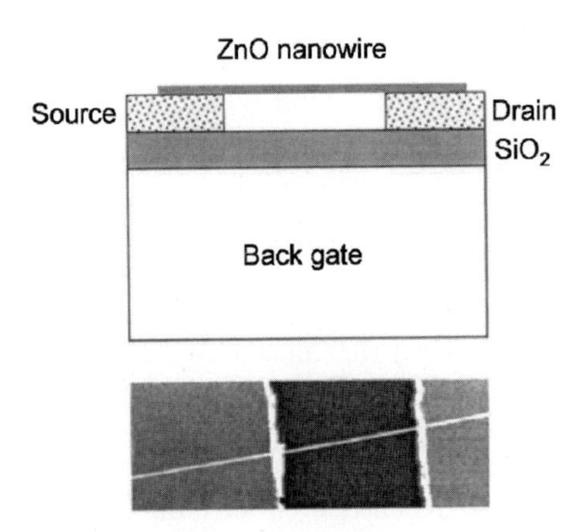

Fig. 5.14. A nanorod embedded between two electrodes

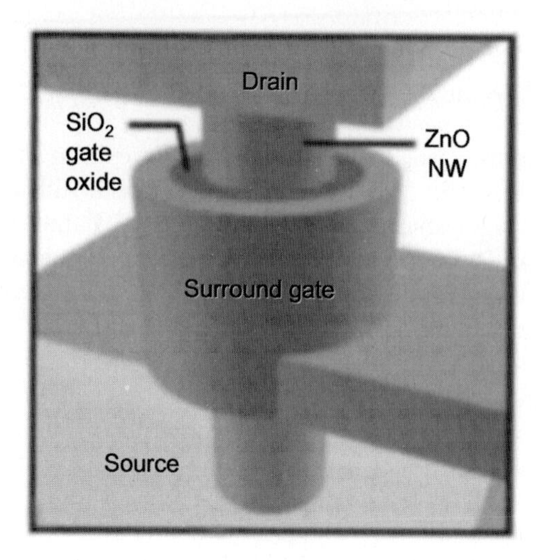

Fig. 5.15. A vertical surround-gate nanorod FET structure

was also reported, where the diameter of the ZnO nanorod was 35 nm. Nanorod transistors are promising for highly integrated circuits with low power consumption.

Field Emitters

Vertically aligned ZnO nanorods with high aspect ratio are effective for field emission of electrons [72], but the large electron affinity may be a handicap. Recently excellent field emission characteristics have been seen. For example, a field emission current density of 2.5 mA cm^{-2} at an electric field of 2.5 V μm^{-1} were reported for GaN nanorods because of their geometrical configuration, good crystalline quality, together with their small electron affinity [73]. Electron field emitters are becoming important for vacuum electronics, electron beam lithography, and flat panel displays.

5.2 Novel AlGaN/GaN Heterostructure Devices

5.2.1 GaN-Based High Temperature Operating Hall Devices (A. Sandhu)

In 1890 Edwin Hall discovered that magnetic fields distort the equipotential lines in current-carrying conductors. This "Hall effect," is now the basis of a multimillion dollar industry. Hall sensors are ubiquitous, being used in vending machines, modern electronics equipment for controlling DC motors in VCRs, CD-ROM drives and more recently even for reducing the power consumed by electric motors in household washing machines [74–76].

Recent applications of Hall effect devices include scanning Hall probe microscopy (SHPM) for magnetic imaging ferromagnetic domains [77–79] and more recently biosensing platforms incorporating superparamagnetic labels [80–83].

Conventional Hall sensors fabricated using silicon, InAs, and InSb show stable operation for temperatures up to about 125°C but their performance is extremely poor above 200°C due to the onset of intrinsic conduction as well as the physical degradation of the semiconducting materials themselves at elevated temperatures.

However, there are niche demands for high temperature Hall sensors including the localized measurement of the Curie temperature of ferromagnetic materials, the automotive industry and aerospace [84].

Wide bandgap semiconductors could be potentially used for producing high temperature Hall sensors because of their large bandgap, high electron saturation velocity, and excellent thermal stability. The potential of silicon carbide (SiC) for high temperature Hall sensors has been reported using 4H-SiC [85]. However, the need for precise control of dopants and the large thickness of the conducting layers severely limits the sensitivity and stability at high temperatures of SiC Hall devices.

AlGaN/GaN heterostructures with a two-dimensional electron gas (2DEG) at the interface, show promise as high sensitivity micro-Hall sensors for use at high temperatures. The AlGaN/GaN structure combination offers the advantages of wide bandgap materials that are stable at high temperatures as well as an extremely thin conducting layer, i.e., the 2DEG, for high magnetic sensitivity. Here we describe the fabrication and electrical characteristics of micro-Hall sensors fabricated using AlGaN/GaN 2DEG heterostructures for operation at temperatures up to 600°C.

Design Parameters for Hall Sensors

Hall Effect

In its simplest manifestation, in the Hall effect a voltage is generated transversely to the current flow direction in a conductor when a magnetic field is applied perpendicularly to the flow of current. This transverse voltage is the Hall voltage (V_H) and its magnitude is equal to

$$V_H = R_H \frac{IB}{d} = \frac{IB}{ned} \mathrm{G} \tag{5.1}$$

where, I is the drive current, B the magnetic field, d the sample thickness, n the electron (hole) density, e (1.602×10^{-19} C) the elementary charge, and G a geometrical parameter. R_H is known as the Hall coefficient. From (5.1) it can be seen that a large Hall voltage is obtained for thin samples with a low carrier concentration. The Hall effect is used for quantitative measurement of magnetic field intensity. High sensitivity Hall effect sensors are made using platelets of semiconductors such as Si, GaAs, and InSb.

Magnetic Sensitivity

The signal (S) to noise (N) ratio of a Hall device can be defined as,

$$\frac{S}{N} = \frac{V_{\mathrm{H}}}{V_{\mathrm{N}}} = \frac{R_{\mathrm{H}} I B}{\sqrt{4 k_{\mathrm{B}} T R_{\mathrm{S}} \Delta f}} \tag{5.2}$$

The main noise component is due to Johnson noise, $V_{\mathrm{noise}} = (4k_{\mathrm{B}}\, T R_{\mathrm{S}} \Delta f)^{1/2}$, where k_{B}, is the Boltzmann constant, R_{S}, the series resistance $= \frac{l}{\mu e n w d} M$ (M is a geometrical factor), Δf, the measurement bandwidth and T the absolute temperature.

From this relationship, the minimum field sensitivity, B_{min}, can deduced as being,

$$B_{\mathrm{min}} = \frac{\sqrt{4 k_{\mathrm{B}} T R_{\mathrm{S}} \Delta f}}{R_{\mathrm{H}} I} = \frac{\sqrt{4 k_{\mathrm{B}} T M d \Delta f}}{v_{\mathrm{sat}} \sqrt{e n \mu} G(w/l) \sqrt{w l}} \tag{5.3}$$

where, ν_{sat} = saturation velocity and μ = carrier mobility.

For $\frac{l}{w} = 2$, G(l/w) = 0.4, under weak fields, $M = 1$, then (5.3) can be rewritten as,

$$B_{\mathrm{min}} = \frac{\sqrt{4 k_{\mathrm{B}} T R_{\mathrm{S}}}}{v_{sat} w} \tag{5.4}$$

From (5.3), it can be seen that a high carrier mobility and saturation velocity and low sheet resistance are important material design parameters for high sensitivity Hall devices [86].

Fabrication and Characterization of AlGaN/GaN Micro-Hall Sensors

The high temperature Hall sensors were fabricated using AlGaN/GaN heterostructures grown on sapphire substrates by molecular beam epitaxy (MBE) under growth conditions reported previously [87]. The AlGaN/GaN micro-Hall sensors were fabricated by a three-step photolithography process involving (1) formation of the mesa and active "cross" patterns by argon ion milling; (2) thermal and electron beam evaporation of Ti/Al/Ni/Au ohmic contacts; and (3) rapid thermal annealing in a nitrogen atmosphere at 830°C for 90 s. Figure 5.16 shows a typical AlGaN/GaN Hall sensor. The sensing "cross-region" is 50 × 50 μm.

The electrical properties of the micro-Hall sensors were measured from 20 to 600°C using a specially constructed high vacuum chamber.

The Hall sensors were used to measure the Curie temperature of barium ferrite permanent magnets and integrated into a scanning Hall probe microscope for imaging of magnetic domains of ferromagnetic garnet crystals.

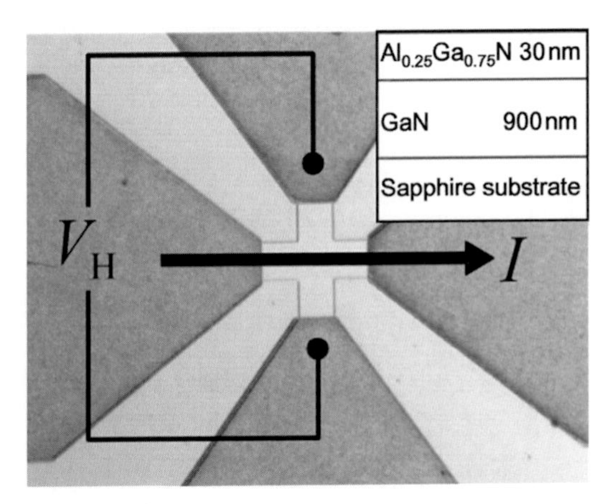

Fig. 5.16. Typical AlGaN/GaN Hall sensor. The sensing "cross-region" is 50×50 μm

Electrical Characteristics and Applications

The room temperature electron mobility and sheet carrier concentration of the 2DEG induced at the heterointerface were $850\,\text{cm}^2\,\text{Vs}^{-1}$ and $7.7 \times 10^{12}\,\text{cm}^{-2}$, respectively.

Figure 5.17 shows typical results of the variation of the Hall voltage output with measurement temperature. The Hall voltage has a linear dependence on the external applied field at each measurement temperature up to 0.50 T for drive currents up to 0.3 mA. This linearity is of critical importance for device applications and our results show that AlGaN/GaN Hall sensors function as expected up to 600°C.

Figure 5.18 shows the variation of the magnetic field sensitivity (B_{min}) of a 100 × 100 μm AlGaN/GaN Hall sensor with temperature. Temperature dependent measurements of the carrier concentration and device resistance showed that the B_{min} increased because of an increase of the series resistance rather than from significant changes in carrier concentration. Nevertheless, the B_{min} measured at 600°C of 0.35 G $(\text{Hz})^{-1/2}$ shows the high sensitivity of the devices even at these elevated temperatures.

Figure 5.19 shows the results of Curie temperature measurements on the barium ferrite using a 100×100 μm AlGaN/GaN Hall sensor. The T_{c} of 358°C is in excellent agreement with published data obtained by VSM measurements.

Figure 5.20a shows the magnetic field image on the surface of a ferromagnetic garnet crystal obtained at 25°C, by scanning Hall probe microscopy (SHPM) using a 2 × 2 μm AlGaN/GaN Hall probe (Fig. 5.20b). The scan area was 25×25 μm and black and white regions correspond to surface field variations of ±81 G into and out of the plane of the paper. These results agree with our previous experiments conducted using GaAs-2DEG and InSb Hall probes and confirm that the 2 μm AlGaN/GaN Hall probe has high sensitivity [79]. A high temperature scanning Hall probe is being developed.

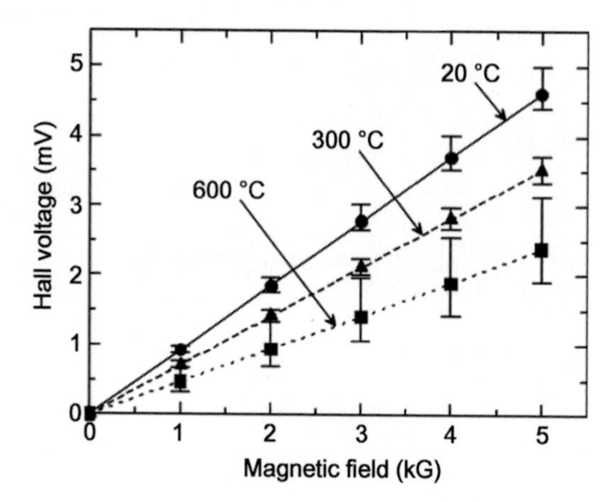

Fig. 5.17. Variation of the Hall voltage output with measurement temperature

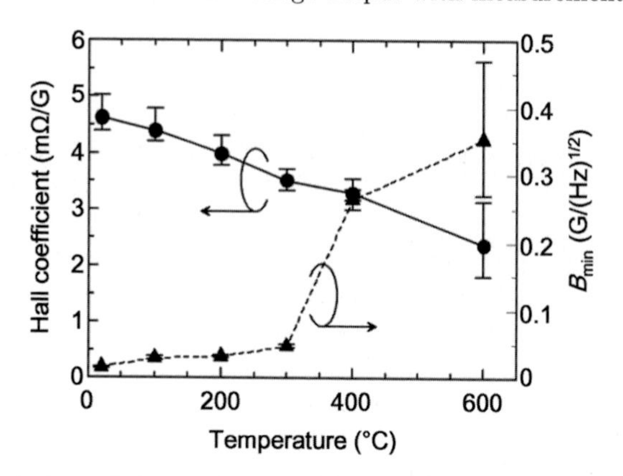

Fig. 5.18. Variation of the magnetic field sensitivity (B_{min}) of a $100 \times 100\,\mu\text{m}$ AlGaN/GaN Hall sensor with temperature

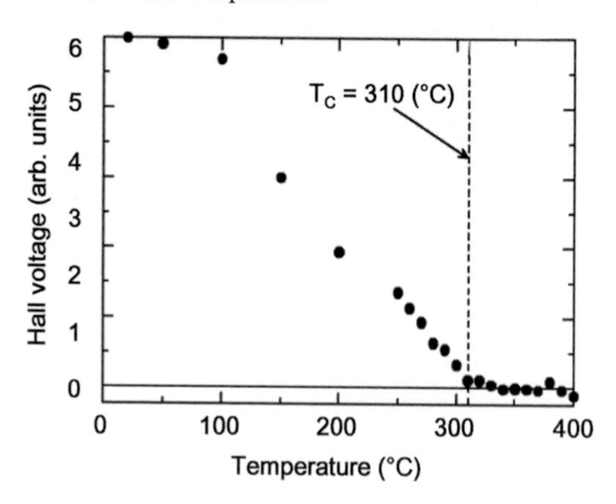

Fig. 5.19. Curie temperature measurements on barium ferrite using a $100 \times 100\,\mu\text{m}$ AlGaN/GaN Hall sensor

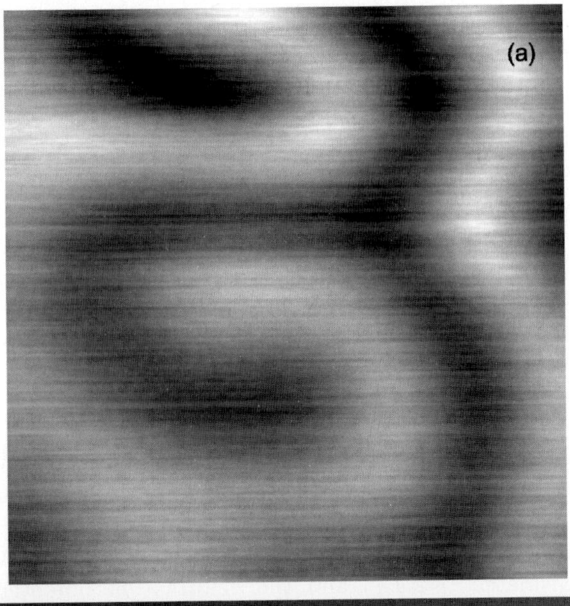

Fig. 5.20. (**a**) Magnetic field image on the surface of a ferromagnetic garnet crystal obtained at 25°C, by scanning Hall probe microscopy (SHPM) using a $2 \times 2\,\mu$m AlGaN/GaN Hall probe (**b**)

5.2.2 GaN-Based Intersubband Transition Optical Switches (N. Suzuki)

An ultrafast semiconductor all-optical switch is one of the key devices for future photonic networks. However, switching operation above 100 Gb/s is difficult in conventional semiconductor all-optical switches based on interband transitions, since the response time is limited by the carrier lifetime (>10 ps). The intersubband transitions (ISBT) in semiconductor quantum wells (QW) are utilized for infrared optoelectronic devices such as quantum cascade lasers (QCL) [88] and quantum well infrared photodetectors (QWIP) [89]. They are also applicable to ultrafast optical switches, because the intersubband (ISB) relaxation time is less than 10 ps [90]. ISB relaxation time in GaN/AlGaN quantum wells (MQW) was predicted to be as short as about 100 fs at $\lambda = 1.55\,\mu\text{m}$ [91,92]. Near-infrared ISBTs in GaN/Al(Ga)N MQW have been realized by some groups [93–97]. The GaN-based ISBT all-optical modulators/switches [98–101] have been proven to be faster than ISBT switches based on InGaAs [102] or II–VI materials [103].

Basic Concept

We utilize saturable absorption of the ISBT in GaN/AlN MQWs, where the GaN wells are heavily doped with Si. Figure 5.21a shows a conduction band diagram of a GaN/AlN QW grown along the c-axis on a sapphire substrate. The conduction band discontinuity was assumed to be 1.76 eV, considering the lattice mismatch (~2.4%) and the deformation potential. The piezoelectric effect and the spontaneous polarization cause strong built-in fields (>5 MV cm^{-1}), which raise the effective barrier height [104]. A near-infrared ISBT is realized when the well width is about 2 nm. Most of the electrons in the well are located in the ground subband, and absorb light resonant to the ISBT. The absorption coefficient for the light with an optical angular frequency of ω is given by

$$\alpha(\omega) = \frac{\mu_{21}^2(N_1 - N_2)}{2c_0\varepsilon_0 n} \frac{\hbar\omega_{21}/\tau_{\text{ph}}}{(\hbar\omega_{21} - \hbar\omega)^2 + (\hbar/\tau_{\text{ph}})^2}, \tag{5.5}$$

where $\hbar$ is Planck's constant divided by 2π; c_0 is the velocity of light in vacuum; ε_0 is the electric permittivity in vacuum; n is the index of refraction;

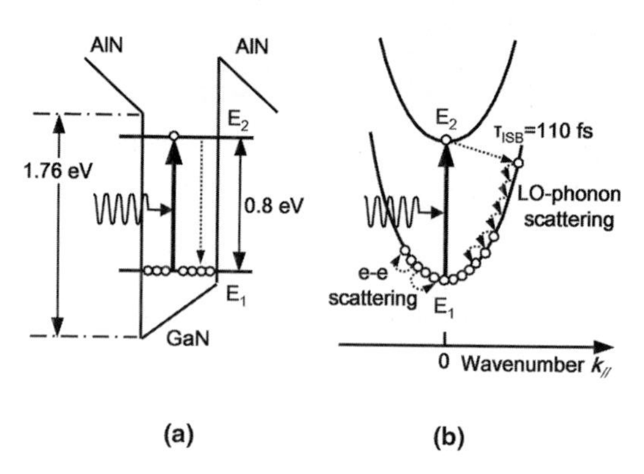

Fig. 5.21. (**a**) Conduction band diagram of a GaN/AlN QW grown along the c-axis. (**b**) Relaxation processes of electrons excited by ISB absorption

N_i is the carrier density of the i-th subband; $\hbar\omega_{21}$ is the transition energy; and $\tau_{\rm ph}$ is the dephasing time. The transition dipole moment between the i-th and the j-th subbands is given by [105]

$$\mu_{ji} = e\int \phi_j(z)^* z\phi_i(z)dz, \tag{5.6}$$

where e is the electron charge; z is the ordinate normal to the well; and $\phi_{\rm i}(z)$ is the envelope function of the i-th subband. Because of the symmetry, the ISBT occurs only for light with the electric field normal to the well. Transitions between the subbands with the same parity are forbidden in a symmetric well. The transition between the adjacent subbands is much stronger than the other transitions. The absorption saturates when almost half of the electrons are excited by strong light.

Carrier relaxation processes are shown schematically in Fig. 5.21b. At room temperature, the dominant intersubband relaxation process is longitudinal optical (LO) phonon scattering. The intersubband LO-phonon scattering rate can be approximated by [106]

$$W_{21} = \frac{e^2\omega_{\rm LO}}{8\pi\,\hbar}\left(\frac{2m^*}{E_1}\right)^{1/2}\left[\frac{1}{\varepsilon_\infty} - \frac{1}{\varepsilon_{\rm s}}\right]\left[\frac{1}{4-(\hbar\omega_{\rm LO}/E_1)} + \frac{1}{12-(\hbar\omega_{\rm LO}/E_1)}\right], \tag{5.7}$$

where $\hbar\omega_{\rm LO}$ is the LO-phonon energy; m^* is the effective mass; ε_∞ and $\varepsilon_{\rm s}$ are the optical and the static dielectric permittivities, respectively; and E_1 is the energy of the ground subband. The solid, dashed, and dotted lines in Fig. 5.22 show $\tau_{\rm ISB}(= 1/W_{21})$ for GaN, CdS, and InGaAs, respectively. More

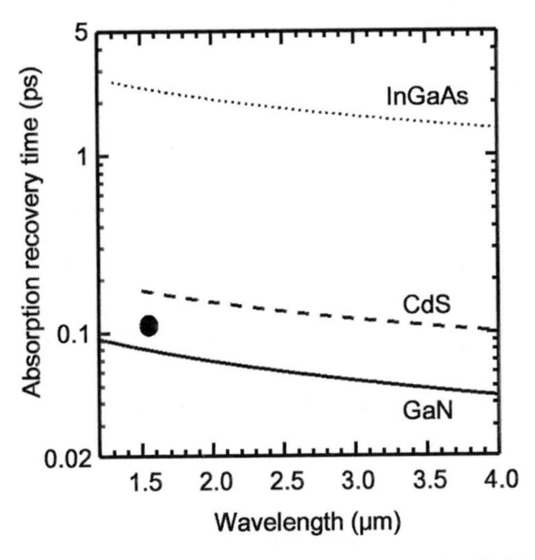

Fig. 5.22. Intersubband LO-phonon scattering time in GaN (*solid line*), CdS (*dashed line*), and InGaAs (*dotted line*) as functions of wavelength. *Circle* is the measured absorption recovery time in a GaN waveguide [13]

rigorous calculation suggested that τ_{ISB} in GaN is 109 fs at $\lambda = 1.55\,\mu$m [92]. Because of the band nonparabolicity, the energy separation of the two subbands is smaller for larger wavenumber $k_{\parallel}$. Hence, electrons scattered to the high energy state of the ground subband do not contribute to the absorption of the signal light until they lose energy by emitting several LO phonons. This carrier cooling process takes longer than τ_{ISB}. However, the electron–electron scattering (thermalization) time (τ_{ee}) is about one order of magnitude shorter than τ_{ISB}. Therefore, the absorption recovery time at a shorter wavelength, which represents the transition for $k_{\parallel} \sim 0$, is dominated by τ_{ISB}. For the transition near the Fermi level (absorption at a longer wavelength), slower recovery due to carrier cooling is observed [107]. The circle in Fig. 5.22 denotes the absorption recovery time measured in a GaN waveguide [100]. The short τ_{ee} also causes a short dephasing time ($\tau_{ph} \sim 10$ fs), which leads to a broad homogeneous linewidth ($\sim$100 meV). It enables us to allocate different wavelengths to the signal and the control pulses with a pulse width of about 100 fs.

Saturation intensity at the absorption peak wavelength is given by [108]

$$I_S = \frac{c_0 \varepsilon_0 n \hbar^2}{2\mu_{21}^2 \tau_{ISB} \tau_{ph}}. \tag{5.8}$$

There is a trade-off between response time and switching energy. Although the saturation intensity in GaN ISBT switches tends to be high, GaN withstands a high-power excitation. Two-photon absorption (TPA), which may destructively interfere with ISBT, is negligible in wide-gap GaN.

Transition Wavelengths in MOCVD-Grown and MBE-Grown GaN/Al(Ga)N MQWs

The thick and thin lines in Fig. 5.23 show the well width dependencies of the transition wavelengths calculated assuming a constant built-in field in the GaN well for AlN and $Al_{0.65}Ga_{0.35}N$ barriers, respectively [104]. The circles and the triangles represent the measured data for GaN/AlN MQWs grown by molecular beam epitaxy (MBE) and for GaN/AlGaN MQWs grown by metal–organic chemical deposition (MOCVD) [109], respectively. The effective field strengths in the well were in the range of 5–10 MV cm^{-1} for GaN/AlN MQWs and 0–3 MV cm^{-1} for GaN/$Al_{0.65}Ga_{0.35}N$ MQWs, respectively.

The shortest transition wavelengths reported so far were 1.08 μm for MBE-grown samples [94] and 2 μm for MOCVD-grown samples [110], respectively. Reciprocal space mappings of the X-ray diffraction data suggested that the MOCVD-grown GaN/$Al_xGa_{1-x}N$ MQWs are grown nearly coherently to the underlying GaN layer. Therefore, thicker and higher barriers lead to inferior crystal quality. The crystal quality is improved to some extent by inserting thicker GaN intermediate layers in GaN/Al(Ga)N MQWs. Even so, TEM images suggested that abruptness of the interface is inferior to that of MBE-grown samples. Furthermore, heavy Si-doping ($N_d > 10^{19}$ cm^{-3}) is difficult by MOCVD. The dislocation density in MBE-grown GaN/AlN MQWs

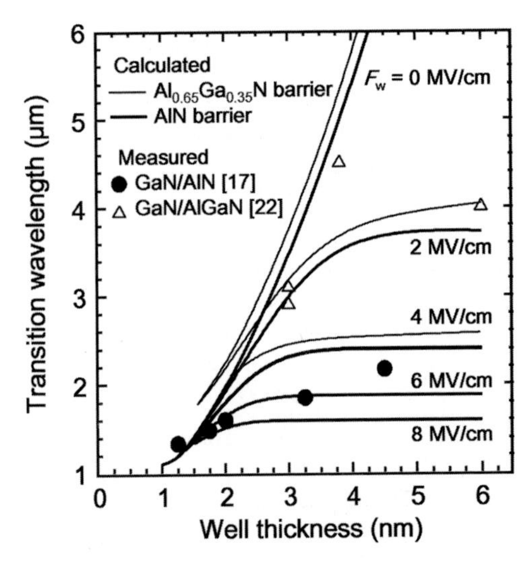

Fig. 5.23. Intersubband transition wavelengths as functions of thickness and built-in field in GaN wells [104]. The *thick* and *thin lines* show the transition wavelength for GaN/AlN MQW and GaN/$Al_{0.65}Ga_{0.35}N$ MQW, respectively, calculated assuming a constant built-in field in the well. The *circles* and the *triangles* represent the measured data for MBE-grown GaN/AlN MQW [104] and for MOCVD-grown GaN/AlGaN MQW [109], respectively

is one order of magnitude higher than that in MOCVD-grown samples. However, the dislocations tend to merge together with the growth of MQWs and the in-plane lattice constant approaches that of an AlGaN layer of the average composition [111]. It allows MBE-growth of uniform GaN/AlN MQWs with a large number of wells.

Ultrafast ISBT Optical Switches

An ISBT is allowed only for TM-mode light in waveguides. Unfortunately, high-density ($>10^{10}\,\mathrm{cm}^{-2}$) edge dislocations run perpendicular to the substrate in nitride layers grown directly on a sapphire substrate by MBE. Electric charges trapped by the dislocations cause excess polarization-dependent loss (PDL) for the TM mode [112]. Actually, a PDL of 8–15 $\mathrm{dB\,mm^{-1}}$ was observed in GaN waveguides without MQWs. The excess PDL depended on both the dislocation density and the carrier density, but was almost independent of the optical pulse energy. Reduction in dislocations was crucial in order to observe the saturation of ISB absorption in nitride waveguides.

Ultrafast optical modulation was achieved for the first time in a rib waveguide (WG-A) [98], where the dislocations were reduced by a multiple intermediate layer (MIL) [113] consisting of ten pairs of unintentionally-doped GaN (40 nm) and AlN (10 nm). Then, 10-dB switching was achieved in

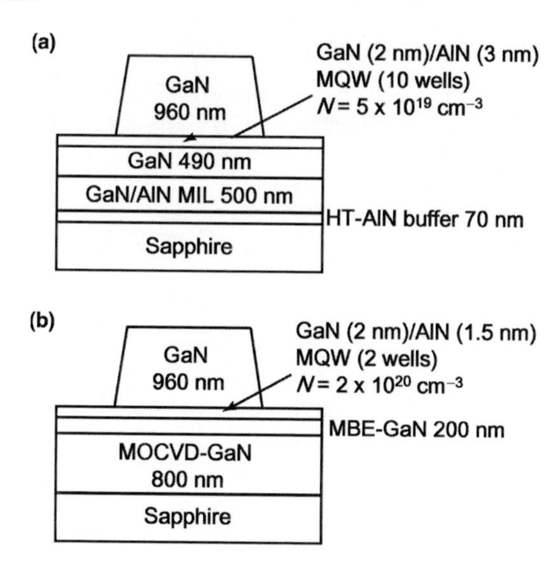

Fig. 5.24. Cross-sectional structure of (**a**) WG-A [98] and (**b**) WG-B [100]

WG-B [100] where the dislocations were reduced further ($1.7 \times 10^9\,\mathrm{cm}^{-2}$) by utilizing a 0.8-μm-thick MOCVD-grown GaN template. Fig. 5.24 shows the cross-sectional structure of the rib waveguides where the light is confined by high index contrast between the nitride layers and the surroundings. In WG-A, the MQW consisted of ten pairs of Si-doped ($5 \times 10^{19}\,\mathrm{cm}^{-3}$) GaN (2 nm) and AlN (3 nm) layers. The absorption peak wavelength and the linewidth were 1.75 μm and 120 meV, respectively. In WG-B, the number of wells N_w was reduced to two and the thickness of the AlN barriers was reduced to 1.5 nm to suppress generation of edge dislocations in the MQW layer. To compensate for the reduction in N_w, the Si-doping level was increased to $2 \times 10^{20}\,\mathrm{cm}^{-3}$. In both waveguides, the upper GaN layer ($\sim$1 μm) was etched to form mesas by Cl_2-based electron cyclotron resonance reactive ion beam etching (ECR-RIBE). The width of the mesa was 1–2 μm, and the length was about 400 μm. The chips were AR-coated and housed in a module with fiber pigtails.

The measured absorption saturation characteristics for TM pulses are shown in Fig. 5.25 [98,100]. The wavelength and the width of the signal pulses were 1.55 μm and 130 fs, respectively. TE-mode insertion loss including the coupling loss was 8.1 and 6.6 dB for WG-A and WG-B, respectively. From the PDL at $\lambda = 1.3$ μm, the excess PDLs were estimated to be 9.3 dB and 4.7 dB for WG-A and WG-B, respectively. In WG-A (open circles), the TM insertion loss was saturated by 5 dB at a fiber input pulse energy of 120 pJ. In WG-B (solid circles), 10-dB saturation was achieved at a pulse energy of 101 pJ.

In WG-A, the 1.55-μm signal pulses were modulated by 1.7-μm, 120-pJ control pulses as shown in Fig. 5.26a [98]. The FWHM of the response was 360 fs, which was limited by the convolution of the pulses. The control pulse width was broadened to 230 fs in the 1-m fiber pigtail due to dispersion and

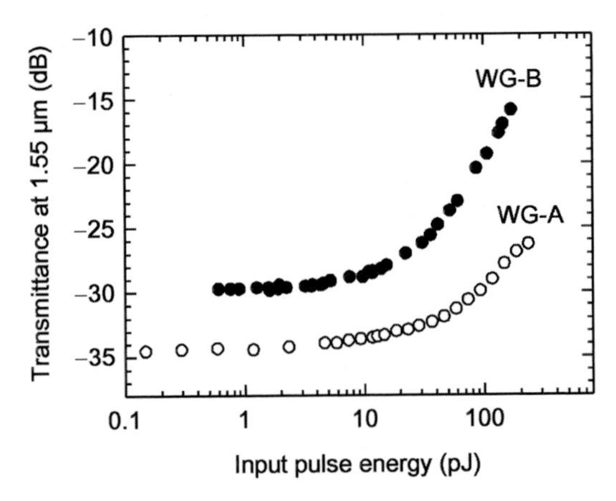

Fig. 5.25. Absorption saturation characteristics of WG-A (*open circles*) [98] and WG-B (*solid circles*) [100]

nonlinearity. Rather small modulation ratio (∼2.4 dB) is due to the excess PDL. As shown in Fig. 5.26b, pattern-effect free optical modulation corresponding to 1.5 $\mathrm{Tb\,s^{-1}}$ was achieved in WG-A [99]. In WG-B, the 1.55-μm signal pulses were switched by 11.5 dB by 1.7-μm, 150 pJ control pulses as shown in Fig. 5.26c. The gate width was reduced to 230 fs, since the broadening of the control pulse was suppressed by shorter fiber pigtails (50 cm). In addition to the fast recovery ($\tau_{\mathrm{ISB}} \sim 110$ fs), however, a slow recovery component ($\tau \sim 2$ ps) was observed. It caused pattern effects at above 500 $\mathrm{Gb\,s^{-1}}$ [101]. The slow recovery is considered to be caused by tunneling of excited electrons to the adjacent bulk GaN layer.

The 10-dB saturation pulse energy was reduced to 90 pJ in an AlN rib waveguide. It is attributable to stronger carrier confinement. The shortest ISB absorption wavelength (∼1.35 μm) in waveguides was also realized, because the built-in field in the wells is stronger than that in GaN waveguide [114].

Challenges and Prospects

Theoretical calculation [115] predicted that the 10-dB switching pulse energy will be about 10 pJ in improved rib waveguides without excess loss. The switching energy will be further reduced (∼1 pJ) in ideal nitride photonic wire waveguides with a cross-section of 0.8×0.8 μm. Reductions in the coupling loss and the scattering loss are crucial in such small waveguides. Even if the switching energy is reduced to 1–10 pJ, the power consumption amounts to 1–10 W at 1 THz. Nitride semiconductors can endure such power consumption as long as they are formed on a material with a good thermal conductivity. When 1-Tb/s OTDM technology matures in the future, the function of a photonic node will be integrated on a planar lightwave circuit (PLC) where optical

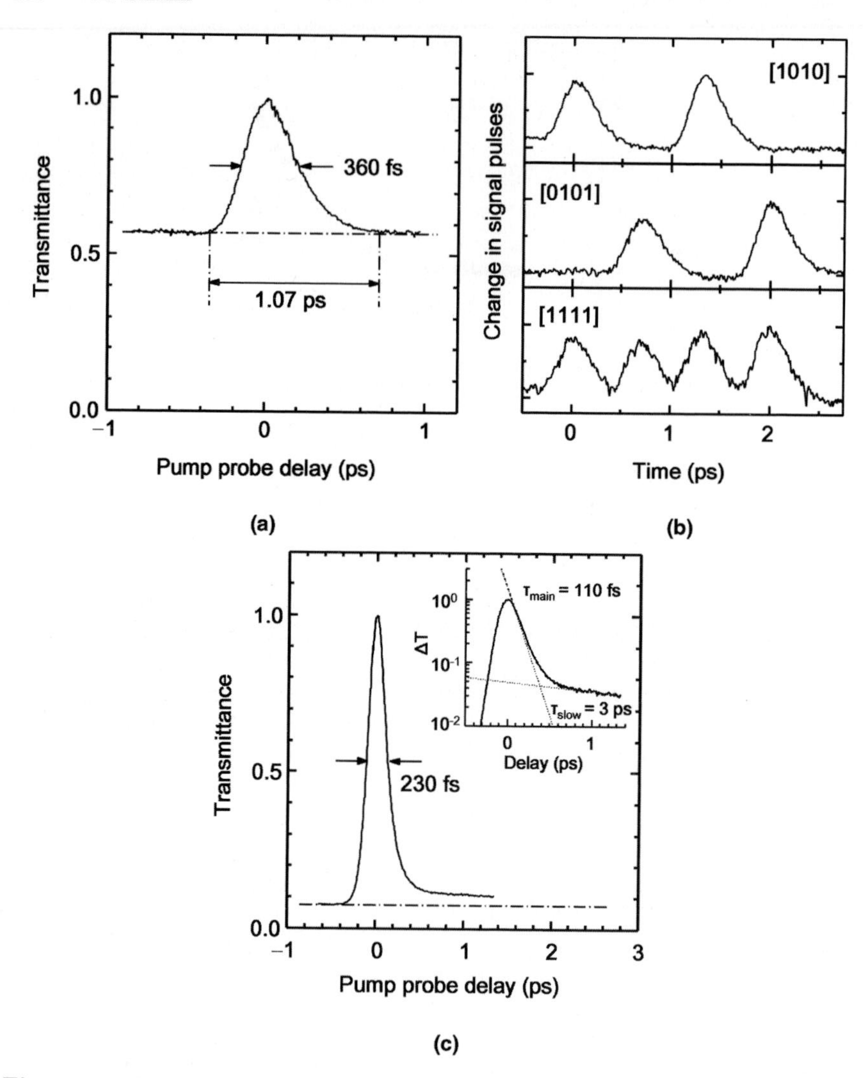

Fig. 5.26. Ultrafast response of GaN/AlN ISBT switches. (**a**) Pump (1.7 μm) and probe (1.55 μm) response of WG-A [98]. (**b**) Response of WG-A for four optical pulses with a 0.67-ps interval (corresponding to 1.5 THz) [99]. (**c**) Pump and probe response of WG-B [100, 101]

devices are synchronized by optical clock pulses. Polarization dependency of ISBT will not be a crucial problem on such a PLC.

Development of near-IR QCLs [116] and QWIPs [96, 117] pose big challenges. The main difficulties in realizing QCLs are the large voltage drop (~1 V per stage) and large current density ($J \propto {\tau_{\mathrm{ISB}}}^{-1}$). In the case of an optically-pumped intersubband amplifier, a pumping power of greater than 5 W will be required to obtain a sufficient gain [118]. In QWIPs, ultrafast ISB relaxation disturbs fast carrier transport and reduces the efficiency. Inter-sublevel transitions in quantum dots (QD) [117,119] will not remove the polarization dependency, because the aspect ratio of QDs is much smaller than unity. However, the dephasing time is considered to be longer because the electron–electron scattering probability will be reduced in QDs. It may relax the difficulties concerning QCLs. Longer dephasing time is also effective for reducing the switching energy in the ISBT optical switch, although the influence on the response time is not yet clear. Control of the dot size and the field uniformity will be crucial for realizing high-performance QD devices.

5.2.3 Nitride Photocatalysis (K. Ohkawa)

Hydrogen is a promising clean energy for fuel cells and so on. Molecular H_2 must be separated from hydrogen-containing compounds such as water and hydrocarbons. Direct photoelectrolysis by solar power is a good method of producing molecular hydrogen from aqueous water [120, 121]. A glass of water (200 cc) contains 270 l of H_2 gas at room temperature. To split aqueous water in a photoelectrochemical cell, the conduction band-edge potential of a semiconductor electrode must be lower than that of the hydrogen-evolving half-reaction ($E(H^+/H_2) = 0.0–0.059$ pH V vs. normal hydrogen electrode (NHE)) and its valence band-edge potential must be higher than that of the oxygen-evolving half-reaction ($E(O_2/H_2O) = +1.23 - 0.059$ pH V vs. NHE) [122]. Some semiconductors, such as $SrTiO_3$ and ZrO_2, were reported to have the ability to split aqueous water without the need to apply external bias [121]. The GaN was also reported to have band-edge potentials that make it applicable as the material of a photoelectrode for splitting aqueous water [123–125]. Recently, hydrogen generation from water by using GaN photocatalyst has been successful [126]. Nitrides are promising materials for photocatalyst due to the following points:

1. Bandgap control in the range of 0.6–6.2 eV
2. p- and n-type conduction
3. Chemical stability

The first point gives the possibility of high efficiency. Most photocatalysis materials are oxides and they have a wide bandgap, which gives sensitivity in the ultraviolet (UV) region. Nitride properties make it possible to have sensitivity from the UV to infrared regions. The second point enables hydrogen generation from the working-electrode side using p-type conduction. Oxides usually show n-type conduction. Oxidation occurs on the surface in the case of an n-type working electrode. The oxidation reaction corrodes most materials. Using a p-type material, a reduction reaction occurs on the surface, necessary for long-life time operation. The third point of chemical stability is also important for stable operation. To confirm these merits, recent photoelectrochemical properties of nitrides are described here [126–131].

Experimental Set-up

A nitride layer on a sapphire substrate was used as the working electrode. Nitride layers were grown by metalorganic vapor-phase epitaxy (MOVPE) using trimethylgallium (TMGa), trimethylindium (TMIn), trimethylaluminum (TMAl), and ammonia (NH_3) as precursors [132]. Conduction control was realized by Si doping for n-type or by Mg doping for p-type. The MOVPE system enables growth of AlGaInN layers with various bandgap. An Ohmic contact was made on the surface at the periphery of the nitride samples. The photoelectrochemical cell is constructed to be similar to the Hoffman apparatus to be able to collect the generated gases [126]. A schematic diagram of the cell is shown in Fig. 5.27. The aqueous contact area of the nitride working electrode was identical to the light-illuminated area, and was 9.7 mm in diameter. The counter and reference electrodes were made of Pt, and Ag/AgCl/NaCl (sodium-saturated silver–chloride electrode (SSSE); electrode potential E(AgCl/Ag) = +0.212 V vs. NHE), respectively. The light source was a 150 W Xe-lamp. We used several kinds of electrolytes such as 1 mol L^{-1} HCl (pH = 0.1), 0.5 mol L^{-1} H_2SO_4 (pH = 0.4), 0.5 mol L^{-1} Na_2SO_4 (pH = 6.7), 1 mol L^{-1} NaCl (pH = 7.7) a mixture of 0.01 mol L^{-1} NaOH with 0.5 mol L^{-1} Na_2SO_4 (pH = 11.8), 1 mol L^{-1} NaOH (pH = 13.5), and 1 mol L^{-1} KOH (pH = 14) at room temperature. A potentiostat and a gas-chromatograph were used to evaluate the electrochemical properties and to measure the generated gas composition, respectively.

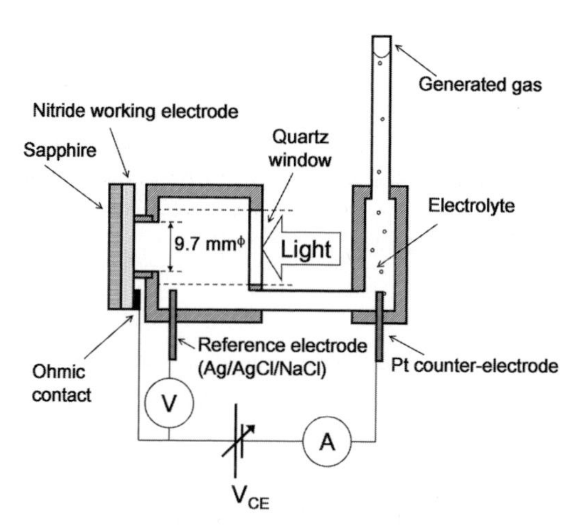

Fig. 5.27. Schematic diagram of the photoelectrochemical cell. The working electrode, reference electrode (for the measurement of electrochemical properties), and counterelectrode were a nitride wafer, Ag/AgCl/NaCl, and Pt, respectively. The nitride working electrode bias vs. Pt counterelectrode is indicated as V_{CE}

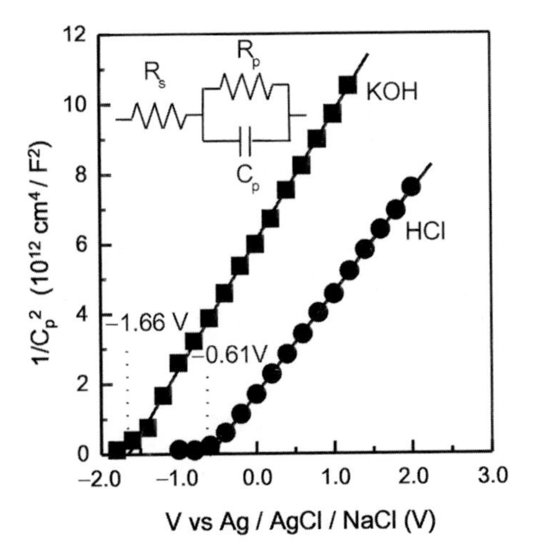

Fig. 5.28. Mott-Schottky plots for n-type GaN in aqueous 1 mol L^{-1} HCl and 1 mol L^{-1} KOH solutions. The *bold lines* show the least-squares fit. The flat-band potentials in HCl and KOH were −0.61 and −1.66 V vs. SSSE, respectively. The inset shows the equivalent circuit for capacitance calculation

Band-edge Potentials of Nitrides

The band-edge potentials of GaN are evaluated by Mott-Schottky plots [129]. Those for n-type GaN in 1 mol L^{-1} HCl and 1 mol L^{-1} KOH solutions are shown in Fig. 5.28. The inset shows the equivalent circuit from impedance fitting. We considered the capacitance (C_p) and resistance (R_p) in the semiconductor/electrolyte interface and the series resistance (R_s) of the electrolyte and semiconductor. The frequency range and amplitude of impedance measurements were 0.1–20,000 Hz and 20 mV, respectively. The flatband potentials of n-type GaN in HCl and KOH were −0.61 V and −1.66 V vs. SSSE, respectively, as obtained from Fig. 5.28.

The band-edge potentials of GaN were calculated from the measured flatband potential Φ_{fb}. For the n-type material, the conduction band-edge potential Φ_c is

$$\Phi_c = \Phi_{fb} - \frac{kT}{q} \ln\left(N_c^*/n\right), \tag{5.9}$$

where k is the Boltzmann constant, T is the absolute temperature, q is the elemental charge, n is the net carrier concentration, and N_c^* is the conduction band effective density of states. N_c^* is expressed as:

$$N_c^* = 2\left(\frac{2\pi m_e^* kT}{h^2}\right)^{3/2}, \tag{5.10}$$

where m_e^* is the electron effective mass and h is Planck's constant. For GaN, $m_e^* = 0.25 m_0$, where m_0 is the electron mass [133]. Using the GaN bandgap

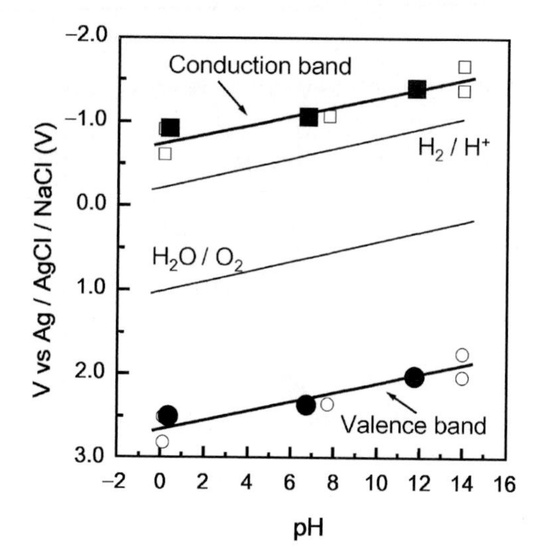

Fig. 5.29. Conduction and valence band-edge potentials of GaN as functions of pH *Solid* and *open* marks indicate p- and n-type GaN, respectively. The *thick solid lines* of the potentials were fitted from the data. The potentials for H_2 and O_2 gas-evolution half reactions are also plotted as *thin solid lines*

of 3.42 eV [134], we calculated the conduction and valence band-edge potentials of n- and p-type GaN as shown in Fig. 5.3. The band-edge potentials in Fig. 5.29 were measured not only using 1 mol L^{-1} HCl and 1 mol L^{-1} KOH but also using various electrolytes with different pH values. The electrode potentials of hydrogen and oxygen generation half reactions [135] are described in Fig. 5.29. It is found that the band-edge potentials of GaN sandwich the electrode potentials of hydrogen and oxygen generation half reactions [129]. The conduction and valence band-edge potentials, respectively, were obtained in our experiments to be

$$\begin{aligned}\Phi_c &= -0.73 - 0.055\,pH\,(\text{V vs. SSSE}) \\ &= -0.52 - 0.055\,pH\,(\text{in V vs. NHE}), \end{aligned} \tag{5.11}$$

$$\begin{aligned}\Phi_v &= +2.69 - 0.055\,pH\,(\text{V vs. SSSE}) \\ &= +2.90 - 0.055\,pH\,(\text{in V vs. NHE}), \end{aligned} \tag{5.12}$$

where Φ_v is the valence band-edge potential [129]. These results are approximately the same as previous results [123–125]. From these results, the band-edge potentials of HCl and KOH were almost the same as those of other solutions at the same pH.

The hole effective mass used was $0.80\,m_0$ in the calculation. The band-edge potentials of p-type GaN at various pH values are also plotted in Fig. 5.29 with the potentials of n-type GaN. It is found that the band-edge potentials of p-type GaN are identical with those of n-type. By contrast, the photocurrent method implied that the band-edge potentials of the p-type material were

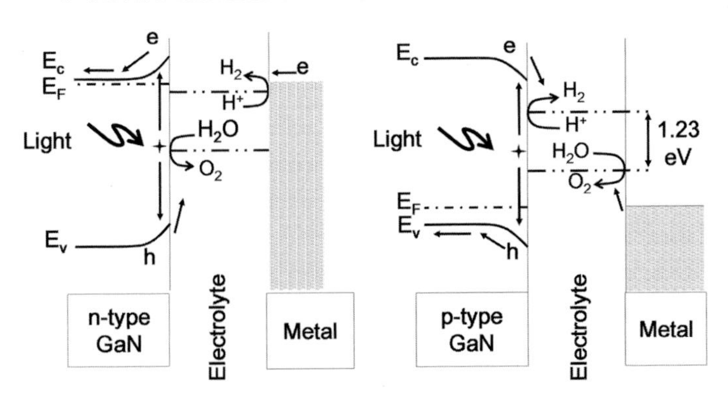

Fig. 5.30. Surface band structures of n-type and p-type GaN during photoelectrolysis. Metals are connected with each semiconductor through Ohmic contacts

approximately 0.4–0.7 V positive in potential compared with the n-type band-edge potentials [125]. It is known that band-edge potentials for p- and n-types are the same in many semiconductors [121]. Thus, we think that the potentials obtained by our capacitance measurements are closer to the "true" values.

Surface band structures of n-type and p-type GaN in an electrolyte bend up or down, respectively, then form a depletion layer on the surface as shown in Fig. 5.30. The relation between band-edge potentials of GaN and the electrode potentials of hydrogen and oxygen generation half reactions is identical with that in Fig. 5.29. In the case of n-type, light absorbed in the depletion layer causes electron–hole pairs. Electrons go inside the semiconductor, but holes move to its surface due to the band bending. Holes can oxidize water generating O_2 gas. On the other hand, electrons move to the Pt counterelectrode through the Ohmic contact. Electrons reduce H^+ ions to H_2 gas at the Pt counterelectrode. The oxidation and reduction reactions easily happen, because holes and electrons contribute oxygen and hydrogen generation half reactions without energy barrier both at the n-type GaN and at the metal surfaces, respectively. The opposite phenomenon occurs in the case of p-type GaN as can be seen from Fig. 5.30. Basically, it is possible to choose an oxidation or reduction reaction at the semiconductor surface by the choice of n-type or p-type semiconductor.

Since the bandgap of GaN is wide, the absorption of solar light is a small amount in the UV region. The InGaN alloys are promising because their bandgaps are narrower than that of GaN. $In_xGa_{1-x}N$ ($x = 0.02$ and 0.09) layers were grown on (0001) sapphire substrates by MOVPE. A Si-doped n-type $In_xGa_{1-x}N$ layer (0.1 μm, $n \geq 10^{18}$ cm^{-3}) was grown on an unintentionally doped GaN layer (1.6 μm) followed by a low-temperature GaN buffer layer for each sample. A Si-doped n-type GaN layer (5.3 μm, $n = 3 \times 10^{18}$ cm^{-3}) on an undoped GaN layer (0.8 μm) followed by a low-temperature GaN buffer layer was also used as a reference.

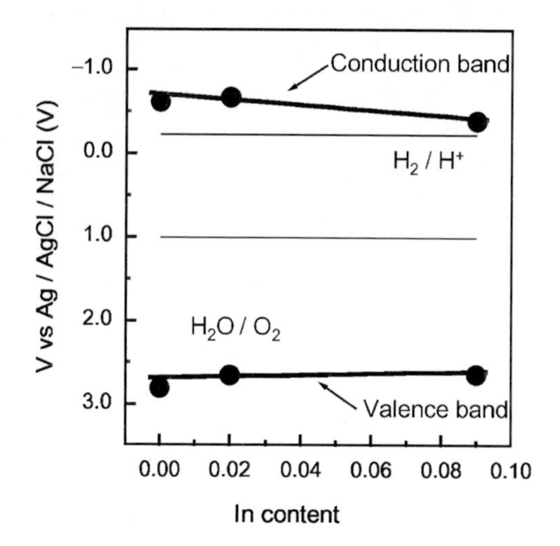

Fig. 5.31. Band-edge potentials vs. Ag/AgCl/NaCl of $In_xGa_{1-x}N$ in aqueous $1\,mol\,L^{-1}$ HCl electrolyte as functions of In composition. The potentials of H_2 and O_2 evolving half-reactions are also plotted in the graph

The bandgap energies for GaN, $In_{0.02}Ga_{0.98}N$, and $In_{0.09}Ga_{0.91}N$ defined from the photoluminescence peaks were obtained to be 3.4 eV, 3.3 eV, and 3.0 eV at room temperature, respectively. Band-edge potentials were evaluated from a flatband potential for each sample using the Mott-Schottky plot [128]. The In composition dependences of the band-edge potentials in $1\,mol\,L^{-1}$ HCl are shown in Fig. 5.31. The conduction band-edge potential increased with In composition. The potential of $In_{0.09}Ga_{0.91}N$ was close to the potential of H_2 evolving half-reaction. This potential position indicates that some external bias is required to generate H_2 gas using an $In_{0.09}Ga_{0.91}N$ working photoelectrode. On the other hand, the valence band-edge potential did not change greatly with In composition.

The relationships between photocurrent density and the applied bias to n-type GaN and $In_xGa_{1-x}N$ working electrodes in 1 mol L^{-1} HCl electrolyte are shown in Fig. 5.32a. The current density was defined as the current divided by the aqueous contact area of a working electrode. The bias of a working electrode (V_{CE}) was measured from the Pt counterelectrode. The direction of current was defined as positive flowing from a working electrode surface to an electrolyte. The photocurrent density at $V_{CE} = 0.0$ V increased with decreasing conduction band-edge potential for a working electrode; that is, the sample order was GaN $>$ $In_{0.02}Ga_{0.98}N > In_{0.09}Ga_{0.91}N$. However, the photocurrent density using an $In_{0.09}Ga_{0.91}N$ working electrode was the largest of the three samples at around $V_{CE} = 1.6$ V. The order for the photocurrent density was $In_{0.09}Ga_{0.91}N > In_{0.02}Ga_{0.98}N >$ GaN, which was opposite that at $V_{CE} = 0.0$ V. These results indicate that not only the light absorption efficiency but also the conduction band-edge potential determines

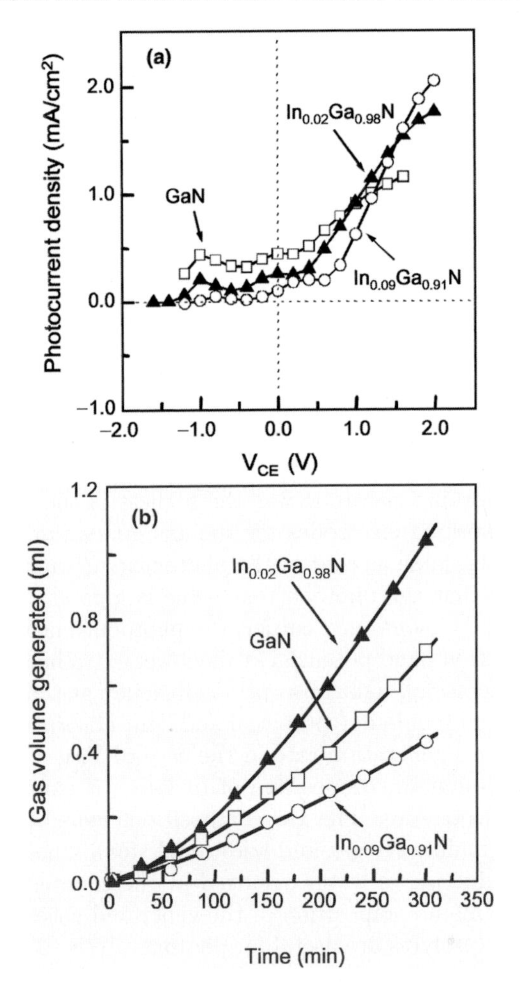

Fig. 5.32. (**a**) Relationships between the current densities and the n-type GaN and $In_xGa_{1-x}N$ working photoelectrode potentials (V_{CE}) from Pt counterelectrode in aqueous HCl. (**b**) Time dependences of the gas generation at Pt counterelectrode by GaN and $In_xGa_{1-x}N$ photoelectrolysis under the external bias of $V_{CE} = +1.0\,V$

the photocurrent density. Light absorption efficiency is one of the factors for the ability to generate photoinduced carriers. This efficiency is increased with In composition of $In_xGa_{1-x}N$. The conduction band-edge potential defines the motive force of electron-related reactions; that is, the lower potential enhances the reduced reaction. The difference between the conduction band-edge potential and the hydrogen-evolving half-reaction is the motive force of the hydrogen-reduced reaction in our experiments. This difference decreased with increasing In composition; thus, the photocurrent was probably decreased with increasing In composition in the case of zero bias. On the other hand, the motive force of the reduced reaction is sufficient when sufficient bias is applied to a working electrode; therefore, the light absorption ability was probably a dominant factor in defining the photocurrent. The higher photocurrent requires the lower conduction band-edge potential and the higher light absorption efficiency from these results. Since the characteristics of the photoelectrode samples were not the same, the In compositions and/or net donor concentration would affect the photocurrent results.

Hydrogen Gas Generation by Photoelectrolysis Effect

We selected $V_{CE} = +1.0\,V$ to get sufficient current for gas generation. Photocurrents in $1\,mol\,L^{-1}$ HCl electrolyte were not so much changed in time. The time dependences of gas generation for these $In_xGa_{1-x}N$ samples in HCl are shown in Fig. 5.32b. The gas generation rate at the counterelectrode using a $In_{0.02}Ga_{0.98}N$ working electrode was the highest of the three samples. The sample order of working electrodes for the gas generation rate at the counterelectrode was the same as that of the photocurrent values at $V_{CE} = 1.0\,V$. This result shows that the photocurrent value is a good indicator of the gas generation ability. As mentioned earlier, the photocurrent was increased with decreasing conduction band potential at the lower extra bias of a working electrode and with increasing light absorption efficiency at the higher extra bias. Both the conduction band-edge potential and light absorption efficiency probably affected the gas generation rate in the case of $V_{CE} = 1.0\,V$. These indicate that the optimum In composition of $In_xGa_{1-x}N$ exists for each applied voltage of photoelectrolysis. The gas compositions generated from the counterelectrode were mainly H_2 gas and additional small amounts of O_2 and N_2. It is clear that reduction of water occurred at the counterelectrodes because hydrogen was the main composition of the generated gases.

Conductive electrolytes are useful for photocatalysis to produce H_2 gas, so we compared different pH solutions such as 1 mol L^{-1} HCl, $1\,mol\,L^{-1}$ NaCl, and $1\,mol\,L^{-1}$ KOH [131]. The external bias of $V_{CE} = +1.0$ V was applied in the cases of KOH and HCl electrolytes. It should be noted that this bias is lower than that required for the splitting of water, that is, the ideal bias is 1.23 V [135] as shown in Fig. 5.30. The bias for the NaCl electrolyte was $V_{CE} = +1.4\,V$ because of the low photocurrent density. The time dependence of the photocurrent densities and the gas volume generated at the Pt counterelectrode are shown in Fig. 5.33a. It is worth noting that H_2 gas has been generated even by an NaCl solution. This can be one of the important methods of producing H_2 energy from seawater and sunshine.

Figure 5.33b shows that the photocurrent in KOH rapidly decreased after 150 min, but that in HCl and NaCl were stable. The weight losses of the n-type GaN working electrode after 300 min were 1.98 mg in KOH, 0.05 mg in HCl, and within the measurement error in NaCl. The large weight loss in KOH means that anodic oxidation and corrosion occurred at the GaN working electrode, that is [129],

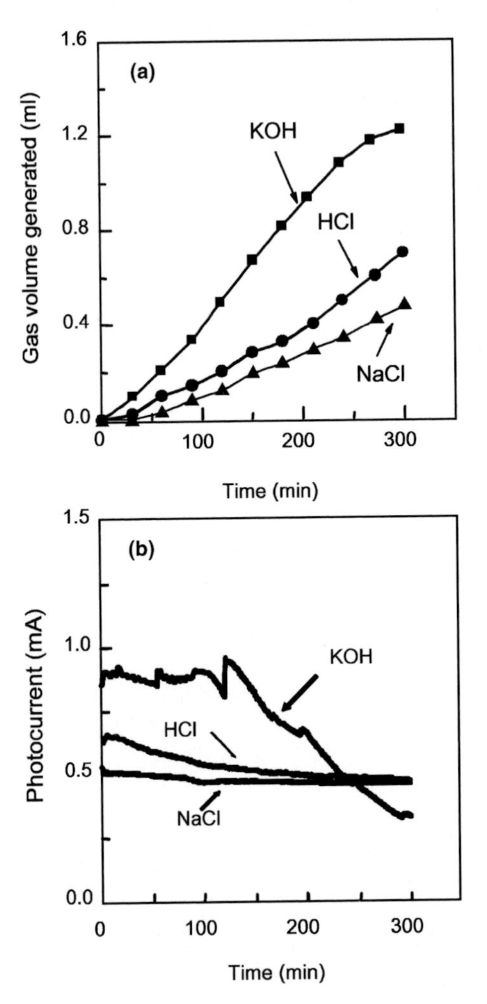

Fig. 5.33. Time-dependences of (**a**) gas volume generated at the Pt counterelectrode and (**b**) current density. The n-type GaN working photoelectrode was under illumination and in aqueous 1 mol L^{-1} HCl, 1 mol L^{-1} NaCl, and 1 mol L^{-1} KOH. The external bias (V_{CE}) was +1.0 V for HCl and KOH, and was +1.4 V for NaCl

$$2GaN(s) + 6h^{+} = 2Ga^{3+} + N_2(g). \quad (5.13)$$

The current for corrosion of the GaN working electrode was as high as 68% for KOH but only 2% for HCl of the total current charge measured. The stability of GaN electrodes in HCl and NaCl suggests that some other reactions, which were neither oxygen generation nor GaN corrosion, probably occurred at the GaN surface. The possible reaction at the GaN surface in HCl and NaCl is oxidation of Cl^- considering the results of the electrolysis using a Cl^--ion containing electrolyte [129].

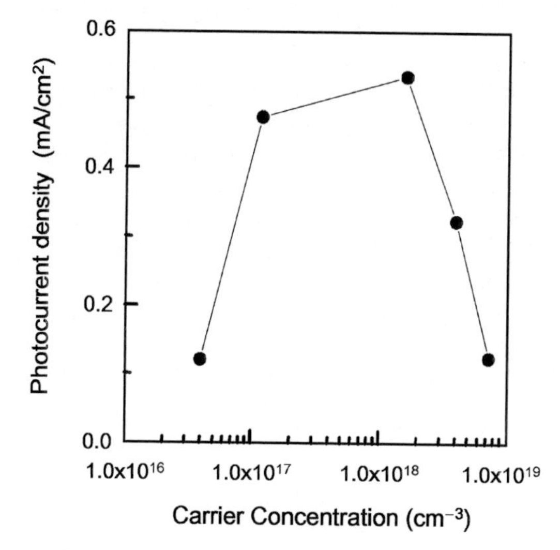

Fig. 5.34. Relationship between photocurrent density at zero-bias ($V_{CE} = 0.0$ V) and carrier concentration of n-type GaN

In order to improve the H_2 gas generation rate at zero bias, we need more photocurrent. Figure 5.34 shows the photocurrent at zero-bias vs. carrier concentration of n-type GaN. It is found that the samples with the carrier concentration range 10^{17}–10^{18} cm^{-3} show higher photocurrents than the others. The number of electron–hole pairs excited by light is the same in all the samples, but the recombination rates of electron–hole pairs in GaN are different. There are two factors affecting the photocurrent; the first factor is thickness of the depletion layer, and the second factor is series resistance. The thickness of the depletion layer depends on carrier concentration. Higher concentration causes a thinner depletion layer. If the depletion layer is thinner than the photoabsorption region (a few 10^2 nm), electron–hole pairs generated outside of the depletion layer recombine easily. Therefore the thicker depletion layer is favorable to separate electrons and holes. The thicker depletion layer, however, causes higher resistivity of a sample because of its lower carrier concentration. The lower resistivity is necessary to get photocurrent through the sample and the Ohmic contact. Thus, the photocurrent has the optimum condition for carrier concentration. The photocurrent density became more than 0.5 mA cm^{-2} after this optimization then H_2 generation at zero bias was successful. The photocurrent was 0.37 mA, and the light power irradiated was 84 mW. The photocurrent causes H_2 gas. Therefore, the conversion efficiency from light power to hydrogen energy is defined as [136]:

$$\eta = \frac{1.23\,[V] \times photocurrent\,[mA]}{light - power\,[mW]}. \tag{5.14}$$

Fig. 5.35. Large-scale experiment using a ϕ50-mm GaN photocatalyst. H_2 bubbles are generated at the Pt wire used as a counterelectrode

We obtained conversion efficiency of 0.5% by nitride photocatalyst at zero bias. Considering the fact that GaN can absorb only 3% of the Xe lamp spectrum, the conversion efficiency from the light energy absorbed is as high as approximately 20%.

Outlook

Application of nitrides as photocatalysts has just started. Fundamental properties of nitride photocatalyst are studied by using a small piece of GaN layer. In practice, a wider area will be favorable as can be seen from the solar panel. Figure 5.35 is a photograph of a lot of H_2 generation from a Pt counter-electrode using a 50-mm GaN sample illuminated by a 500-W Xe lamp. The amount of H_2 gas makes us think nitride photocatalyst promising. Nitrides are therefore attractive materials not only in the field of light emission but also in the field of photocatalysis. Nitrides are relatively safety materials. Nitrides can have sensitivity to the visible light region, and therefore nitride photocatalyst will have high efficiency to convert clean solar energy to clean hydrogen energy.

References

1. C.H. Bennett and G. Brassard: Proceedings of the IEEE International Conference on Computers System and Signal Proceedings, Bangalore, India, 1984, 175
2. P.D. Townsend, J.G. Rarity, and P.R. Tapster: Elec. Lett. 29 (1993) 634.
3. A. Muller, T. Herzog, B. Huttner, W. Tittel, H. Zbinden, and N. Gisin: Appl. Phys. Lett. 70 (1997) 793

4. R. Loudon: The Quantum Theory of Light, 3rd ed., Oxford University Press, Oxford, New York, 2000
5. B. Lounis and M. Orrit: Rep. Prog. Phys. 68 (2005) 1129
6. H.J. Carmichael: Phys. Rev. Lett. 55 (1985) 2790
7. J. McKeever, A. Boca, A.D. Boozer, R. Miller, J.R. Buck, A. Kuzmich, and H.J. Kimble: Science 303 (2004) 1992
8. B. Lounis and W.E. Moerner: Nature 407 (2000) 491
9. C. Kurtsiefer, S. Mayer, P. Zarda, and H. Weinfurter: Phys. Rev. Lett. 85 (2000) 290
10. X. Brokmann, G. Messin, P. Desbiolles, E. Giacobino, M. Dahan, and J.P. Hermier: New J. Phys. 6 (2004) 99 and references therein
11. C. Santori, D. Fattal, J. Vuckovic, G.S. Solomon, and Y. Yamamoto: New J. Phys. 6 (2004) 89 and references therein
12. Z. Yuan, B.E. Kardynal, R.M. Stevenson, A.J. Shields, C.J. Lobo, K. Cooper, N.S. Beattie, D.A. Ritchie, and M. Pepper: Science 295 (2002) 102
13. K. Sebalda, P. Michler, T. Passow, D. Hommel, G. Bacher, and A. Forchel: Appl. Phys. Lett. 81 (2002) 2920
14. P. Michler, A. Imamoglu, M.D. Mason, P.J. Carson, G.F. Strouse, and S.K. Buratto: Nature 406 (2000) 968
15. C. Santori, S. Gotzinger, Y. Yamamoto, S. Kako, K. Hoshino, and Y. Arakawa: Appl. Phys. Lett. 87 (2005) 051916
16. S. Strauf, P. Michler, M. Klude, D. Hommel, G. Bacher, and A. Forchel: Phys. Rev. Lett. 89 (2002) 177403
17. A. Muller, P. Bianucci, C. Piermarocchi, M. Fornari, I.C. Robin, R. André, and C.K. Shih: Phys. Rev. B 73 (2006) 081306
18. S. Kimura, H. Kumano, M. Endo, I. Suemune, T. Yokoi, H. Sasakura, S. Adachi, S. Muto, H.Z. Song, S. Hirose, and T. Usuki: Jpn J. Appl. Phys. 44 (2005) L793
19. T. Yokoi, S. Adachi, H. Sasakura, S. Muto, H.Z. Song, T. Usuki, and S. Hirose, Phys. Rev. B 71 (2005) 041307
20. H. kumano, S. Kimura, M. Endo, I. Suemune, H. Sasakura, S. Adachi, S. Muto, H.Z. Song, S. Hirose, and T. Usuki: Physica E 32 (2006) 144
21. E. Waks, E. Diamanti, and Y. Yamamoto: New J. Phys. 8 (2006) 4
22. R. Hanbury-Brown and R.Q. Twiss: Nature 177 (1956) 27
23. S. Kimura, H. Kumano, M. Endo, I. Suemune, T. Yokoi, H. Sasakura, S. Adachi, S. Muto, H.Z. Song, S. Hirose, and T. Usuki: Phys. Stat. Sol. (c) 2 (2005) 3833
24. M. Yoshizawa, A. Kikuchi, M. Mori, N. Fujita, and K. Kishino: 23rd International Symposium on Compound Semiconductor, 25-2, St. Petersburg, Russia, Sept. 1996 and Intitute of Physics Conference Series No. 155, 1997, pp. 187–190
25. M. Yoshizawa, A. Kikuchi, M. Mori, N. Fujita, and K. Kishino: Jpn J. Appl. Phys. 36 (1997) L459–L462
26. M. Yoshizawa, A. Kikuchi, N. Fujita, K. Kushi, H. Sasamoto, and K. Kishino: 2nd International Conference on Nitride Semiconductors, F2-4, pp. 496–497, Tokushima, Japan, 27–31 October 1997, and J. Cryst. Growth 189/190 (1998) 138–141
27. K. Hiruma, M. Yazawa, T. Katsuyama, K. Ogawa, K. Haraguchi, M. Koguchi, and H. Kakibayashi: J. Appl. Phys. Lett. 77 (1995) 447–462
28. T. Hamano, H. Hirayama, and Y. Aoyagi: Jpn J. Appl. Phys. 36 (1997) L286–L288

29. M.T. Björk, B.J. Ohlsson, C. Thelander, A.I. Persson, K. Deppert, L.R. Wallemberg, and L. Samuelson: Appl. Phys. Lett. 81 (2002) 4458–4460
30. M.H. Huang, S. Mao, H. Feick, H. Yan, Y. Wu, H. Kind, E. Weber, R. Russo, and P. Yang: Science 292 (2001) 1897–1899
31. A. Kikuchi, M. Kawai, M. Tada, and K. Kishino: Jpn J. Appl. Phys. 43 (2004) L1524–L1526
32. H.M. Kim, Y.H. Cho, H. Lee, S.I. Kim, S.R. Ryu, D.Y. Kim, T.W. Kang, and K.S. Chung: Nano Lett. 4 (2004) 1059–1062
33. M.A.S-Garcia, E. Calleja, E. Monroy, F.J. Sanchez, F. Calle, E. Munoz, and R. Beresford: J. Cryst. Growth 183 (1998) 23–30
34. E. Calleja, M.A.S-Garcia, F.J. Sanchez, F. Calle, E. Munoz, S.I. Molina, A.M. Sanchez, F.J. Pacheco, and R. Garcia: J. Cryst. Growth 201/202 (1999) 296–317
35. J. Ristic, M.A. S-Garcia, J.M. Ulloa, E. Calleja, J. S-Paramo, J.M. Calleja, U. Jahn, A. Trampert, and K.H. Ploog: Phys. Status Solidi B 234 (2002) 717–721
36. T. Araki, Y. Chiba, M. Nobata, Y. Nishioka, and Y. Nanishi: J. Cryst. Growth 209 (2000) 368–372
37. H.M. Kim, D.S. Kim, D.Y. Kim, T.W. Kang, Y.H. Cho, and K.S. Chung: Appl. Phys. Lett. 81 (2002) 2193–2195
38. V.V. Mamutin, V.A. Vekshin, V.N. Jmerik, V.V. Ratnikov, V.Y. Davydov, N.A. Cherkashin, S.V. Ivanov, G. Pozina, J.P. Bergman, and B. Monemar: Proceedings of the International Workshop on Nitride Semiconductors, IPAP Conf. Series 1, 2000, pp. 413–416
39. A. Kikuchi, K. Yamano, M. Tada, and K. Kishino: Phys. Stat. Sol. (b) 241 (2004) 2754–2758
40. H.-J. Choi, J.C. Johnson, R.R. He, F. Kim, P. Pauzauskie, J. Goldberger, R.J. Saykally, and P.D. Yang: J. Phys. Chem. B 107 (2003) 8721–8725
41. H.M. Kim, T.W. Kang, and K.S. Chung: Adv. Mater. 15 (2003) 567–569
42. A. Kikuchi and K. Kishino: Technical Program with Abstracts of 45th Electronics Materials Conference, V5, Salt Lake City, USA, 25–27 June 2003, p. 61
43. A. Kikuchi, K. Yamano, M. Tada, and K. Kishino: The International Workshop on Nitride Semiconductors (IWN-04), P13.6, Pittsburgh, Pennsylvania, USA, 19–23 July 2004
44. K. Kishino and A. Kikuchi: International Conference on Nitride Semiconductors, Bremen, Germany, 28 August–2 September, 2005
45. T. Baba, K. Inoshita, H. Tanaka, J. Yonekura, M. Ariga, A. Matsutani, T. Miyamoto, F. Koyama, and K. Iga: J. Lightwave Technol. 17 (1999) 2113–2120
46. T. Hino, S. Tomiya, T. Miyajima, K. Yanashima, S. Hashimoto, and M. Ikeda: Appl. Phys. Lett. 76 (2000) 3421–3423
47. M.H. Huang, S. Mao, H. Feick, H. Yan, Y. Wu, H. Kind, E. Weber, R. Russo, and P. Yang: Science 292 (2001) 1897
48. K. Maejima, M. Ueda, Sz. Fujita, and Sg. Fujita: Jpn J. Appl. Phys. 42 (2003) 2600
49. Z.W. Pan, Z.R. Dai, and Z.L. Wang: Science 291 (2001) 1947
50. B.P. Zhang, N.T. Binh, K. Wakatsuki, Y. Segawa, Y. Yamada, N. Usami, M. Kawasaki, and H. Koinuma: Appl. Phys. Lett. 84 (2004) 4098

51. W.I. Park, G.C. Yi, M.Y. Kim, and S.J. Pennycook: Adv. Mater. 15 (2003) 526
52. G. Kipshidze, B. Yavich, A. Chandolu, J. Yun, V. Kuryatkov, I. Ahmad, D. Aurongzeb, M. Holtz, and H. Temkin: Appl. Phys. Lett. 86 (2005) 033104
53. H.W. Seo, Q.Y. Chen, M.N. Iliev, L.W. Tu, C.L. Hsiao, J.K. Mean, and W.K. Chu: Appl. Phys. Lett. 88 (2006) 153124
54. S.J. An, W.I. Park, G.C. Yi, Y.J. Kim, H.B. Kang, and M. Kim: Appl. Phys. Lett. 84 (2004) 3613
55. S.W. Kim, Sz. Fujita, and Sg. Fujita: Appl. Phys. Lett. 81 (2002) 5036
56. T. Yatsui, T. Kawazoe, M. Ueda, Y. Yamamoto, M. Kourogi, and M. Ohtsu: Appl. Phys. Lett. 81 (2002) 3651
57. S.W. Kim, Sz. Fujita, and Sg. Fujita: Jpn J. Appl. Phys. 42 (2003) L568
58. S. Tanaka, M. Iwai, and Y. Aoyagi: Appl. Phys. Lett. 68 (1996) 4096
59. X.Q. Shen, S. Tanaka, S. Iwai, and Y. Aoyagi: Appl. Phys. Lett. 72 (1998) 344
60. M. Miyamura, K. Tachibana, and Y. Arakawa: Appl. Phys. Lett. 80 (2002) 3937
61. P. Chen, S.J. Chua, Y.D. Wang, M.D. Sander, and C.G. Fonstad: Appl. Phys. Lett. 87 (2005) 143111
62. K. Hoshino, S. Kako, and Y. Arakawa: Appl. Phys. Lett. 85 (2004) 1262
63. E.S. Jang, J.Y. Bae, J. Yoo, W.I. Park, D.W. Kim, G.C. Yi, T. Tatsui, and M. Ohtsu: Appl. Phys. Lett. 88 (2006) 023102
64. M.C. Jeong, B.Y. Oh, M.H. Ham, and J.M. Myoung: Appl. Phys. Lett. 88 (2006) 2102105
65. K. Keem, H. Kim, G.T. Kim, J.S. Lee, B. Min, K. Cho, M.Y. Sung, and S. Kim: Appl. Phys. Lett. 84 (2004) 4376
66. S. Tanaka, J.S. Lee, P. Ramvall, and H. Okagawa: Jpn J. Appl. Phys. 42 (2003) L885
67. M. Kang, J.S. Lee, S.K. Sim, B. Min, K. Cho, G.T. Kim, M.Y. Sung, S. Kim, and H.S. Han: Jpn J. Appl. Phys. 43 (2004) 6868
68. Q.H. Li, Y.X. Liang, Q. Wan, and T.H. Wang: Appl. Phys. Lett. 85 (2004) 6389
69. Y. Zhang, K. Yu, D. Jiang, Z. Zhu, H. Geng, and L. Luo: Appl. Surf. Sci. 242 (2005) 212
70. Q. Wan, Q.H. Li, Y.J. Chen, T.H. Wang, X.L. He, J.P. Li, and C.L. Lin: Appl. Phys. Lett. 84 (2004) 3654
71. H.T. Ng, J. Han, T. Yamada, P. Nguyen, Y.P. Chen, and M. Meyyappan: Nano Lett. 4 (2004) 1247
72. Y. Zhang, K. Yu, S. Ouyang, and Z. Zhu: Mat. Lett. 60 (2006) 522
73. T. Yamashita, S. Hasegawa, S. Nishida, M. Ishimaru, Y. Hirotsu, and H. Asahi: Appl. Phys. Lett. 86 (2005) 082109
74. E.H. Hall: Am. J. Math. 2 (1879) 287
75. G. Kida: III-Vs Review 16 (2003) 24
76. P. Ripka (ed.): Magnetic Sensors and Magnetometers, Artech House, Norwood, MA, 2001
77. A.M. Chang, H.D. Hallen, L. Harriott, H.F. Hess, H.L. Kao, J. Kwo, R.E. Miller, R. Wolfe, J. Vanderziel, and T.Y. Chang: Appl. Phys. Lett. 61 (1992) 1974
78. A. Oral, S.J. Bending, and M. Henini: Appl. Phys. Lett. 69 (1996) 1324
79. A. Sandhu, I. Shibasaki, and A. Oral: Microelectron. Eng. 73–74 (2004) 524

80. P.A. Besse, G. Boero, M. Demierre, V. Pott, and R. Popovic: Appl. Phys. Lett. 80 (2002) 4199
81. K. Togawa, H. Sanbonsugi, A. Sandhu, M. Abe, H. Narimatsu, K. Nishio, and H. Handa: J. Appl. Phys. 99 (2006) 08P103
82. A. Sandhu and F. Handa: IEEE Trans. Magn. 41 (2005) 4123
83. A. Lapicki, H. Sanbonsugi, T. Yamamura, N. Matsushita, M. Abe, H. Narimatsu, H. Handa, and A. Sandhu: IEEE Trans. Magn. 41 (2005) 4134
84. P.G. Neudeck, R.S. Okojie, and L.Y. Chen: Proceedings IEEE 90 (2002) 1065
85. J.L. Robert, S. Contreras, J. Camassel, J. Pernot, S. Juillaguet, L. Di Cioccio, and T. Billon: in Silicon Carbide and Related Materials 2001, Parts 1 and 2, Proceedings, Materials Science Forum, Vol. 389-3, Trans Tech Publications Ltd, Zurich-Uetikon, 2002, pp. 1435–1438
86. G. Boero, M. Demierre, P.A. Besse, and R.S. Popovic: Sens. Actuator A-Phys. 106 (2003) 314
87. M. Higashiwaki and T. Matsui: Jpn J. Appl. Phys. Part 2 - Lett. Express Lett. 44 (2005) L475
88. F. Capasso, R. Paiella, R. Martini, R. Colombelli, C. Gmachl, T.L. Myers, M.S. Taubman, R.M. Williams, C.G. Bethea, K. Unterrainer, H.Y. Hwang, D.L. Sivco, A.Y. Cho, A.M. Sergent, H.C. Liu, and E.A. Whittaker: IEEE J. Quantum Electron. 38 (2002) 511
89. B.F. Levine: J. Appl. Phys. 74 (1993) R1
90. S. Noda: in Femtosecond Technology: From Basic Research to Application Prospects, (T. Kamiya, F. Saito, O. Wada, and H. Yajima, eds.), Springer, Berlin Heidelberg New York, 1999, pp. 222–233
91. N. Suzuki and N. Iizuka: Jpn J. Appl. Phys. 36 (1997) L1006
92. N. Suzuki and N. Iizuka: in Physics and Simulation of Optoelectronic Devices VI, Proceedings SPIE, Vol.3283, (M. Osinski, P. Blood, and A. Ishibashi eds.), SPIE, Bellingham, 1998, pp. 614–621
93. H.M. Ng, C. Gmachl, S.N.G. Chu, and A.Y. Cho: J. Crystal Growth 220 (2000) 432
94. K. Kishino, A. Kikuchi, H. Kanazawa, and T. Tachibana: Appl. Phys. Lett. 81 (2002) 1234
95. N. Iizuka, K. Kaneko, and N. Suzuki: Appl. Phys. Lett. 81 (2002) 1803
96. D. Hofstetter, S.-S. Schad, H. Wu, W.J. Schaff, and L.F. Eastman: Appl. Phys. Lett. 83 (2003) 572
97. A. Helman, M. Tchernycheva, A. Lusson, E. Warde, F.H. Julien, Kh. Moumanis, G. Fishman, E. Monroy, B. Daudin, Le Si Dang, E. Bellet-Amalric, and D. Jalabert: Appl. Phys. Lett. 83 (2003) 5196
98. N. Iizuka, K. Kaneko, and N. Suzuki: Electron. Lett. 40 (2004) 962
99. N. Iizuka, K. Kaneko, and N. Suzuki: in 2004 IEEE LEOS Annual Meeting Conference Proceedings IEEE, Rio Grande, Puerto Rico, Vol. 2, 2004, pp. 665–666
100. N. Iizuka, K. Kaneko, and N. Suzuki: Opt. Exp. 13 (2005) 3835
101. N. Iizuka, K. Kaneko, and N. Suzuki: in Abstract of the International Quantum Electronics Conference 2005 and the Pacific Rim Conference on Lasers and Electro-Optics 2005, IEEE, Tokyo, 2005, pp. 1279–1280
102. T. Akiyama, N. Georgiev, T. Mozume, H. Yoshida, A.V. Gopal, and O. Wada: Electron. Lett. 37 (2001) 129
103. R. Akimoto, B.S. Li, K. Akita, and T. Hasama: Appl. Phys. Lett. 87 (2005) 181104

104. N. Suzuki, N. Iizuka, and K. Kaneko: Jpn J. Appl. Phys. 42 (2003) 132
105. L.C. West and S.J. Eglash: Appl. Phys. Lett. 46 (1985) 1156
106. A.K. Ridley: in Hot carriers in semiconductor nanostructures: Physics and applications (J. Shah, ed.), Academic, Boston, 1992, pp. 17–51
107. J. Hamazaki, H. Kunugita, K. Ema, A. Kikuchi, and K. Kishino: Phys. Rev. B 71 (2005) 165334
108. K.L. Vodopyanov, V. Chazapis, C.C. Phillips, B. Sung, and J.S. Harris, Jr.: Semicond. Sci. Technol. 12 (1997) 708
109. N. Suzuki and N. Iizuka: Jpn J. Appl. Phys. 38 (1999) L363
110. S. Nicolay, E. Feltin, J.-F. Carlin, M. Mosca, L. Nevou, M. Tchernycheva, F.H. Julien, M. Ilegems, and N. Grandjean: Appl. Phys. Lett. 88 (2006) 151902
111. K. Kaneko, N. Iizuka, and N. Suzuki: Presented at 44th Electronic Materials Conference, Santa Barbara, 2002, M7
112. N. Iizuka, K. Kaneko, and N. Suzuki: J. Appl. Phys. 99 (2006) 093107
113. A. Kikuchi, T. Yamada, S. Nakamura, K. Kusakabe, D. Sugihara, and K. Kishino: Jpn J. Appl. Phys. 39 (2000) L330
114. A. Kumtornkittikul, N. Iizuka, N. Suzuki, and Y. Nakano: Presented at Conference on Lasers and Electro-Optics, IEEE, 2006, CWD4
115. N. Suzuki, N. Iizuka, and K. Kaneko: IEICE Trans. Electron. E88-C (2005) 342
116. A. Ishida, K. Matsue, Y. Inoue, H. Fujiyasu, H.-J. Ko, A. Setiawan, J.-J. Kim, H. Makino, and T. Yao: Jpn J. Appl. Phys. 44 (2005) 5918
117. A. Vardi, N. Akopian, G. Bahir, L. Doyennette, M. Tchernycheva, L. Nevou, F.H. Julien, F. Guillot, and E. Monroy: Appl. Phys. Lett. 88 (2006) 143101
118. N. Suzuki: Jpn J. Appl. Phys. 42 (2003) 5607
119. Kh. Moumanis, A. Helman, F. Fossard, M. Tchernycheva, A. Lusson, F.H. Julien, B. Damilano, N. Grandjien, and J. Massies: Appl. Phys. Lett. 82 (2003) 868
120. A. Fujishima and K. Honda: Nature 238 (1992) 37
121. For example, A.J. Nozik and R. Memming: J. Phys. Chem. 100 (1996) 13061
122. For example, M. Pourbaix: Atlas of Electrochemical Equilibria in Aqueous Solutions, 2nd ed., National Association of Corrosion Engineers, Houston, 1974, p. 97
123. I.M. Huygens, K. Strubbe, and W.P. Gomes: J. Electrochem. Soc. 147 (2000) 1797
124. S.S. Kocha, M.W. Peterson, D.J. Arent, J.M. Redwing, M.A. Tischler, and J.A. Turner: J. Electrochem. Soc. 142 (1995) L238
125. J.D. Beach, R.T. Collins, and J.A. Turner: J. Electrochem. Soc. 150 (2003) A899
126. K. Fujii, T. Karasawa, and K. Ohkawa: Jpn J. Appl. Phys. 44 (2005) L543
127. K. Fujii and K. Ohkawa: Jpn J. Appl. Phys. 44 (2005) L909
128. K. Fujii, K. Kusakabe, and K. Ohkawa: Jpn J. Appl. Phys. 44 (2005) 7433
129. K. Fujii and K. Ohkawa: J. Electrochem. Soc. 153(3) (2006) A468
130. K. Fujii, M. Ono, T. Ito, and K. Ohkawa: in Generation, Storage and Fuel Cells (A. Dillon, C. Olk, C. Filiou, J. Ohi, eds.), Proceedings of the Materials Research Society, 885, Warrendale, PA, 2006, 0885-A11-04
131. K. Fujii and K. Ohkawa: Phys. Status Solidi (c) 3 (2006) 2270
132. K. Kusakabe, T. Hara, and K. Ohkawa: J. Appl. Phys. 97 (2005) 043503

133. S. Nakamura and S.F. Chichibu: Introduction to Nitride Semiconductor Blue Lasers and Light Emitting Diodes, p46-c, Taylor and Francis, London and New York, 2000
134. S. Chichibu, T. Azuhata, T. Sota, and S. Nakamura: J. Appl. Phys. 79 (1996) 2784
135. M. Pourbaix: Atlas of Electrochemical Equilibria in Aqueous Solutions, 2nd ed., National Association of Corrosion Engineers, Houston, 1974, pp. 97–105
136. S.U.M. Khan, M. Al-Shahry, and W.B. Ingler Jr.: Science 297 (2002) 2243

6

Crystal Growth

6.1 Bulk Crystal Growth

6.1.1 SiC (N. Ohtani)

Silicon carbide (SiC) exhibits a peritectic reaction type phase diagram and dissociates into graphite and a silicon melt containing atomic carbon at 2,830°C. Thus bulk SiC cannot be produced by congruent melt growth, which needs a stoichiometric melt of elements. Further, due to the low solubility of carbon in Si, it is difficult to grow large single crystals from the Si melt. Thus vapor phase growth is normally used to produce bulk single crystals of SiC.

Historically, the Acheson process (1891) has been used for synthesis of SiC single crystals. This method, where a mixture of silica and a carbon source is heated to 2,000°C, is used to manufacture polishing powder (abrasive). Later, the Lely method was developed in 1955, where SiC powder is sublimated in a graphite crucible and recrystallized on to the inner walls of crucible. However, it was only possible to produce crystals of 10–15 mm in diameter using these methods; insufficient for the manufacture of semiconductor devices.

Currently, large-sized bulk single crystal SiC is produced by seeded sublimation growth, also known as the *modified Lely method*. The original Lely method had problems with low growth rates and lack of control of nucleation sites during the initial stages of growth. These problems were resolved by Tairov et al., who developed a physical vapor transport growth method of SiC [1]. The growth process takes place inside a semiclosed crucible wherein vapor from sublimated Si and carbon sources is transported over a seed crystal maintained at a slightly lower temperature than the source materials. This results in supersaturation and condensation of the vapor onto the seed and growth of single crystal SiC. The rate of crystal growth depends on the temperature of the sources, and the temperature gradient and pressure of the system. The early crystals grown by Tairov et al. were 14 mm in diameter and currently it is possible to produce 100-mm diameter wafers. Commercial SiC wafers are 2 and 3 in. in diameter.

Computer simulations are being increasingly used for optimizing growth conditions of bulk SiC. Such theoretical approaches are particularly important for this semiconductor system where in situ observation of growth processes is extremely difficult. The main parameters required for the simulation of sublimation growth processes are:

(1) Generation and transmission of heat (induction heating, thermal conduction, radiation, and convection)
(2) Mass transport (diffusion and convection of multielement particle systems such as Si, SiC_2, and Si_2C)
(3) Chemical reactions (interfaces of source-gas, gas-graphite wall and gas-crystal growth surface, and vapor phase)

Simulations combining (1) and (2) are widely used, as well as reports incorporating the parameters in (3) [2].

Modified Lely Method for Manufacture of Bulk Single Crystal SiC

Figure 6.1 shows an example of a modified Lely system used for growing bulk single crystal SiC. A graphite crucible is located inside a double-walled (for water cooling) quartz growth chamber. The relative positions of the high-frequency coil and the base of graphite crucible can be moved vertically and/or rotated for control of the temperature gradient. A turbomolecular pump (TMP) is used for evacuating the growth vessel and a conductance valve for controlling the pressure. The pump is used to remove residual gas

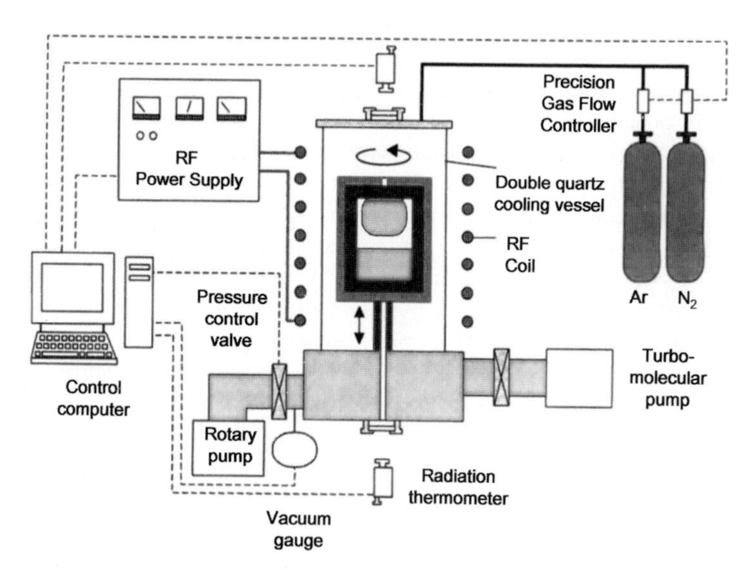

Fig. 6.1. Example of a modified Lely system used for growing bulk single crystal SiC

prior to growth and the conductance valve to control the internal pressure during growth. High-purity Ar is used for controlling the transport of the source gases and high-purity nitrogen is used for n-type doping. The flow of the gases is controlled using high precision mass flow controllers. The internal pressure of the growth vessel is measured in real time with a diaphragm vacuum gauge and the signal is read by a computer which is used to automatically control the conductance valve. The crucible temperature is measured using an infrared radiation thermometer and the resulting data fed into the computer control system.

A typical growth process is as follows (1) the TMP is used to remove the residual nitrogen from the growth vessel; (2) the graphite crucible is heated and Ar gas introduced into the vessel and the pressure is increased to ~100 kPa in order to prevent the unwanted initiation of polycrystalline growth (often due to 3C-SiC nucleation) that occurs at low temperature ($<2,000°C$); (3) the crucible temperature is increased to 2,200–2,400°C over a few hours and in order to initiate growth, the Ar pressure is reduced from ~100 kPa to as low as 1 kPa over 5–60 min; (4) the Ar pressure and temperature are kept constant during growth and afterward the temperature is reduced to room temperature over the same time taken to raise it to the growth temperature; and (5) the crystal is removed.

Control of Single Crystal SiC Polytypes

Growth of the desired crystal polytype is not always possible using the modified Lely method even if the seed crystal of desired polytype is used.

The polarity of the growth surface of seed crystals is the most important parameter for controlling the polytype. The (0001)Si and $(000\bar{1})$C surfaces, which have different chemical properties, exist on SiC{0001} surfaces. Table 6.1 shows the polytypes generated by 6H and 4H seed crystals and their surface polarities [3]. In the case of (0001)Si surface growth, 15R crystals are often generated but 6H-SiC is dominant and 4H-SiC is never produced. However with growth on the $(000\bar{1})$C surface, both the 4H and 6H crystals are generated. In many cases a mixture of both is observed but, in contrast to growth on the (0001)Si surface, it is found that growth on $(000\bar{1})$C is dominated by crystals with the same polytype as the seed crystals. That is, the possibility of

Table 6.1. Polytypes generated by 6H and 4H seed crystals and their surface polarities

seed crystal		crystal growth
polytype	plane polar	generation polytype
6H	$(000\bar{1})$C	4H, 6H
	(0001)Si	6H (15R often)
4H	$(000\bar{1})$C	4H (dominating)
	(0001)Si	6H (15R often)

6H-SiC growth on 6H seed crystals is greater than 4H and, conversely, 4H-SiC is more probable than 6H growth on 4H-SiC seeds. The 15R-SiC growth is observed on both surface polarities but 15R is not a dominant polytype.

Transition metal impurities such as Sc and Ce are not readily incorporated into SiC but these impurities are known to affect the formation of polytypes. On the other hand, SiC can be heavily doped with N and Al, which are also known to affect the formation of polytypes. The transition metals are thought to change the surface energy of growth nuclei, which affects the polytype of grown crystals.

Polytypes have also been found to be affected by the degree of supersaturation and crystal stoichiometry, and the optimization of these growth parameters enables the reproducible growth of single 4H or 6H polytype crystals.

Control of the Electrical Properties of Bulk SiC Crystals

Low resistivity crystals are required for LED and power devices, while high resistivity (semi-insulating) substrates are needed for high-frequency devices to reduce parasitic capacitance and for effective device isolation. Nitrogen is widely used for n-type doping. Figure 6.2 shows the variation of SiC resistivity with nitrogen partial pressure in the Ar atmosphere [4]. Nitrogen atoms incorporated in single crystal SiC act as donors at an energy level about 100 meV below the conduction band edge. As seen from Fig. 6.2, it is possible to produce SiC crystals with a resistivity in the mΩ cm range above nitrogen partial pressures of 1 kPa. However, using the modified Lely method it is only possible to produce p-type SiC with a minimum resistivity of $\sim$1 Ω cm.

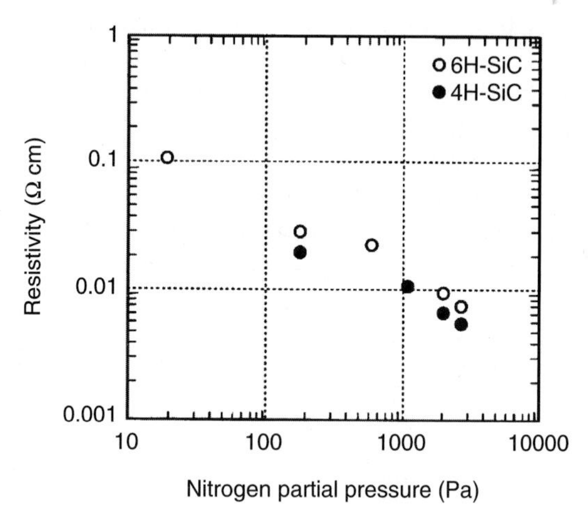

Fig. 6.2. Variation of SiC resistivity with nitrogen partial pressure in the Ar atmosphere

Recently, semi-insulating single crystal SiC has become increasingly important for high-frequency device applications. The initial reports were on vanadium (V) doping of 6H-SiC single crystals [5], where growth by the modified Lely method produced material with a resistivity of $10^{15}\,\Omega\,\text{cm}$. Vanadium acts as a deep donor in p-type SiC and a deep acceptor in n-type materials, in both cases compensating residual acceptors and donors and leading to high resistivity material. The problem with this technology is that due to the large difference in the vapor pressures of SiC and V, it is difficult to uniformly dope SiC with V during growth. Also, precipitates are generated when vanadium doping exceeds a concentration of $3 \times 10^{17}\,\text{cm}^{-3}$, which results in degradation of the SiC crystal quality.

Semi-insulating SiC has also been produced by compensation of residual acceptors and donors by intrinsic point defects such as vacancies. SiC crystals with resistivity of $\sim 10^{8}\,\Omega\,\text{cm}$ have also been reported [6].

Dislocations in Bulk Single Crystal SiC

The effect of dislocations (such as micropipes) on device performance is of concern. Dislocations in SiC are defined in terms of the directions they extend with threading dislocations in the *c*-axis and basal plane dislocations that extend in directions approximately perpendicular to the growth direction. Threading dislocations determine the breakdown voltage of SiC devices (particularly at voltages greater than 2,000 V), and basal plane dislocations are responsible for degradation of the forward bias characteristics of SiC bipolar devices and reliability of the gate oxides of SiC MOSFETs. The basal plane of single crystal SiC is the most easily activated slip plane, and there have been reports of basal plane dislocations being generated during and after crystal growth due to thermal stress. On the other hand, threading dislocations extend in the *c*-direction and have screw or edge characters. Threading edge dislocations often form low-angle grain boundaries in SiC crystals [7].

Micropipes, hollow tubes extending along the growth direction with diameters of 1–3 μm, are the largest problem to resolve in SiC crystals because they cause degradation of the blocking voltages of high power devices. These defects are hollow core dislocations as theoretically predicted by Frank in 1951 [8]. Figure 6.3 shows a growth spiral (40×40 μm AFM image) responsible for a micropipe on the surface of 6H-SiC single crystal. The black pit seen in the middle of the image is a micropipe that is a threading screw dislocation with a huge Burgers vector. The spiral step height corresponding to the Burgers vector of the micropipe is 13.5 nm, which is nine times the *c*-axis lattice constant ($c = 1.512$ nm) of 6H-SiC.

As described above, basal plane dislocations in SiC are introduced during or after crystal growth due to thermal stress. The slip band of basal plane dislocations has been observed and Frank–Reed type dislocation multiplication in SiC crystals was reported [9].

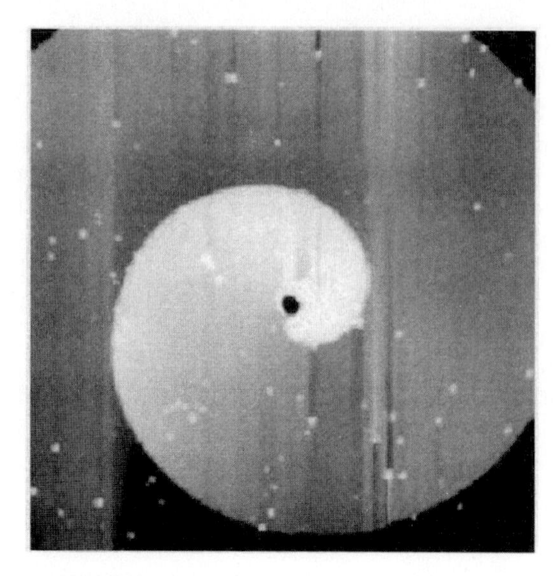

Fig. 6.3. Growth spiral (40 × 40 μm AFM image) responsible for a micropipe on the surface of 6H-SiC single crystal

On the other hand, most threading dislocations in single crystal SiC are grown in type dislocations and arise due to growth instabilities such as foreign polytype inclusions. Threading dislocations generated during the initial stages of growth are produced in pairs, and their densities have been reported to have a well-defined correlation with the stacking fault density [10].

In modified Lely growth, {0001} plane SiC single crystal platelets or wafers are usually used as the seed crystal and growth occurs parallel to the $\langle 0001 \rangle$ c-direction. However, as described above, growth parallel to the c-axis leads to the foreign polytype inclusions and generation of micropipes. A solution to these problems is growth of bulk SiC along directions perpendicular to the c-axis [11], namely, $[1\bar{1}00]$ and $[11\bar{2}0]$, which has been reported to yield crystals without micropipes.

The structure of dislocations depends on the direction of crystal growth. Figure 6.4 shows the X-ray topographic images of 6H-SiC grown in the $[1\bar{1}00]$ direction [12]. The image for the diffraction vector perpendicular to the growth direction (i.e., the $[\bar{1}\bar{1}20]$ direction, Fig. 6.4a) shows stripes with strong contrast. But a featureless image is seen for vectors parallel to the $[1\bar{1}00]$ growth direction (Fig. 6.4b). The X-ray images imply that basal plane dislocations in the crystal extend parallel to $[1\bar{1}00]$ and are aligned in the $\langle 0001 \rangle$ directions.

Table 6.2 is a comparison of the densities of dislocations and stacking faults formed during growth parallel and perpendicular to the c-axis. It can be seen that it is possible to completely suppress the formation of micropipes and threading dislocations in the c-direction for SiC growth perpendicular to the c-axis. However, the table also shows that basal plane stacking faults are

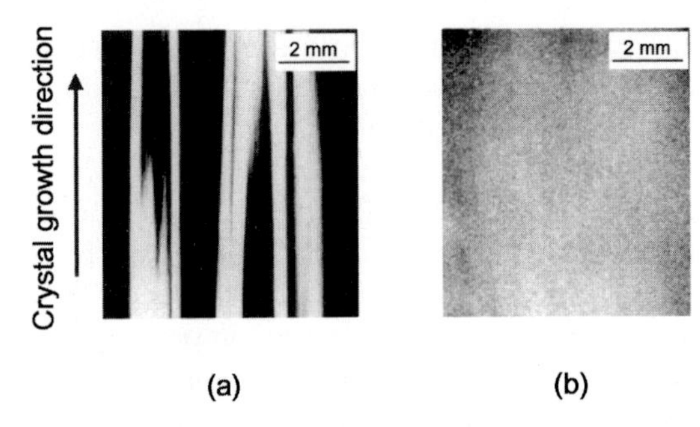

Fig. 6.4. The X-ray topographic images of 6H-SiC grown in the $[1\bar{1}00]$ direction. The image for the diffraction vector perpendicular to the growth direction (i.e., the $[\bar{1}\bar{1}20]$ direction, (**a**)) shows stripes with strong contrast. But a featureless image is seen for vectors parallel to the $[1\bar{1}00]$ growth direction (**b**)

Table 6.2. Comparison of the densities of dislocations and stacking faults formed during growth parallel and perpendicular to the c-axis

	growth along c-axis	growth perpendicular to c-axis
growth direction	[0001] or $[000\bar{1}]$	$[11\bar{2}0]$ or $[1\bar{1}00]$
micropipe density (cm^{-2})	1–10^2	0
threading dislocation density (cm^{-2})	10^3–10^4	0
basal plane dislocation density (cm^{-2})	10^4–10^5	10^4–10^5
stacking fault density (cm^{-1})	<10	10^2–10^3

easily formed during the growth [12], which leads to leakage currents in power devices. The use of off angle seed crystals has been proposed to resolve this problem [13]. Furthermore, repeated growth perpendicular to the c-axis has recently been reported to significantly reduce all types of dislocations [14].

6.1.2 Diamond (H. Sumiya, T. Nakamura)

The synthesis of diamond can be divided into two main methods: high pressure–high temperature (HP–HT) and chemical vapor phase deposition (CVD). The former includes the solubility difference and temperature-gradient method using metal solvents (catalysts) and direct conversion method and shock-wave compression method. The most effective method to grow large and good quality diamond single crystals is the temperature-gradient method under static high pressure.

Diamond Synthesis by Temperature-Gradient Method

This method was developed in the early 1970s by General Electric (GE) and used to synthesize a 1 carat (200 mg, 5 mm diameter) diamond crystal

[15, 16]. The same method was recently used to produce an almost perfect, 1-cm diameter, 7–8 carat single crystal diamond [17].

Figure 6.5 shows the components of a system used for synthesizing diamond at high pressure using the temperature differences between a carbon source in the high temperature area, a small diamond seed in the low temperature area, and Fe, Ni, and Co metal solvents (catalysts) in the intermediate temperature gradient. There is a temperature difference of 20–50°C between the carbon source and seed. Figure 6.6 shows the temperature and pressure region for growing diamond single crystals. By selection of the appropriate growth conditions it is possible to synthesize large (several millimeters), high-quality diamonds without inclusions.

If the growth rate is above a certain value then the solvent becomes incorporated into the crystal (inclusions) and at much higher growth rates polycrystalline (skeletal crystal) growth occurs. The growth rate limit for high-quality crystals without inclusions depends on the type of solvent and synthesis temperature. Also, changes in the pressure and temperature

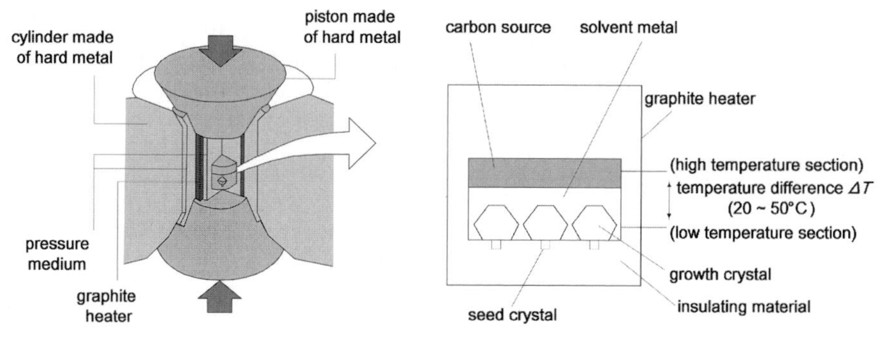

Fig. 6.5. The method of diamond synthesis (Temperature Gradient Method)

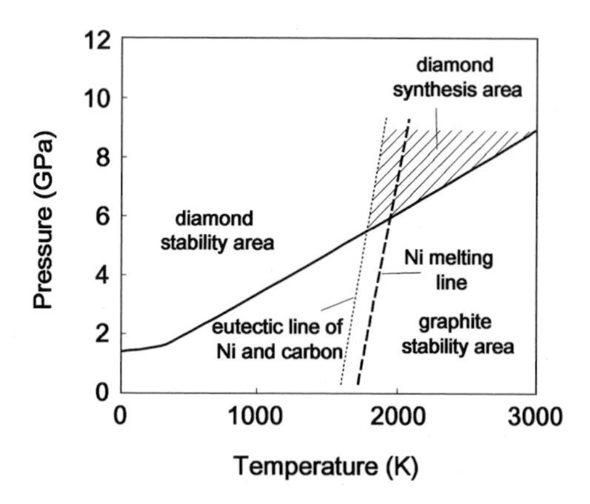

Fig. 6.6. Phase diagram for diamond synthesis (with Ni solvent)

during synthesis also results in the generation of polycrystalline defects and inclusions. During the early days of development, a growth rate of $\sim$1 mg h^{-1} was used to prevent inclusions, but it took about 1 week to grow a 1 carat crystal [15] and this method was not cost-effective for industrial applications. However, advances in industrial equipment led to improvements in the growth rates, and in 1985 it became possible to mass produce 1 carat Ib-type (with nitrogen impurities of less than 100 ppm) diamonds [18]. Furthermore, further refinements in growth procedures enabled the synthesis in 100–200 h of large Ib-type rectangular platelet diamonds with sides of 10–12 mm (5–9 carats) [19].

Improvement of the Purity of Diamonds

Diamonds produced by high pressure synthesis are yellow or amber in color. This is due to the presence of $\sim$100 ppm of nitrogen-isolated substitutional impurities. Such crystals are called *Ib-type diamonds.* The presence of these nitrogen impurities leads to absorption in the infrared and ultraviolet spectral regions. Synthesis of colorless, transparent diamond crystals (known as *IIa-type*) is very difficult compared with the production of Ib.

Nitrogen impurities arise due to residual air gaps in the synthesis chamber or nitrogen in the solvent. Thus it is necessary to remove all residual nitrogen during the synthesis. A solution to this problem was the addition of Al, Ti, and Zr (elements with high affinity with nitrogen) to the metal solvent in order to getter nitrogen [15]. In particular the group IV elements Ti and Zr are able to reduce the nitrogen concentration in crystals to $\sim$0.1 ppm. Figure 6.7 shows the ultraviolet spectra of diamond crystals synthesized by

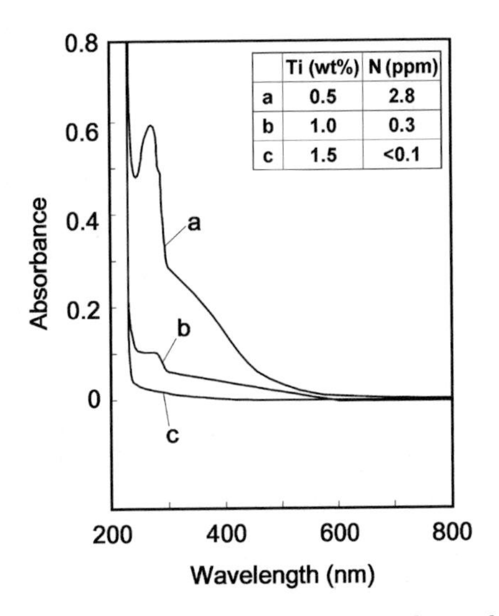

Fig. 6.7. Ultraviolet spectra of diamond crystals synthesized by the temperature difference method using Ti for gettering nitrogen

the temperature difference method using Ti for gettering nitrogen [20]. The absorption peak at 270 nm is due to nitrogen and the absorption coefficient at this wavelength is used to determine the quantity of nitrogen in the crystals. For high-purity IIa-type diamonds, the nitrogen is almost completely removed for a Ti content more than 1.5 wt% and absorption at 270 nm is not observed [20]. However, the addition of Ti and Zr can result in the formation of TiC and ZrC carbide compounds which degrade the quality of diamonds even at very low growth rates. In the case of Ti, the formation of carbides is suppressed by the addition of Cu and Ag to the solvent, which decompose TiC carbides. The use of this method has enabled synthesis of IIa-type diamonds at growth rates of 3 mg h^{-1} [20]. The incorporation of boron and nickel are also known to adversely affect the quality of diamonds.

Figure 6.8 is a comparison of the absorption spectra of synthetic IIa- and Ib-type diamonds, and natural Ia and IIa diamonds. The synthetic IIa-type is transparent from the UV to IR regions. These excellent optical properties find applications as FT-IR windows and compression cells as well as for windows in high pressure anvils.

Improvement of Crystalline Quality

Natural diamonds have many crystalline defects reflecting their growth history in the interior of the earth. In contrast, there are few defects in diamonds produced by the temperature difference method. In particular, high-purity IIa-type diamonds have extremely high crystalline quality. Figure 6.9 shows

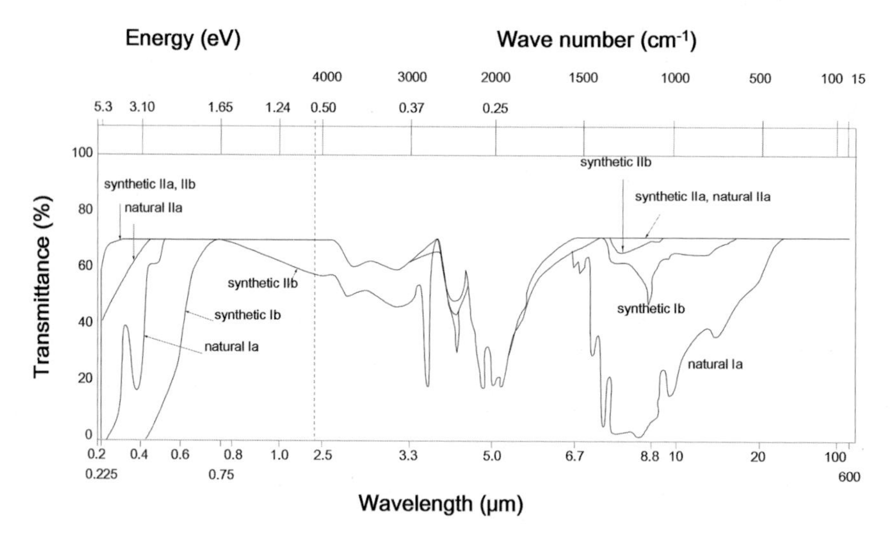

Fig. 6.8. A comparison of the absorption spectra of synthetic IIa- and Ib-type diamonds, and natural Ia- and IIa-type diamonds

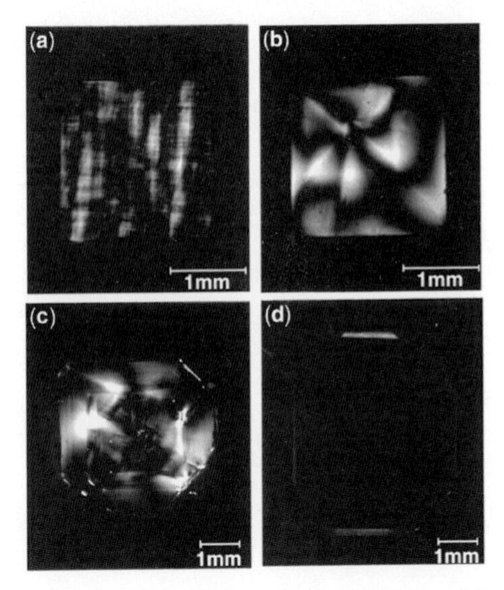

Fig. 6.9. Polarization microscope images of various types of diamonds: (**a**) natural type IIa, (**b**) natural type Ia, (**c**) synthetic type Ib, (**d**) synthetic type IIa

Table 6.3. FWHM of X-ray rocking curves and Raman spectra of diamond crystals

	synthetic IIa	synthetic Ib	natural IIa	natural Ia
natural abundance (%)	–	x	1–2	
nitrogen content (ppm)	<0.1	10–120 (dispersed)	<1	∼1000 (aggregated)
internal strain observed by polarizing microscopy	none or little	much (radial, stripe)	much (tatami)	much (radial, stripe)
defects observed by X-ray topography	some (line, plane)	many (line, stripe)	very many (tatami)	many (line, stripe)
FWHM of 004 rocking curve ($CuK\alpha_1$, arcsec)	4–6	6–20	200–2,500	7–60
FWHM of Raman spectra (cm^{-1})	1.6–1.8	1.8–2.6	2.0–2.5	3.2–3.8

polarized microscope images of a selection of diamonds. The image of the synthetic IIa-type diamond does not show mosaic structures and stress radiation lines, besides no stress corresponding to the nitrogen impurity fluctuations of Ib-type crystals. Table 6.3 shows the full-width at half maximum (FWHM) of X-ray rocking curves (XRCs) and Raman spectra of diamond crystals [21,22]. The internal stress and defect content of synthetic IIa-type diamond are thus much less than those of natural diamond and synthetic Ib-type crystals.

Thus the crystal quality of synthetic IIa-type diamond is excellent, but there are needle-like dislocations radiating from the seed crystal similar to the results shown in the X-ray topography of Fig. 6.10a; needle-like dislocations

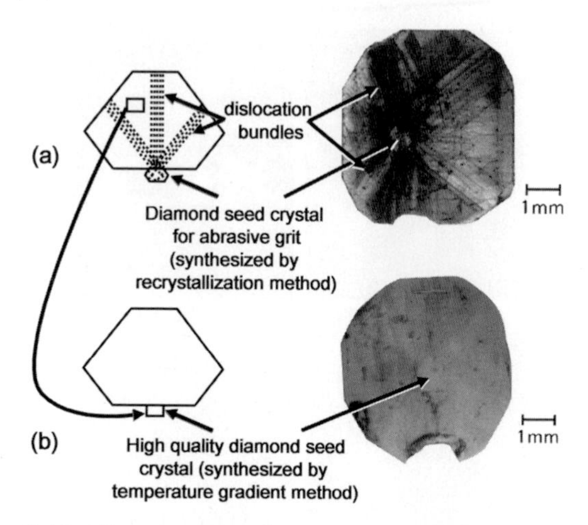

Fig. 6.10. X-ray topographic images of synthetic diamond

are often seen in diamonds grown by the temperature difference method. Usually 0.5 mm synthetic diamond grits (synthesized by the solubility difference method) are used as seeds, which are thought to act as origin of many defects including the aforementioned needle-like defects. To resolve this problem, seed crystals with extremely small densities of defects (synthetic IIa-type diamond) have been used to synthesize high-quality IIa-type diamond [22]. Figure 6.10b shows the IIa-type crystals obtained where hardly any needle-like defects can be seen [23].

Synthesis of Large 1 cm Sized IIa-Type Diamond

The growth of high-quality, large-sized diamond over a period of 200 h has been attempted [24]. The results are shown in Fig. 6.11. For growth times less than 60–70 h, high-quality crystals are obtained for growth rates of $\sim$3 mg h^{-1} and, with increasing growth time, high-quality crystals are obtained even for faster growth rates. For growth times greater than 100 h, high-quality crystals are obtained even for growth rates greater than 6 mg h^{-1}. Inclusions are present for slower growth rates but these are thought to be due to slight changes in the synthesis temperature.

As described above, it is possible to synthesize 7–8 carat (1-cm diameter), high-purity IIa-type diamond crystals by optimization of the solvent composition, nitrogen gettering, synthesis temperature, and synthesis time at high growth rates of 6–7 mg h^{-1} [24]. This is the largest sized, high-purity synthetic IIa-type diamond ever reported and the growth rate is much higher than other reports of IIa-type growth with rates of 1.5–1.8 mg h^{-1} [16, 25].

The large synthetic IIa-type diamond crystals shown in Fig. 6.12 are being used in synchrotron X-ray diffraction systems such as SPring-8 [26].

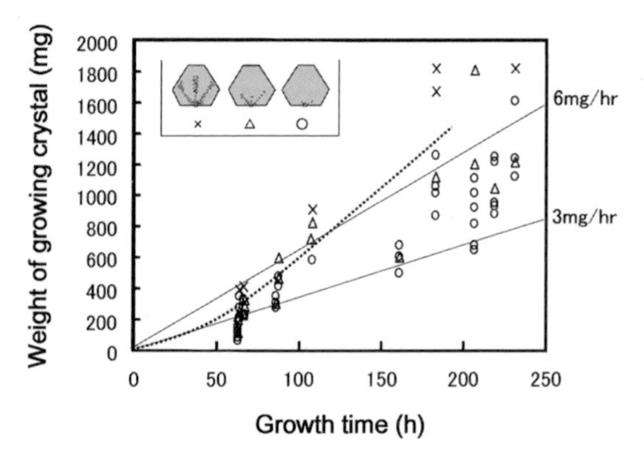

Fig. 6.11. The growth of high-quality, large-sized diamond over a period of 200 h

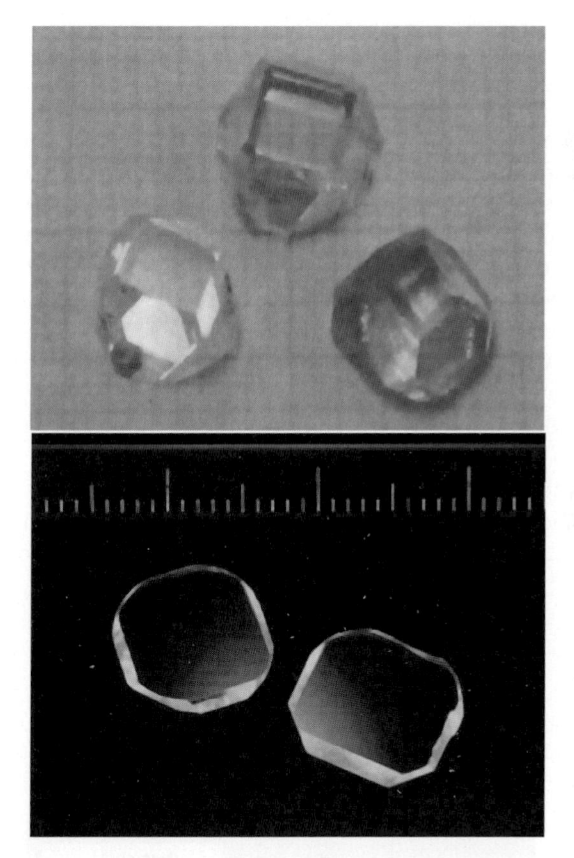

Fig. 6.12. Diamond crystal plates made of large size and high-quality synthetic IIa type diamond

6.1.3 GaN (F. Hasegawa, S. Sarayama)

It is not possible to grow bulk GaN by melt growth due to the high dissociation vapor pressure of nitrogen. Poland's high pressure laboratory grew GaN at 1–2 GPa nitrogen pressure but it was only possible to produce crystals of 17×17 mm [27, 28] which are too small for industrial applications.

The only method practically used now for the growth of bulk GaN is halide vapor phase epitaxy (HVPE). This section also describes growth from solution using a sodium flux.

Halide Vapor Phase Epitaxy

Maruska et al. reported the first growth of relatively high quality crystals of GaN by HVPE and clarified them to have a direct bandgap 3.4 eV [29]. The reaction was

$$2\text{Ga} + 2\text{HCl} \rightarrow \text{GaCl} + \text{H}_2, \quad (6.1)$$

$$2\text{GaCl} + 2\text{NH}_3 \rightarrow 2\text{GaN} + 3\text{H}_2. \quad (6.2)$$

The use of a low temperature buffer layer enabled growth of thick GaN crystals on sapphire substrates by HVPE. The problem was how to remove the sapphire substrate. One method is shown in Fig. 6.13, where the laser light irradiated from the sapphire side is absorbed by the GaN which results in the separation of the GaN/sapphire interface [30]. Although it was possible to separate 2-in. substrates (wafers) using this method, there have not been reports on its use for industrial applications due to poor yield. Another method for separating GaN and the sapphire substrate involves evaporation of a thin layer of Ti onto the sapphire, growth of GaN by HVPE where voids are formed at the interface, which enables separation; this is known as the "void-assisted separation method" (VAS) [31]. However, there have not been any reports of device fabrication using this method.

On the other hand, there have been reports by Sumitomo Electric Industries on the fabrication of bulk GaN substrates by HVPE growth of GaN on

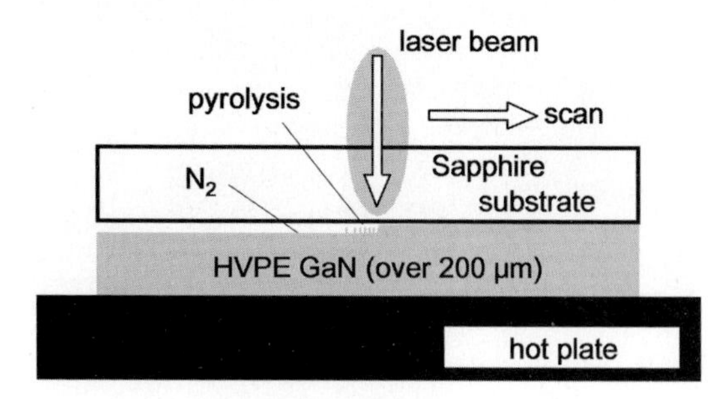

Fig. 6.13. The method of separating GaN, grown by HVPE, from sapphire substrate by laser radiation

GaAs substrates [32]. Such substrates have been sold commercially but the price is thought to be extremely high at about one million Yen per wafer.

These GaN substrates were grown by (1) depositing a 0.1-μm layer of SiO_2 onto a GaAs (111)A substrate; (2) opening 2-μm holes at the top of hexagonal peaks on the SiO_2; (3) growth of a GaN buffer layer at 500°C; and (4) high temperature growth at 1,030°C [33]. As shown in Fig. 6.14, initially selective growth occurs but it coalesces and eventually forms a uniform layer of bulk GaN. The GaAs substrate is removed by aqua regia solution and both sides of the GaN are polished to yield transparent 2-in. substrates as shown in Fig. 6.15. The FWHM of the XRC and the etch pit density were reported to be 100 arcsec and $5 \times 10^5\ \mathrm{cm}^{-2}$, respectively. These substrates were used for the growth and fabrication of laser diodes which showed excellent characteristics compared with GaN devices grown on sapphire substrates.

However, as shown in Fig. 6.16a, GaAs substrates are known to be eroded by NH_3 during GaN growth by HVPE with or without the presence of an SiO_2 layer. The Sumitomo report does not show the state of the GaAs substrate after growth of ∼100 μm of GaN. It is not known whether in fact the substrate remains after growth or if an undisclosed method is used to prevent the GaAs being eroded during growth. Further, as shown in Fig. 6.16b, erosion of GaAs substrates from NH_3 can be prevented by covering the entire substrate with GaN and then growing an intermediate GaN layer at 800–900°C [34, 35].

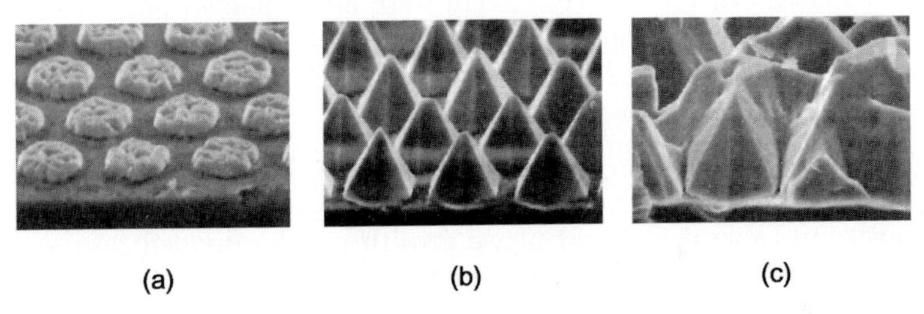

Fig. 6.14. SEM images of GaN with SiO_2 mask having 2-μmφ windows grown by HVPE on GaAs(111)A template

Fig. 6.15. Bulk GaN substrate grown by HVPE

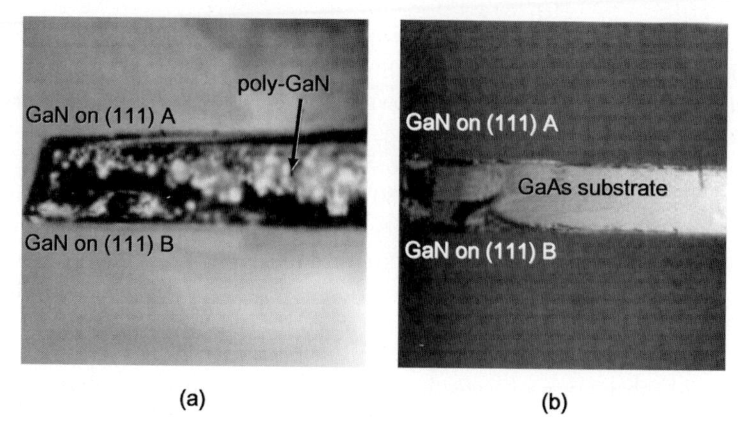

Fig. 6.16. Cross-section of GaN grown by HVPE at 1,000°C on GaAs template

GaN Growth by Sublimation

Sublimation growth of GaN is carried out using a semiclosed space, as shown in Fig. 6.1, but the difference with SiC growth is that it is necessary to suppress dissociation of GaN, and NH_3 gas is used in addition to nitrogen. However, it is only possible to grow several mm square pellets of GaN by this method due to depletion of the sources. There were many papers published on this growth method in the 1960s and 1970s, but there is hardly any interest now due to the lack of distinct advantages over HVPE.

Liquid Phase Growth

As shown in Table 6.4, liquid phase growth can be divided into melt and solution growth, where the latter can be further divided into the high pressure, flux and ammonothermal methods. Solution growth enables growth of high quality crystals with low densities of defects but the maximum crystal size is small. Due to the high partial pressure of nitrogen in GaN, it can only exist in a molten state at very high temperatures and pressures that are not industrially practical. Thus GaN is dissolved into a solvent and GaN crystals nucleate and grow from a supersaturated state.

Melt Growth

The melting point of GaN had been assumed to be 2,300°C at 6 GPa [27]. Recently, in situ X-ray diffraction measurements showed GaN to exist in a molten state at 2,220°C (6 GPa) and that 100-μm sized GaN crystals could be formed by slowly cooling the melt [36]. Since the partial pressure of nitrogen in GaN is as high as 0.1 MPa (almost atmospheric pressure) even at 900°C. Therefore, below 6-GPa GaN decomposes into either the GaN crystals,

Table 6.4. Methods for growth of GaN

method		characteristics
molten liquid growth		cooling by degrees under hot and high nitrogen pressure, more than 2,200°C and 6 GPa, respectively. 100 μm GaN is grown
liquid solution growth	high pressure liquid solution method	melting nitrogen into metal Ga under high temperature and high pressure, 1,600°C and 1–2 GPa, respectively. Crystal of mm order is grown
	flux method	several mm GaN is grown, the mixed molten liquid consists of Ga and alkaline metal of Na, K, etc., under a nitrogen pressure <10 MPa
	stable heat method	melting GaN material into super- or subcritical NH_3, and it recrystallizes. Crystal size is 100 μm – 1 mm

metallic Ga, and nitrogen gas or a state where nitrogen is dissolved in the Ga metal. However, above 6 GPa and 2, 220°C, GaN does not dissociate and melts:

$$\text{Pressure} < 6\,\text{GPa} \qquad \text{GaN(crystal)} \Leftrightarrow \text{Ga} + {}^1\!/_2\text{N}_2, \tag{6.3}$$

$$\text{Pressure} \geq 6\,\text{GPa} \qquad \text{GaN(crystal)} \Leftrightarrow \text{GaN(liquid)}. \tag{6.4}$$

In these equations, GaN is stable on the LHS and unstable on the RHS. Thus it is possible to grow GaN crystals by carefully controlling the equilibrium conditions toward the LHS. Melt growth yields GaN crystals of only about 100 μm in size.

High Pressure Solution Growth

This method involves dissolving nitrogen into molten Ga at 1,600°C/1−2 GPa followed by the nucleation and growth of GaN crystals from a supersaturated state produced by either slowly reducing the temperature or by increasing the pressure of the melt. By cooling at 2–4°C min^{-1} from initial conditions of 1,200–1,600°C, it is possible to produce n-type single crystal GaN platelets as large as 17 mm (after 150 h) [27], with carrier densities in the range 10^{18}–$10^{20}\,\text{cm}^{-3}$. At constant pressure and 1,500°C, the crystals are brownish-red and at lower temperatures they are transparent. This is because more nitrogen defects are generated at high temperatures. Spontaneous nucleation growth leads to crystals with low densities of dislocations in the range 10^3–$10^6\,\text{cm}^{-2}$.

It is also possible to grow crystals by keeping the temperature constant and varying the nitrogen pressure. Inoue et al. [37] reported the growth of

12-mm platelets at 1,475°C in 16 h with defect densities of $< 10^5\,cm^{-2}$. The necessity for large equipment capable of applying the large pressures is the main problem with implementing this method.

Flux Method

As shown in Fig. 6.17, in this method, nitrogen is dissolved into a molten mixture of Na and Ga and $(Ga_xN_y)\cdot(Na_z)$ soluble species are formed that enable precipitation of GaN crystals at lower temperature and low pressure. It is also possible to achieve similar results using potassium [38].

This method was first reported by Yamane et al., where they mixed and heated NaN_3 and Ga [39]. The mixture was heated to temperatures between 600 and 800°C, resulting in 500-μm sized GaN crystals. They found that during growth NaN_3 was being dissociated into Na and N_2 and that the growth rate and size of the crystals could be increased by independently supplying N_2 and Na.

Figure 6.18 is a comparison of the pressure and temperature of the high pressure solution and flux methods. The region for stable growth of GaN is clearly defined. Compared with the high pressure solution method, the flux method enables growth of GaN crystals at lower temperature and pressure (800°C and less than 10 MPa).

As shown in Fig. 6.19, the growth of GaN by the flux method can be divided into three regions. If the nitrogen pressure corresponding to the temperature is too high (region (1)) then GaN precipitates too quickly which results in the growth of many nucleation sites. If the temperature is higher than the most appropriate one for a particular pressure (region (3)) then no precipitation occurs and there is no growth. Growth conditions in region (2) are the most suitable for growth of GaN crystals [40].

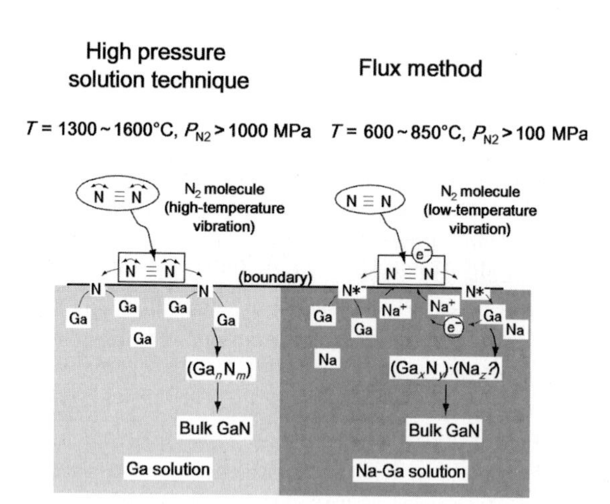

Fig. 6.17. Crystal growth models for high pressure solution and flux methods

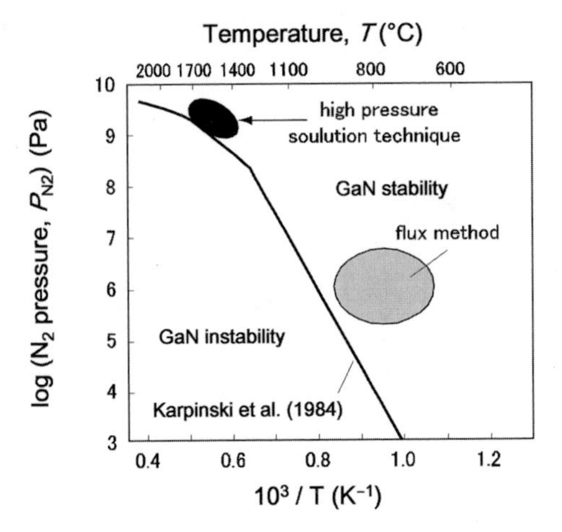

Fig. 6.18. Comparison between the high pressure solution technique and the flux method within a pressure–temperature phase stability diagram

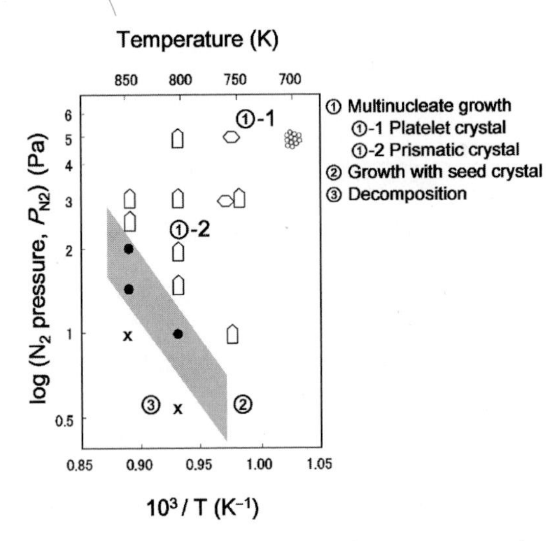

Fig. 6.19. The relation between pressure and temperature about flux method

Figure 6.20 shows optical images of GaN crystals grown by the flux method which typically have dislocation densities of $\sim 10^5\,\mathrm{cm}^{-2}$.

Ammonothermal Method

This method involves dissolving GaN into super- or subcritical ammonia followed by recrystallization [41]. It is possible to grow GaN crystals up to several hundred micrometers in size after several days of growth [42]. GaN crystals grown by the ammonothermal method on HVPE free-standing GaN substrates

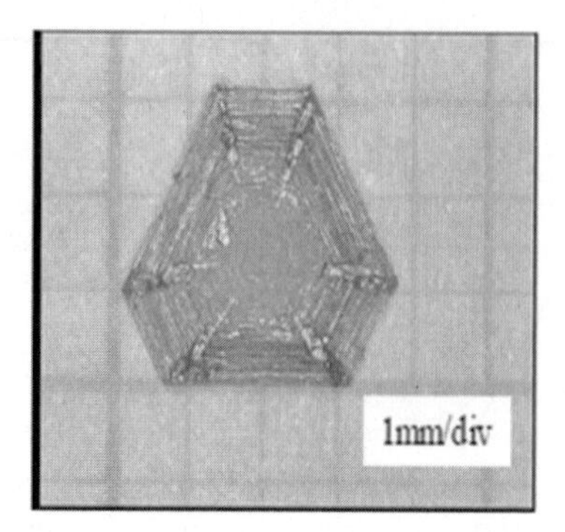

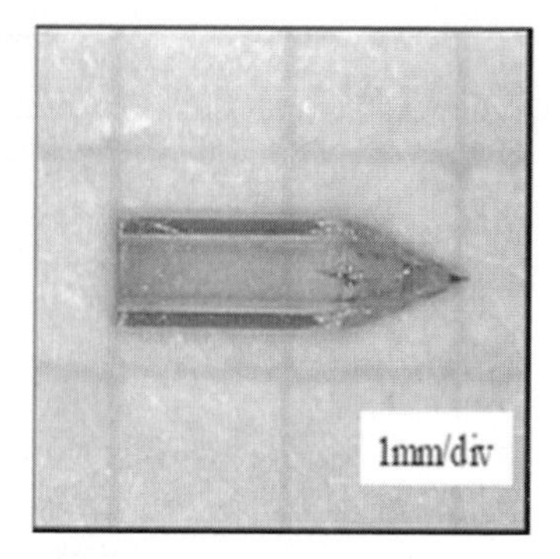

Fig. 6.20. The examples of GaN crystals grown by the flux method

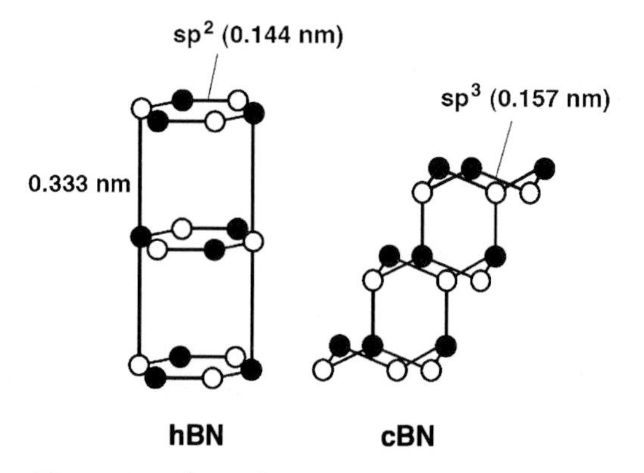

Fig. 6.21. Crystal structure of hBN and cBN

with dislocation densities of $\sim 10^7\,\mathrm{cm}^{-2}$ were found to have increased density of dislocations of $\sim 10^9\,\mathrm{cm}^{-2}$ [42].

6.1.4 BN (T. Taniguchi, K. Watanabe)

The periodic table of the elements shows that boron nitride (BN) is the combination occupying the highest position among the so-called "III–V group compound semiconductors." Hexagonal BN (hBN) and cubic BN (cBN) are the representative crystal structures of BN, as shown in Fig. 6.21. Hexagonal BN is chemically and thermally stable and has been widely used as an electrical insulator and heat-resistant material for many years. The crystal structure of hBN is composed of stacks of two-dimensional chicken-wire layers of boron and nitrogen atoms coupled by sp^2 bonds. The interaction between the layers is much less than that between the boron and nitrogen within each layer. The distance between nearest-neighbor atoms is 0.144 nm in the plane with the chicken-wire layers and 0.333 nm between the layers. This anisotropic crystal structure characterizes the mechanical and electronic properties of hBN. A single crystal has surfaces with predominant cleavage at c facets along the layers and is easily collapsed by mechanical perturbations, which cause the layers to glide along the surface. This is the reason that this material is used as a lubricant. In addition, this quasi-two-dimensional structure dominates the two-dimensional band structures and intriguing optical properties of hBN as described later originate from this anisotropy. Cubic BN, which is a high-density phase, is a material with a hardness next to diamond. In practical applications, it is an indispensable material for tools used in machining ferrous metal materials [43, 44]. The crystal structure of cBN is a typical zinc blende structure whose mechanical properties are characterized by strong sp^3 bonding between boron and nitrogen atoms. Its band structure is similar to that of diamond, namely it is an indirect bandgap material with a wide bandgap energy of 6.3 eV, which has attracted much attention for its potential application to semiconductor and optical devices. Synthesizing p- and n-type crystals by adding appropriate impurities has made it possible to fabricate p–n junctions and ultraviolet (UV)-light-emitting diodes [45–47].

A cBN single crystal can typically be grown at a high pressure (HP) of 5 GPa and a high temperature (HT) of 1,400°C using the temperature-gradient method first developed by a GE research group for diamond growth [48–50]. The size of the cBN crystals grown with this method is on the order of a few millimeters, and the crystal morphology depends upon the solvent and growth conditions. Apparently, cBN crystallizes through precipitation from BN-dissolved solvents at HP–HT, where a solvent of alkali metal boron nitrides has commonly been used [51, 52]. Although it is understood that the key to obtaining large crystals of high-quality is to control the supersaturation of BN solute in the solvent, the process of growing a well-facetted large single crystal is still being studied.

Although the intrinsic nature of the cBN crystal should be colorless, as seen in pure diamond, commercially available cBN powder abrasives always exhibit a variety of colors. Grown cBN crystals have also been reported to be amber to yellow in color [48–50]. The typical amber color of the crystals represents imperfection of the crystals, exhibiting electrical conductivity of an n-type semiconducting nature, which is probably affected by oxygen and carbon impurities [53]. By using a barium–boron nitride system as a growth solvent, nearly colorless cBN crystals with high electrical resistivity can be obtained [50, 54]. Notably, a band edge emission representing the 6.3 eV indirect bandgap was observed in colorless cBN crystals [55]. In the conventional process, however, the growth rate of cBN crystals is extremely small as compared to diamond growth [50, 56]. Some breakthrough that achieves a high growth rate and impurity control is needed to realize the promising practical potential of cBN. This will be the subject of future research.

Synthesis of cBN thin films has also been the subject of study for several decades. Plasma-assisted chemical vapor deposition (CVD), laser ablation,

ion-beam deposition, and sputtering are all common techniques used for this purpose. The films fabricated by many of these processes are nanocrystalline and contain a mixture of hBN (sp^2 bonding) and cBN (sp^3 bonding) phases. Among the variety of approaches, a breakthrough in obtaining cBN thin films that should be noted was one using a fluorine-assisted plasma CVD process [57].

Adequate attention has not been paid to obtaining hBN single crystals of quality good enough for various evaluations. In particular, research focusing on light-emitting properties has been inadequate. Recently however, high-purity single crystals of hBN were synthesized under high pressure, in work that succeeded in educing hBN's interesting properties. High-quality hBN single crystals were fabricated using the aforementioned HP crystal growth process for colorless cBN. Although hBN, as the less dense phase, does not need to be grown under HP, HP conditions are advantageous because they utilize the growth solvent of the barium–boron nitride system without causing decomposition at HT.

A temperature-gradient method was used in the hBN growth experiments. As a BN source, hBN was heat-treated at 2,100°C for 2 h in a flow of nitrogen gas to remove oxygen impurities. After the heat treatment, the oxygen content of the hBN source was measured to be about 0.06 wt%. The solvent used was barium–boron nitride. A molybdenum (Mo) sample chamber was used for the growth experiments. The assembled cell was compressed to 5.5 GPa and then heated to between 1,500 and 1,750°C. The holding time varied from 20 to 80 h. After HP–HT experiments, the recovered Mo sample chamber was dissolved using hot aqua regia to obtain the grown crystal.

As shown in Fig. 6.22, the high-purity hBN single crystals obtained under high pressure display a clear hexagonal idiomorphology, are colorless, and have high transparency. According to a secondary ion mass spectrometry

Fig. 6.22. High-purity hBN crystals produced under high pressure

study, oxygen and carbon impurity levels are both on the order of less than $10^{18}\,\mathrm{cm}^{-3}$.

The pure single crystals grown by the HP–HT method show high luminous efficiency in the deep UV region. Figure 6.23 shows an example of the luminescence spectrum measured by a UV-optimized cathodoluminescence (CL) system. A single peak in the 215-nm band dominates the spectral range from the visible to the UV region. Optical absorption spectrum and CL studies at various temperatures found a bandgap of 5.97 eV and a large exciton binding energy of 149 meV [55, 58].

Owing to the high luminous efficiency, some of the evidence for laser action excited by an accelerated electron beam was observed to compose an optical cavity consisting of reflections from the back and front surfaces of a cleaved crystal. Figure 6.24 shows an example of room temperature laser oscillation spectra from an hBN single crystal. The amplified spectral feature was observed at the etalon cavity mode [58].

Judging from the results of experiments, hBN's luminous properties apparently show that it is a direct bandgap material. However, almost all theoretical studies based on first-principle calculations predict that hBN is an indirect bandgap material [59–61]. A detailed understanding of hBN's optical nature, from both the experimental and theoretical points of view, is absolutely necessary for designing future devices such as high efficiency light-emitting devices.

Impurity doping and control of electric conduction for hBN will be an important challenge to develop high efficiency UV emissions for the future. It goes without saying that accomplishing this is indispensable for achieving a deep UV-emitting device using a p–n junction. Because of hBN's insulating

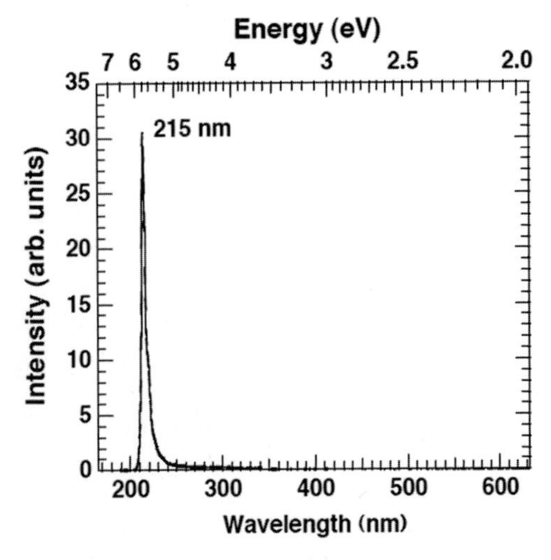

Fig. 6.23. CL spectrum of HP–HT single crystals

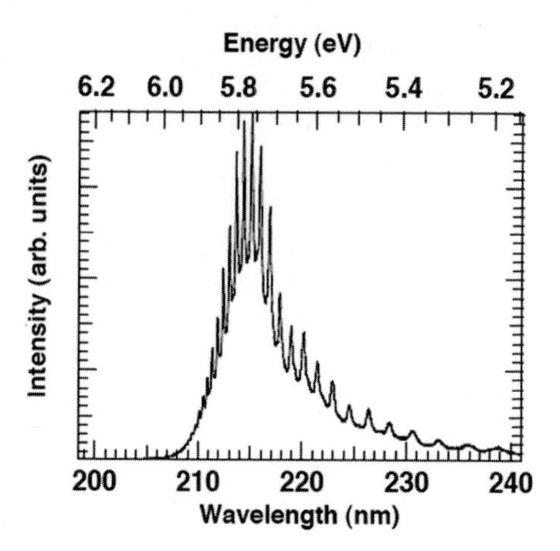

Fig. 6.24. Room temperature laser oscillation spectrum from hBN single crystal

properties, obtaining conductive hBN by doping is difficult but should be possible. From the viewpoint of the synthesis process, establishing thin film growth technology may also be the key issue for practical application.

The compound boron nitride shows a variety of characteristic physical, mechanical, and electronic properties, but its chemical and thermal stabilities are particularly outstanding. In the past few decades, considerable effort, now beginning to bear fruit, was put into obtaining pure crystals in order to elucidate intrinsic band edge properties. Both boron nitrides (hBN and cBN) were found to have wide bandgap energies with wavelengths around 200 nm, and hBN in particular was found to have high luminous efficiency in the wavelength region. These properties open up a new research field for application to robust semiconductor devices operating under extreme conditions and for deep UV-light-emitting devices.

6.1.5 ZnSe (S. Fujiwara, T. Nakamura)

The melting points (as wells as dissociation pressures at the melting points) of ZnSe and ZnO are high. Furthermore, due to the high reactivity of group VI element gases, it is not possible to synthesize these compounds by melt growth. In particular, ZnSe undergoes a phase transition from the wurtzite to zinc blende structure at ~1,430°C which is below its melting point of 1,520°C. Thus it is extremely difficult to produce high-quality crystalline substrates for optical emission devices. For these reasons, ZnSe and ZnO crystals are produced by methods enabling low temperature growth. In the case of ZnSe, large, 2-in. single crystal wafers have been grown by sublimation method, known as *physical vapor transport* (PVT). However, impurity doping is not possible by this method, which severely limits its applications. Iodine doping is possible by the chemical vapor transport (CVT) method, which enables growth of conducting crystals, but it is difficult to make large area crystals reproducibly, although 1-in. crystals have recently been reported.

Crystals grown from solution are still too small and contain defects. Further, centimeter-sized ZnSe crystals have been grown by solid-phase-recrystallization method where polycrystalline ZnSe is annealed in a Se atmosphere. However, the quality of the crystals is not as good as those grown from vapor phase and it is difficult to dope the crystals.

ZnO crystals are mainly produced from vapor phase or solution where 2-in. ZnO crystals have been grown by CVT and the hydrothermal method. As for conductivity, generally ZnO crystals show n-type conductivity originating in oxygen vacancies.

Growth of ZnSe Crystals

Physical Vapor Transport

The PVT method exploits the sublimation of ZnSe [62, 63] as shown the following equation:

$$\begin{aligned} &\mathrm{ZnSe(S)} \Leftrightarrow \mathrm{Zn(g)} + \frac{1}{2}\mathrm{Se_2(g)} \\ &K(T) = p_{Zn} {p_{Se_2}}^{1/2} \end{aligned} \tag{6.5}$$

Here, K is the temperature-dependent equilibrium constant. Growth (Fig. 6.25) involves (a) placing polycrystalline ZnSe crystals (source) at one end of a quartz glass ampoule and seed crystals at the other end; (b) raising the temperature of the source ZnSe to ~1,000°C, which leads to sublimation of ZnSe according to (1); (c) sublimed gases are transported to the seed crystals whose temperature is a few tens of degrees less than the source; and (d) ZnSe crystals grow on the seeds. The growth principle is simple but, as described below, the reproducible growth of large, high-quality crystals requires considerable ingenuity.

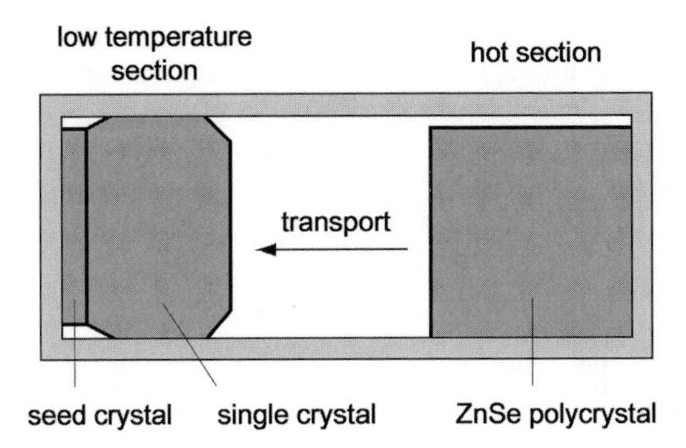

Fig. 6.25. The basic concept of the sublimation method

If PVT growth of ZnSe crystals is grown as shown in Fig. 6.25 using a simple closed ampoule, then the growth rate will be different from growth to growth. This is because the composition of Zn and Se_2 is not controlled by (1) and the composition of the gas in the ampoule fluctuates considerably. This problem is resolved by attaching a reservoir of Zn or Se to directly control the partial pressure of Zn and Se_2 or by the use of a quasi-open ampoule where the partial pressure of the gasses is controlled by exhausting some of the source gas during growth [64].

Since the critical sheer stress required for propagation of dislocations in ZnSe is small, it is difficult to grow high-quality crystals when the ZnSe is in contact with the crucible walls. Also, compared with the CVT process described later, crystal twins arise easily in the PVT method. Thus growth without contact with the ampoule is necessary. In response to this, the Markov method [65] is widely used for noncontact growth of ZnSe where, as shown in Fig. 6.26, there is a gap between the pedestal holding the seed crystal and ampoule wall. Of course a method for controlling the gas composition is required.

Figure 6.27 shows a PVT grown 2-in. diameter ZnSe single crystal grown by Sumitomo Electric Industries. The dislocation density of this crystal is $\sim 10^4\,\mathrm{cm}^{-2}$ and can be grown reproducibly. However, it is not possible to dope the crystals with n-type impurities during growth and it is difficult to produce crystals exhibiting sufficient electrical conductivity. This is the main disadvantage of the sublimation method. Diffusion of Al into ZnSe wafers has been found to yield conducting crystals [66] although there are still problems due to nonuniform doping.

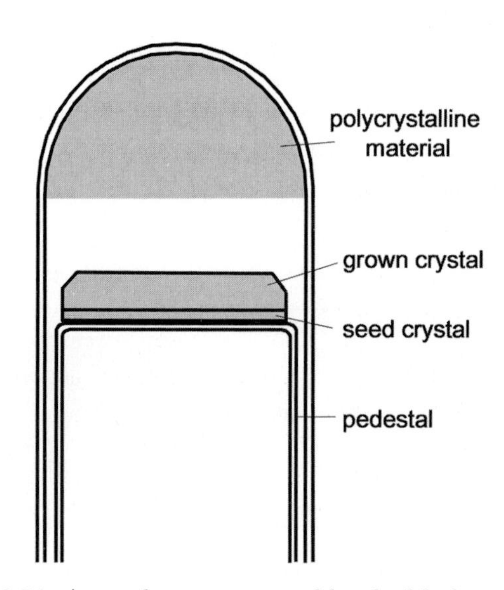

Fig. 6.26. Ampoule structure used by the Markov method

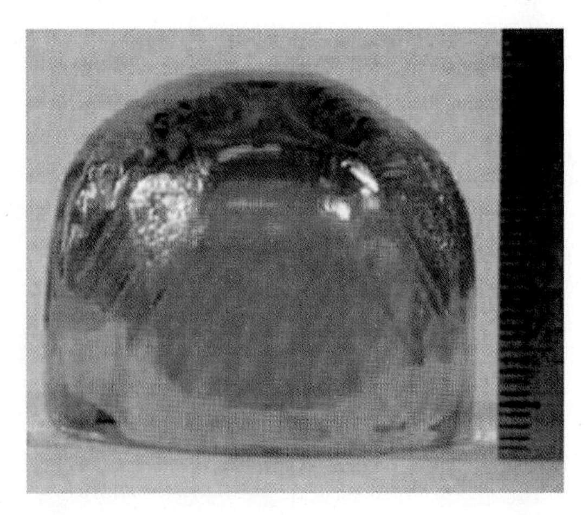

Fig. 6.27. The large single crystal of ZnSe obtained by the PVT (presented by Mr. Namikawa, Sumitomo Electric)

Chemical Vapor Transport

This growth method uses the reaction between iodine and ZnSe [62, 63], as expressed by the following equation:

$$\begin{aligned} &ZnSe(S) + I_2(g) \Leftrightarrow ZnI_2(g) + \frac{1}{2}Se_2(g) \\ &K(T) = \frac{p_{ZnI_2} {p_{Se_2}}^{1/2}}{p_{I_2}} \end{aligned} \tag{6.6}$$

Polycrystalline ZnSe and the seed crystals are put into an ampoule as well as the appropriate weight of iodine. After evacuating the ampoule, the polycrystalline source is raised to 900°C, and then gaseous iodine reacts with the source ZnSe, producing a gaseous mixture of ZnI_2 and Se_2. This gas is transported to the seed crystal, which is at a slightly lower temperature (850°C), where ZnSe crystals begin to grow. Typically, about 5 mg cm^{-3} of iodine are used and the ampoule pressure is ∼2 atm.

It is relatively easy to produce 10-mm diameter crystals by the CVT method. However, it is extremely difficult to produce 1-in. crystals because gas convection becomes stronger inside large ampoules which results in unstable growth.

In order to reduce the effects of gas convection, Fujiwara et al. developed the rotating CVT method and used it to grow large ZnSe crystals with diameters greater than 1 in. [67]. In this method, the growth ampoule maintained in a horizontal position is rotated at constant velocity around its central axis. Convection currents of the gas are sufficiently suppressed at rotations of 60 rpm. Also, it is possible to reduce the dislocation density to less than 10^4 cm^{-2} by careful suppression of thermal stress during the growth and cool-down steps. Figure 6.28 shows a large ZnSe crystal grown by CVT.

Fig. 6.28. A large single crystal of ZnSe obtained by CVT

ZnSe crystals grown by CVT contain ~200 ppm of iodine, the transport agent. In as-grown crystals, iodine, which acts as a donor, is completely compensated by Zn vacancies leading to high resistivity material. However, annealing ZnSe in a Zn atmosphere results in crystals with resistivities of ~0.05 Ω cm.

At the present time, in terms of size, the PVT method is a step ahead of CVT for the growth of single crystal ZnSe. CVT has advantages regarding doping and the ability to produce larger crystals will enable a wider range of applications for this method of ZnSe growth.

6.1.6 Crystal Growth of Bulk ZnO Single Crystal (I. Niikura, Y. Kashiwaba)

Zinc oxide (ZnO) is a wide bandgap semiconductor with great potential for a variety of commercial applications such as acoustic wave devices, UV photodetectors, light-emitting diodes, laser diodes, and high-frequency electronic devices. Single crystal ZnO with good quality is needed for the realization of these applications.

Single crystals used for electronic devices can be commercially grown in a variety of ways such as melt growth for Si and GaAs, vapor transport growth for SiC, and solution growth for quartz. Bulk ZnO single crystal growth has also been developed based on these three methods or modified methods.

Melt Growth of ZnO

The melting point of ZnO is as high as 1,977°C. Since ZnO decomposes into atomic components near the melting point at atmospheric pressure, flux methods [68–70] and a pressurized melt method [71–73] have been developed for melt growth. A pressurized melt method for growth of ZnO has been developed by Cermet, Inc. (Atlanta, GA, USA), and ZnO wafers are now being produced commercially by using this method. This process uses a modified Bridgman configuration. By radio frequency induction heating of a water-cooled crucible in a high pressure apparatus to prevent the decomposition of ZnO, high-purity ZnO starting material is heated to the melting point (1,977°C) and a melt of ZnO is formed. The liquid ZnO is crystallized using growth rates ranging from 5 to 20 mm h^{-1}. The entire melting and containment process is carried out in a controlled gas atmosphere ranging from 1 to over 100 atm. Wafers of 2 in. in size are obtained from ingots of 5.5 in. in size. Crystals grown with a growth rate of approximately 10 mm h^{-1} have demonstrated etch pit densities of approximately 10^4 cm^{-2}. The FWHM of the X-ray ω rocking curve of the (0002) reflection peak is 49 arcsec. The resistivity of this wafer is typically 0.09 Ω cm. The electrical properties of ZnO obtained by this method vary from those of a semiconductor to those of a semi-insulator and may be able to be tailored for various applications by adjusting the stoichiometry of the compound or appropriate doping. This method has a very fast growth rate and is scalable to larger dimensions (100 mm) compared with other existing processes.

Chemical Vapor Transport Growth of ZnO

CVT growth [74–84] consists of reduction, transport, and reoxidation. ZnO powder reacts with hydrogen to become gas-phase Zn and H_2O in one region of the furnace. Furthermore, gas-phase Zn is transported by hydrogen and an inert gas and is reoxidized by reaction with oxygen to become ZnO in another region of the same furnace. Thus, ZnO grows on the seeds placed in the reoxidation region. The reaction gases H_2 and O_2 and the carrier gas are introduced into the apparatus through separate tubes. The temperatures of the reduction and reoxidation regions are appropriately controlled to optimize the chemical reactions and the transportation. The chemical reactions at the source and growth surfaces have been investigated and it has been shown that the pressures of hydrogen, water vapor, and other gases are effective for the transport and growth process. The case of hydrogen as the transporting agent is mentioned above, but the selection of a transporting agent is very important in the CVT method. The effectiveness of various transporting agents such as halogen, hydrogen halide, ammonium halide, hydrogen and carbon for ZnO growth has been investigated. However, the influence of oxygen and water absorbed on ZnO powder as the starting material make the growth phenomena complicated. Eagle Picher Technologies, LLC (Miami, OK) have developed a CVT method for growth of ZnO single crystals by using hydrogen as the transporting agent and using ZnO seeds as substrates [80–83]. In this seeded chemical vapor transport (SCVT) growth method, they used high-purity ZnO powder, which was self-produced. It is interesting that this highly pure powder may play an important role in the crystal growth, although there is no detailed report about the process.

Single crystals of ZnO of 2 in. in diameter are now being produced commercially by this SCVT method (ZN Technology, Inc., Brea, CA, USA) [84]. FWHM of the X-ray ω rocking curve for the (0002) plane of this ZnO crystal is ~30 arcsec. The total impurity level was measured as ~0.5 ppm by glow discharge mass spectrometry, and the ZnO crystals were found to have mainly Si, N, and C impurities. The purity of this ZnO crystal is much better than that of ZnO crystals obtained by using other existing methods. The technique to prevent twinning in ZnO crystals seems to be important.

Hydrothermal Growth of ZnO

Crystal growth by the hydrothermal method [85–98] is carried out in a supercritical state solution. A hydrothermal method for growth of quartz crystal has been established, and large quartz SAW wafers of 6 in. in diameter have been commercially produced.

Figure 6.29 shows the internal structure of the autoclave used for the growth of ZnO single crystals. Sintered ZnO polycrystals as the starting material are put in an aqueous solution of mineralizers of LiOH (1 mol l^{-1}) and KOH (3 mol l^{-1}). Basicity of the mineralizer solution for ZnO crystal growth is higher than that for artificial quartz growth. Therefore, a closed Pt inner container is used to prevent impurity incorporation from the inner surface of the autoclave. The crystal growth zone and sintered ZnO dissolution zone were designed to be separated by a Pt baffle plate in the Pt inner container. Sintered ZnO was put into the bottom of the Pt inner container, and seed

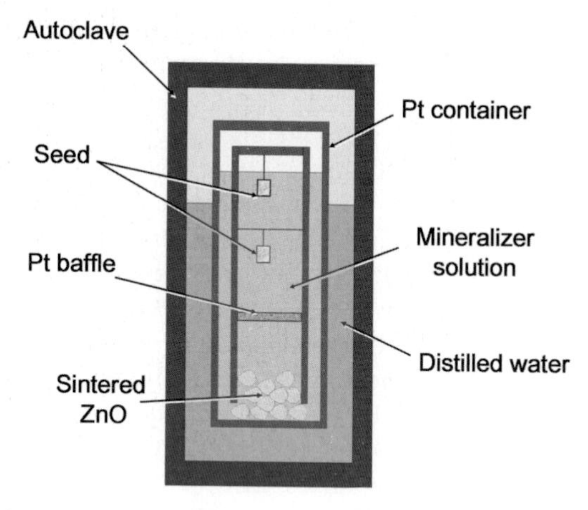

Fig. 6.29. Hydrothermal crystallization furnace for single crystal growth of ZnO

crystals were suspended in the growth zone by Pt wires. The Pt inner container was filled with an aqueous solution of mineralizers and sealed. The amount of this solution was determined from the relationships between temperature and pressure in the inner container. The Pt inner container was put into a pressure-resistant autoclave. An appropriate quantity of distilled water was supplied in the space between the autoclave and the Pt inner container for pressure balance. ZnO single crystals were grown at temperatures of 300–400°C and pressures of 80–100 MPa.

The main growth parameters of the hydrothermal method are the temperatures of the crystal growth and dissolution zones, the difference in these temperatures, the inner pressure, the mineralizer concentration, the shape of the seed crystals, and the shape of the baffle plate. The design of the baffle plate is important to control the solution flow of the convection. The quality of a ZnO single crystal changes with changes in these growth parameters.

Figure 6.30 shows a photograph of a ZnO single crystal grown by the hydrothermal method (Tokyo Denpa Co., Ltd, Japan) [95,96]. The diameter of the crystal is more than 2 in. A schematic drawing of the growth sectors of the crystal is shown in Fig. 6.31. There are five sectors named +c, +p, m, −p, and −c. The faces of the +c and −c sectors are terminated with zinc atoms and oxygen atoms, respectively. Although the crystal of 2 in. in size seems pale green, this color originates from the region near the seed in the −c sector, and the +c sector of the crystal is colorless. The transmissivity of visible light was more than 80% and the reflectivity was about 15% in the wafers of the +c sector. By ICP-MS analysis, incorporation of Fe and Al was observed and they were concentrated in the region close to the seed in the −c sector. The concentrations of Fe and Al were low except in this area of the crystal. It is thought that the pale green color of the −c sector crystal is related to Fe

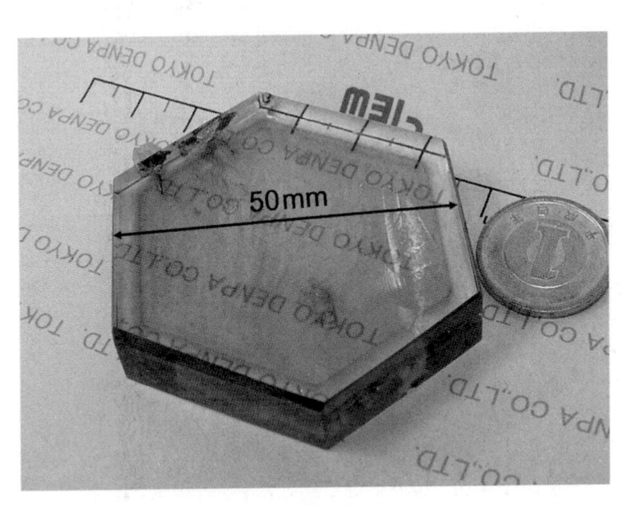

Fig. 6.30. Photograph of a ZnO single crystal grown by the hydrothermal method (Tokyo Denpa Co., Ltd, Japan)

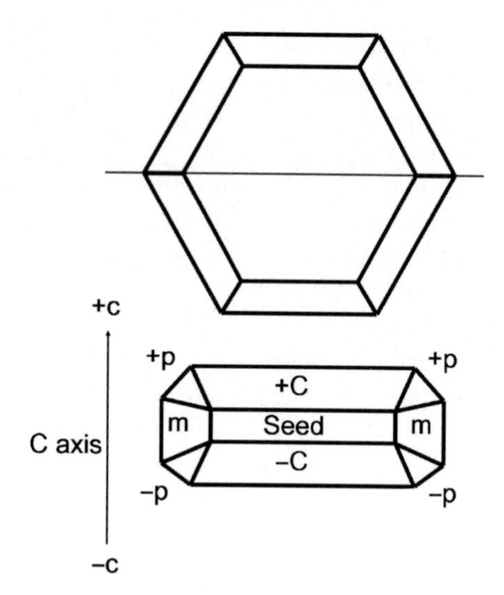

Fig. 6.31. A schematic drawing of the growth sectors of the ZnO crystal shown in Fig. 6.30

and Al impurities. Incorporated concentrations of Li and K originating from the mineralizer were several ppm for Li and under 0.2 ppm for K. The ratio of c- and m-direction growth rates depends on the Li ion concentration in the mineralizer [94]. Optimization of the growth rate in each direction is needed to obtain ZnO crystals with good crystallinity.

Wafers of 2 in. in diameter and 10-mm square wafers with thickness of 0.5 mm were sliced from bulk crystals and polished. After 4 h of annealing at 1,100°C in an atmosphere of 1 atm., the surface damage caused by the chemical mechanical polishing (CMP) was removed completely, leaving atomically flat- and terrace-like features on both the O- and Zn-face as shown by the AFM images in Fig. 6.32.

The X-ray ω rocking curve of the (0002) reflection peak of ZnO single crystals grown by the hydrothermal method is shown in Fig. 6.33. The FWHM is 18 arcsec for the ZnO crystals. The spot area of the X-ray reciprocal space map of (11–24) reflection shown in the inset of Fig. 6.33 is very small. These results show that the crystallinity of the ZnO single crystals grown by the hydrothermal method is better than that of crystals grown by other methods.

Figure 6.34 shows the photoluminescence (PL) spectrum of the hydrothermally grown crystal measured at 5.2 K. Strong UV emissions from the band edge were observed at about 3.36 eV. However, green emission originating from oxygen vacancies was not observed. Therefore, a ZnO single crystal grown by the hydrothermal method has high crystallinity.

Figure 6.35 shows electrical characteristics of a ZnO single crystal wafer of 2 in. in size. The distributions of resistivity, mobility, and carrier concentration are $380\,\Omega\,\text{cm} \pm 15\%$, $200\,\text{cm}^2\,\text{V}^{-1}\,\text{s}^{-1} \pm 10\%$, and $8 \times 10^{13}\,\text{cm}^{-3} \pm 20\%$, respectively. The uniformity of electrical properties in one wafer is good.

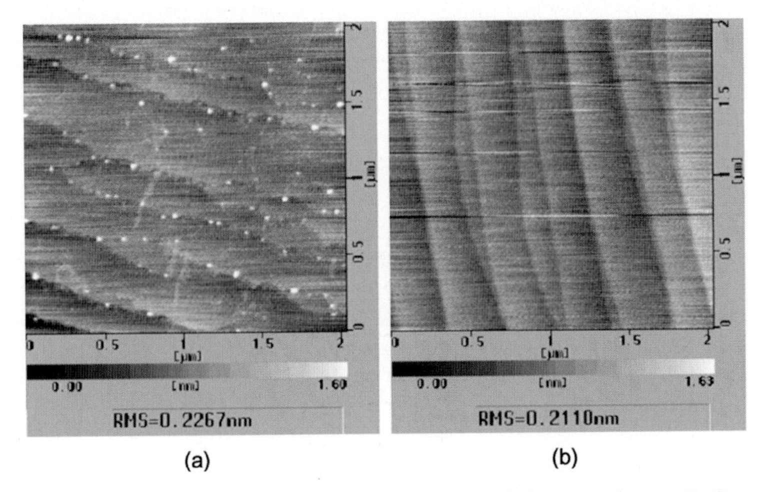

Fig. 6.32. AFM images of the oxygen (**a**) and zinc (**b**) faces of 2-in. ZnO surfaces

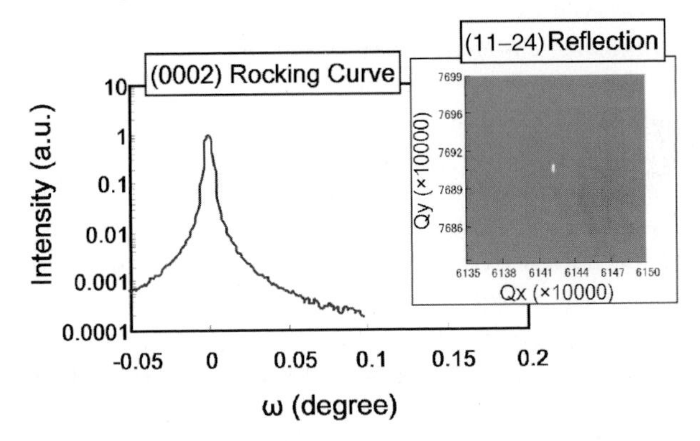

Fig. 6.33. The X-ray ω rocking curve of the (0002) reflection peak of ZnO single crystals grown by the hydrothermal method

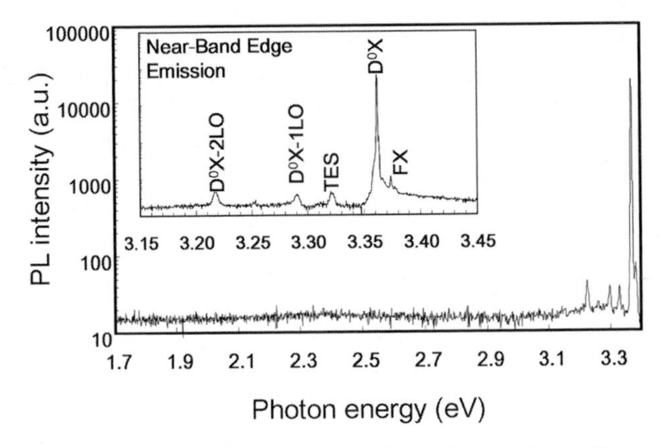

Fig. 6.34. Photoluminescence spectrum of the hydrothermally grown crystal measured at 5.2 K

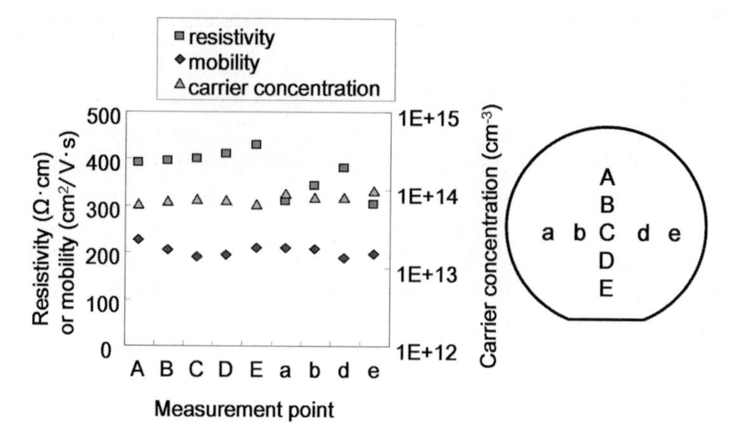

Fig. 6.35. Electrical characteristics of a 2-in. ZnO single crystal wafer

The EPD observed by chemical etching with HCl (0.7%) aqueous solution at 60°C for 5 min was less than 80 cm^{-2} on both the Zn- and O-face.

The rate of the hydrothermal growth of ZnO is as slow as ~0.2 mm day^{-1}, but it is comparable to the rate of ~0.5 mm day^{-1} for industrialized quartz crystal growth. A Pt liner is costly, but it is forever reusable, so this growth method for ZnO is commercially feasible.

XRD analysis showed that the crystallinity of ZnO crystals grown by the hydrothermal method is better than that of crystals grown by other existing methods, but the effects of Li and K impurity concentrations on the characteristics of various devices have not yet been elucidated. Methods for reducing the concentrations of Li and K will be developed in the near future.

The crystal growth methods mentioned above all have some merits and some demerits, and all of them are still under development. It is expected that the quality of bulk ZnO single crystals will be improved in the near future and that improvement in quality will be accelerated by progress in the fabrication of devices. Now GaN devices form a big market, but in the development stage of GaN devices many efforts had be made to overcome the mismatches of lattice constant and thermal expansion coefficient induced by using sapphire or SiC as substrates. On the other hand, the development of ZnO devices is free from these problems by using ZnO single crystals as the substrate and may be expected to have a better chance of success in a shorter time.

6.2 Epitaxial Growth

6.2.1 Epitaxial Growth of SiC (T. Kimoto)

Homoepitaxial Growth of 4H and 6H

Hexagonal 4H- and 6H-SiC can be grown homoepitaxially on SiC{0001} substrates. Commercial products are available with diameters of 2–3 in. and at 4-in. SiC wafers are under development by the sublimation method. However, cubic 3C-SiC undergoes a phase change to a hexagonal polytype at high temperatures thus it is difficult to grow bulk 3C-SiC. In the early days, liquid phase epitaxy (LPE) was used where carbon was transported over molten silicon in a graphite crucible [99]. Typical growth temperatures are 1,650–1,700°C, and growth rates are 5–20 μm h^{-1}. High-quality SiC can be grown by LPE but surface morphology and high-purity are still issues to be resolved and CVD has become the most widely used method for epitaxial growth of SiC [99, 100]. Other methods include molecular beam epitaxy (MBE), where the low growth rate of ~0.2 μm h^{-1} is of concern [101].

Step-Controlled Epitaxy

The CVD growth of SiC is carried out using SiH_4 and C_3H_8 (or C_2H_4) source gases transported by hydrogen at substrate temperatures of 1,500–1,600°C [102]. The growth rates are typically 2–15 μm h^{-1}. To date, it has been necessary to grow homoepitaxial 4H- and 6H-SiC on SiC{0001} substrates at high temperature (above 1,800°C) in order to obtain high-quality material. Growth at lower temperatures results in formation of 3C-SiC in the films. In 1987, it was found that homoepitaxial 6H-SiC{0001} films could be grown at the relatively low temperatures of ~1,500°C by using substrates which had been cut a few degrees off the main surface plane. This method was named the "step-controlled epitaxy" [102, 103]. Figure 6.36 shows growth modes on "just" and off-cut substrates. Deposition on SiC{0001} just substrate surfaces proceeds by two-dimensional nucleation growth and includes different polytypes (the dominant one being 3C-SiC). However, with SiC{0001} off-cut substrates, the presence of atomic steps results in growth in the horizontal direction (step-flow growth) where homoepitaxy proceeds without generation of other polytypes. "The steps in the off-cut substrate control the epitaxial growth so that only the desired polytype results."

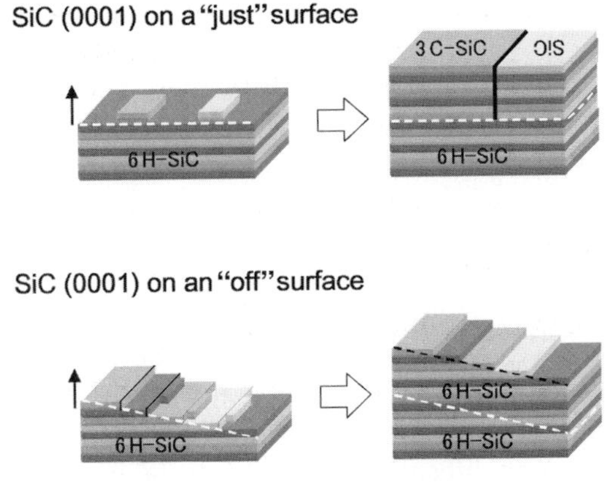

Fig. 6.36. Growth mode on "just" and "off" planes of 6H-SiC{0001}

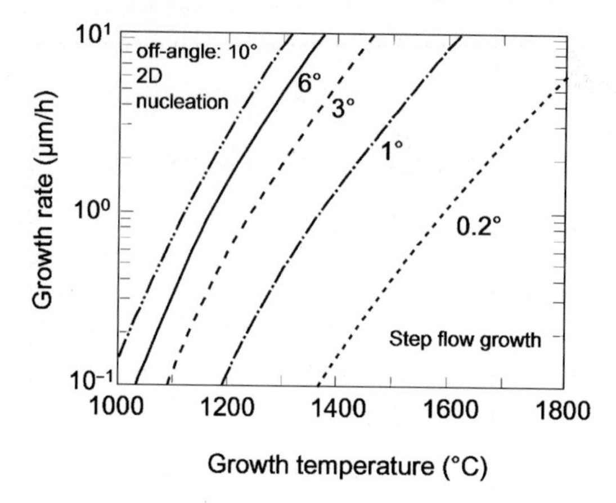

Fig. 6.37. Critical conditions of 6H-SiC homoepitaxial growth (step-flow) and 3C-SiC growth (two-dimensional nucleation)

Figure 6.37 shows the critical conditions for the growth of 6H- or 3C-SiC deduced using the Burton–Cabrera–Frank (BCF) theory [104]. It can be seen that by varying the substrate temperature, growth rate, and off angle, the growth mode can be varied to be either two-dimensional site growth (left upper region) or step-flow (right lower region). For example, at 1,500°C and a growth rate of 5 μm h^{-1}, an off angle greater than 1° is required for step-flow growth (homoepitaxy). Conversely, it is necessary to use growth rates greater than 0.5 μm h^{-1} to produce step-flow at 1,200°C using substrates with off angles greater than 3°.

Figure 6.38 shows the variation of growth rate with the C/Si ratio [102]. If the C_3H_8 flow rate is varied at a constant SiH_4 flow rate, then the C supply dominates at low C_3H_8 flow rates (low C/Si ratio) and Si supply dominates at high C_3H_8 flow rates (high C/Si ratio). Low C/Si ratios (excess Si) lead to the formation of surface defects induced by Si droplets while at high C/Si ratios (excess carbon) macrosteps are readily formed. That is, a high-quality surface morphology needs stoichiometric conditions.

High-Purity Growth

The unintentional incorporation of nitrogen during growth of SiC creates shallow levels that result in n-type (10^{15}–10^{16} cm^{-3}) crystals. The optimization of the C/Si ratio has been found to reduce the background donor density [105]. Figure 6.39 shows the variation of the donor density of undoped 4H-SiC layers grown on 4H-SiC{0001} with C/Si ratio. At high C/Si ratios, on SiC(0001)Si surfaces, it is possible to grow high-purity n-type SiC layers with a background impurity concentration of 5×10^{12} cm^{-3}. This is thought to be because the coverage of the surface with carbon atoms increases under excess carbon conditions, making it more difficult for nitrogen to be incorporated into carbon

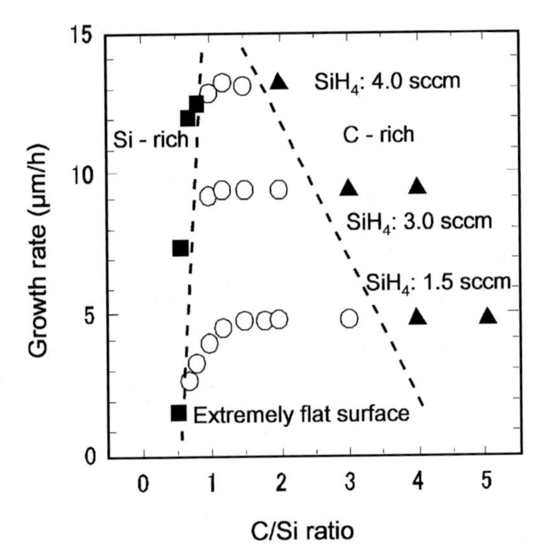

Fig. 6.38. C/Si ratio dependence of growth rate of 4H-SiC(0001) deposited by CVD

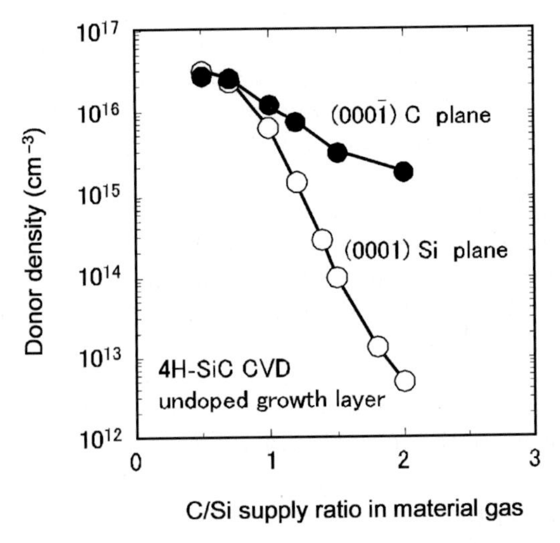

Fig. 6.39. Dependence of donor density on the C/Si ratio for undoped 4H-SiC growth layer on 4H-SiC{0001} off substrate

sites. Such conditions are also effective for reducing the donor density in low pressure CVD. The donor density on $(000\bar{1})$C surfaces is relatively high and the C/Si ratio does not have a significant effect on impurity incorporation. However, under conditions of low pressure, high temperature, and high C/Si ratio, it is possible to reduce the donor density to $\sim 5 \times 10^{14}\,\text{cm}^{-3}$.

Figure 6.40 shows temperature dependence of the PL spectrum of a 50-μm thick, high-purity SiC layer [106]. At low temperature, the strong emission

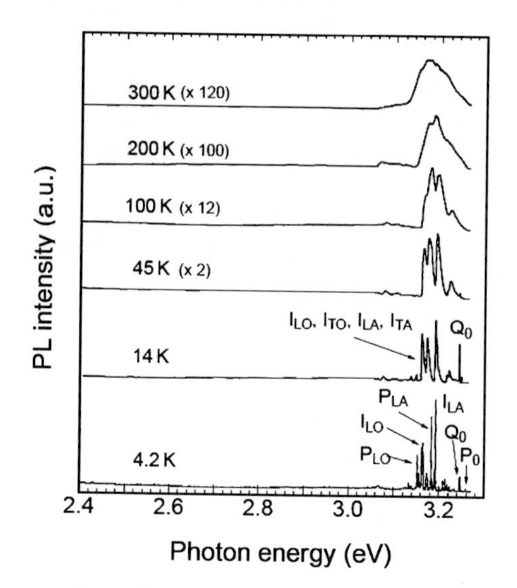

Fig. 6.40. Temperature dependence of PL spectrum of high-purity N-doped 4H-SiC growth layer (thickness 50 μm)

from bound excitons (intrinsic nitrogen donors) and free excitons is clearly visible. Emission from nitrogen–aluminum related donor–acceptor transitions, titanium, boron, or point defects is not observed. The intrinsic nitrogen donor bound exciton peak intensity decreases with increasing measurement temperature and above 50 K free exciton emission dominates. Free exciton emission is observed even at room temperature. The mobility of high-purity 4H-SiC (carrier density $\sim 10^{14}$ cm^{-3}) at room temperature is 981 cm^2 V^{-1} s^{-1} and at 42 K a value of 46,200 cm^2 V^{-1} s^{-1} has been achieved [106].

Control of Defects

Figure 6.41 shows a typical DLTS spectrum of an n-type 4H-SiC layer. The DLTS measurements were made using a Ni/SiC Schottky barrier at 100–820 K. Traps at energies of $E_c - 0.65$ eV and $E_c - 1.55$ eV were detected that were found to be independent of the growth conditions, crystal orientation, and impurity density. The $E_c - 0.65$ eV trap is known as the "$Z_{1/2}$ center" and the latter the "$EH_{6/7}$ center." However, the microscopic structure of these centers is still not clear [107]. Since the forbidden gap of 4H-SiC at room temperature is 3.26 eV, the $EH_{6/7}$ is a midgap level. Figure 6.42 shows the variation of the density of the $Z_{1/2}$ and $EH_{6/7}$ centers in 4H-SiC(0001) layers with C/Si ratio. It is noteworthy that the density of both centers decreases to $\sim 1 \times 10^{11}$ cm^{-3} at high C/Si ratios. On the other hand, DLTS measurements of p-type 4H-SiC layers show traps in the lower half of the forbidden gap at E_v+1.49 eV [108].

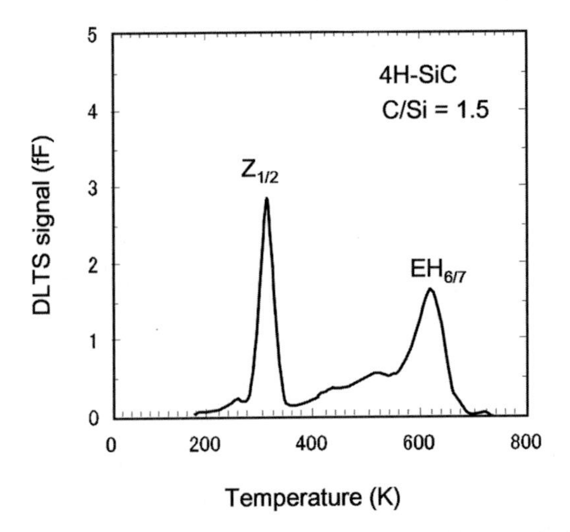

Fig. 6.41. Typical DLTS spectrum of N-doped 4H-SiC growth layer deposited by CVD

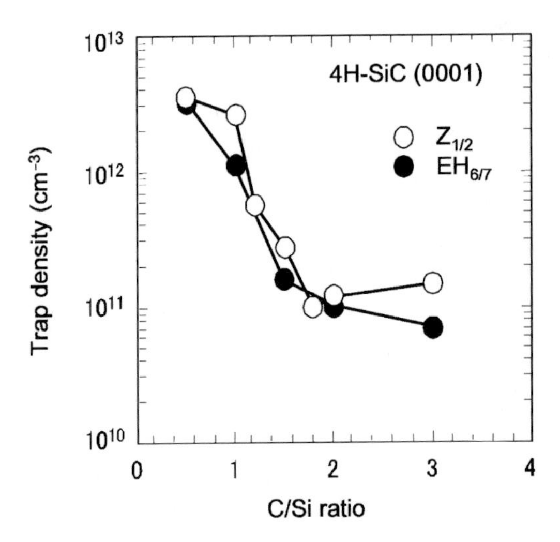

Fig. 6.42. C/Si ratio dependence of $Z_{1/2}$ center and $EH_{6/7}$ center density in 4H-SiC(0001) growth layer deposited by CVD

The existence of micropipes in 4H- and 6H-SiC substrates has been widely reported [109], including the use of epitaxial growth for filling micropipes in substrates [110, 111]. Figure 6.43 shows variation of micropipe filling efficiency with C/Si ratio where the samples were prepared by KOH etching and polishing. It can be seen that micropipes are easy to fill under excess Si conditions and it is easier to fill $(000\bar{1})$C surfaces than (0001)Si planes. It is possible to achieve 99% filling under optimized conditions.

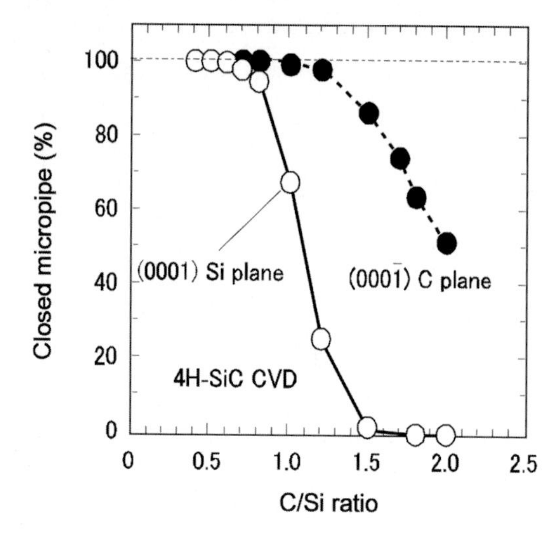

Fig. 6.43. C/Si ratio dependence of closed micropipe in 4H-SiC(0001) growth layer deposited by CVD

Recently, basal plane defects that exist on SiC{0001} surfaces have been found to degrade the forward I–V characteristics of bipolar devices and the insulating properties of thermal oxides [112,113]. Usually, there are $\sim 10^4\,\text{cm}^{-2}$ basal dislocations in SiC substrates but there are typically only $\sim 10^2\,\text{cm}^{-2}$ in the epitaxial layer itself [114].

Impurity Doping

In spite the wide bandgap of SiC, it is relatively easy to produce both p-(Al) and n-type (N) crystals. In the CVD growth of SiC, nitrogen (N_2) is used as an n-type dopant and $Al(CH_3)_3$ (TMA) for producing p-type films. Figure 6.44 shows the nitrogen doping of 4H-SiC(0001) surfaces. The donor density of the epitaxial layer is proportional to the nitrogen flow rate with a maximum density of $\sim 3 \times 10^{19}\,\text{cm}^{-3}$ which corresponds to a resistivity of $\sim 0.005\,\Omega\,\text{cm}$. Under excess Si conditions, nitrogen doping characteristics are almost independent of the surface polarity but under excess carbon conditions, the incorporation efficiency of the (0001)Si surface is low but that of the (000$\bar{1}$)C surface is high.

Figure 6.45 shows the Al doping characteristics of growth on SiC, (0001)Si and (000$\bar{1}$)C surfaces. The acceptor density of the layer increases with the amount of TMA added and it is possible to produce $\sim 10^{21}\,\text{cm}^{-3}$. However, the incorporation of Al on (000$\bar{1}$)C surfaces is low. In low doped crystals, the mobility of holes in 4H-SiC is 120 and $\sim 100\,\text{cm}^2\,\text{V}^{-1}\,\text{s}^{-1}$ in 6H-SiC. The ionization of Al in SiC is relatively large at 190 meV in 4H-SiC and 240 meV in 6H-SiC, and the hole density is approximately one order of magnitude less than the acceptor density. However, it is possible to produce p-type 4H-SiC with doping of 10^{20}–$10^{21}\,\text{cm}^{-3}$ and resistivity of $\sim 0.02\,\Omega\,\text{cm}$.

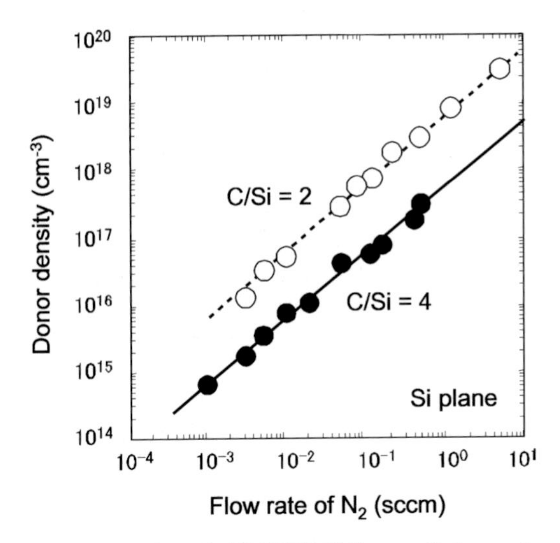

Fig. 6.44. Properties of N-doped 4H-SiC(0001) growth layer deposited by CVD

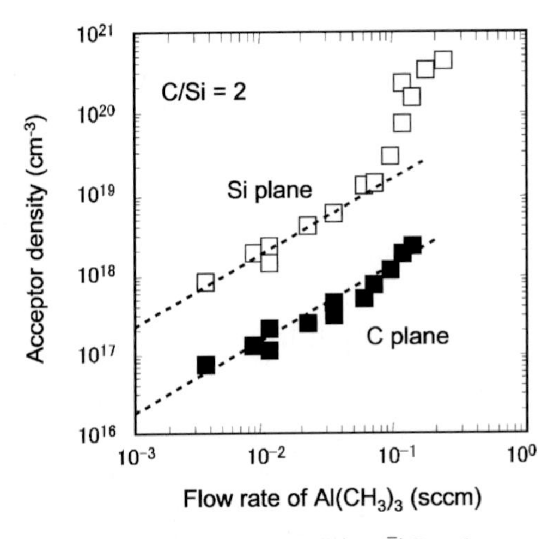

Fig. 6.45. Properties of Al-doped 4H-SiC(000$\bar{1}$)Si plane and (000$\bar{1}$)C plane deposited by CVD

Heteroepitaxial Growth of 3C-SiC

Layers of 3C-SiC can be grown on Si substrates, which also have a cubic crystal structure. However, it is difficult to produce high-quality heteroepitaxial structures using SiC and Si due to the large lattice mismatch (20%) and the large difference in thermal expansion coefficients (8%) between these crystals. However, the carbonization (carbonized buffer) of the Si surface prior to growth enables high-quality growth of 3C-SiC on Si substrates [115, 116]. The buffer layer is usually formed by increasing the temperature from room temperature to 500°C while introducing C_3H_8 into the growth reactor and then increasing the temperature to 1,200–1,350°C, which leads the carbonization of the Si surface. The carbonized buffer not only serves to reduce the effects of the lattice mismatch between the substrate and epilayer, but also this process prevents (1) diffusion of Si atoms from the Si substrate to its surface, which results in excess Si supply; and (2) suppression of void formation. For CVD, the typical substrate temperatures and growth rates are 1,300–1,350°C and 1–4 $\mu m\,h^{-1}$, respectively [115–117]. Growth of SiC-polar crystals on nonpolar Si(001) substrates leads to the formation of antiphase domains (APDs) [118, 119]. Recently, there have been reports on growth on silicon on insulator (SOI) substrates [120].

In the growth of 3C-SiC on silicon substrates, the undoped SiC growth layer is n-type with a carrier concentration of 10^{15}–10^{16} cm^{-3} due to nitrogen incorporation. It is relatively easy to produce n-type material using nitrogen and ammonia, but it is difficult to produce p-type SiC greater than 10^{19} cm^{-3} due to surface roughness. The bandgap and breakdown voltage of 3C-SiC is less than 4H-SiC but its MOS interface characteristics are excellent with potential applications for low voltage devices and sensors.

Problems associated with 3C-SiC heteroepitaxy films include high-density of structural defects, wafer bowing and cracks. It is possible to grow single crystal SiC using a carbonized buffer layer but problems related to misfit/layer dislocations due to the lattice mismatch and device degradation still remain to be resolved.

Methods for reducing APD and defects in 3C-SiC include the use of substrates with ultra-small one-dimensional undulation structures [121, 122].

6.2.2 GaN/Si Epitaxy (T. Egawa)

Due to the lack of suitable homosubstrates, InGaN-based LEDs and AlGaN/GaN HEMTs have been fabricated on sapphire or SiC substrate. The low thermal conductivity and insulating properties make sapphire less perfect, and the widely used SiC blocks are highly expensive. On the other hand, the use of Si as a substrate offers many advantages such as low cost, good thermal conductivity, large size wafer, and the possibility of integration of Si electronics on the same chip. Moreover, the fabrication process of LEDs on Si becomes easier than that of LEDs on sapphire because one of the ohmic contacts can be made from the backside through a conductive Si substrate. As shown in Fig. 6.46, we need to overcome the problems caused by large mismatches in lattice constants (16.9%) and in thermal expansion coefficients (57%) in order to grow high-quality GaN on Si. Previously reported GaN-based LEDs and HEMTs on Si have been grown with a low temperature thick AlN buffer layer (~750°C, 8–30 nm) and an AlGaN/GaN strained-layer superlattice (SLS) [123]. The low temperature-grown (LT) buffer layer technique was developed for the growth of epitaxial GaAs on Si [124]. It has been called as *two-step growth technique* and widely used in

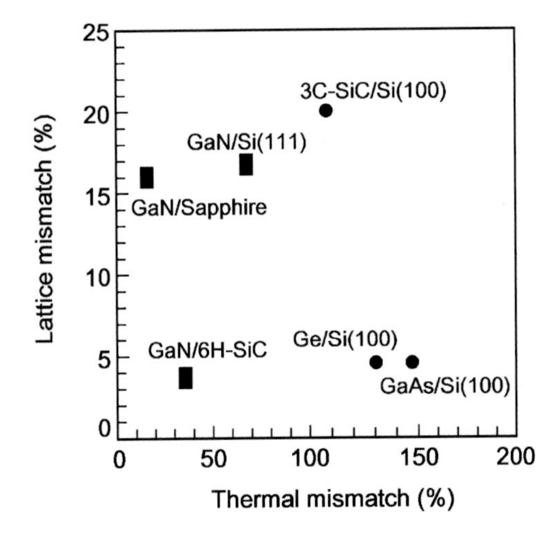

Fig. 6.46. Thermal and lattice mismatches in the growth of heteroepitaxy

the highly mismatched system such as the growth of GaN on sapphire. It was also proven that the SLS is very effective in improving the crystal quality of GaAs on Si. However, the LEDs on Si with a thick LT-AlN buffer layer suffer from both high operating voltage and high series resistance, which result from the insulating AlN layer and the large band offset at the AlN/Si interface. In this study, the InGaN-based LEDs and the AlGaN/GaN HEMTs on Si are reported using the high temperature-grown AlN/AlGaN intermediate layers (HT-AlN/AlGaN ILs).

InGaN-Based LEDs on Si

Figure 6.47 shows the schematic cross-sectional structure of the MOCVD-grown MQW LEDs on n-Si(111) substrate. After thermal cleaning at 1,130°C in H_2 ambient to remove oxide from Si surface, an AlN layer, a 20-nm thick n-$Al_{0.27}Ga_{0.73}N$ layer, 20 pairs of AlN (5 nm)/GaN (20 nm) multilayers at 1,080°C, a 0.2-μm thick n-GaN layer, an InGaN MQW structure, a 20-nm thick p-$Al_{0.15}Ga_{0.85}N$ layer, and a 0.2-μm thick p-GaN layer were deposited subsequently. The active layer consists of 3-nm thick $In_xGa_{1-x}N$ wells and 5-nm thick $In_{0.01}Ga_{0.99}N$ barrier layers. The typical In-content of the wells are 17 and 23% for the blue and green LEDs, respectively. In order to study the effect of the thickness of the AlN layer on the operating voltage, the 3- and 120-nm thick AlN layers were used in this study.

Ni/Au thin transparent metals and Ni/Au (12/100 nm) p-type ohmic metals were deposited on the p-GaN layer and annealed at 610°C for 3 min in N_2 ambient. The n-type ohmic contact was made from the backside through Si substrate using AuSb/Au (18/100 nm) annealed at 380°C for 1 min in N_2 ambient. For a comparison of the LED characteristics, the LED with the same

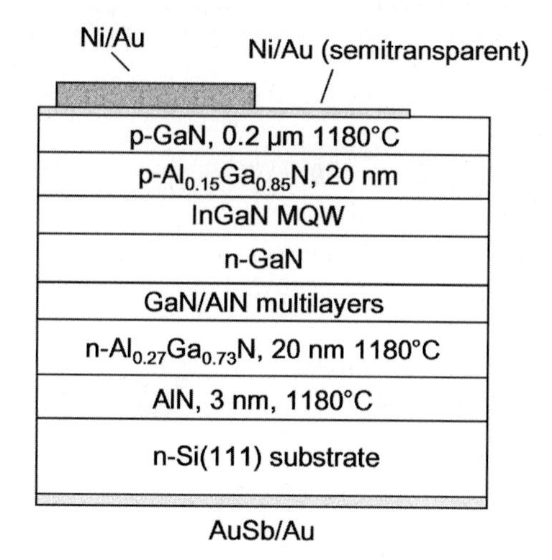

Fig. 6.47. Schematic cross-sectional structure of InGaN MQW LED on Si

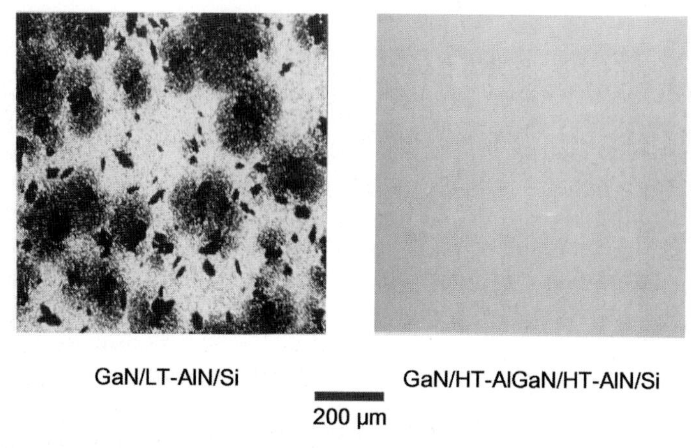

Fig. 6.48. Comparison of surface morphology of GaN on Si using LT-AlN buffer layer and HT-AlN/AlGaN ILs

MQW structure was also fabricated on c-plane sapphire, in which both the ohmic contacts were made on the top side with a mesa structure formed by reactive ion etching. Thus, the fabrication process of LED on Si is simple when compared to that of LEDs on sapphire. For light output power measurements, the detector was placed 10 mm above the surface of the LED chip. Cross-sectional transmission electron microscopy (TEM) observation was carried out to study the microscopic structure of the device.

Figure 6.48 shows the comparison of the surface morphology of the 1-μm thick GaN layers on Si using the conventional LT-AlN buffer layer (two-step growth technique) and the HT-AlN/AlGaN ILs. The GaN on Si grown by the LT-AlN buffer layer showed a cloudy surface morphology. However, the sur-

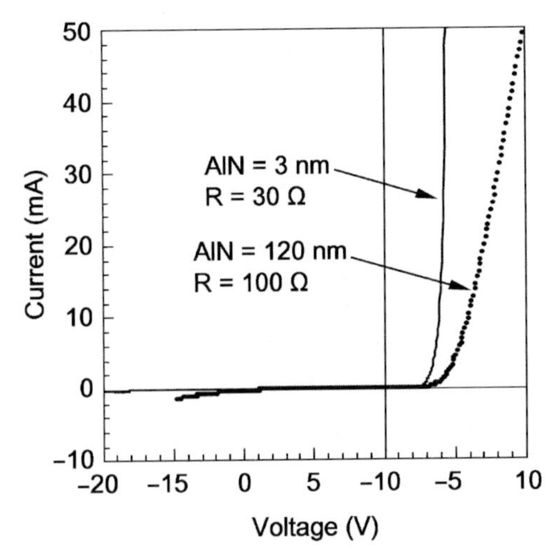

Fig. 6.49. Comparison of I–V characteristics of LEDs on Si with the different thickness of AlN layer

face morphology was improved using the HT-AlN/AlGaN ILs, because the Si substrate was protected from meltback etching by the HT-AlN/AlGaN ILs [125]. The FWHM values of XRC symmetric (0004) and asymmetric (20–24) reflections were 670 and 1,535 arcsec, respectively.

Figure 6.49 shows the I–V characteristics of blue LEDs grown on Si. The operating voltage of 7.0 V and the series resistance of 100 Ω were obtained at the forward current of 20 mA for the LED on Si with the thick AlN layer. On the contrary, the operating voltage and the series resistance were reduced to 3.8–4.1 V and 30 Ω, respectively, when the thin AlN layer was used. The same LED grown on sapphire showed an operating voltage of 3.8 V and a series resistance of 25 Ω at a forward current of 20 mA. The band diagram at the AlN/Si interface was evaluated using the X-ray photoelectron spectroscopy (XPS) measurement. As shown in the Fig. 6.50, the valence and conduction band discontinuities at the AlN/Si interface were evaluated to be 2.8 ± 0.4 and 2.3 ± 0.4 eV, respectively [126]. Moreover, the AlN layer is an insulator with a bandgap of 6.2 eV. When an n-contact is made on the backside of n-Si substrate, the electrons are injected into the active layer from the n-Si substrate side through the AlN layer. There is a large barrier to the electron injection due to the presence of thick AlN layer. On the contrary, when the AlN layer is thin below a critical value, the formation of the tunnel junction between the n-AlGaN layer and the n-Si substrate will reduce the operating voltage of the LED greatly.

Figure 6.51 shows the comparison of L–I characteristics of the LEDs on sapphire and Si substrates under highly injected current. The wavelengths at 20 mA were 470 and 475 nm for the LEDs on sapphire and Si substrates, respectively. The output power from the LED on sapphire increased with the

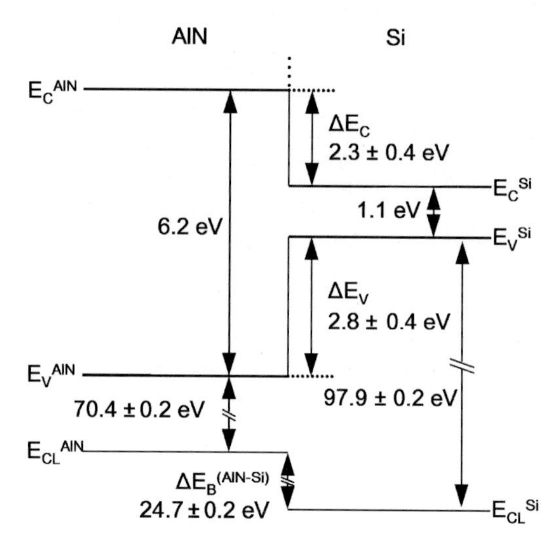

Fig. 6.50. Schematic energy band diagram at AlN/Si interface

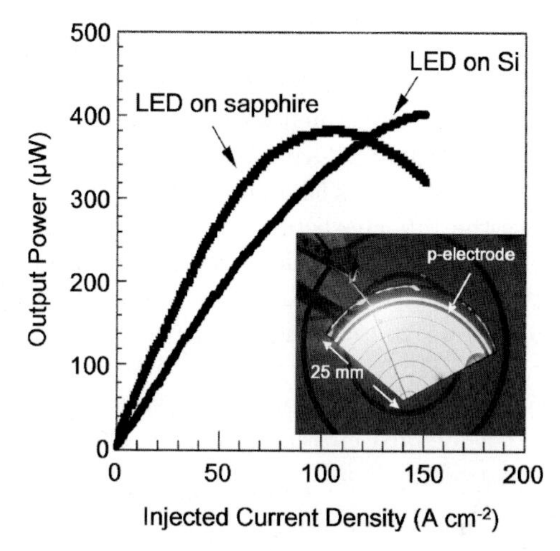

Fig. 6.51. Comparison of L–I characteristics of LEDs on sapphire and Si substrates under highly injected current. The detector was set 10 mm above the surface of LED chip. *Inset* shows the demonstration of emission from a large area green LED on Si with thin AlN layer under injected current of 800 mA, corresponding to current density as low as 120 mA cm^{-2}

injected current density up to 100 A cm^{-2}, and then decreased with increasing the injected current density. The thermal conductivity of sapphire is as low as 0.35 W cm^{-1} K^{-1}. Therefore, the saturation of the output power of the LED on sapphire is due to the heating effect. The output power from the LED on Si was lower than that on sapphire below the 120 A cm^{-2} because of the absorption of emission by Si substrate. However, the output power from the LED on Si saturated at the injected current density of 150 A cm^{-2} because of higher thermal conductivity of Si substrate (1.5 W cm^{-1} K^{-1}). Thus, the LED on Si exhibited the better characteristic than that on sapphire under highly injected current. Some LED chips were molded by epoxy resin. The optical output power of the molded blue LEDs on Si was as high as 1.5–2 mW at 20 mA. Taking into account of the absorption of partial emission by Si substrate, it is believed that the high-quality active layer is grown on Si substrate in spite of its highly mismatched system. The inset in Fig. 6.51 shows the emission morphology of a large area green (505 nm) LED on Si with a thin AlN layer under the injected current of 800 mA (120 mA cm^{-2}). Although the injected current density was as low as 120 mA cm^{-2}, a uniform emission has been observed for the large area. The onset of emission was observed at 12 mA, corresponding to the injected current density as low as 1.8 mA cm^{-2}.

In order to study the reliability of the LED on Si, the aging tests were performed at 27 and 80°C under automatic current control (ACC). Figure 6.52 shows the reliability results of the blue LED on Si with the thin AlN layer. The output power was constant during the 1,000 h aging at 27°C. At 80°C, there was a decrease in the output power at initial stage, which was as low as 6%. After that, the output power of LED almost remained constant. Therefore, it is proven that the superior characteristics of Si substrate are of great advantage to the characteristics of LED built on it [127, 128].

The dark spot densities were measured to be 5×10^8 and 4×10^9 cm^{-2} for the n-GaN layers on sapphire and Si, respectively, by using cathodoluminescence technique. In spite of high dark spot density in the GaN layer on Si, the LED on Si with the thin HT-AlN/AlGaN ILs exhibited a relatively low operating voltage, a good stability, and a uniform emission under a low injected

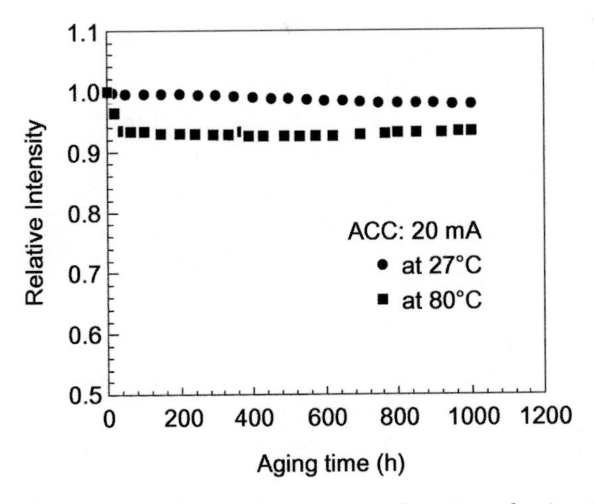

Fig. 6.52. Variation of optical output power as a function of aging time measured at 27 and 80°C

current density as explained above. TEM analysis was carried out to examine the microscopic structure of our sample. Figure 6.53a shows a cross-sectional TEM image for the overall LED structure, and Fig. 6.53b is an enlarged picture near the active layer. As shown in Fig. 6.53, the V-defects were initiated at the threading dislocations during MQW growth. The V-defect has been observed in the InGaN MQW structure on sapphire [129]. It is noteworthy that the pyramid-shaped structures are formed during the growth of last few QWs, and no threading dislocation was observed in these pyramid-shaped structures. The QWs near the top of pyramid-shaped structures become the quantum-dot-like structure. The dots are ~10 nm in diameter and ~3.8 nm in height. Moreover, the valleys between pyramid-shaped structures are filled with p-AlGaN and p-GaN layers during growth at high temperature due to en-

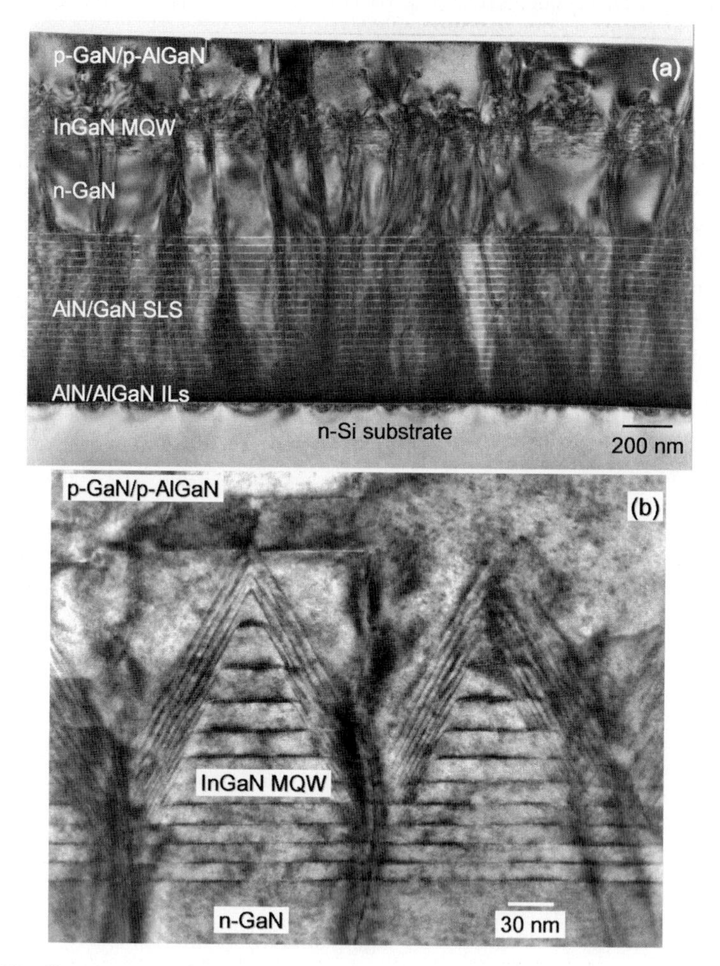

Fig. 6.53. Cross-sectional TEM images for (**a**) overall structure and (**b**) near active layer of InGaN MQW LED on Si with the thin AlN layer

hanced surface migration of reactants. Because the resistivities of both p-GaN and p-AlGaN layers are relatively high, the holes could be effectively injected into the QWs near the top of pyramid-shaped structure through the shortest current passage from the p-electrode. The most likely reason for excellent characteristics in GaN-based LED on Si is that the dislocation-free pyramid-shaped (quantum-dot-like) structure contributes to the emission. Further study is necessary to clarify the effect of pyramid-shaped structures on the LED characteristics.

The problem with the LED on Si is the absorption of partial emission by the Si substrate. However, the combination of the selective etching of Si and the metal-to-metal bonding was effective in improving the LED characteristics, such as 49% increase in the optical output power and the reduction of the operating voltage to 3.6 V [130].

AlGaN/GaN HEMTs on Si

The AlGaN/AlN/GaN HFETs were grown on 4-in. Si(111) substrate by MOCVD. The schematic structure of the AlGaN/GaN HFETs on Si is shown in Fig. 6.54. All the layers were grown at 1,130°C. The device structure consists of the 100-nm thick AlN layers, the 40-nm thick $Al_{0.26}Ga_{0.74}N$, 20 pairs of GaN/AlN (20/5 nm) multilayers, 1-μm thick *i*-GaN layer, 1-nm thick AlN spacer layer, and *i*-$Al_{0.26}Ga_{0.74}N$ (25 nm) top layer. The modified AlGaN/AlN/GaN structure, which employs the thin AlN spacer layer between AlGaN and GaN layers, showed higher 2DEG properties than those of the conventional AlGaN/GaN structure. The values of electron mobility and sheet carrier density of the 2DEG were 1,711 cm^2 V^{-1} s^{-1} and 1.5×10^{13} cm^{-2},

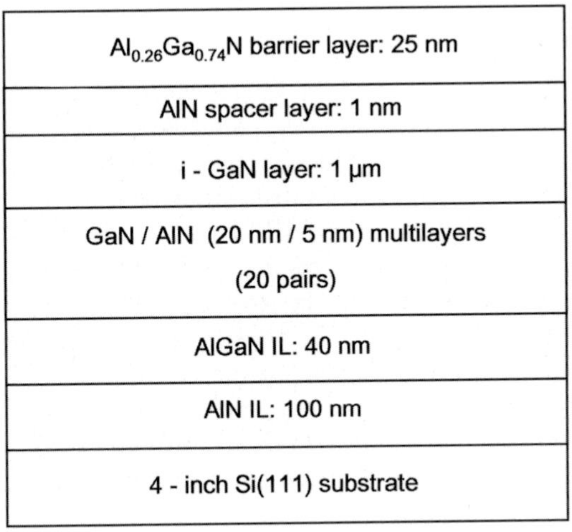

$Al_{0.26}Ga_{0.74}N$ barrier layer: 25 nm
AlN spacer layer: 1 nm
i - GaN layer: 1 μm
GaN / AlN (20 nm / 5 nm) multilayers (20 pairs)
AlGaN IL: 40 nm
AlN IL: 100 nm
4 - inch Si(111) substrate

Fig. 6.54. Schematic cross-sectional structure of AlGaN/GaN HEMT on Si

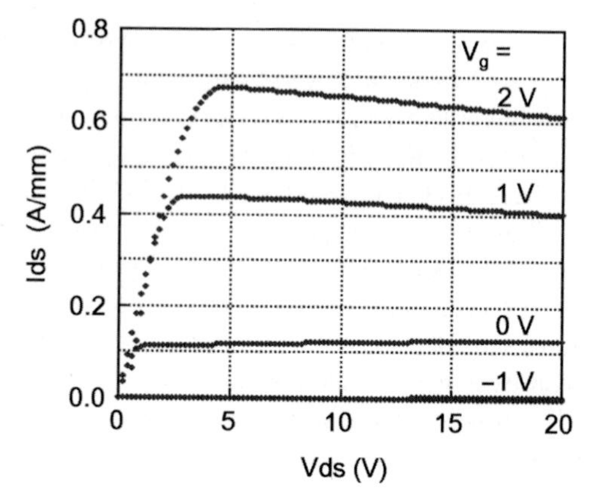

Fig. 6.55. DC characteristic of 1.0-μm gate-length AlGaN/GaN HEMT on Si

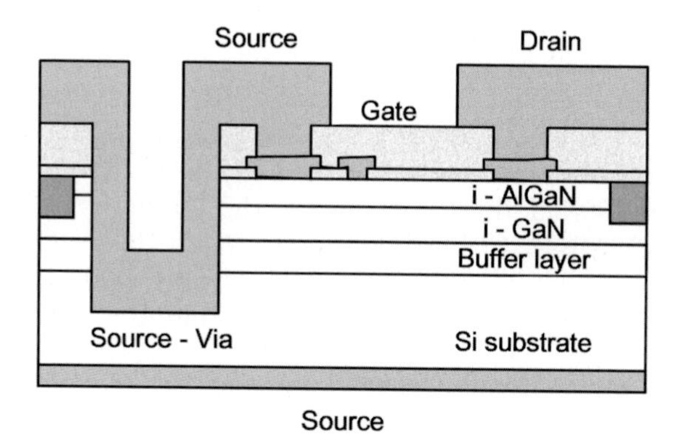

Fig. 6.56. Schematic cross-sectional view of AlGaN/GaN HEMT with SVG structure

respectively. A cross-sectional TEM micrograph showed the smooth interfaces were obtained at the GaN/AlN multilayers and many dislocations originating from the AlN/Si interface propagated into the upper layer due to the highly mismatched system. The dislocation density was estimated to be $4–8 \times 10^9\,\text{cm}^{-2}$ for the AlGaN/GaN HFET on Si, which was one order higher than that on sapphire. Figure 6.55 shows the DC characteristic of the device with the gate-length of 1.0 μm. The device showed the maximum drain current density (I_{Dmax}) of $670\,\text{mA mm}^{-1}$ and the maximum extrinsic transconductance (g_{mmax}) of $323\,\text{mS mm}^{-1}$. The current gain cut-off frequency (f_T) and a maximum frequency of oscillation (f_{max}) were 56 and 115 GHz, respectively, for 0.2-μm gate-length. The AlGaN/GaN HEMTs on Si exhibited the reduced self-heating effect due to high thermal conductivity of Si when compared to HEMTs on sapphire [131, 132].

Recently, the high power AlGaN/GaN HFET has been demonstrated on the conductive Si substrate with the source-via grounding (SVG) structure [133]. Figure 6.56 shows the schematic cross-sectional view of AlGaN/GaN HEMT with SVG structure. In SVG structure, the source is ohmic-contacted to the conductive Si through the surface via-hole. The backside of the substrate is grounded. The SVG structure plays three important roles. First, it reduces the source interconnection resistance. Second, the source parasitic inductance is also significantly reduced because of the eliminated source wires and bonding. Finally, the SVG structure relieves the electric field between the drain and gate because the grounded substrate can act as a backside field plate. These benefits result from the use of Si as the substrate. The device with the gate-length of 1.0 μm exhibited a g_{mmax} of 200 mS mm^{-1}, I_{Dmax} of 400 mA mm^{-1}, very low specific on-state resistance of 1.9 mΩ cm^2, the high off-state breakdown voltage of 350 V, and the current handing capability of 150 A. In addition to these excellent characteristics, the subnanosecond switching t_r of 98 ps and t_f of 96 ps with the current density as high as 2.0 kA cm^{-2} were achieved.

The HT-AlN/AlGaN ILs and AlN/GaN multilayers have been used for the growth of high-quality GaN on Si substrate. The LED at 20 mA was reduced to as low as 3.8–4.1 V due to the formation of tunnel junction between the n-AlGaN layer and n-Si substrate when the high temperature AlN layer is reduced to 3 nm. Because Si has a better thermal conductivity than sapphire, the optical output power of the LED on Si saturates at a higher injected current density. When the injected current density was higher than 120 A cm^{-2}, the output power of the LED on Si was higher than that of LED on sapphire. The LED also exhibited the good reliability and a uniform emission from a large size wafer. Cross-sectional TEM observation indicated that the active layers of these LEDs consist of the dislocation-free pyramid-shaped (quantum-dot-like) structure.

The 1-nm-thick AlN spacer layer was effective in improving the 2DEG characteristics. The AlGaN/GaN HEMT on Si exhibited the I_{Dmax} of 670 mA mm^{-1}, g_{mmax} of 323 mS mm^{-1}, f_T of 56 GHz, and f_{max} of 115 GHz. The device also showed excellent characteristics for high power applications.

6.2.3 Diamond (H. Okushi)

Homoepitaxy Growth

It is difficult to produce n-type diamond films. However, recently there have been reports on the use of microwave plasma CVD growth for the homoepitaxial growth of high-quality, atomically flat diamond films on diamond substrates [134].

Microwave Plasma CVD Growth

Step-flow growth on off substrates is necessary to produce flat, high-quality single crystal films. In the case of diamond, the microwave plasma method, using a mixture of methane and hydrogen can be used for the growth of high-quality films [135, 136] where the growth mode strongly depends on the methane density (CH_4/H_2 ratio) and the orientation of the substrate. Figure 6.57 shows the variation of the surface morphology of plasma CVD diamond films grown on off angle substrates with CH_4/H_2 ratio. The solid dots indicate anomalous growth particles/hillocks (region (I): NC); the solid triangles, macroscopic step bunching; and open triangles, surfaces when macroscopic step bunching and anomalous growth particles/hillocks are not observed simultaneously, corresponding to regions when growth is dominated by the step-flow mode (region (II): NC+B). The solid squares are regions where the surface is atomically flat with CH_4/H_2 less than 0.1% and off angle less than 1.5° (region (III)). However, even if the CH_4/H_2 ratio is less than 0.1%, atomically flat surfaces are not observed if the off angle is greater than 1.5°, the region shown by the open squares (IV) [137].

During synthesis of the atomically flat surfaces under extremely low methane concentration growth conditions (III), the diamond surface is etched by hydrogen atoms present in the plasma. Also, the degree of hydrogen etching is dependent on the substrate off angle, with greater roughness at larger off angles. The density of atomic steps increases with increasing off angle, which results in pronounced differences in the etch rates at step edges (different from

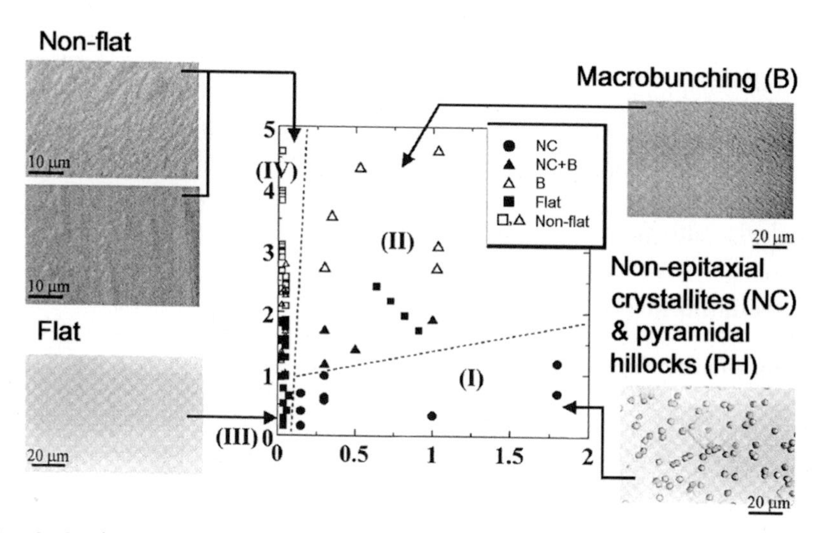

Fig. 6.57. A map of surface morphologies as a function of both the misorientation angles and the CH_4/H_2 ratios. The *dotted lines* in the figure are a guide to the eye and divided into four regions, I–IV

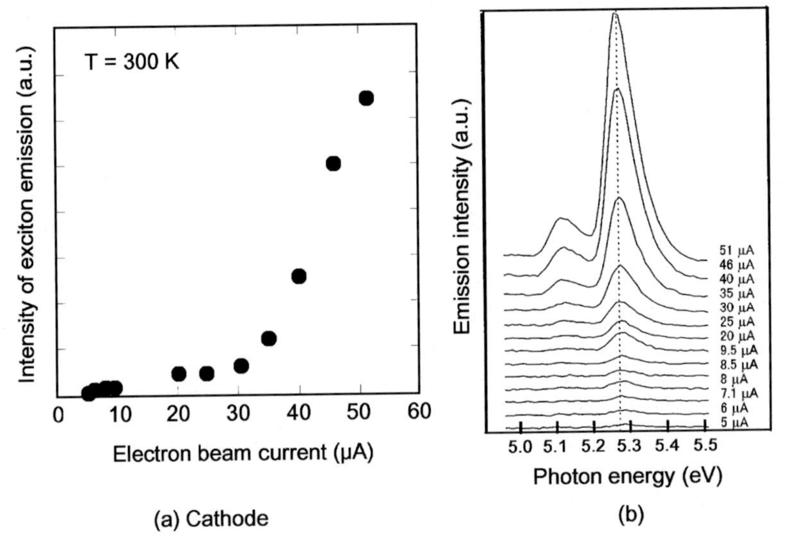

Fig. 6.58. Relationship from cathodoluminescence (CL) between exciton emission intensity and excited electron beam current of high-quality diamond thin film deposited by CVD at room temperature (**a**). Observed CL emission spectrum (**b**). This peak position did not shift also in nonlinear area in (**a**). These main and satellite peaks indicate free exciton with transverse optical phonon and side band of this phonon, respectively

the (001) direction) and step surfaces ((001) surface) and hence the observation of the surface morphology variation with off angle [136].

Optical Properties of Diamond Films

Atomically flat diamond thin films grown as described above exhibit room temperature cathode-luminescence (CL) due to free excitons at 235 nm (5.27 eV) (Fig. 6.58.). The emission intensity increases nonlinearly with increasing electron beam current [138]. Figure 6.58(b) shows the actual CL spectra observed in these measurements. The strong CL emission at 235 nm (free exciton with transverse optical phonon) and the sideband at 242 nm are not observed from the Ib diamond substrate itself.

The binding energy of free excitons in diamond is large (80 meV) due to its smaller dielectric constant compared with other semiconducting materials. Thus a high-density of free excitons (critical free exciton density $\sim 6 \times 10^{19}\,\mathrm{cm}^{-3}$) can exist in diamond films even at room temperature. Conventional CVD diamond thin films exhibit emission in the visible spectrum due to the presence of dislocations and impurities, but only free exciton emission is observed from the atomically flat diamond films described above [138], which underscores their high crystalline quality. However, the extremely low growth rate of microwave plasma CVD limits its applications for the synthesis of practical device structures.

Impurity Doping

Fabrication of electron devices requires control of the p- and n-type conduction (impurity density) of diamond thin films. At present, both p- and n-type diamond films have been realized but there are still many issues to resolve in order to produce low resistivity diamond films.

Boron has been successfully used for producing p-type diamond films during CVD growth [139] irrespective of the substrate orientation. In particular, homoepitaxial films grown on (001) surfaces have been reported with room temperature hole densities and mobilities of $10^{15}\,\mathrm{cm}^{-3}$ and $2{,}000\,\mathrm{cm}^2\,\mathrm{V}^{-1}\,\mathrm{s}^{-1}$, respectively [140]. At present, it is possible to reproducibly produce boron-doped films with hole concentrations from 10^{14} to $10^{17}\,\mathrm{cm}^{-3}$ and carrier mobilities of $\sim 1{,}000\,\mathrm{cm}^2\,\mathrm{V}^{-1}\,\mathrm{s}^{-1}$. These electrical properties are similar to those of Si films doped with boron and phosphorus.

However, the problem is that the room temperature acceptor ionization efficiency is low because the energy level of boron acceptors is 0.37 eV above the valence band edge, with the density of free holes being two to three orders of magnitude less than the boron-doping density. Boron-doped films with resistivities as low as $\sim 10^{-3}\,\Omega$ cm have been produced.

Doping for n-type conduction has proved to be extremely difficult. Phosphorus doping of homoepitaxial diamond films on (111) substrates was a major breakthrough in 1997 [141]. In this case, the phosphorus forms a donor level at 0.58 eV below the conduction band edge, which makes the growth of low resistivity films extremely difficult compared with p-type films. Hall effect measurements of these films showed the electron mobility to be $600\,\mathrm{cm}^2\,\mathrm{V}^{-1}\,\mathrm{s}^{-1}$ at a phosphorus doping density of $10^{16}\,\mathrm{cm}^{-3}$ and in experiments on producing high doping density, an electron density of $10^{13}\,\mathrm{cm}^{-3}$ was reported [142]. However, these results were obtained using (111) substrates and phosphorous doping is extremely difficult on (001) surfaces.

For practical applications, the (111) crystals are extremely hard, making it difficult to polish them and, apart from cleaving the crystals, it is extremely difficult to produce atomically flat surfaces. On the other hand, (001) surface can be mechanically polished and etched for fabrication of ultra-small structures and n-type doping of (001) films is the only practical alternative at this stage. Recently, phosphorus doping of (001) surfaces yielded films with electron mobilities of $400\,\mathrm{cm}^2\,\mathrm{V}^{-1}\,\mathrm{s}^{-1}$ [143]; a value comparable with (111) films and offering tremendous potential for diamond electron devices.

Heteroepitaxial Growth

Heteroepitaxial growth of diamond on substrates that are not diamond is vital for synthesis of large area diamond films. There have been reports on covering nondiamond substrates with diamond particles for the synthesis of polycrystalline diamond films. However, more recently, there have been attempts to grow high-quality diamond films on Pt [144] and Ir [145] substrates, representing significant advances in this field. But it is still necessary to collect basic data about the electrical properties of heteroepitaxial films for comparison with homoepitaxial materials. The optical applications of diamond films are being studied [146].

6.2.4 Epitaxial Growth of GaN (K. Onabe)

Role of Low Temperature Buffer Layers and Advances in Epitaxial GaN Growth Technology

The epitaxial growth of high-quality GaN (1986) was the key breakthrough leading to development of nitride-based, optoelectronic devices. That is, the inclusion of a low temperature buffer layer on sapphire substrates during MOCVD growth is a critical step for producing high-quality GaN thin films [147]. During the early days of development, the lack of bulk GaN meant that the c-surface of sapphire (0001) substrates, which have a 16% lattice mismatch, was widely used for growth of GaN. The AlN or GaN low temperature buffer layers assume the same orientation as the underlying substrate and provide a high-density of growth sites which recrystallize as the temperature is increased and at high temperatures, lateral growth of GaN is enhanced by the recrystallized buffer layers [148–151]. As a result, "islands" that exist during the initial stages of high temperature GaN growth coalesce to eventually form flat growth surfaces. Figure 6.59 illustrates the main stages involved during the growth of GaN on a low temperature AlN buffer layer [149]. The buffer layer is deposited at 500°C and growth is carried out at $\sim$1,000°C. Figure 6.60 shows the main components of a typical MOVPE system [147,148]. Trimethylgallium (TMG) and trimethylaluminum (TMA) are used as source gases for Ga and Al, respectively, and ammonia (NH_3) as a source of nitrogen. The substrate crystal is placed onto the susceptor and the substrate temperature is increased by RF heating. Crystal growth progresses due to pyrolysis reactions near the substrate.

Currently, single crystal GaN thin films are also grown by HVPE and MBE, and other substrates are SiC, Si, and GaAs [152]. Growth on these lattice-mismatched substrates almost always involves the use of buffer layers. In MBE, Ga and Al are evaporated as molecular beams from heated Knudsen cells and nitrogen is produced by dissociating N_2 gas in the form of an RF- or ECR-plasma to generate atomic nitrogen that serves as the nitrogen source.

Reduction of Dislocations by Epitaxial Lateral Overgrowth Technology

The dislocation density of GaN layers grown by MOVPE (on AlN or GaN buffer layers) is 10^8–$10^{10}\,cm^{-2}$, which is two to three orders of magni-

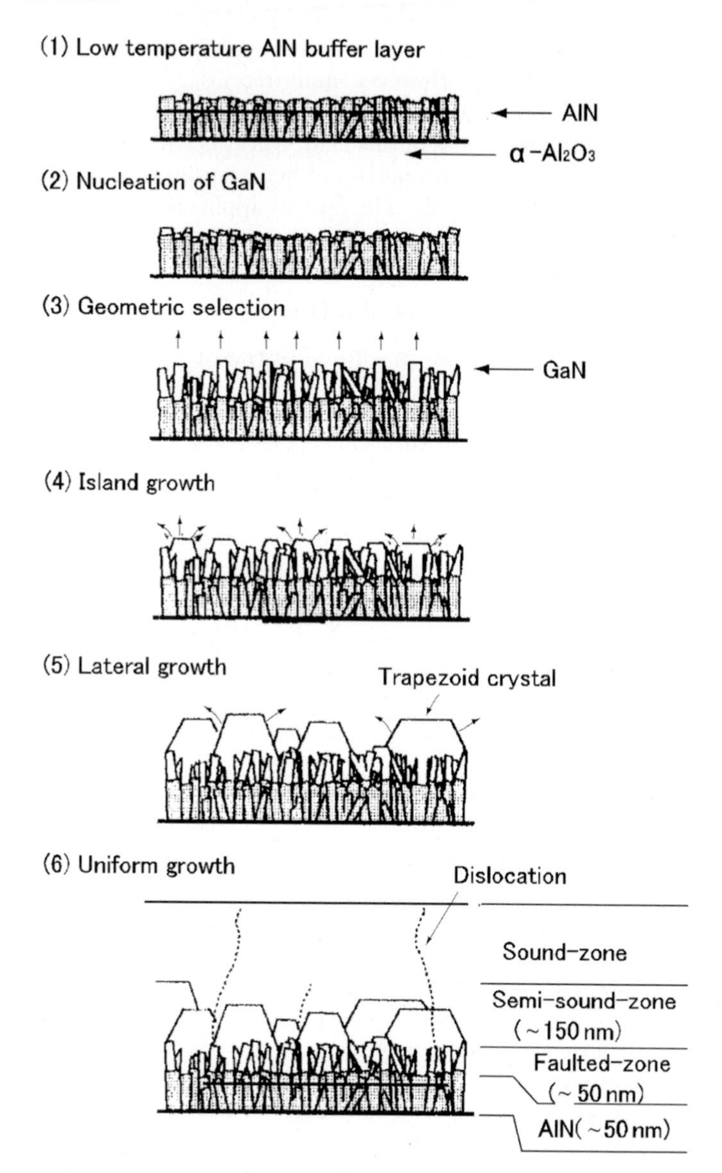

Fig. 6.59. Growth process of GaN single crystal on low temperature AlN buffer layer [149]

tude larger than the required for use as part of device structures such as lasers. Reduction of threading dislocations (defects originating at the substrate/epilayer interface up through the growing crystal) is the major issue for device applications. Threading dislocations have been reduced by one order of magnitude by introducing low temperature AlN and GaN interlayers [153]. Epitaxial lateral overgrowth (ELO) method is widely used for reducing the density to 10^6–10^7 cm^{-2} for GaN thin films grown by HVPE [154, 155] and

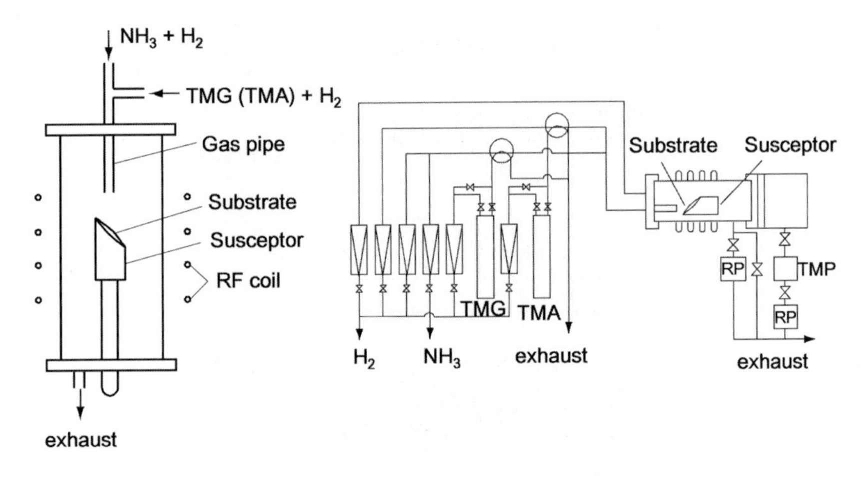

(a) Vertical reactor [147] (b) Horizontal reactor and gas supply piping [148]

Fig. 6.60. Main components of a typical MOCVD system

MOVPE [156–159]. HVPE can be used to grow GaN films at rates as high as 10–100 μm h^{-1} for producing films several hundred micrometers thick, which are then removed from the underlying sapphire for use as substrates in the growth of homoepitaxial GaN [160,161]. MOVPE is used to produce substrate layers for directly growing laser structures [162, 163].

The facet-initiated ELO (FIELO) method is based on HVPE growth. As described by Usui et al. [154], the substrate consists of a 1–1.5 μm thick GaN layer grown by MOVPE on c-surface of sapphire (0001) patterned with SiO_2 stripes defined by photolithography. The stripes are designed to be in the $\langle 11\bar{2}0\rangle$ or $\langle 1\bar{1}00\rangle$ directions with the mask regions 1–4 μm wide and a repetition period of 7 μm. GaN films are grown on these patterned substrate surfaces at 1,000°C by HVPE. In HVPE, reactions between GaCl and NH_3 are used for crystal growth, where GaCl is produced by a reaction between metallic Ga and HCl gas upstream of the reactor at 850°C. At a GaCl partial pressure of 5.2×10^{-3} atm, an NH_3 partial pressure of 0.26 atm and a total flow rate of 3.8 slm, the growth rate was reported to be 50 μm h^{-1}. Figure 6.61 shows the evolution of the ELO process of GaN when the strip is in the $\langle 11\bar{2}0\rangle$ direction. Triangular facets with $\{1\bar{1}01\}$ planes are seen after initial growth in the (0001) direction. Next, growth occurs in the horizontal direction over the masked areas while maintaining the facet growth areas and then the adjacent regions coalesce. After 30 min, flat growth layers are obtained. In the $\langle 1\bar{1}00\rangle$ direction, the $\{11\bar{2}2\}$ planes become facets but the same growth process occurs as in the $\langle 11\bar{2}0\rangle$ direction. Figure 6.62 is a cross-sectional TEM image of a 140-μm thick GaN layer [164]. Dislocations near the sapphire propagate through the MOVPE GaN region perpendicularly but they bend over sharply in the horizontal direction near the mask region of the HVPE GaN layer. Most of the bending dislocations occur at the initial facet surface. Dislocations extending in the horizontal direction have a tendency to propagate in the vertical

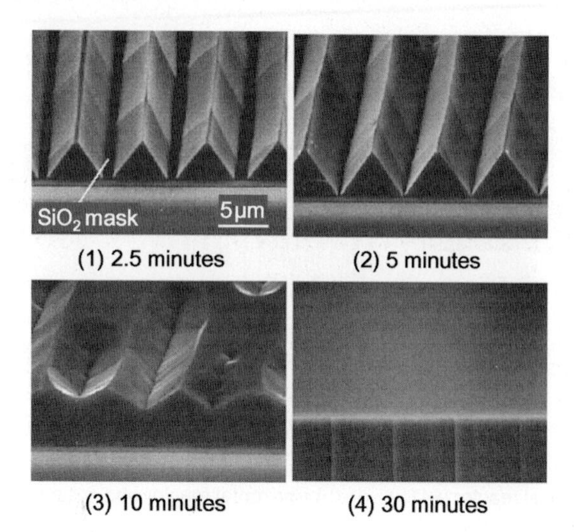

Fig. 6.61. Growth process of FIELO GaN evaluated by SEM. The stripes are in the $\langle 11\bar{2}0\rangle$ direction [154]

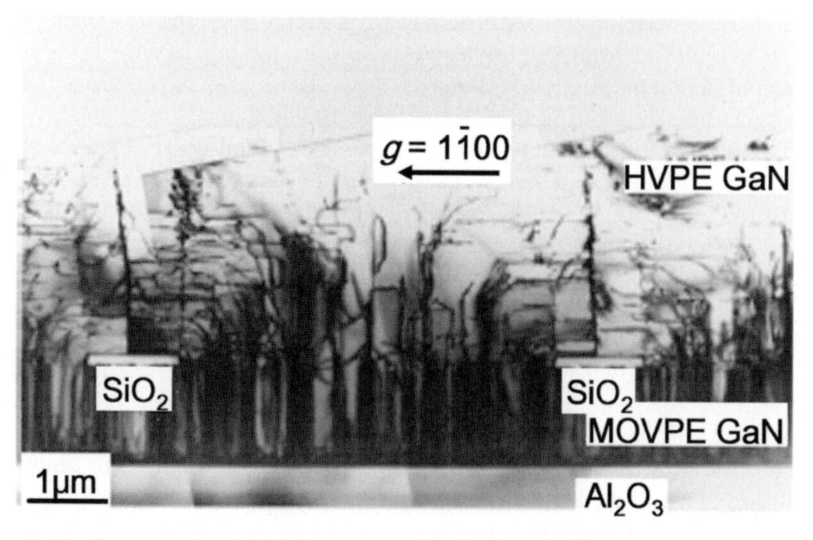

Fig. 6.62. Cross-sectional TEM image of FIELO GaN $\langle 11\bar{2}0\rangle$ (diffraction vector $\boldsymbol{g}$) [164]

direction again in the coalescence growth region above the center of the mask. However, dislocations in other regions are concentrated in a thickness ~5 μm from the HVPE growth layer interface. The density of threading dislocations near the surface is $\sim 6\times 10^{7}\,\mathrm{cm}^{-2}$, compared with a value of $10^{9}\,\mathrm{cm}^{-2}$ in the MOVPE layer.

Threading dislocations grown on sapphire (0001) surfaces arise parallel to the [0001] (c-axis) direction. For hexagonal crystal structures, three types

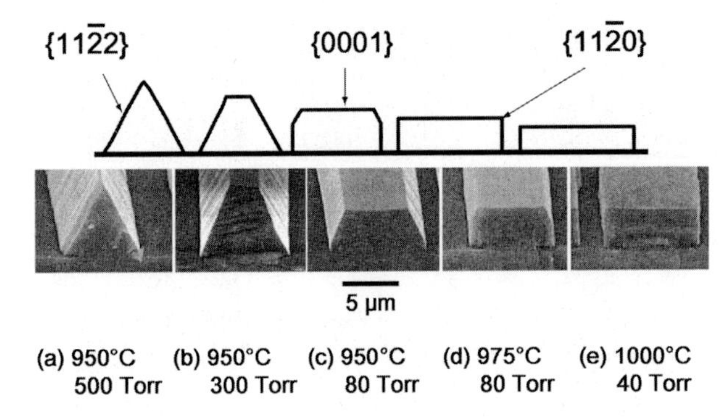

Fig. 6.63. Growth temperature and pressure dependence of facet formation of ELO GaN deposited by MOVPE. The direction of stripes is $\langle 1\bar{1}00\rangle$ [156]

of Burgers vector, **b**, are possible, $\mathbf{a} = \langle 11\bar{2}0\rangle/3$, $\mathbf{c} = \langle 0001\rangle$, and $\mathbf{a}+\mathbf{c} = \langle 11\bar{2}3\rangle/3$, corresponding to edge dislocations, screw dislocations, and mixed dislocations as threading dislocations along the c-axis. In GaN, edge and mixed dislocations are dominant [164].

In MOVPE work, Hiramatsu et al. [156, 157] developed the "facet-controlled ELO" method (FACELO) to reduce the density of threading dislocations, where the shape of the GaN grown horizontally from the $\langle 1\bar{1}00\rangle$ SiO_2-striped windows on (0001) GaN (Fig. 6.63) was controlled by appropriate selection of growth temperature and reactor pressure. Here, TMG and NH_3 were used as source gases and H_2 as a carrier gas. The growth temperature was 950–1,050°C, the pressure was 40–500 Torr, and the growth time was 30 min. For a width of 5 μm and periodicity of 10 μm, at low temperature and high pressure, growth yields pyramidal features with $\{11\bar{2}2\}$ side walls. In contrast, growth at high temperature and low pressure results in a $\{11\bar{2}0\}$ surface perpendicular to the substrate surface. On the other hand, for strips in the $\langle 11\bar{2}0\rangle$ direction, pyramidal growth with $\{1\bar{1}01\}$ planes generally results. The formation of these facets has been explained by consideration of the polarity and dangling bond density required for stable surfaces [157]. That is, for stripes in the $\langle 1\bar{1}00\rangle$ direction, use of low temperature, high pressure, and high III/V growth conditions during MOCVD growth results in $\{11\bar{2}2\}$ (nitrogen-polarity/surface) stable surfaces, and at high temperature and low pressure, the (0001) (Ga surface) and $\{11\bar{2}0\}$ (nonpolar) surfaces, which have a lower dangling bond density than $\{11\bar{2}2\}$ surfaces, are found to be stable. For stripes in the $\langle 11\bar{2}0\rangle$ direction, the $\{1\bar{1}01\}$ surface (nitrogen-polarity) is thought to be stable. In the HVPE method, pyramidal growth always occurred in the openings, but this is thought to be due to HVPE being an atmospheric pressure growth method.

FACELO offers two ways of reducing propagation of dislocations, both of which employ a two-step change in the temperature and pressure during growth on striped masks [156]. An example of the first method has the stripes in the $\langle 1\bar{1}00\rangle$ direction (open width 3 μm; mask width 7 μm; thickness 80 nm)

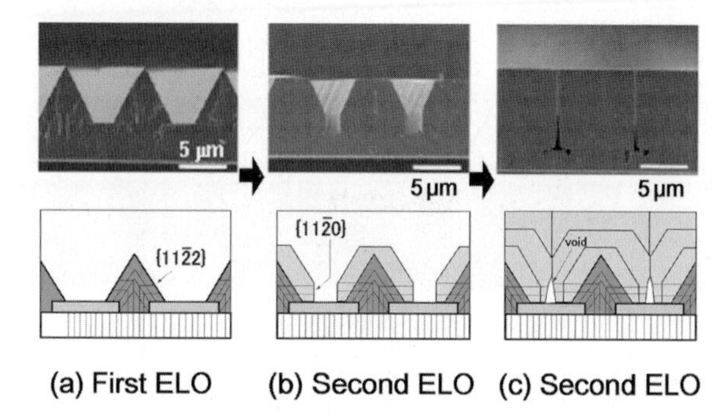

Fig. 6.64. SEM images and schematic models of FACELO GaN via Model B: (**a**) is the first ELO (950°C, 45 min); (**b**) is the second ELO (950°C, 45 min + 1,050°C, 30 min); (**c**) is the second ELO (950°C, 45 min + 1,050°C, 105 min). *Bar* indicates 5 μm [156]

and in the first step, a rectangular cross-section is grown at 1,000°C and 80 Torr (high temperature, low pressure conditions) for 30 min. During the next step, growth is continued at 1,050°C and 500 Torr (high temperature, high pressure conditions) for 90 min, resulting in lateral growth, coalescence and flat (0001) surfaces. In this method, only dislocations propagating in the vertical direction above the mask openings remain and the average dislocation density on the mask is reduced to $\sim 10^7$ cm^{-2}. An example of the second method is shown in Fig. 6.64 for stripes in the $\langle 1\bar{1}00 \rangle$ direction (opening width 5 μm; mask width 5 μm), where 90-min growth at 950°C and 500 Torr (low temperature, high pressure) produces pyramidal-shaped structures. At this stage, the propagation direction of threading dislocations, extending from the open areas, changes to the horizontal direction due to the $\{11\bar{2}2\}$ facet. In the second step, growth is carried out at 1,050°C and 500 Torr (high temperature, high pressure) for 105 min, but the $\{11\bar{2}0\}$ surface becomes larger with increasing growth time. The lateral growth eventually coalesces resulting in growth of flat (0001) surface growth and completion of the FACELO. The appearance of voids due to insufficient supply of source gases at these regions is a characteristic of this method. The dislocation density at the surface is 10^6 cm^{-2}, which is much less than the $\sim 6 \times 10^8$ cm^{-2} in the GaN substrate layer.

In the MOVPE method, PENDEO (Latin meaning to "hang up") growth is used to reduce threading dislocations by employing lateral growth [158, 159]. The effectiveness of this method was first demonstrated by the growth of GaN on 6H-SiC substrates [158] and later on Si [159] and sapphire substrates. Figure 6.65 illustrates the concept of PENDEO. An AlN buffer layer is deposited on a 6H-SiC(0001) substrate followed by growth of a GaN layer; then a SiN_x film is deposited and deep dry etching used to leave SiN_x stripes in the $\langle 1\bar{1}00 \rangle$ direction (etching down to the SiC substrate), preparing the substrate for PENDEO growth. The GaN $\{11\bar{2}0\}$ surfaces appear on the side walls of the stripe sections, known as "seed columns." The growth of GaN

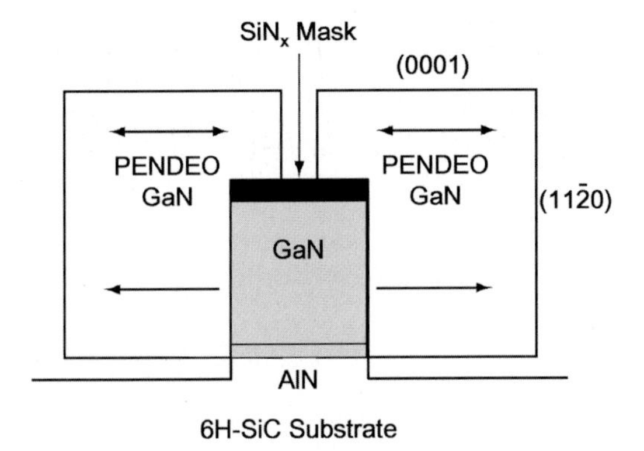

Fig. 6.65. PENDEO growth method [158]

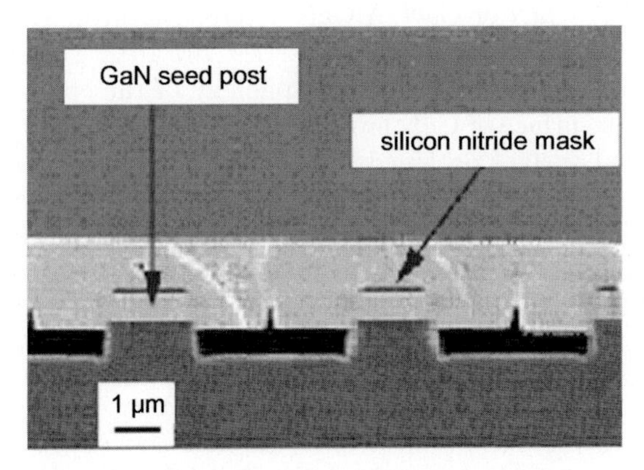

Fig. 6.66. Cross-sectional SEM image of PENDEO GaN [159]

proceeds laterally from the side walls and is followed by growth vertical to the (0001) surface until it reaches the SiN_x film mask where growth proceeds laterally over the mask. Since growth does not occur on the SiC substrate, voids remain between the PENDEO growth region and the SiC substrate. Figure 6.66 shows an SEM image of 2-μm thick GaN layer grown by the PENDEO method [159]. The seed column consists of a 1-μm thick GaN layer, a 100-nm SiN_x mask, and 2-μm seed columns separated by 3 μm. The growth temperature for PENDEO is 1,050–1,100°C and pressure is 40 Torr (high temperature, low pressure), which does not lead to growth on either SiC or SiN_x. Since the PENDEO growth layer is not in direct contact with the substrate, there is no strain associated with heterostructure growth. This is observed in a slight high energy shift of the low temperature PL emission spectrum compared with the GaN layer [159]. Use of maskless PENDEO growth has enabled reduction of dislocation densities to $\sim 10^4\,\mathrm{cm}^{-2}$ [159].

The ELO method cannot be used in MBE growth because in nonequilibrium, ultra high vacuum conditions, the differences in the surface energies between facets are too small to affect the growth process. Optimization of the buffer layer during MBE leads to flat GaN surfaces with dislocation densities of $\sim 10^9\,\mathrm{cm}^{-2}$ [165]. Shen et al. [166] reported on the use of RF-MBE for the growth of an extremely flat, 1.2-μm thick GaN layer on sapphire (0.5° off) at 700°C via a 300-nm thick AlN high temperature (700°C) buffer layer.

Control of the Polarity of the Epitaxial Layer

As shown in Fig. 6.67, wurtzite GaN grown on sapphire (0001) surfaces has either a Ga or N surface ("polarity"). For Ga-polarity, three of the four Ga bonds point to the substrate and one in a perpendicular direction toward the surface; N-polarity is the opposite. The polarity of the growth layer affects the emission of quantum wells and the density of the two-dimensional electron gas of GaN/InGaN or GaN/AlGaN heterostructures, and in many cases the Ga-polarity has more favorable properties [167,168]. The effects of surface polarity are observed in the surface morphology of the epilayers. Figure 6.68 shows the growth surface of GaN layers grown by MOVPE but compared with the extremely flat surface obtained with Ga surfaces, nitrogen leads to rough surfaces [167]. Growth on mixed polarity surfaces leads to cloudy surfaces. Also, Ga-polarity surfaces are resistant to etching in KOH and NaOH solutions, but N-polarity surfaces are easily etched, which enables identification

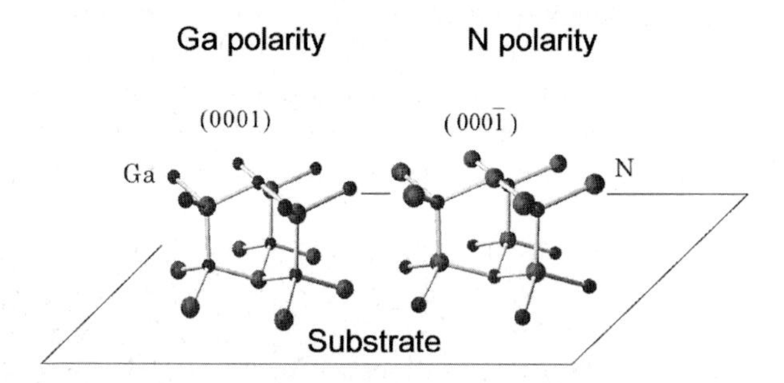

Fig. 6.67. Ga- and N-polarity in wurtzite structure GaN

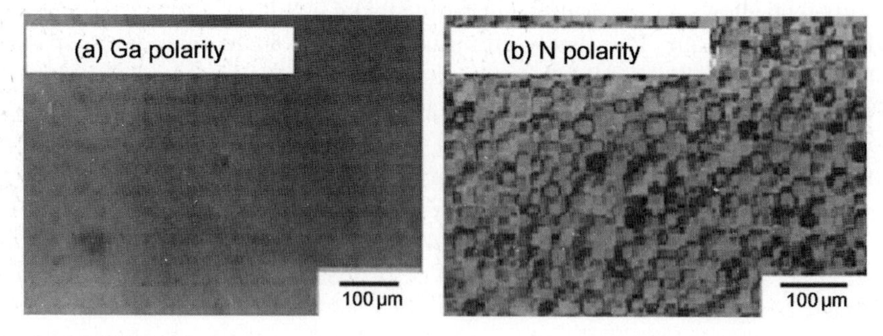

Fig. 6.68. Optical microscope images of Ga- and N-polar surface of GaN deposited by MOVPE [167]

of the type of surface polarity. The polarity of surfaces has been clarified by use of coaxial impact-collision ion scattering spectroscopy (CAICISS) [167].

Sumiya and Fuke [167, 169] used CAICISS to investigate the polarity of GaN layers grown on sapphire (0001) by MOVPE. XPS can be used to check that the thermal treatment of sapphire substrates at 1,000°C in flowing H_2 leads to Al-rich surfaces due to the expulsion of oxygen. The Ga (Al)-polarity of GaN and AlN buffer layers grown directly on such surfaces is always the same and, further, first-principle calculations show that the interface structure is stable [170]. On the other hand, if surface H_2 treatment is followed by exposure to NH_3 above 600°C, then nonuniform AlO_xN_{1-x} is formed on the surface. GaN or AlN layers grown on AlO_xN_{1-x} have been shown to have stable nitrogen surface polarity and if low temperature GaN buffer layers are grown at 600°C on nitrogen-processed surfaces, then mixed Ga- and N-polarity regions are found to grow. Figure 6.69 shows the variation of the half width (ω mode) of (0002) X-ray diffraction peaks from 1-μm GaN layers (1,040°C) with the thickness of GaN and AlN low temperature buffer layers (600°C) [167]. Thin buffer layers yield GaN- with N-polarity and both polarities are seen as the thickness is increased, eventually changing to only Ga-polarity for larger thicknesses. A Ga-polar surface can be obtained by (1) H_2 treatment; (2) start TMA flow to prevent nitridization; and (3) start growth by introducing NH_3 [167]. Also, a process involving deep nitridization of the sapphire substrate, and a combination of TMA flow and producing a high temperature AlN buffer before introducing the NH_3, has been reported to be effective in producing Ga surfaces [171].

In the growth of GaN by MBE, the sapphire substrate is "nitridized" prior to growth using nitrogen radicals generated by an RF-plasma. The nitrogen radicals react strongly with sapphire surfaces and nitridization is possible

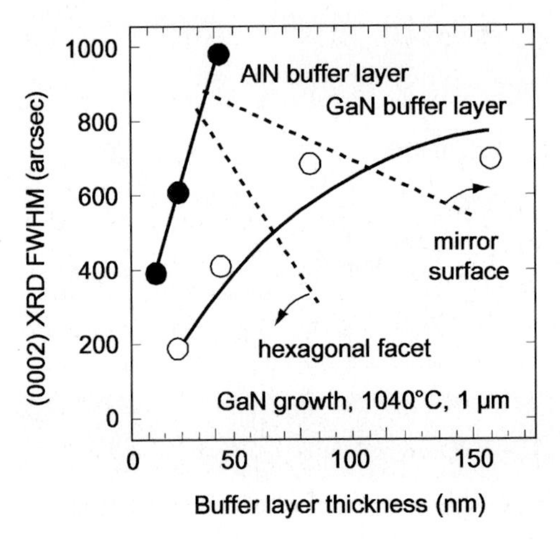

Fig. 6.69. The (0002) XRD FWHM as dependent on the thickness of GaN and AlN low temperature buffer layer about MOVPE-grown GaN films [167]

even at temperatures as low as 200°C resulting in the formation of an AlN layer on the sapphire substrate. The stability of AlN layers on sapphire has been studied using first-principle calculations [170], which show that the GaN layers grown directly on nitridized sapphire substrates at high temperatures (~700°C) always have nitrogen-polarity [172]. When low temperature GaN and AlN (~500°C) buffers are used, the polarity of the GaN layer depends on the buffer layer thickness, and at a thickness of 15 nm the GaN surface has nitrogen-polarity and the AlN surface has Ga-polarity [173]. Both buffers have nitrogen-polarity near the substrate but it changes to Ga (Al) with increasing buffer layer thickness; thus the final polarity of the GaN surface is determined by the thickness of the buffer layer. High temperature AlN buffer layers have Ga-polarity [173]. Furthermore, incorporation of two-atomic layers of Mg, Al, or Ga metals on GaN surfaces with nitrogen-polarity enables the inversion of the GaN growth layer to Ga-polarity [174]. The same results are found if GaN is grown on nitridized sapphire substrates that have undergone TMA treatment [175]. In MBE growth, the polarity of GaN surfaces can be determined by analysis of the RHEED patterns. Figure 6.70 shows the temperature and V/III dependence of the surface reconstruction of Ga and nitrogen surfaces of GaN {0001} [176]. For Ga-polarity surfaces, the 2×2, 5×5, 6×4, and 1×1 surfaces appear dependent on the temperature and V/III ratio, and for nitrogen surfaces, the 1×1, 3×3, 6×6, and $c(6 \times 12)$ structures are seen.

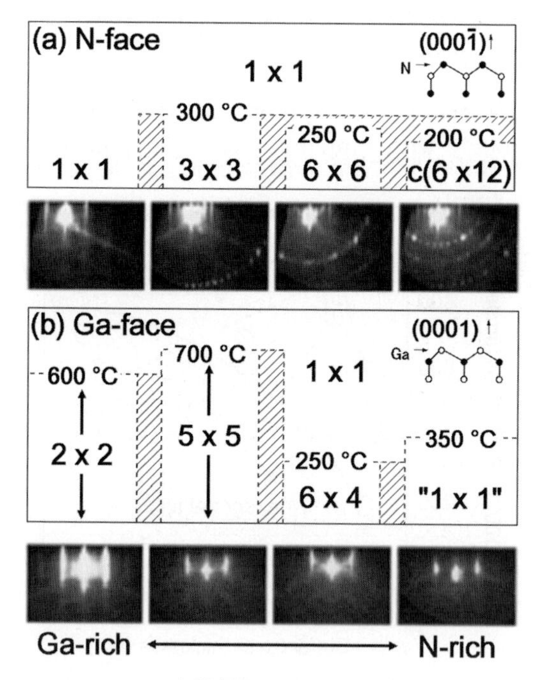

Fig. 6.70. The temperature and V/III ratio dependence of surface reconstruction about {0001} face of GaN [176]

There have also been reports on the control of surface polarity of 6H-SiC and Si(111) surfaces. For 6H-SiC, a low temperature AlN buffer is used and Ga-polarity results for Si surfaces. For Si(111) surfaces, hexagonal column structures associated with nitrogen-polarity are observed for surfaces without AlN buffer layers, and with the presence of AlN buffers, flat surfaces are obtained reflecting the existence of Ga-polarity [167, 168].

Growth Along Nonpolar Orientations

Recently, it has been found that growth on $(11\bar{2}0)$ (A-plane) or $(10\bar{1}0)$ (M-plane) GaN surfaces leads to the formation of quantum well structures without the internal electric field. Growth on sapphire $(10\bar{1}2)$ (R-plane) substrates enables the growth of GaN $(11\bar{2}0)$ (A-plane), where there is a +16% lattice mismatch along the a-axis and only +1% along the c-axis. There has been a report of the MBE growth of A-plane GaN/AlGaN MQW structures where the emission characteristics showed elimination of the internal field and nonexistence of the (0001)-oriented growth [177]. Also MOVPE was used to grow A-plane GaN/AlGaN and GaN/InGaN MQW LEDs exhibiting similar effects [178, 179]. Also, when ELO is used in MOVPE, the density of threading dislocations has been reduced to $10^7\,\mathrm{cm}^{-2}$ [180]. When 6H (4H)-SiC substrates are used, then with the use of $(11\bar{2}0)$ (A-plane) or $(10\bar{1}0)$ (M-plane) it is possible to grow GaN $(11\bar{2}0)$ (A-plane) or $(10\bar{1}0)$ (M-plane), respectively. Reports of MOVPE growth have shown the GaN surfaces to be rough and to have threading dislocation densities of $\sim 10^{10}\,\mathrm{cm}^{-2}$ [180, 181].

6.2.5 InGaN and InGaN/GaN Heterostructures (K. Tadamoto)

MOVPE Growth of Epitaxial InGaN Ternary Alloys

InGaN ternary alloys are used as emission layers in a wide range wavelength of light emitting devices. The first report on the growth of InGaN by MOVPE was by Nagamoto et al. [182] in 1989. The first device application was reported by Nakamura et al. [183] in 1993 using p-GaN/n–InGaN/n–GaN/sapphire double heterostructures (DHs). Growth of InGaN is one of the most important key technologies for nitride semiconductor devices.

Epitaxial growth of AlInGaP-based phosphorus and arsenic compound semiconductors can be expressed in a simple form by the following series of reaction equations:

$$\mathrm{Ga(CH_3)_3} + \frac{3}{2}\mathrm{H_2} \longleftrightarrow \mathrm{Ga(g)} + 3\mathrm{CH_4} \tag{6.7}$$

$$\mathrm{VH_3} \longleftrightarrow \frac{1}{4}\mathrm{V_4} + \frac{3}{2}\mathrm{H_2} \tag{6.8}$$

$$\frac{1}{4}\mathrm{V_4} + \mathrm{Ga(g)} \longleftrightarrow \mathrm{GaV} \tag{6.9}$$

In the reaction shown in (6.7), metal–organic materials such as trimethylgallium (source of the group III element) decompose near the growth temperature and exist as metallic vapor (Ga(g)). The group V element containing compounds PH_3 and AsH_3 also decompose near the growth temperature and exist as group V molecules in dimer and tetramer (6.8). Epitaxial growth proceeds due to reactions of gaseous metals and group V molecules (6.9) with the hot semiconductor surfaces [184].

The reactions for the growth of nitrides are

$$\mathrm{NH_3 + Ga(g) \overset{K_{GaN}}{\longleftrightarrow} GaN + \frac{3}{2}H_2} \tag{6.10}$$

$$\mathrm{NH_3 + In(g) \overset{K_{InN}}{\longleftrightarrow} InN + \frac{3}{2}H_2} \tag{6.11}$$

with equilibrium constants of the form:

$$\mathrm{K_{GaN}} = \frac{\mathrm{a_{(GaN)}P_{(H_2)}}^{3/2}}{\mathrm{P_{(Ga)}\,P_{(NH_3)}}} \tag{6.12}$$

$$\mathrm{K_{InN}} = \frac{\mathrm{a_{(InN)}P_{(H_2)}}^{3/2}}{\mathrm{P(In)\,P_{(NH_3)}}} \tag{6.13}$$

where $\mathrm{a_{(GaN)}}$ is the activity of GaN and $P(NH_3)$ is the partial pressure of NH_3. Figure 6.71 shows the temperature dependence of the equilibrium constants of the main compound semiconductors obtained from thermodynamic theory [185]. As shown in this diagram, the reasons for difficulties in the

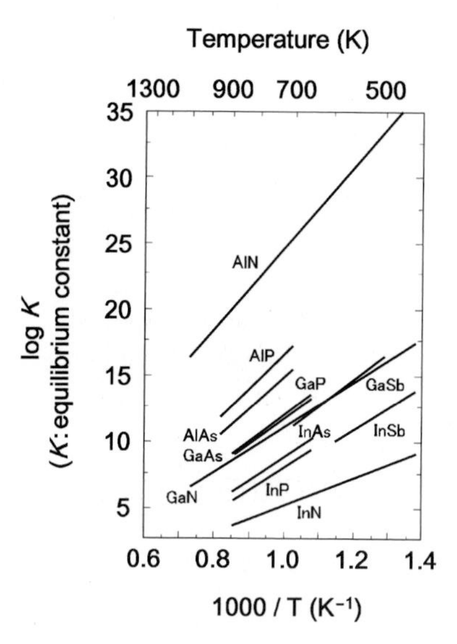

Fig. 6.71. The temperature dependence of equilibrium constant by main preparation reaction (the thermodynamic calculations of equilibrium constant)

epitaxial growth of InGaN are (1) the extremely small equilibrium constant of the main InN formation reactions; and (2) that the main formation reaction is accompanied by the generation of hydrogen (6.11). Thus it is necessary to remove hydrogen from the growth system and nitrogen is used as a carrier gas. Also, in order to increase the equilibrium constant of the formation reaction, the growth temperature must be set at ∼800°C, which is less than that for GaN growth.

Figure 6.72 shows the theoretically calculated variation of InN alloy composition ratio (InN in InGaN) with gas-phase composition [185, 186]. It can be seen that the InN alloy ratio strongly depends on the growth temperature with high InN composition films becoming difficult to grow at high growth temperatures. This data also shows the existence of a miscibility gap under certain growth conditions. The alloy composition ratio is also strongly affected by the hydrogen partial pressure and growth rate.

Generally, MOVPE growth occurs under mass transport limited conditions. Under such conditions the effect of growth temperature on the alloy composition is minimal and the desired alloy composition can be controlled by varying the ratio of the source gases only. The growth of InGaN is fundamentally mass transport limited but as described above, the actual InN composition ratio differs from the expected value. This is because the large equilibrium partial pressure of In cannot be ignored.

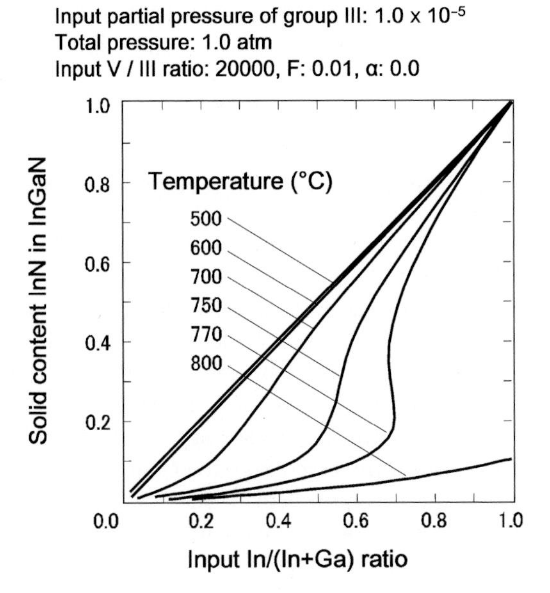

Fig. 6.72. Solid composition as a function of input In mole ratio (in calculations input partial pressure of group III sources: 1.0×10^{-5} atm., total pressure: 1.0 atm., H_2 partial pressure: 0.01, V/III: 20,000, of the effective mole fraction of decomposed NH_3: 0)

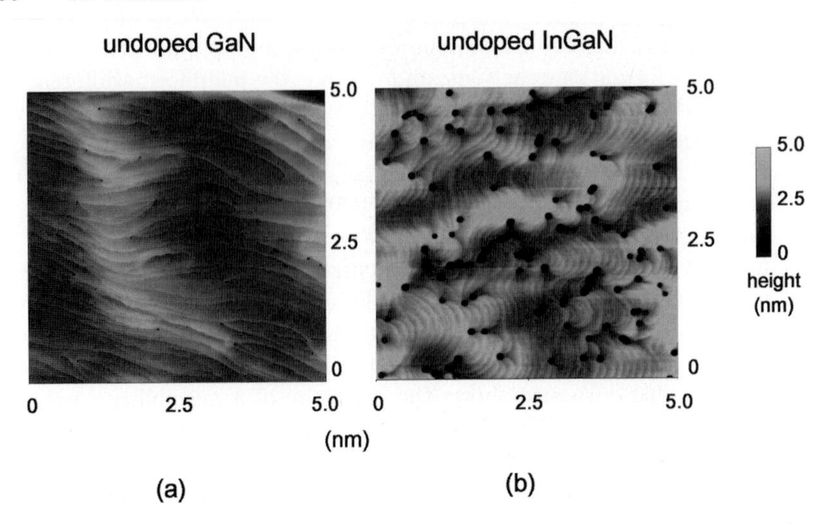

Fig. 6.73. SPM image of GaN (**a**) and InGaN/GaN heterostructure (**b**)

Figure 6.73 contains atomic force microscopy (AFM) images of the surface of GaN (a) and InGaN/GaN MQW structures obtained using the tapping mode. The surface of GaN shows extremely small pits due to dislocations terminating at the surface and step-flow growth. The height of the steps is 0.5 nm, corresponding to the c-axis length. In contrast, the surface of InGaN/GaN MQW structures exhibit helical morphology, which is thought to be due to use of low growth rates. Changes in the growth modes are thought to lead to fluctuations in the indium composition of the InGaN layers [187, 188].

Growth of Epitaxial InGaN/GaN Heterostructures

InGaN has a direct band structure and is used as the well layer of quantum well light-emitting devices. The barrier layer is usually GaN or InGaN with a bandgap larger than the well layer. Generally, InGaN/GaN means a well/barrier structure. The emission wavelength is controlled by varying the InN composition ratio, i.e., through altering the InGaN bandgap. Growth of InGaN on GaN occurs coherently up to a critical thickness (lattice matched to the GaN a-axis), resulting in the generation of biaxial-compressed strain along the a-axis and tensile strain in the c-axis direction [189]. These strains result in the generation of a piezoelectric field in the c-axis direction of the InGaN layers. Furthermore, the piezoelectric field causes a gradient in the band structure along the c-axis that has the effect of reducing the emission recombination energy of injected electron–hole pairs (the emission wavelength shifts to higher wavelengths), which is called the quantum-confined Stark effect [189, 190].

Figure 6.74 shows a cross-section TEM image of an InGaN/GaN (well width ∼3 nm) multiquantum well heterostructure. It can be seen that since

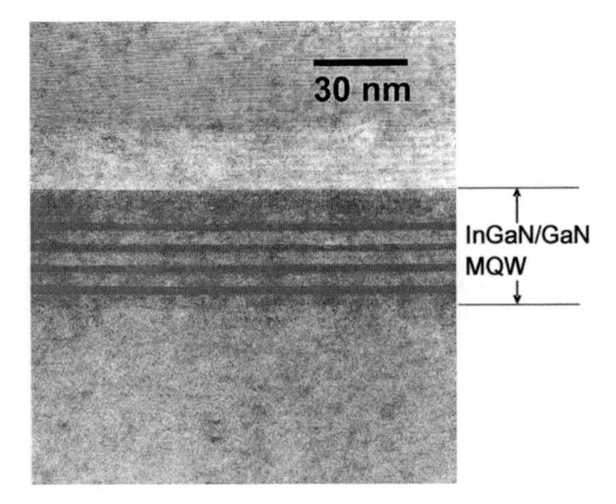

Fig. 6.74. Cross-section TEM image of InGaN/GaN heterostructure (MQW)

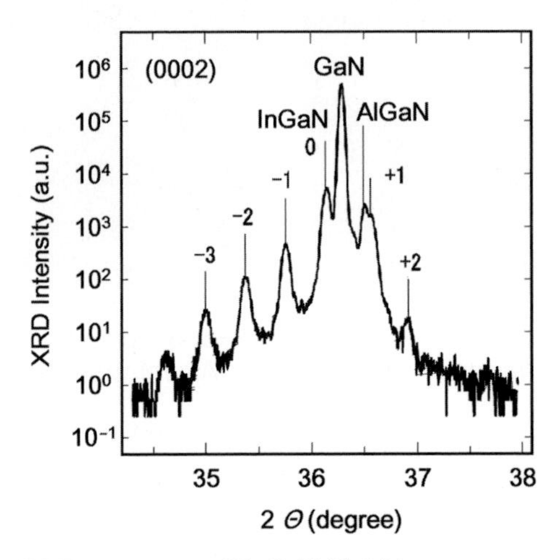

Fig. 6.75. XRD spectrum of InGaN/GaN heterostructure (MQW)

the MQW has a superlattice structure, its XRD spectrum contains satellite peaks (Fig. 6.75). The main peak is due to diffraction from GaN and the satellite peaks are from InGaN, AlGaN, and the MQW structure. The thickness of the quantum well and barrier layer and InGaN composition can be calculated from the separation of the satellite peaks.

Spatial Distribution of the In Composition of InGaN/GaN Heterostructures

The In compositional fluctuation and phase separation of the InGaN well layer is well known [188]. Recently, it has become possible to observe directly the microscopic compositional fluctuations using cathode-luminescence, scanning near-field optical microscopy (SNOM), and confocal laser microscopy [188, 191–193]. Figure 6.76 shows the submicron spatial variation of the PL intensity of an InGaN/GaN SQW structure wafer measured using a JSACO Corp NSF-300 SNOM system [193]. From measurements of the ratio of the PL intensity and peak wavelength, it is found that regions with high In composition exhibit higher PL intensity, while the peak is shifted to higher wavelengths. These compositional fluctuations are thought to be due to the presence of dislocations [188]. They are formed at regions of low potential and cause localization of electron–hole pairs and excitons; their existence strongly affects the high optical emission of InGaN/GaN heterostructures. Figure 6.77 shows the macroscopic PL intensity distribution across a wafer of an InGaN/GaN MQW structure (peak wavelength 405 nm) measured using a PL mapping system (ACCENT, RPM SIGMA, He–Cd laser excitation source). The PL

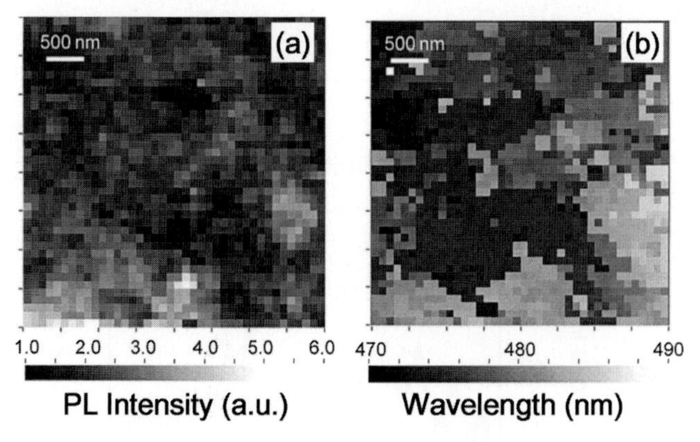

Fig. 6.76. PL mapping of InGaN/GaN heterostructure (SQW) by SNOM measurement

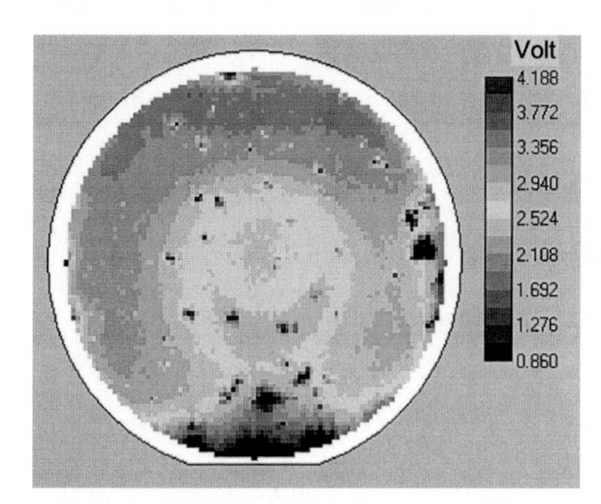

Fig. 6.77. PL mapping of epitaxial substrate for LED (2 in.) bearing InGaN/GaN (MQW) heterostructure

intensity and peak wavelength exhibit concentric distributions due to the rotation of the substrate during growth. As described later, the PL intensity depends on the crystalline quality of the MQW structure as well as on the quantum well width. The peak wavelength depends on the InN composition ratio and well width. These macroscopic PL measurements are important for development of light-emitting devices.

Internal Quantum Efficiency and Emission Wavelength (InN Composition Ratio) of InGaN/GaN Heterostructures

The internal quantum efficiency of LEDs is defined as the ratio of the number of photons generated inside a device (MQW structure) due to radiative recombination to the number of injected electrons. Figure 6.78 shows the reported external quantum efficiency of InGaN/GaN-based LEDs as a function of wavelengths [194, 195]. The label A represents devices fabricated on patterned sapphire substrates [196, 197]; B and C are two SiC-based devices from Cree (B = XThin, C = XBright) [198]; D, a device fabricated on patterned sapphire substrates with mesh p-type electrodes [199]; and E, a device structure with no GaN layer incorporated [200]. The external quantum efficiency is determined by the product of the internal quantum efficiency and light extraction efficiency which is affected by the device structure. Since all the devices (A–E) have different structures, it is not easy to directly compare the results. However, it is thought that the variation of the internal quantum efficiency with emission wavelength – i.e., trends related to the InN alloy composition – is clearly shown in these results. From these results, the internal quantum efficiency of InGaN/GaN LEDs is highest (and similar in magnitude) around from 400 to 460 nm. Although it is difficult to accurately calculate the

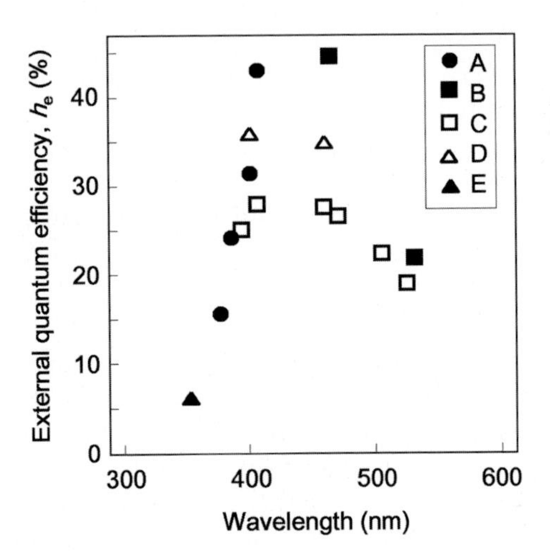

Fig. 6.78. The dependence of emission wavelength of external quantum efficiency of InGaN LED

internal quantum efficiency, the value for device A (peak wavelength 405 nm) was estimated to be 54% based on the temperature dependence of the PL characteristics [201]. The external quantum efficiencies are seen to decrease in the short and high wavelength regions. The small value at short wavelengths is due to decreases of the InGaN/GaN band offset and optical absorption due to the GaN layer. In the high wavelength region, the small efficiencies are mainly due to decreases in the internal quantum efficiency.

As described above, during InGaN growth, the partial pressure of hydrogen should be decreased as low as possible and the films should be grown at low temperatures in order to increase the equilibrium constant of the formation reaction. In the case of GaN, poor crystalline growth occurs if there is insufficient hydrogen flow [202]. However, high-quality InGaN crystals cannot be grown even with low hydrogen flow rates. Also, the difference of optimal growth temperatures between InGaN and GaN is another problem to grow high-quality InGaN/GaN heterostructure. Generally, insufficient atomic migration during growth at low temperatures results in poor crystal quality. Low temperature growth also leads to increased impurity incorporation and hence increases in the density of nonradiative impurities.

Increase in the InN alloy composition leads to an increase of the InGaN lattice constant, which consequently leads to increases in the strain in the InGaN layers and eventually increases in the quantum-confined Stark effect. Assuming a quantum well width of 3 nm, the piezoelectric fields generated in quantum wells at various wavelengths are: green (530 nm) 1.2 MV cm^{-1}; blue (460 nm) 0.8 MV cm^{-1}; and near-ultraviolet (400 nm) 0.3 MV cm^{-1} [203]. Thus, this quantum-confined Stark effect is thought to be one reason for reductions in the internal quantum efficiency in the long wavelength region. For LEDs emitting ~400 nm, blue shifts are not observed with increasing injection currents, which is thought to be because this piezoelectric effect is blocked out by the standard structure of device [196]. Further improvement in the internal quantum efficiency of InGaN LEDs will necessitate further studies on the growth of InGaN/GaN heterostructures, design of device structures that minimize piezoelectric effects and development of doping technology.

6.2.6 AlGaN, AlN, and AlGaN/GaN Heterostructures (K. Hiramatsu)

Al-Based Nitride Semiconductors

The direct transition bandgap energy of $Al_xGa_{1-x}N$ can be changed from 3.4 to 6.2 eV by varying the mole fraction of GaN and AlN. This alloy can be used for fabricating ultra-short wavelength LEDs, LDs, and photodetectors. For such applications, it is necessary to develop technology for the growth of high-quality, high AlN mole fraction $Al_xGa_{1-x}N$ layers with thicknesses of several 100 nm. In 1988, Koide et al. succeeded in growing high-quality $Al_xGa_{1-x}N$ layers on sapphire substrates using a low temperature AlN buffer layer [204]. Then, Itoh et al. reported (1) the crystalline quality of $Al_xGa_{1-x}N$

layers to be degraded with increasing AlN mole fraction; (2) improvements in the crystalline quality of $Al_xGa_{1-x}N$ by growth at high temperatures on GaN; and (3) the appearance of cracks above the elastic limit due to the lattice mismatch between $Al_xGa_{1-x}N$ and GaN [205]. In 1999, Amano et al. succeeded in the growth of a 1-μm thick, crack-free $Al_xGa_{1-x}N$ for all AlN mole fractions by inserting a low temperature intermediate layer on GaN [206]. The low temperature buffer layer was also found to suppress the propagation of spiral dislocations [207]. However, there are also reports on the use of high temperature AlN buffer layers prior to $Al_xGa_{1-x}N$ growth. Ohba et al. used a high temperature AlN during the growth of 1.8-μm thick crack-free $Al_{0.2}Ga_{0.8}N$ deposited by MOVPE [208]. Tensile stress in a layer of $Al_xGa_{1-x}N$ is reduced when grown on high temperature AlN layer with the potential of reducing the generation of cracks but due to difficulties in the growth of high-quality AlN, there is sparse experimental data to support such expectations. Thus, there have been no detailed reports on dislocations and crystalline properties of GaN and $Al_xGa_{1-x}N$ grown on AlN substrates. This section describes the epitaxial growth AlN and growth of GaN and AlGaN on epitaxial AlN layers.

Growth of Epitaxial AlN

Shibata et al. used MOVPE to grow 0.5–2.0 μm thick AlN epilayers on 2-in. sapphire (0001) and 6H-SiC substrates without low temperature buffer layers [209, 210]. Figure 6.79 shows the AFM images of AlN grown on sapphire and 6H-SiC substrates. In both cases, 0.2-nm atomic steps corresponding to one molecular layer can be seen in the c-axis direction. The surface roughness (Ra) is less than 0.2 nm. These results confirm the possibility of producing high-quality AlN epilayers for device applications.

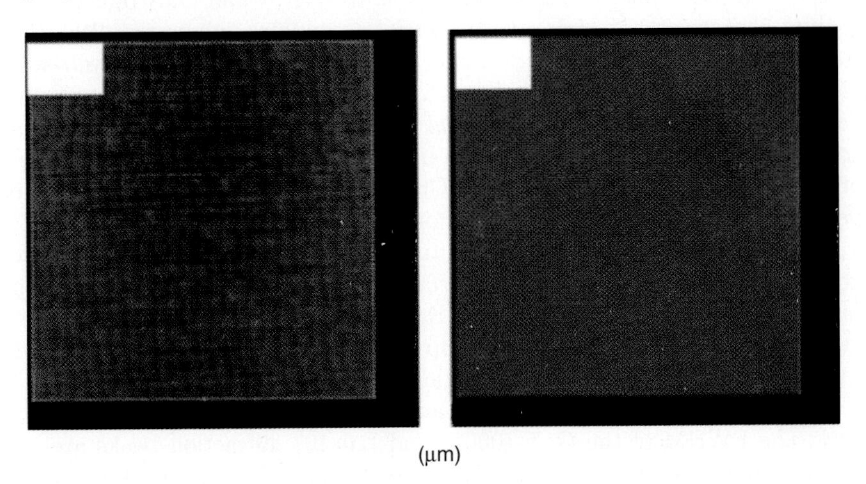

Fig. 6.79. AFM image of AlN epitaxial layer (**a**) on sapphire substrate and (**b**) 6H-SiC substrate

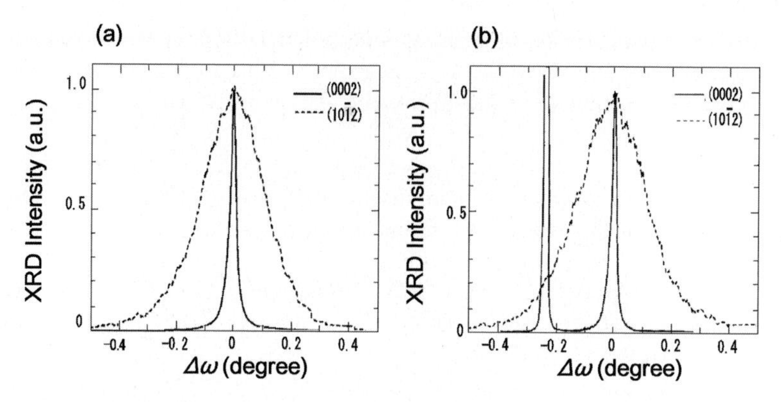

Fig. 6.80. XRD rocking curve of AlN epitaxial layer (**a**) on sapphire substrate and (**b**) 6H-SiC substrate

Figure 6.80 shows the XRD rocking curves for AlN epilayers grown on sapphire and 6H-SiC substrates. For sapphire, the FWHM of the diffraction peaks from the (0002) and (10–12) planes are 80 and 850 arcsec, respectively. For 6H-SiC the FWHM of (0002) and (10–12) are 56 and 880 arcsec, respectively. The FWHM of (0002) is extremely narrow but in contrast, that of the (10–12) is relatively large which indicates that the AlN film has small fluctuations in the c-axis component and that the rotational component has larger fluctuations.

Growth of GaN on AlN

It is well known that the use of low temperature AlN and GaN buffer layers enables the growth of high-quality GaN on sapphire substrates [211, 212]. However, use of low temperature buffer layers still results in GaN layers with dislocations densities of 10^8–$10^{10}\,\mathrm{cm}^{-2}$. On the other hand, GaN films grown on epitaxial AlN show better electrical and optical characteristics than films grown on standard low temperature buffers [213]. However, the crystalline properties or dislocation density of GaN grown on epitaxial AlN have not been reported to date. Recent results are described below.

Figure 6.81 shows an AFM ($5 \times 5\,\mu\mathrm{m}$) image of a 1.7-μm thick GaN layer grown on epitaxial AlN [209]. The surface is atomically flat and atomic steps can be seen. The surface roughness (Ra) is 0.36 nm over this $5 \times 5\,\mu\mathrm{m}$ area. Further, the height of the atomic step is 0.2 nm, which is equal to $c/2$ and corresponds to the thickness of a single atomic layer.

Figure 6.82 shows typical (0002) and (10–10) XRCs for GaN grown on AlN. The FWHM of the GaN (0002) and (10–10) diffraction peaks are 170 and 410 arcsec, and for AlN the FWHM are 40 and 1,800 arcsec. These results show that fluctuations in the c-axis direction of these structures are less than those grown using low temperature buffer layers. Fluctuation of the c-axis rotation component has also been decreased.

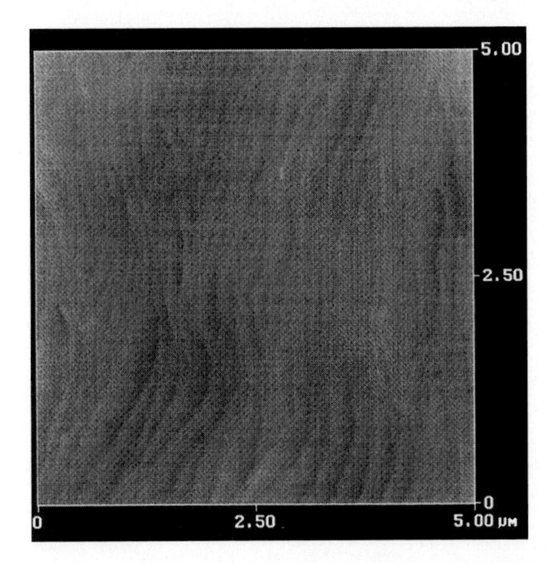

Fig. 6.81. AFM image of GaN on AlN epitaxial layer

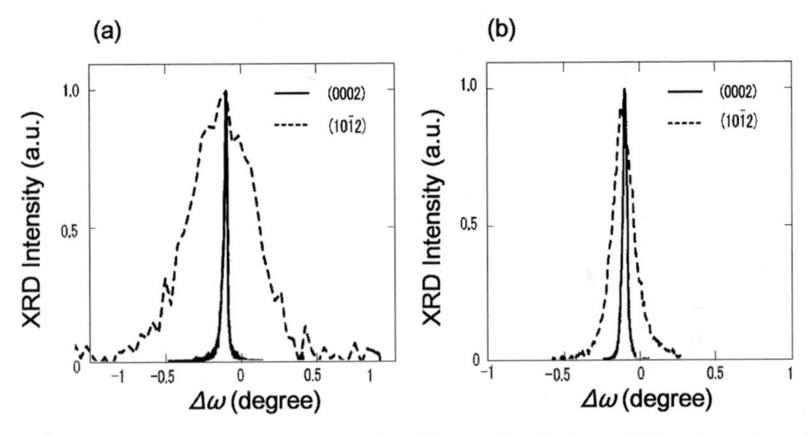

Fig. 6.82. XRD rocking curve of (**a**) AlN and (**b**) GaN on AlN epitaxial layer

Figure 6.83 shows the CL of GaN grown on epitaxial AlN. The dislocation density was estimated to be $3.8 \times 10^8\,\mathrm{cm}^{-2}$ from the dark spot density, which is similar or less than values measured in GaN films grown on low temperature buffers [214, 215]. Further, the dislocation density of the underlying AlN was $10^{10}\,\mathrm{cm}^{-2}$ which indicates a two orders of magnitude decrease of dislocations in the GaN layer.

Growth of AlGaN on AlN

Growth of high mole fraction of AlN is difficult because of the generation of cracks arising from the release of strain between GaN and $\mathrm{Al_xGa_{1-x}N}$ layers

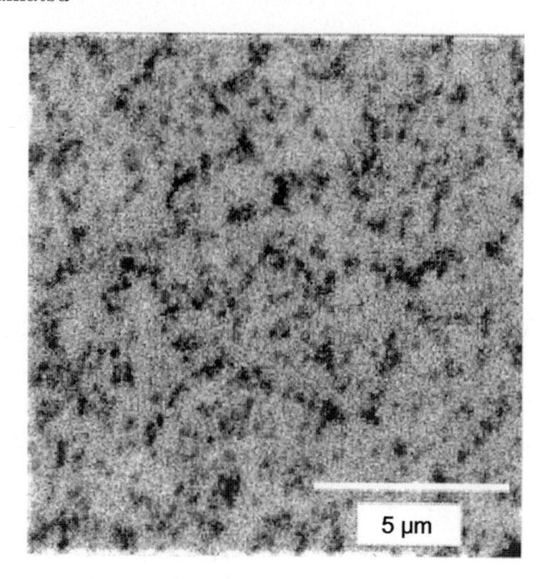

Fig. 6.83. Surface CL image of GaN on AlN epitaxial layer

when GaN is grown on $Al_xGa_{1-x}N$ [205]. Thus the fabrication of devices with AlGaN/GaN structures necessitates careful control of the AlN composition and layer thickness. The growth of a low temperature AlN layer on GaN followed by deposition of AlGaN has been reported as a solution this problem [206]. The use of this method enables growth of crack-free, high-quality AlGaN on GaN. However, the growth of AlGaN on epitaxial AlN has also been reported [208]. The use of an AlN underlayer for AlGaN is expected to prevent and reduce the density of cracks in the AlGaN layer. This section reviews the properties of high-quality AlGaN grown on epitaxial AlN/sapphire.

Here, the growth of high-quality $Al_xGa_{1-x}N$ using epitaxial AlN/sapphire substrates is described. Figure 6.84 shows SEM and AFM ($1 \times 1\,\mu m$) images of crack-free $Al_xGa_{1-x}N$ ($0.4 < x < 0.8$) layers grown between 1,120 and 1,220°C. The AFM images show the appearance of tiny pits at a growth temperature of 1,120°C but increasing the temperature to above 1,140°C results in atomically flat surfaces.

Figure 6.85 shows the variation of the (0002) XRC peak from $Al_xGa_{1-x}N$ with Al mole fraction. The data from Amano et al. [206] is also included for comparison. The FWHM of $Al_xGa_{1-x}N$ ($0.4 < x < 0.8$) is less than 200 arcsec indicating a reduction in c-axis fluctuations.

An issue still to consider is that the dislocation density of $Al_xGa_{1-x}N$ with AlN mole fraction greater than 0.5 is $10^9\,cm^{-2}$ and must be reduced. ELO is known as an effective means of reducing the density of threading dislocations in GaN films. However, for Al-based group III nitride semiconductors, the Al reacts with SiO_2 masks thus making selective growth extremely difficult. A solution to this problem was studied by producing undulations in the epitaxial AlN underlayer and growing $Al_xGa_{1-x}N$ on top.

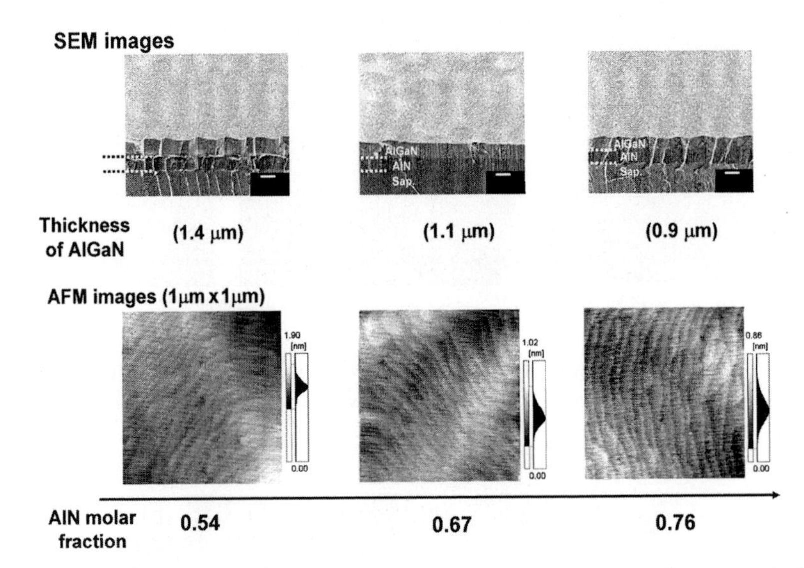

Fig. 6.84. AFM image and surface AFM image of $Al_xGa_{1-x}N$ ($0.4 < x < 0.8$) on AlN epitaxial layer

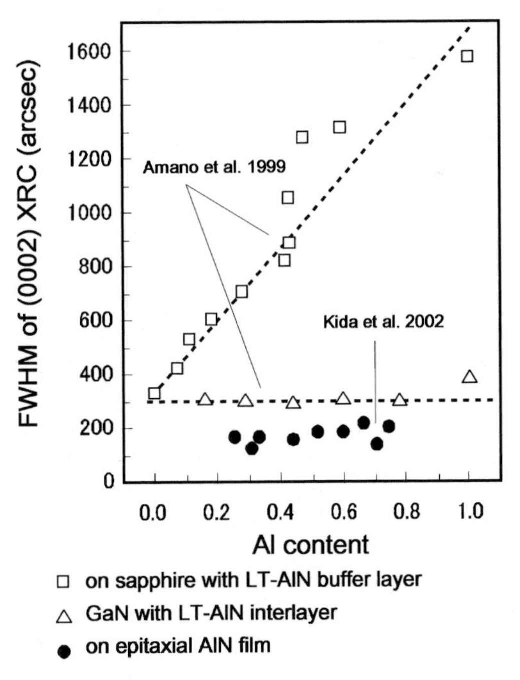

Fig. 6.85. FWHM value of XRD of AlGaN (0002) diffraction

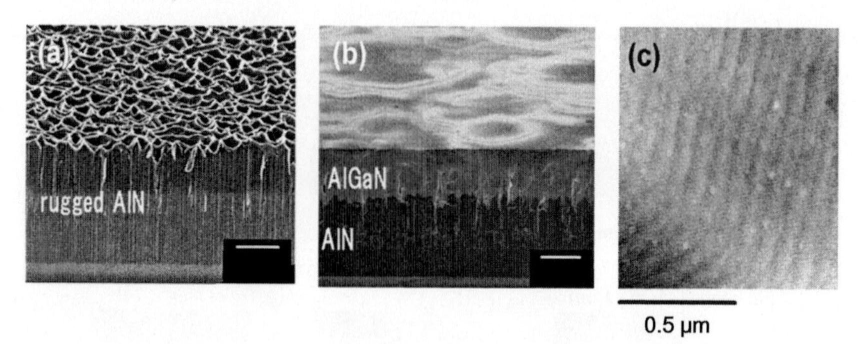

Fig. 6.86. (**a**) SEM image of convexo-concave AlN substrate, (**b**) bird's-eye SEM image of AlGaN on convexo-concave AlN substrate, and (**c**) AFM image of AlGaN surface on convexo-concave AlN substrate

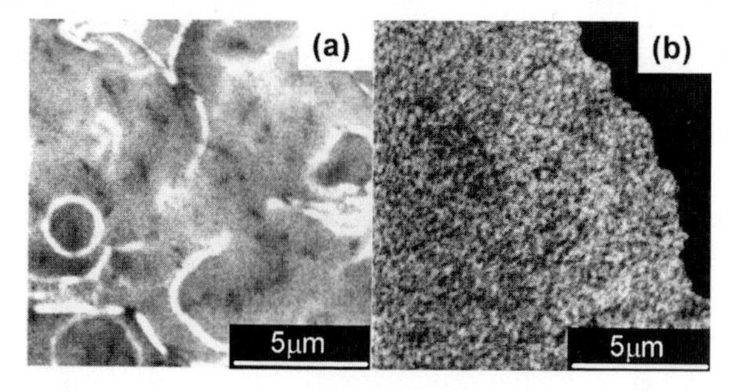

Fig. 6.87. (**a**) Surface CL image of AlGaN on convexo-concave AlN substrate and (**b**) surface CL image of AlGaN on flat AlN substrate

Figure 6.86 shows SEM images of the substrates used and AFM images of the resulting $Al_xGa_{1-x}N$ films [216]. The SEM image is of a 5.0-μm thick $Al_xGa_{1-x}N$ layer where shallow islands can be seen although the overall surface is flat. X-ray diffraction showed the AlN mole fraction to be high at 0.48. The AFM image shows tiny pits although atomic steps were observed indicating the excellent flatness of the $Al_{0.48}Ga_{0.52}N$ surface.

Figure 6.87 shows the CL of AlGaN grown on modified epitaxial AlN substrates [217]. The dislocation density was estimated to be $5.5\times10^8\ cm^{-2}$, which was less than AlGaN films grown on flat epitaxial AlN substrates ($>10^9\ cm^{-2}$). Thus modified AlN substrates are effective in reducing the dislocation density of AlGaN layers.

This section described the growth and crystalline properties of Al-based nitride semiconductors. It is possible to grow crack-free, low dislocation density GaN and AlGaN layers using epitaxial AlN underlayers. Further improvements in crystal quality are expected to be possible by epitaxial growth of AlGaN and GaN on bulk AlN crystals. At the present time, there is increasing activity on the growth of bulk single crystal AlN [218], which is expected to contribute to the fabrication of high performance Al-based nitride optical devices.

6.2.7 InN and Related Semiconductors (A. Yoshikawa)

Energy Bandgap of InN

InN has the smallest bandgap amongst III–V nitride semiconductors. Also, InN is used for the fabrication of GaN-based multiquantum well structures (MQW) that emit blue and green light. That is, InGaN ternary alloys are combined with GaN barrier layers to construct MQW device structures. However, InGaN is prone to phase separation due to the large differences in the lattice constants of GaN and InN as well as in their chemical properties, which lead to the formation of higher In composition quantum dots rather than uniform layers under increased indium compositions. Such dots exhibit high emission efficiencies, which are good for applications. However, the nonuniformity of this alloy makes it difficult to estimate the true bandgap of InN itself.

On the other hand, since InN has the largest electron mobility amongst nitride semiconductors, it has tremendous potential for use in ultrahigh speed electronic devices employing two-dimensional electron gas structures. The use of MBE has enabled the growth of high-quality InN on GaN substrates at low temperature.

Estimates of the bandgap of InN were initially based on measurements on polycrystalline or microcrystalline thin films produced by sputter deposition and the resulting 1.9 eV value [219, 220] held sway for many years. However, in 2001–2002, it became possible to grow high-quality single crystal InN films, which were reported to have a bandgap between 0.75 and 0.9 eV [221–226]. Recently, the bandgap has been shown to be ~0.64 eV [227].

Some of the discrepancies of the bandgap of InN have been attributed to increases of the bandgap due to oxygen incorporation, effective increases in the optical bandgap due to the high background electron density (Burstein–Moss effect) and measurements being carried out on ultra thin films.

Estimates of the bandgap can be obtained from optical absorption/reflectance and band edge photoluminescence measurements. Figure 6.88 shows the optical absorption/reflectance characteristics of InN films grown by MBE in ~2003 [228]. The optical bandgap is seen to be ~0.7 eV, but this is larger than the actual value due to the large background electron density of $\sim 8 \times 10^{18}\,\mathrm{cm}^{-3}$.

Epitaxial Growth of InN

As shown in Fig. 6.89, the equilibrium vapor pressure of nitrogen on InN is extremely high thus necessitating growth at low temperatures of between 500 and 600°C [229]. On the other hand, the vapor pressure of In is only 10^{-6} Torr even at 600°C, which leads to the formation of growth inhibiting In

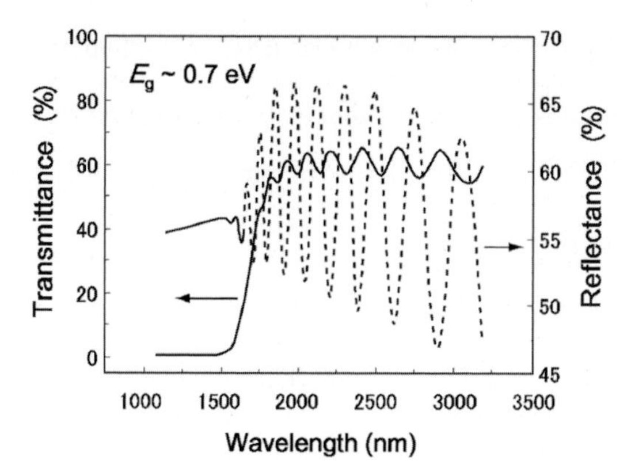

Fig. 6.88. The optical transmittance and reflectance of InN [228]

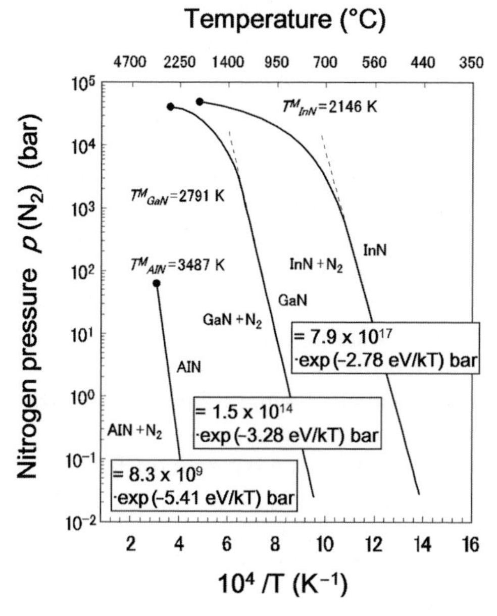

Fig. 6.89. The temperature dependence of equilibrium vapor pressure of nitrogen on InN [229]

droplets during deposition. While, as with other nitrides, at high V/III ratios the surface diffusion of In is reduced and the surface flatness and crystallinity degrades. That is, the stoichiometry of the growing surface must be controlled with high precision. Further, the low growth temperature for InN leads to another problem of large mismatch in epitaxy temperatures among other III-nitrides when fabricating MQW structures for device application. That is, the growth of GaN, AlN, and ternary alloys at the same low temperature is difficult.

A comparison of MBE [230, 231] and MOCVD [232] techniques for the growth of high-quality InN thin films shows that, in MBE, the nitrogen used as the nitrogen source is extremely chemically active, enabling low temperature growth; but, in MOCVD, the ammonia nitrogen source is thermally stable and does not decompose at low temperatures, necessitating use of extremely high and impractical V/III ratios of greater than 10,000. Further, there is an additional problem with ammonia (NH_3) that the hydrogen produced as a by-product during decomposition attacks the InN layer itself. Thus MBE is considered better suited for growth of InN.

Effect of Crystal Polarity

The properties of epitaxial InN grown by MBE depend on the polarity of the crystalline surfaces. Generally, GaN and AlN are grown on +c polarity surfaces where the film deposition progresses by the step-flow mode, and the surface becomes atomically flat resulting in high-quality crystalline epilayers. It is generally thought that the growth of many nitride semiconductors, including InN, must be grown on cation polarity surfaces, i.e., In surfaces in the case of InN. Figure 6.90 shows the temperature dependence of the growth rate on both polarities [233, 234], where growth was carried out on either the Ga or nitrogen surface of GaN grown on the c-surface of a sapphire substrate under stoichiometric but slightly nitrogen-rich conditions. The growth rate steeply decreases for In and nitrogen surface beyond 500 and 600°C, respectively. These results show that a window of suitable growth temperatures is 100°C higher in nitrogen-polarity due to higher nitrogen incorporation rate. Figure 6.91 shows the step-flow-like flat surface morphology of InN thin films grown on nitrogen surfaces [233]. It is possible to obtain similar surface morphologies on In surfaces but the optimal growth temperature window is narrow. Recently it was reported by the authors that step-flow growth having

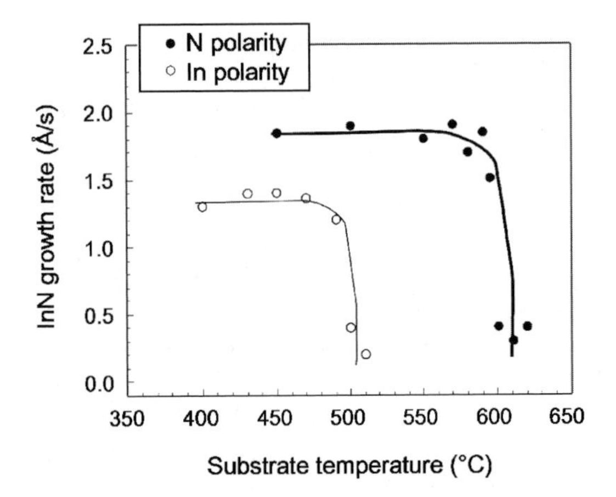

Fig. 6.90. The dependence of InN growth rate by MBE on polarity [233, 234]

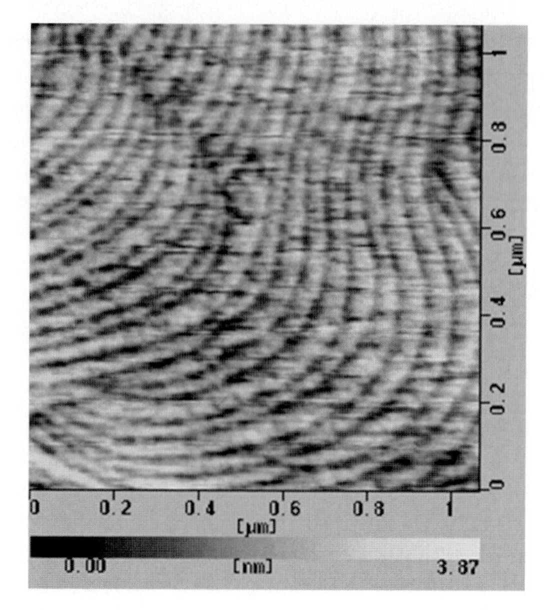

Fig. 6.91. AFM image of N-polarity InN surface [233]

exactly one monolayer step height can be more easily achieved in In-polarity InN epitaxy on Ga-polarity GaN template with a dislocation density less than $10^8\,\mathrm{cm}^{-2}$. The rms roughness was as small as 0.27 nm in $2 \times 2\,\mu\mathrm{m}^2$ and less than 1 nm even for $10 \times 10\,\mu\mathrm{m}^2$.

The room temperature electron density and mobility of typical high-quality InN films are $10^{18}\,\mathrm{cm}^{-3}$ and $2{,}000\,\mathrm{cm}^2\,\mathrm{V}^{-1}\,\mathrm{s}^{-1}$, respectively [235]. The reduction of the background electron density (the lowest background electron density is still $\sim 3\times10^{17}\,\mathrm{cm}^{-3}$) and control of p-type doping are still issues to consider.

Growth of InN-Based Ternary Alloys

The application of InN films for optoelectronic devices will require the formation of quantum well structures which necessitate a deeper understanding of the physical properties of ternary InGaN and InAlN. The discovery of the smaller than expected InN bandgap led to widespread studies of the properties of InGaN [236, 237]. Figure 6.92 shows the variation of the bandgap of ternary InGaN with composition. The bowing parameter is seen to be in the range 1.4–2.5 eV. However, there are very few reports on InAlN alloys with high InN compositions, but there has been a report on its growth over all InN compositions without phase separation [238]. Figure 6.93 shows the variation of the bandgap of InAlN alloys with Al content. The bowing parameter is ~4.96 eV. These results show that the (AlN–GaN–InN) nitride alloy system incorporating InN can be used to cover the optical spectrum from deep UV to near-infrared, the optical telecommunications waveband.

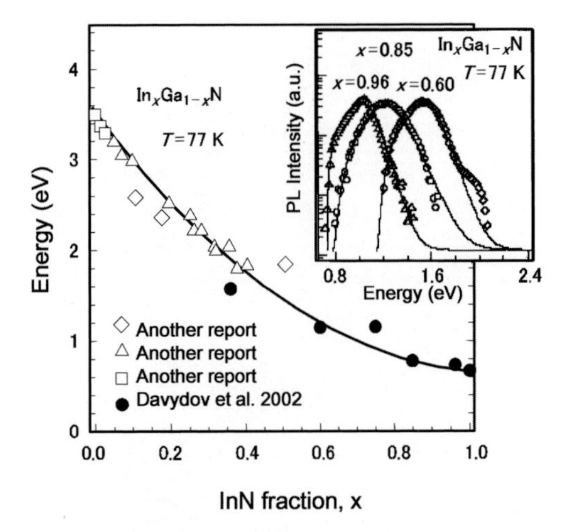

Fig. 6.92. Variation of the optical bandgap of InGaN on InN composition [236]

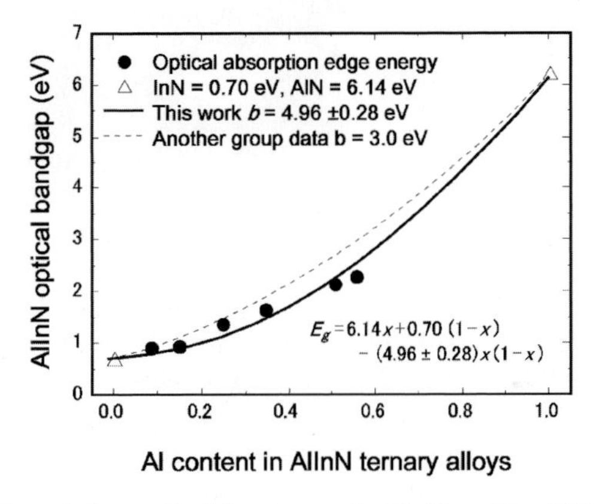

Fig. 6.93. Variation of the optical bandgap of AlInN on the AlN composition [238]

InN-Based Multiquantum Well Structures

The main issues in the growth of single or multiple quantum well structures using InN as a well are the large lattice mismatches between the composite materials and their optimal epitaxial temperatures. There is a lattice mismatch of ~11% between InN and GaN, and it is difficult to grow high-quality quantum well structures if the GaN must be grown at the same low temperature as InN [239]. High-quality InN/InGaN MQWs were grown using InGaN barriers with an In composition of 20–30% as shown in the TEM of Fig. 6.94. Such structures exhibited X-ray diffraction third-order satellite peaks which confirm their high crystalline quality [240, 241].

Changing the barrier material from InGaN to InAlN, it is possible to maintain the difference in the bandgap with InN as well as reducing the lattice

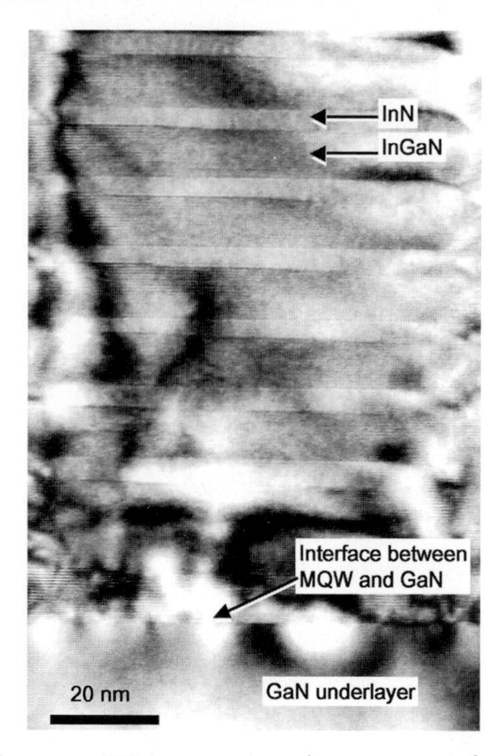

Fig. 6.94. TEM image of InN/InGaN MQWs [241]

mismatch. This is an important factor in the growth of InN-based MQWs with large conduction band discontinuities as used in the development of optical devices with intersubband electron transitions in the quantum wells. Figure 6.95 shows the X-ray diffraction characteristics from InN/InAlN MQWs with an Al composition of 30%. Again, third-order satellite peaks are observed, indicative of good quantum well interfaces.

Low-Dimensional InN-Based Quantum Structures (Quantum Dots and Nanocolumns)

The 11% lattice mismatch between InN and GaN leads to the generation of compressive stress in InN films grown on GaN. For such combinations of InN and GaN, the growth occurs three dimensionally with the formation of quantum dots and nanocolumns (crystalline columns with diameters on the nm scale). Figure 6.96 shows the lattice relaxation effects for InN grown on the nitrogen-polarity surface of GaN under excess nitrogen stoichiometry conditions (V/III ratio of ∼3) taking the InN deposition rate as a parameter [242]. The in-plane lattice is independent of the deposition rate and matches the GaN lattice for one monolayer InN coverage, but for larger thicknesses, there is a sudden change in the lattice relaxation and the

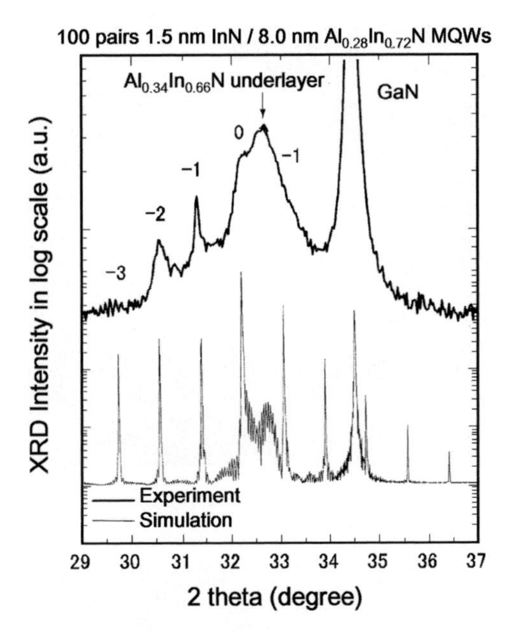

Fig. 6.95. X-ray diffraction of InN/AlInN MQWs [238]

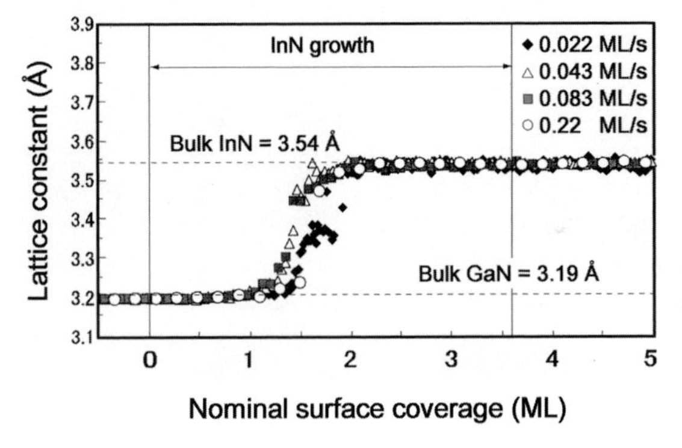

Fig. 6.96. Properties of lattice relaxation of InN epitaxial growth on N-polar GaN [242]

lattice constant becomes that of the free-standing InN layer. That is, the InN layer is elastically strained for one monolayer but for larger thicknesses the strain is released and three-dimensional growth starts.

Further deposition of InN under excess nitrogen conditions results in the formation of quantum dots. Figure 6.97 is an AFM image of quantum dots formed by the growth of InN for 4 min on the c-surface of nitrogen-polarity GaN substrates off-cut by approximately 1° [242]. Typically, the quantum dots have a diameter of 20 nm and height of 1–2 nm. Growth on Ga-polarity GaN surfaces under excess nitrogen conditions (to minimize the diffusion of

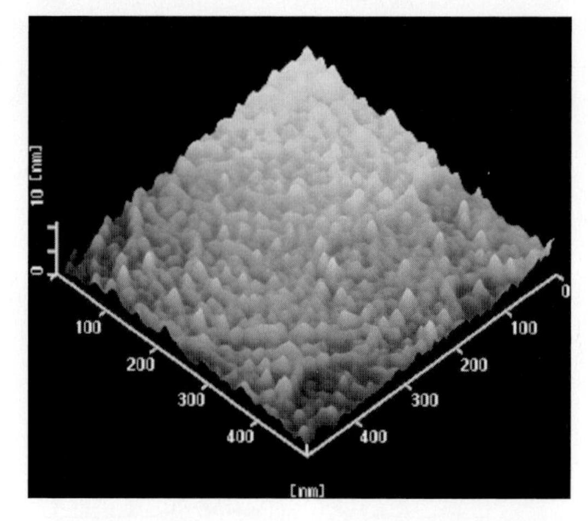

Fig. 6.97. Growth of InN dots on N-polar GaN template [242]

Fig. 6.98. Growth of In-polar InN nanocolumns on Ga-polar GaN template [243]

In on the GaN surface) results in the formation of nanocolumns as shown in Fig. 6.98 [243]. The columns typically have a diameter of ∼170 nm, a height of ∼1,180 nm, and a density of $4 \times 10^8\,\mathrm{cm}^{-2}$.

Further, the authors have recently proposed and demonstrated the first successful achievement of a novel structure InN/GaN MQW consisting of one monolayer and fractional monolayer InN well insertion in GaN matrix under In-polarity growth regime. Since the critical thickness of InN epitaxy on c-plane GaN is about one monolayer and also the growth temperature for one monolayer InN insertion can be remarkably highered, the proposed MQW structure can avoid/reduce new generation of misfit dislocation, resulting in high quality MQW-structure nature in principle.

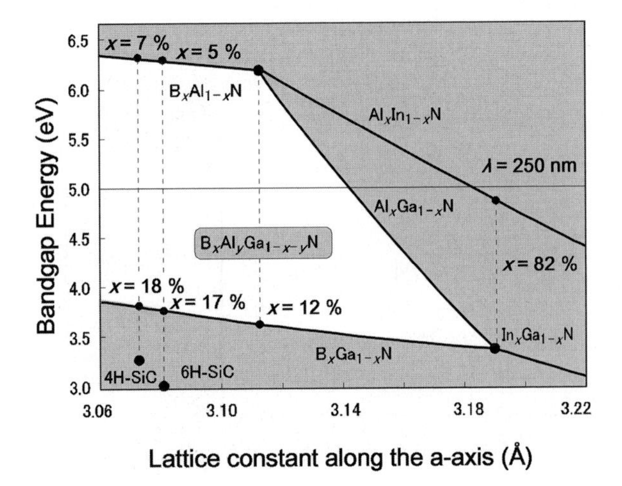

Fig. 6.99. Relationship between lattice constant along the a-axis and bandgap energy of GaN, AlN, and BAlGaN

6.2.8 Lattice-Matched Quaternary BAlGaN Alloys

Wide Range of Emission Wavelengths Using Quaternary BAlGaN Semiconductors

The first GaN/AlGaN UV LEDs grown on SiC substrate were fabricated using Low-Pressure MOVPE [244]. Then AlGaN quaternary lattice matched to 4H- or 6H-SiC was proposed as wide bandgap semiconductor covering 200–360 nm spectral region, as shown in Fig. 6.99 [245]. It is necessary to use 17–18% boron (B) to achieve lattice-matched BGaN structures on SiC. However, the boron composition would enable reduction to ∼ 12% as the quaternary was grown on AlN substrates or free standing AlN template.

Epitaxial Growth of Quaternary BAlGaN Alloys (H. Kawanishi)

The atomic radii of boron and nitrogen are small at 54.9 and 79.5 pm, respectively. Such small atomic radii make crystal growth difficult due to the miscibility gap of the crystals. However, thermodynamic analysis indicates that growth should be possible by precise control of the growth conditions and there has been a report on the low pressure MOCVD growth of BAlGaN multiquantum well structures, as shown in Fig. 6.100 [246]. The AlN/GaN multi buffer layer structure was found to be important for controle and realization of high-quality crystals [247]. The actual structure consists of an AlN/GaN multi buffer layer structure, a strain free AlN layer and finally the BAlGaN alloy. Figure 6.101 shows the room temperature emission spectrum from this sample where ultraviolet emission is demonstrated at 250 nm [248]. Figure 6.102 shows the relationship between the residual strain and emission intensity. The residual strain can be reduced by increasing the Al content, which results in increases of the emission intensity.

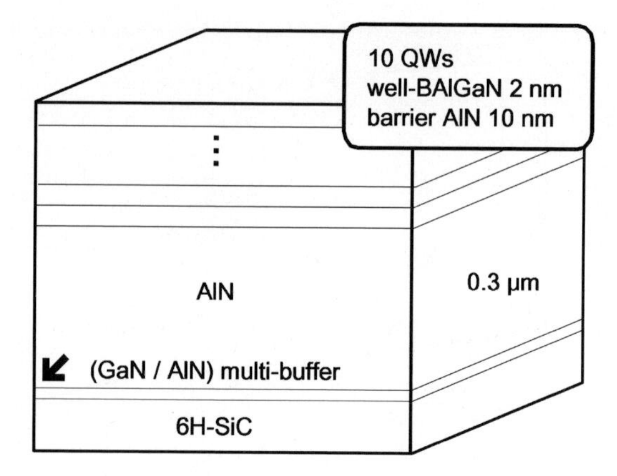

Fig. 6.100. BAlGaN MQWs structure

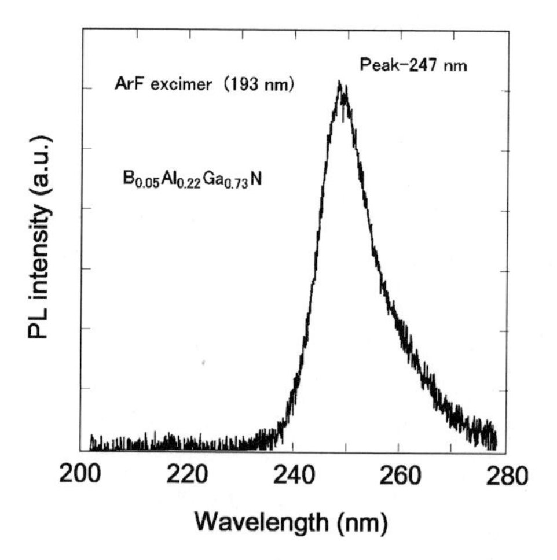

Fig. 6.101. PL spectrum from BAlGaN at room temperature

However, further improvement in the crystalline quality of the BAlGaN alloys is necessary for device applications. In particular, the purity of the metal organic source for boron must be improved.

Refractive Indices of BAlN and BGaN in the UV and Deep UV Range

The refractive indices were determined by absolute reflection measurements where the incident light is irradiated almost perpendicular to the sample surface and the resulting multiple reflection spectrum recorded [249]. This method necessitates samples having surface flatness of the order of a wavelength of the incident light and the resulting experimental data can be used to determine the absorption and band edges of the BAlGaN alloy.

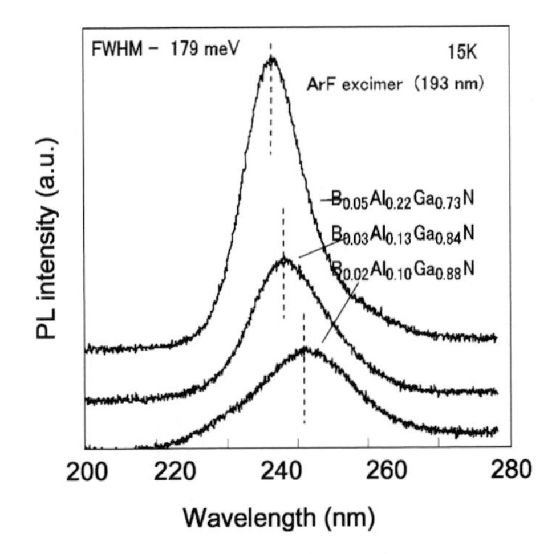

Fig. 6.102. Improvement of PL intensity by lowering residual strain in the case of BAlGaN quaternary materials

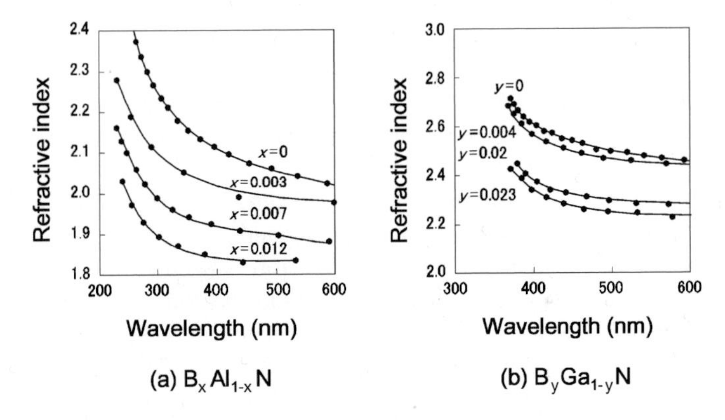

Fig. 6.103. Refractive index change in the region from ultraviolet to deep ultraviolet of BAlN and BGaN (measurement)

Figure 6.103 shows the refractive indices of BGaN and BAlN determined by the aforementioned method in the 200–600 nm wavelength range. It is noteworthy that the addition of a small quantity of boron results in a large variation of the refractive indices, especially for BAlN.

6.2.9 Epitaxial Growth of ZnSe and Related Semiconductors (K. Akimoto)

Growth of ZnSe Crystalline Thin Films

The growth of ZeSe can be summarized as being bulk crystal growth in the 1960s, the use of vapor phase epitaxy in the 1970s, and MOCVD and MBE in

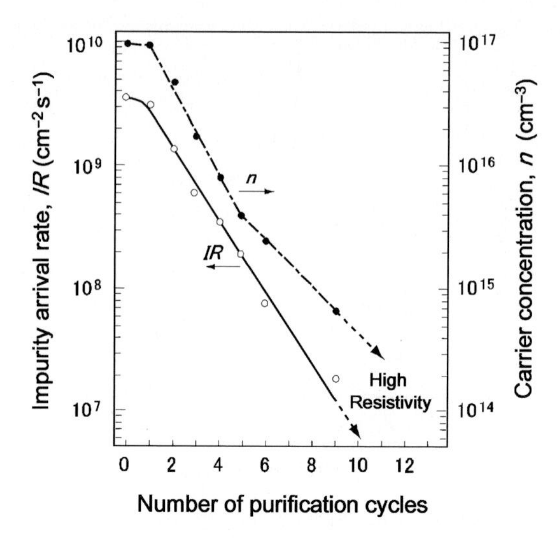

Fig. 6.104. Relation between the number of purification cycle of Se and carrier concentration [144]

the 1980s. In the early 1980s, it was possible to grow ZnSe crystals by MBE and MOCVD with background carrier concentrations of less than $10^{16}\,\mathrm{cm}^{-3}$ and mobilities of greater than $600\,\mathrm{cm}^2\,\mathrm{V}^{-1}\,\mathrm{s}^{-1}$ [251,252]. However, the properties of ZnSe films varied with the purity of the Se source [253]. Figure 6.104 shows the variation of the background carrier density with the number of purification cycles of Se. The carrier concentration is seen to decrease with the number of purification cycles. The origin of the background carriers are not donor-like defects due to self-compensation effects but residual impurities.

Figure 6.105 shows the CL spectra of undoped and Al-doped ZnSe [254]. The excited state of free excitons is observed, indicating the high-purity of the samples. In the figure, the Y-emission line is due to excitons bound to dislocations, which shows that ZnSe films grown on GaAs substrates contain many dislocations.

There have been many studies on optimization of ZnSe growth [255–258]. Figure 6.106 is a map of the stability of ZnSe surfaces with the ratio of Zn and Se pressures. The Zn- and Se-stabilized surfaces can be determined by analysis of RHEED patterns, with the stable configurations for Zn and Se being $c(2 \times 2)$ and (2×1), respectively. The boundary between the Zn- and Se-stabilized surfaces corresponds to the stoichiometric composition of that particular surface. The boundary line is denoted as SZT. Figures 6.107 and 6.108 show the variation of the PL FWHM and carrier mobility with II/VI ratio. It can be seen that high-quality crystals are obtained in a region that has moved from stoichiometry to a Zn-rich growth condition. The growth conditions have to be reoptimized for the growth of n- and p-type crystals.

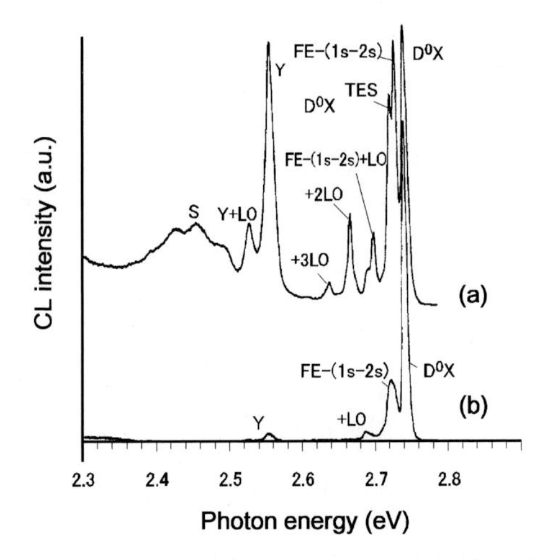

Fig. 6.105. An example of cathode-luminescence from undoped and Al-doped ZnSe

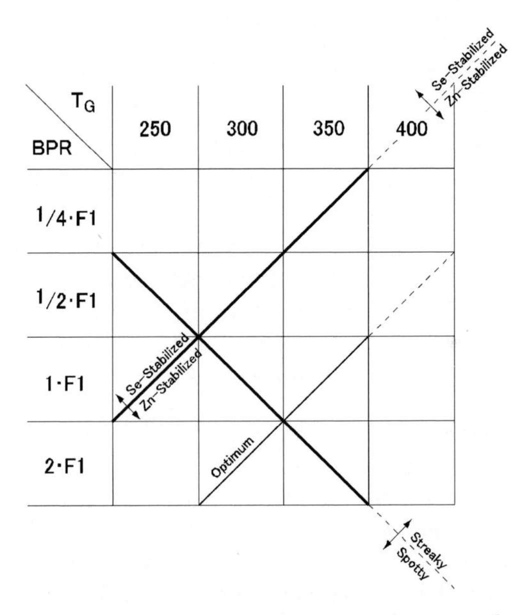

Fig. 6.106. Relationship between ZnSe growth conditions and stabilized surface

Growth of ZnSe on GaAs Substrates

There is a 0.2% lattice mismatch between ZnSe and GaAs substrates leading to strain in the ZnSe epilayer as initially clarified by Mohammed et al. [259]. Figure 6.109 shows the variation of strain in ZnSe with film thickness as determined from PL, TEM, and XRD measurements. These results show that the ZnSe film thickness of about 1 μm at least is needed to reduce the strain negligibly small.

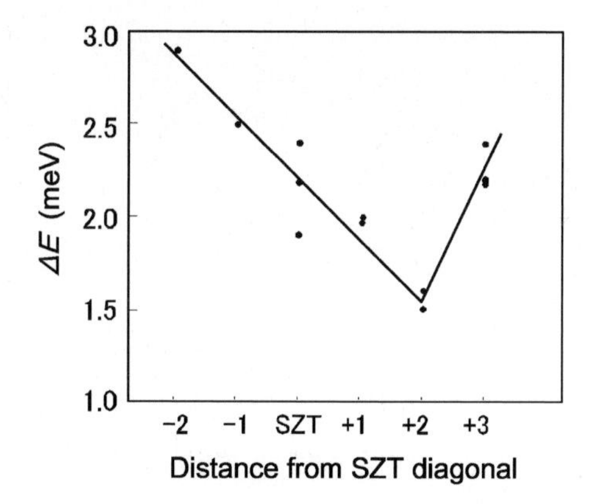

Fig. 6.107. Variation of the FWHM of exciton luminescence with II–VI ratio

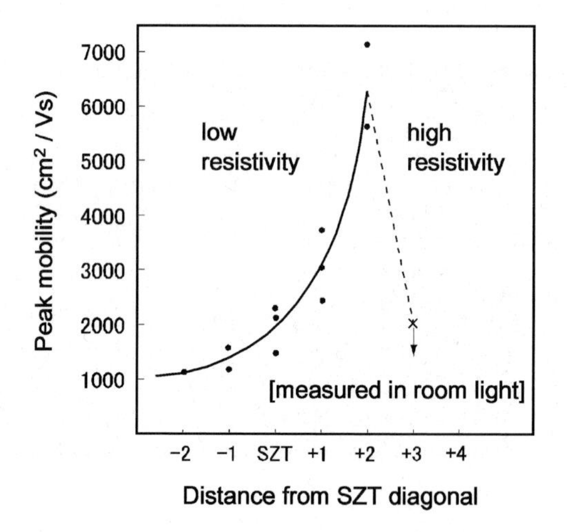

Fig. 6.108. Variation of the carrier mobility with II–VI ratio

Lattice mismatch is also known to affect surface morphology [260]. Figure 6.110 shows the surface of ZnSSe grown on a GaAs substrate as observed using a differential interference contrast microscope. The surface morphology improves for an S composition of 0.06, when the ZnSSe is lattice matched to the GaAs substrate.

Problems with the use of GaAs substrates for growth of ZnSe were predicted by Harrison et al. [261] due to the formation of interface states resulting from differences in the number of valence electrons, changes in the band discontinuity due to the existence of interface dipoles, and the accumulation of carriers. Fortunately, however, these three phenomena were not confirmed experimentally [262–264]. Figure 6.111 shows the C–V characteristics of an n-ZnSe/p-GaAs structure. The solid lines are experimental results and the broken lines are theoretical values calculated assuming no interface states and carrier accumulation. The good agreementbetween both results indicates

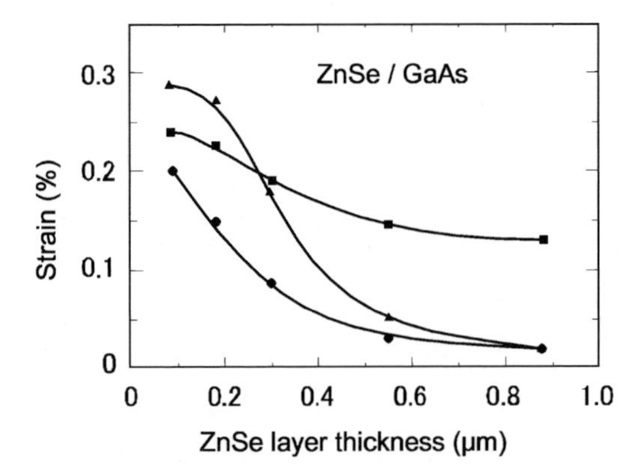

Fig. 6.109. Variation of percent strain as a function of ZnSe layer thickness as measured by PL (*closed circles*), TEM (*closed squares*), and XRD technique (*closed triangles*)

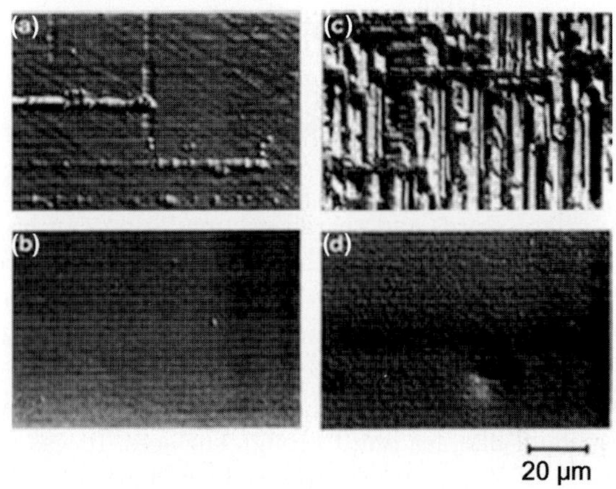

Fig. 6.110. Surface of ZnSSe grown on a GaAs substrate as observed using a differential interference contrast microscope

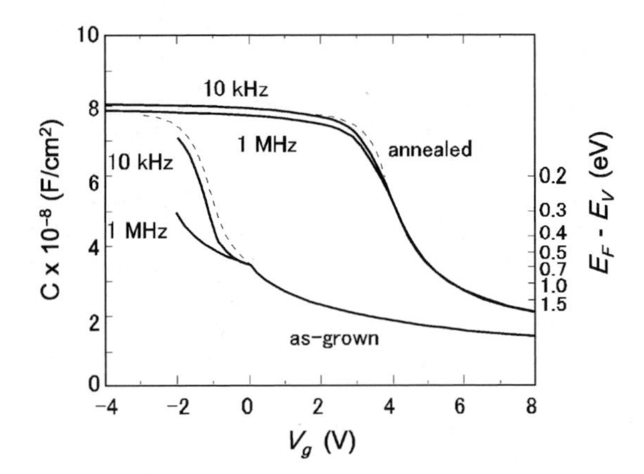

Fig. 6.111. C–V property of n-ZnSe/p-GaAs (*full line* and *dashed line* shows experiment and simulation, respectively)

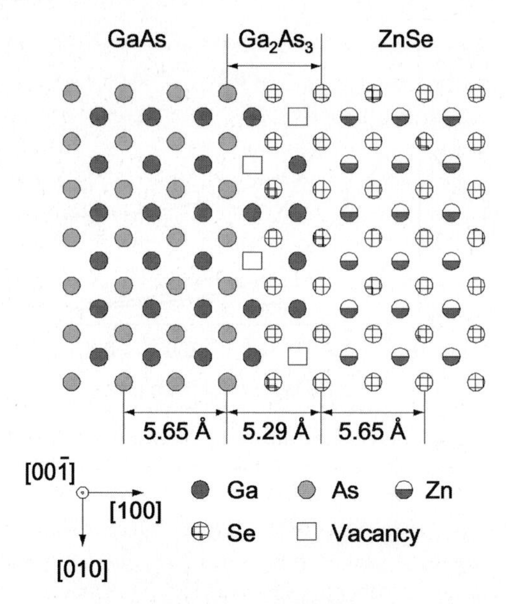

Fig. 6.112. Schematic diagrams of ZnSe/GaAs interface based on TEM images

effects due to interface states and carrier accumulation can be ignored. TEM measurements of the interface show the formation of two layers of Ga_2Se_3. The formation of the Ga_2Se_3 layers is thought to be the reason for absence of interface states and carrier accumulation [265,266]. Figure 6.112 is a schematic representation of the interface cross-section based on TEM images.

As shown in Fig. 6.112, if the Ga_2Se_3 becomes a perfect layer then ZnSe growth becomes island-like, leading to rough surfaces. It was found that two-dimensional growth of ZnSe can be achieved by changing the GaAs surface from Ga- to As-stabilized [267,268]. Interface states and carrier accumulation are still not found in growth on As-stabilized surfaces; the interface consists of As–Zn bonds. The dislocation density of ZnSe films grown on As-stabilized GaAs substrates is low [269], a further reason for use of As-stabilized substrates.

As stabilization is obtained by irradiation with a As molecular beam during thermal cleaning of the GaAs substrate and growth of GaAs under high As partial pressure. It is not possible to use As molecular beams in the ZnSe growth chambers – two chamber systems are required, making the growth system large.

Fundamental Concepts of Band Discontinuities

Chadi [270] and McCaldin [271] led early studies on the band discontinuities at semiconductor heterojunction. The so-called "common anion rule" is based on the top of the valence band edge of semiconductors being due to p-like atomic orbitals. The valence band discontinuity of GaSe and ZnSe has been calculated to be 0.2 eV by tight-binding analysis which generally agrees with experimental values and thus the validity of the common anion rule has been widely accepted. Harrison et al. calculated the state of the interface using the linear combination of atomic orbitals method (LCAO) and showed the importance of including ionicity in the common anion rule [272]. Further, Katnani et al. [273, 274] incorporated electronegativity into calculations and found good agreement with the results of Harrison's method and experimental data. However, although results obtained using Harrison's method can be well explained for only small mismatch between two semiconductors forming a heterostructure, cases of large mismatch lead to large differences between theory and experiment as found by Shih et al. using photoelectron spectroscopy [275,276]. In the 1980s, van de Walle and Cardona found good experimental agreement with calculations made using first-principle simulations and muffin-tin-orbital methods for discontinuities of strained systems [277, 278].

Currently, the Harrison method is widely used for calculating band discontinuities in these semiconductor systems. Figure 6.113 shows the experimentally and theoretically determined band discontinuities of the valence band edge of heterojunctions. Good agreement can be seen.

Quaternary Heterostructure Alloys

Figure 6.114 shows the lattice constant and bandgap energy of II–VI compound semiconductors incorporating Mg and Be [279]. A triangle can be formed by joining ZnS, ZnSe, and ZnTe where this triangle contains ZnS, ZnSe, and ZnTe ternary alloys showing that the possibility of controlling the material properties of these compounds within this region.

As described above, since the band discontinuity is based on the top of the valence band edge, it is necessary to determine the bandgap energy of the alloy

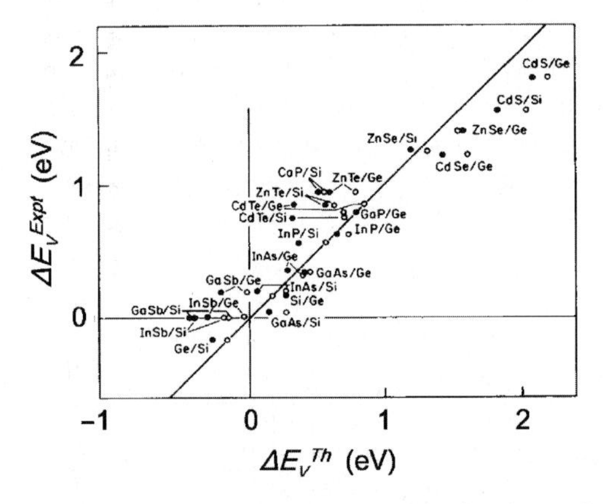

Fig. 6.113. Comparison between Harrison model of valence band discontinuity values and experimental values

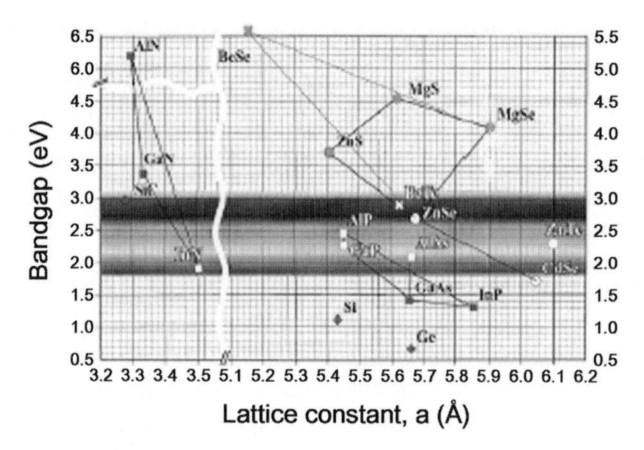

Fig. 6.114. Relation between lattice constant and bandgap energy

Table 6.5. Bowing parameter of group II–III compound semiconductor alloys (eV)

alloy	bowing parameter (eV)
ZnS–Se	0.6
ZnSe–Te	1.23
ZnS–Te	3.00
Zn–CdS	0.3
Zn–CdSe	0.51

in order to calculate the discontinuity of the conduction band. The bandgap energy, $E_g(x)$, as a function of the alloy composition x, can be approximated by the following relationship

$$E_g(x) = E_g(0) + [E_g(1) - E_g(0) - c]x + cx^2,$$

where c is known as the *bowing parameter* and its magnitude can be estimated using the method described by van Vechten et al. [280]. The bowing parameters of II–VI compound semiconductor alloys are shown in Tables 6.5 and 6.6 [281–288]. The band discontinuities of a selection of alloys are described in the following section.

$ZnSe/ZnS_xSe_{1-x}$ (Type I)

The discontinuities at the top of the valence band for $x = 0.12, 0.19, 0.25$ are 100, 121, 170 meV, respectively [289–291]. These values agree well with those obtained by interpolation of the discontinuity for ZnSe and ZnS. It has been shown that the discontinuity of ternary alloy can be estimated by interpolation in case that the lattice mismatch is small [292].

Table 6.6. Band discontinuities of group II–III compound semiconductor alloy heterojunctions

heterojunction	type	composition	ΔE_y	ΔE_g
$ZnSe/ZnS_xSe_{1-x}$		$x = 0.12$	100	~0
	I	$x = 0.19$	121	4
		$x = 0.25$	170	10
$ZnTe/ZnS_xTe_{1-x}$	II			
$ZnSe_xTe_{1-x}/ZnSe$	II			
$Zn_{1-x}Cd_xSe/ZnSe$	I	$x = 0.14$	32	153
		$x = 0.24$	60	180
$Zn_{1-x}Cd_xS/ZnS$	I	$x = 0.22$	107	240
$Zn_{1-x}Cd_xSe/ZnS_ySe_{1-y}$	I	$x = 0.02$, $y = 0.08$	81	305
$Mg_xZn_{1-x}Se/ZnS$	I	$x = 0.5$	600	400
$ZnS/Mg_xZn_{1-x}S$	II	$x = 0.5$	−100	400
$Cd_{1-x}Zn_xSe/MgSe$	I	$x = 0.46$	560	
$Cd_{1-x}Zn_xTe/MgTe$	I	$x = 0.12$	430	
$ZnSe/Zn_xMn_{1-x}Se$	I	$x = 0.23$	20	90
$Zn_{1-x}Cd_xSe/Zn_{1-y}Mn_ySe$	I	$x = 0.15$, $y = 0.16$	30	208
$Zn_{1-x}Cd_xS_ySe_{1-y}/ZnS_{0.84}Se_{0.16}$	I	$x = 0.15$, $y = 0.8$	103	160
$ZnSe/Zn_{1-x}Mg_xS_ySe_{1-y}$	I	$x = 0.15$, $y = 0.21$	$\Delta E_y = 0.6\Delta E_g$ $\Delta E_y = (0.35 - 0.45)\Delta E_g$	

$ZnTe/ZnS_xTe_{1-x}$ (Type II)

Accurate values of discontinuities for x-dependence are not available. The discontinuity of ZnTe and ZnS is 2.15 eV but the effects of mismatch are likely to affect the x-dependence [293].

$ZnSe/ZnSe_xTe_{1-x}$ (Thought to be Type II)

No reliable x-dependence results are available. The discontinuity of ZnSe/ZnTe is thought to be between 0.8 and 1.2 eV [294–297] due to lattice strain.

$Zn_{1-x}Cd_xSe/ZnSe$ (Type I)

The discontinuities of the valence and conduction bands at $x = 0.14$ are 32 and 153 meV, respectively, and for $x = 0.24$ they are 60 and 180 meV. The values for ZnSe and CdSe are 230 and 826 meV, respectively [298]. The interpolation is valid in this system.

$Cd_xZn_{1-x}S/ZnS$ (Type I)

For $x = 0.22$, the valence and conduction band discontinuities are 107 and 240 meV, respectively. The validity of the interpolation is not established since the discontinuity of CdS/ZnS has not been measured [299].

$Zn_{1-x}Cd_xSe/ZnS_ySe_{1-y}$ (Type I)

At x = 0.20 and y = 0.08, the valence and conduction band discontinuities are 81 and 305 meV, respectively. In this system, the addition of Cd to ZnSe reduces the conduction band, and the addition of S leads to a decrease of the valence band, thus both the conduction and valence bands are expected to have large discontinuities. The interpolation is not so effective since lattice mismatch is relatively large [300].

$Mg_xZn_{1-x}Se/ZnS$ (Type I)

In the 1990s, II–VI alloys incorporating Mg were studied for use as the cladding layers of lasers. At x = 0.5, the valence and conduction band discontinuities are 600 and 400 meV, respectively [301].

$ZnS/Mg_xZn_{1-x}S$ (Type II)

For x = 0.5, the valence and conduction band discontinuities are -100 and 400 meV, respectively [301].

$ZnSe/Zn_{1-x}Mn_xSe$ (Type I)

The II–VI compound semiconductors incorporating Mn exhibit ferromagnetic properties and are known as *dilute magnetic semiconductors*. ZnMnSe has a zinc blende structure for Mn < 0.3. The valence and conduction band discontinuities at x = 0.23 are 20 and 90 meV, respectively [302, 303].

$Zn_{1-x}Cd_xSe/Zn_{1-y}Mn_ySe$ (Type I)

For x = 0.15 and y = 0.16, the valence and conduction band discontinuities are 30 and 208 meV, respectively [304]. It has been shown that the common anion rule is valid even in the mixed crystal including transition metals.

Heterojunctions with Quaternaries

The possible quaternary alloys of ZnSe, ZnS, MgSe, and MgS are shown inside the four-sided shape in Fig. 6.114. It is possible to grow lattice-matched structures to GaAs substrates having bandgaps between 2.8 and 4.0 eV.

$ZnSe/Zn_xMg_{1-x}S_ySe_{1-y}$ (Type I)

Figure 6.115 shows changes in the bandgap and lattice constant by varying the composition of ZnMgSSe. For x = 0.21 and y = 0.15, a valence band discontinuity of $0.6{\cdot}\Delta E_g$ has been reported [305].

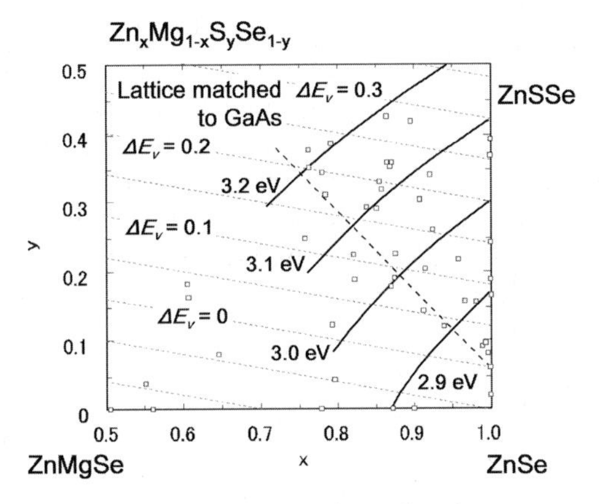

Fig. 6.115. Relations of ZnMgSSe composition, bandgap energy, lattice constant, and valence band discontinuity values with ZnSe

$ZnS_{0.84}Se_{0.16}/Zn_{1-x}Cd_xS_ySe_{1-y}$ (Type I)

$ZnS_{0.84}Se_{0.16}$ can be grown lattice matched to GaAs substrates. For $x = 0.5$ and $y = 0.8$, the valence and conduction band discontinuities are 103 and 160 meV, respectively [306]. Figure 6.116 shows the changes in the bandgap and lattice constant by varying the composition [306].

6.2.10 ZnO and Related Semiconductors (T. Yao, T. Hanada)

The exciton binding energy of ZnO is as large as 60 meV, which makes this semiconductor promising for optical devices [307]. In this section, the growth of ZnO and related semiconductors including MgZnO and CdZnO are discussed.

Growth of ZnO by Molecular Beam Epitaxy

ZnO has a wurtzite (WZ) structure and the (0001) surface (the so-called "c-plane") is widely used. Regarding substrates, the hexagonal sapphire c-surface is readily available, but the c-plane of ZnO ($a = 0.325$ nm) and sapphire ($a/\sqrt{3} = 0.274$ nm) have an 18% lattice mismatch even if the $[2\bar{1}\bar{1}0]$ axis (a-axis) is shifted by 30°. Due to this large lattice mismatch, ZnO is also grown on c-GaN, which enables better lattice matching by first growing c-GaN layers on c-sapphire substrates followed by growth of ZnO on the c-GaN substrates. MgO has a rock salt (RS) crystal structure (cubic) with a triangular lattice structure in the (111) plane that is the same as the hexagonal c-plane. Further, since the lattice constant in the (111) plane ($a/\sqrt{2} = 0.298$ nm) is intermediate between that of the c-surface of ZnO and sapphire, it can be used a buffer layer between the two layers. The most notable effect of using MgO as a buffer layer on sapphire substrates is that single phase ZnO films

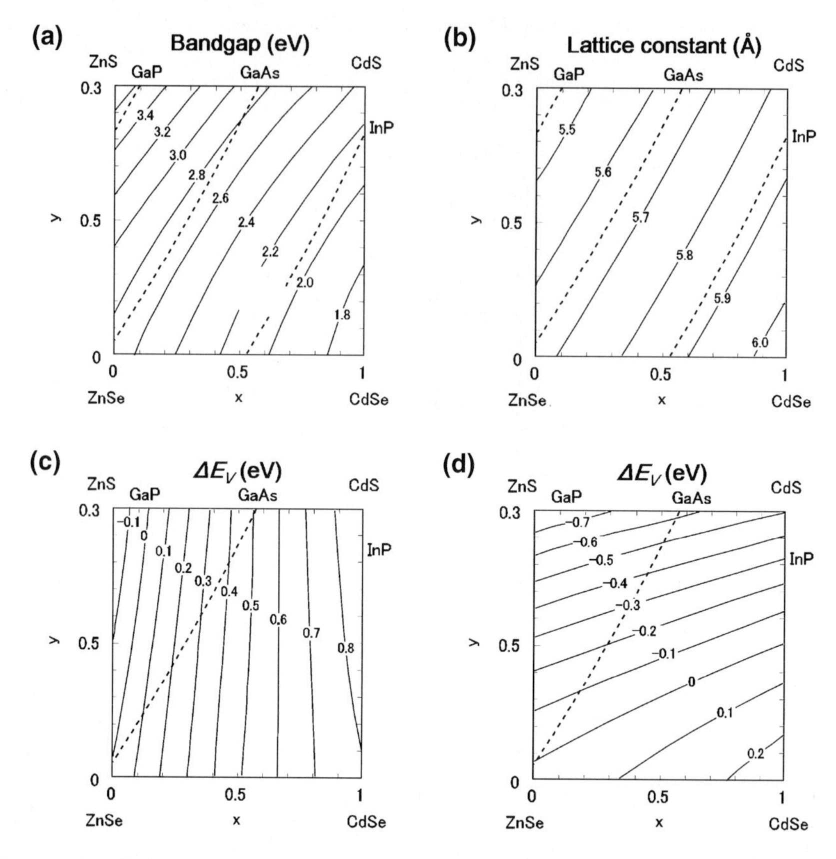

Fig. 6.116. Relations of ZnCdSSe composition, bandgap energy, lattice constant, and band discontinuity values with $ZnS_{0.84}Se_{0.16}$

(with their a-axis shifted by 30°) can be grown without the formation of ZnO domains along the a-axis of sapphire substrates [308].

The growth of ZnO thin films by plasma-assisted molecular beam epitaxy (p-MBE) involves (1) cleaning c-sapphire substrates under an oxygen plasma and high temperature; (2) deposition of MgO buffer layer; and (3) high temperature annealing to produce flat surfaces. The epitaxial relationships between ZnO and MgO films and sapphire substrates are $ZnO(0001)//MgO(111)//Al_2O_3(0001)$, $ZnO[2\bar{1}\bar{1}0]//MgO[1\bar{1}0]//Al_2O_3[1\bar{1}00]$, $ZnO[1\bar{1}00]//MgO[1\bar{2}1]//Al_2O_3[1\bar{2}10]$. ZnO thin films grown on these substrates exhibit (3×3) reconstruction and RHEED oscillations, as shown in Fig. 6.117, indicating two-dimensional growth [308]. The RHEED oscillation amplitude decreases with increasing substrate temperature and oscillations are not observed above 700°C, which indicates a change from two-dimensional island nucleation mode to step-flow mode growth. The crystalline quality of the thin films improves with the transition to step-flow

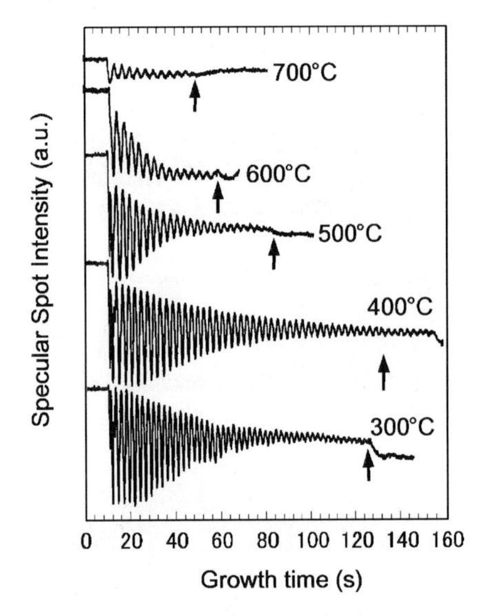

Fig. 6.117. Growth temperature dependence of RHEED specular intensity in initial growth of ZnO

mode, especially with a decrease in the density of screw dislocations. Figure 6.118 shows TEM images of ZnO thin films grown on MgO (buffer)/ c-sapphire substrates. Threading dislocations are seen parallel to the c-axis [309]. Figure 6.118a, c shows screw dislocations (Burgers vector b = $\langle 0001 \rangle$, diffraction point g = (0006)) and Fig. 6.118b, d is edge dislocations (b = 1/3 $\langle 2\bar{1}\bar{1}0 \rangle$, g = $(03\bar{3}0)$). The MgO buffer layer was deposited at 490°C and annealed at 750°C for 5 min (Fig. 6.118a, b) or annealed at 800°C for 25 min (Fig. 6.118c, d). ZnO films were grown at 700°C. The flat MgO buffer layer resulting from high temperature annealing enable a reduction of the density of screw dislocations as well as a reduction of the total dislocation density to $1.9 \times 10^9\,\mathrm{cm}^{-2}$.

Control of the Polarity of ZnO Surfaces

Since the wurtzite structure does not have inversion symmetry and growth on {0001} surfaces occurs (as is also the case for growth on {111} zinc blende surfaces) in units of two-atomic layers, in order to reduce the density of bonds that can be broken with surface termination being on the one hand cationic (0001) and on the other anionic $(000\bar{1})$. In heterostructure growth, these polarities are retained near the junctions.

However, it has been found that it is possible to grow oxygen-terminated (O-polarity) ZnO films on Ga-terminated (Ga-polarity) GaN(0001) surfaces by incorporation of an interface layer with a inversion symmetry structure.

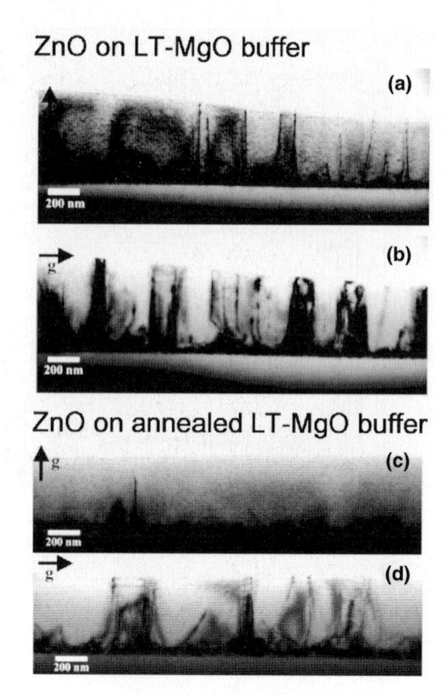

Fig. 6.118. Cross-section TEM images of thin ZnO film on MgO buffer (490°C)/ c-plane sapphire

That is, polarity inverted growth is possible. From the analysis of TEM images it has been found that prior to the growth of ZnO by MBE, if only a Zn beam is irradiated onto a Ga-polarity surface, the interface exhibits a WZ structure, but in the case of only an oxygen plasma beam being irradiated, a Ga_2O_3 interface layer is formed which has a monoclinic crystal structure and inversion symmetry. Convergent-beam electron diffraction (CBED) measurements show that ZnO has Zn-polarity without a Ga_2O_3 interface (Fig. 6.119a) and O-polarity when a Ga_2O_3 is formed (Fig. 6.119b) as compared with simulations for cation-polarity (Fig. 6.119c) [310]. The polarity of the ZnO surface affects and is used to control the reactivity and robustness of the surface; the properties of the underlying heterostructures; and the piezoelectric properties of the thin films. For example, for improving nitrogen doping efficiency, ZnO surfaces with Zn-polarity have been reported to readily incorporate nitrogen [311].

The polarity of ZnO has also been inverted by incorporation of MgO(111) layers (RS structure) between two ZnO layers [312]. Further, the polarity of ZnO layers can be controlled by changing the thickness of MgO buffer layers grown on oxygen-terminated c-sapphire substrates [313]. MgO buffer layers with thicknesses of less than 2.7 nm have a wurtzite c-surface and ZnO layers grown onto such MgO have O-polarity. In contrast MgO buffer layers thicker than 3 nm have (111) rock salt structures and the growth of ZnO results in Zn-polarity.

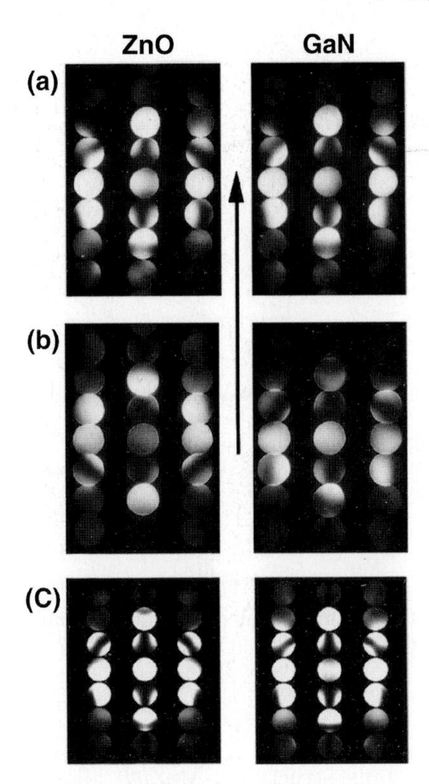

Fig. 6.119. CBED patterns from ZnO and GaN layer of ZnO/GaN/sapphire structure (ZnO-polarity in (**a**) and (**b**) is inverting each other)

MBE Growth of MgZnO and CdZnO Heterostructures

ZnO alloy thin films are produced incorporating the group II elements Mg and Cd. The presence of oxygen in the crystal matrix and the high ionicity of these alloys results in WZ structures for alloys with large Zn composition but MgO and CdO alloys lead to RS structures. The lattice constant and bandgap (E_g) of MgO are $a = 0.421$ nm and 7.8 eV, respectively. For CdO, $a = 0.470$ nm and $E_g = 0.8$ eV (indirect) and 2.3 eV (direct).

Theoretical studies have shown that WZ-MgO (four coordinates; $a = 0.317$ nm; $c = 0.517$ nm) is unstable and RS-MgO (six coordinates; $a = 0.414$ nm) has been predicted to be most stable [314]. Simulations of the structure of bulk CdO indicate that the RS structure ($a = 0.477$ nm) to be most stable but the WZ structure has also been predicted to form easily [315]. The high pressure synthesis of RS-ZnO ($a = 0.428$ nm) has also been reported [316].

Figure 6.120 shows the in-plane lattice constants and bandgaps of $Mg_xZn_{1-x}O$ [317–319] and $Cd_yZn_{1-y}O$ [320] thin films and theoretically

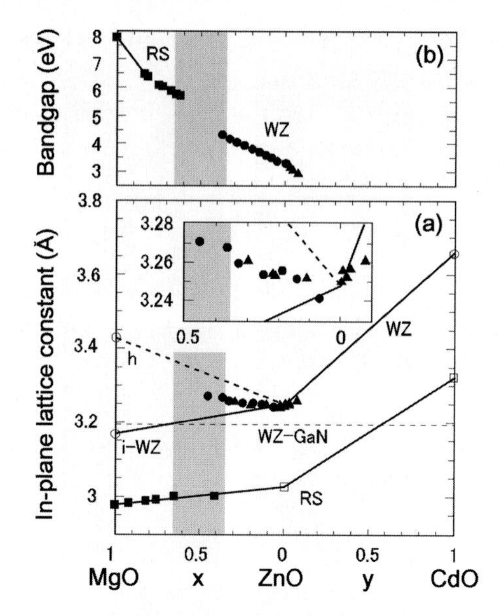

Fig. 6.120. Compositional dependence of bandgap energy and in-plane lattice constant of $Mg_xZn_{1-x}O$ and $Cd_yZn_{1-y}O$ in the case of wurtzite c-plane and rock salt (111) plane

determined values for bulk structures. MgZnO alloys (x maximum of 0.35) grown on c-sapphire substrates by p-MBE and pulsed laser deposition (PLD) are single phase, WZ structures. The a-axis lattice constant increases slightly with solid solubility of MgO and the c-axis lattice constant decreases [317]. Increases of the Mg composition are accompanied by changes in the WZ structure to form layered hexagonal crystals; the internal parameter u is thought to increase. For WZ structures, u is defined as the ratio of the bond length parallel to the c-axis and the lattice constant, c, and ideally $u = 3/8$ for bonds in a regular tetrahedron structure. In the vicinity of WZ structures, the room temperature bandgap increases from the 3.37 eV of ZnO to a maximum of 4.3 eV. For x greater than 0.65, the RS structure is formed and the bandgap changes from 5.7 eV to that of MgO, 7.8 eV. In the intermediate region, shown by the gray region in Fig. 6.120, both structures exist together and the bandgap cannot be determined. In this region, precipitation of grains of MgO and (100) orientation RS-MgZnO have been reported [318]. In growth on RS-MgO(111) buffer layers deposited on c-sapphire substrates, WZ-MgZnO(0001) layers grow after the formation of RS-MgZnO(111) layers. When the two layers form RS(110) and WZ($\bar{1}012$) orientated interfaces then c-axis of the WZ phase is tilted by 10° from the c-axis of the sapphire substrate [319].

For $Cd_yZn_{1-y}O$ alloys with $y = 0.07$ grown on $ScAlMgO_4$ substrates (good lattice matching with c-plane ZnO), the bandgap decreases to 3.0 eV with increasing solubility of Cd and the lattice constants of both the a- and c-axes increase [320]. These results show that in CdZnO alloys, changes in the WZ structure are small. However, due to the large differences in the ionic radii of Cd and Zn, there have only been reports on the epitaxial growth of CdZnO alloys for small Cd compositions.

References

1. Y.M. Tairov and V.F. Tsvetkov: J. Cryst. Growth 43 (1978) 209
2. M. Pons, R. Madar, and T. Billon: in Silicon Carbide: Recent Major Advances (H. Matsunami, G. Pensl, and M.J. Choyke, eds.), Springer, Berlin Heidelberg New York, 2003, pp. 121–136
3. H. Matsunami (ed.): Technology of Semiconductor SiC and Its Application, The Nikkan Kogyo Shimbun, Tokyo, 2003 (in Japanese)
4. K. Onoue, T. Nishikawa, M. Katsuno, N. Ohtani, H. Yashiro, and M. Kanaya: Jpn J. Appl. Phys. Part 1 35 (1996) 2240
5. H.M. Hobgood, R.C. Glass, G. Augustine, R.H. Hopkins, J. Jenny, M. Skowronski, W.C. Mitchel, and M. Roth: Appl. Phys. Lett. 66 (1995) 1364
6. S.G. Müller, M.F. Brady, W.H. Brixius, G. Fechko, R.C. Glass, D. Henshall, H.M. Hobgood, J.R. Jenny, R.T. Leonard, D.P. Malta, A.R. Powell, V.F. Tsvetkov, S.T. Allen, J.W. Palmour, and C.H. Carter: Silicon Carbide and Related Materials. Materials Science Forum, Vol. 389–393, Trans Tech Publications, Zurich–Uetikon, 2002, pp. 23–28
7. N. Ohtani, K. Katsuno, T. Fujimoto, and H. Yashiro: in Silicon Carbide: Recent Major Advances (M.J. Choyke, H. Matsunami, and G. Pensl, eds.), Springer, Berlin Heidelberg New York, 2003, pp. 137–162
8. F. Frank: Acta Crystallogr. 4 (1951) 497
9. S. Ha, M. Skowronski, W.M. Vetter, and M. Dudley: J. Appl. Phys. 92 (2002) 778
10. E.K. Sanchez, J.Q. Liu, M. De Graef, M. Skowronski, W.M. Vetter, and M. Dudley: J. Appl. Phys. 91 (2002) 1143
11. J. Takahashi and N. Ohtani: Phys. Status Solidi B 202 (1997) 163
12. J. Takahashi, N. Ohtani, and M. Kanaya: J. Cryst. Growth 167 (1996) 596
13. N. Ohtani, M. Katsuno, and T. Fujimoto: Jpn J. Appl. Phys. Part 2 42 (2003) L277
14. D. Nakamura, I. Gunjishima, S. Yamaguchi, T. Ito, A. Okamoto, H. Kondo, S. Onda, and K. Takatori: Nature 430 (2004) 1009
15. R.H. Wentorf, Jr.: J. Phys. Chem. 75 (1971) 1833
16. H.M. Strong and R.M. Chrenko: J. Phys. Chem. 75 (1971) 1838
17. H. Sumiya, N. Toda, and S. Satoh: SEI Tech. Rev. 60 (2005) 10
18. A. Hara: Seimitsu Kikai 51 (1985) 1497 (in Japanese)
19. H. Sumiya, S. Satoh, K. Tsuji, and S. Yazu: 31st High Pressure Conference of Japan, Osaka, 19–21 November 1990, Programme and Abstracts, pp. 48–49 (in Japanese)
20. H. Sumiya and S. Satoh: Diam. Relat. Mater. 5 (1996) 1359
21. H. Sumiya, S. Satoh, and S. Yazu: Proceedings of the Joint AIRAPT-16 & HPCJ-38 International Conference on High Pressure Science and Technology. The Proceedings of the High Pressure Conference of Japan, Vol. 7, The Japan Society of High Pressure Science and Technology Kyoto, 1997, p. 960

22. H. Sumiya, N. Toda, Y. Nishibayashi, and S. Satoh: J. Cryst. Growth 178 (1997) 485
23. H. Yamaoka, K. Otomo, D. Hirata, and T. Ishikawa: SR Kagaku Gijutsu Joho 5 (1995) 6 (in Japanese)
24. H. Sumiya, N. Toda, and S. Satoh: J. Cryst. Growth 237–239 (2002) 1281
25. R.C. Burns, J.O. Hansen, R.A. Spits, M. Sibanda, C.M. Welbourn, and D.L. Welch: Diam. Relat. Mater. 8 (1999) 1433
26. M. Yamamoto, T. Kumasaka, and T. Ishikawa: Rev. High Pressure Sci. Technol. 10 (2000) 56 (in Japanese)
27. J. Karpinski, J. Jun, and S. Porowski: J. Cryst. Growth 66 (1984) 1
28. S. Porowski and I. Grzegory: J. Cryst. Growth 178 (1997) 174
29. H.P. Maruska and J.J. Tietjen: Appl. Phys. Lett. 15 (1969) 327
30. M.K. Kelly, R.P. Vaudo, V.M. Phanse, L. Görgens, O. Ambacher, and M. Stutzmann: Jpn J. Appl. Phys. Part 2 38 (1999) L217
31. Y. Oshima, T. Eri, M. Shibata, H. Sunakawa, K. Kobayashi, T. Ichihashi, and A. Usui: Jpn J. Appl. Phys. Part 2 42 (2003) L1
32. Nikkei Sangyo Shimbun, February 18 (2000) 1 (in Japanese)
33. K. Motoki, T. Okahisa, N. Matsumoto, M. Matsushima, H. Kimura, H. Kasai, K. Takemoto, K. Uematsu, T. Hirano, M. Nakayama, S. Nakahata, M. Ueno, D. Hara, Y. Kumagai, A. Koukitu, and H. Seki: Jpn J. Appl. Phys. Part 2 40 (2001) L140
34. F. Hasegawa, M. Minami, K. Sunaba, and T. Suemasu: Jpn J. Appl. Phys. Part 2 38 (1999) L700
35. Y. Kumagai, H. Murakami, Y. Kangawa, and A. Koukitu: Jpn J. Appl. Phys. Part 2 42 (2003) L526
36. W. Utsumi, H. Saitoh, H. Kaneko, T. Watanuki, K. Aoki, and O. Shimomura: Nat. Mater. 2 (2003) 735
37. T. Inoue, Y. Seki, O. Oda, S. Kurai, Y. Yamada, and T. Taguchi: The 1st Asian Conference on Crystal Growth and Crystal Technology (CGCT-1), Sendai, August 29–September 1, 2000, T-B-12, p. 341
38. H. Yamane, M. Aoki, and S. Sarayama: Oyo Butsuri 71 (2002) 548 (in Japanese)
39. H. Yamane, M. Shimada, S.J. Clarke, and F.J. DiSalvo: Chem. Mater. 9 (1997) 413
40. S. Sarayama and H. Iwata: Ricoh Tech. Rep. 30 (2004) 9 (in Japanese)
41. D.R. Ketchum and J.W. Kolis: J. Cryst. Growth 222 (2001) 431
42. T. Hashimoto, K. Fujito, F. Wu, B.A. Haskell, P.T. Fini, J.S. Speck, and S. Nakamura: Jpn J. Appl. Phys. Part 2 44 (2005) L797
43. R.H. Wentorf, Jr., R.C. DeVries, and F.P. Bundy: Science 208 (1980) 873
44. R.C. DeVries: GE Technical Report, 72CRD178 (General Electric Co.), USA, 1972
45. O. Mishima, J. Tanaka, S. Yamaoka, and O. Fukunaga: Science 238 (1987) 181
46. O. Mishima, K. Era, J. Tanaka, and S. Yamaoka: Appl. Phys. Lett. 53 (1988) 962
47. T. Taniguchi, K. Watanabe, S. Koizumi, I. Sakaguchi, T. Sekiguchi, and S. Yamaoka: Appl. Phys. Lett. 81 (2002) 4145
48. O. Mishima, S. Yamaoka, and O. Fukunaga: J. Appl. Phys. 61 (1987) 2822
49. M. Kagamida, H. Kanda, M. Akaishi, A. Nukui, T. Osawa, and S. Yamaoka: J. Cryst. Growth 94 (1989) 261

50. T. Taniguchi and S. Yamaoka: J. Cryst. Growth 222 (2001) 549
51. R.C. DeVries and J.F. Fleischer: J. Cryst. Growth 13–14 (1972) 88
52. O. Fukunaga, S. Nakano, and T. Taniguchi: Diam. Relat. Mater. 13 (2004) 1709
53. T. Taniguchi, T. Teraji, S. Koizumi, K. Watanabe, and S. Yamaoka: Jpn J. Appl. Phys. Part 2 41 (2002) L109
54. T. Taniguchi, K. Watanabe, and S. Koizumi: Phys. Status Solidi A: Appl. Res. 201 (2004) 2573
55. K. Watanabe, T. Taniguchi, and H. Kanda: Phys. Status Solidi A: Appl. Res. 201 (2004) 2561
56. T. Taniguchi and S. Yamaoka: New Diam. Front. Carbon Technol. 10 (2000) 291
57. S. Matsumoto and W.J. Zhang: Jpn J. Appl. Phys. Part 2 39 (2000) L442
58. K. Watanabe, T. Taniguchi, and H. Kanda: Nat. Mater. 3 (2004) 404
59. S. Guerini and P. Piquini: Phys. Rev. B 71 (2005) art. no. 193305
60. B. Arnaud, S. Lebegue, P. Rabiller, and M. Alouani: Phys. Rev. Lett. 96 (2006) art. no. 026402
61. N. Ooi, A. Rairkar, L. Lindsley, and J.B. Adams: J. Phys. Condens. Matter 18 (2005) 97
62. H. Hartmann, R. Mach, and B. Selle: in Current Topics in Material Science (E. Kaldis, ed.), Vol. 9, North-Holland, Amsterdam, 1982, pp. 252–273
63. H. Hartmann et al.: in Recent Development of Bulk Crystal Growth (M. Isshiki, ed.), Research Signpost, Trivandrum, India, 1998, pp. 165–190
64. S. Fujiwara, T. Kotani, K. Matsumoto, and T. Shirakawa: J. Cryst. Growth 169 (1996) 660
65. E.V. Markov and A.A. Davydov: Izv. Akad. Nauk SSSR Neorg. Mater. 7 (1971) 575
66. Y. Namikawa, S. Fujiwara, and T. Kotani: J. Cryst. Growth 229 (2001) 92
67. S. Fujiwara: Nihon Kessho Seicho Gakkaishi 32 (2005) 24 (in Japanese)
68. J.W. Nielsen and E.F. Dearborn: J. Phys. Chem. 64 (1960) 1762
69. K. Oka, H. Shibata, and S. Kashiwaya: J. Cryst. Growth 237 (2002) 509
70. N. Ohashi, T. Sekiguchi, K. Aoyama, T. Ohgaki, Y. Terada, I. Sakaguchi, T. Tsurumi, and H. Haneda: J. Appl. Phys. 91 (2002) 3658
71. J. Nause and B. Nemeth: Semicond. Sci. Technol. 20 (2005) S45
72. J. Nause, S. Ganesan, B. Nemeth, A. Walencia, U. Ozgur, Y. Terada, I. Sakaguchi, T. Tsurumi, and H. Haneda, The 3rd International Workshop on ZnO and Related Materials, Sendai, Japan, 5–8 October 2004, p. 28
73. X. Gu, S. Sabuktagin, A. Teke, D. Johnstone, H. Morkoc, B. Nemeth, and J. Nause: J. Mater. Sci.: Mater. Electron. 15 (2004) 373
74. K.F. Nielsen: J. Cryst. Growth 3–4 (1968) 141
75. Y.S. Park and D.C. Reynolds: J. Appl. Phys. 38 (1967) 756
76. M. Shiloh and J. Gutman: J. Cryst. Growth 11 (1971) 105
77. K. Matsumoto and K. Noda: J. Cryst. Growth 102 (1990) 137
78. J.M. Ntep, S.S. Hassani, A. Lusson, A. Tromson-Carli, D. Ballutaud, G. Didier, and R. Triboulet: J. Cryst. Growth 207 (1999) 30
79. D.C. Look, D.C. Reynolds, J.R. Sizelove, R.L. Jones, C.W. Litton, G. Cantwell, and W.C. Harsch: Solid State Commun. 105 (1998) 399
80. D.B. Eason, G. Cantwell, D.C. Look, D.C. Reynolds, C.W. Litton, and R.L. Jones: International Conference on Solid State Devices and Materials, Nagoya, 17–19 September 2002, G-6-1, # 5010, p. 360

81. D.B. Eason and G. Cantwell: Compound Semicond. 8 (2002) 15
82. D.W. Hamby, D.A. Lucca, M.J. Klopfstein, and G. Cantwell: J. Appl. Phys. 93 (2003) 3214
83. K. Thonke, T. Gruber, N. Teofilov, R. Schonfelder, A. Waag, and R. Sauer: Physica B: Condens. Matter 308 (2001) 945
84. G. Cantwell, H.P. Xin, H.B. Yuan, J.J. Song, C.W. Litton, Y.K. Yeo, and T. Steiner: The 3rd International Workshop on ZnO and Related Materials, Sendai, Japan, 5–8 October 2004, p. 29
85. R.A. Laudise and A.A. Ballman: J. Phys. Chem. 64 (1960) 688
86. R.A. Laudise, E.D. Kolb, and A.J. Caporaso: J. Am. Ceramic Soc. 47 (1964) 9
87. E.D. Kolb and R.A. Laudise: J. Am. Ceramic Soc. 49 (1966) 302
88. E.D. Kolb and R.A. Laudise: J. Am. Ceramic Soc. 48 (1965) 342
89. N. Sakagami and K. Shibayama: Jpn J. Appl. Phys. 20(Suppl.) (1981) 201
90. T. Sekiguchi, S. Miyashita, K. Obara, T. Shishido, and N. Sakagami: J. Cryst. Growth 214 (2000) 72
91. N. Sakagami, M. Yamashita, T. Sekiguchi, S. Miyashita, K. Obara, and T. Shishido: J. Cryst. Growth 229 (2001) 98
92. W.J. Li, E.W. Shi, W.Z. Zhong, and Z.W. Yin: J. Cryst. Growth 203 (1999) 186
93. M. Suscavage, M. Harris, D. Bliss, P. Yip, S.Q. Wang, D. Schwall, L. Bouthillette, J. Bailey, M. Callahan, D.C. Look, D.C. Reynolds, R.L. Jones, and C.W. Litton: MRS Internet J. Nitride Semicond. Res. 4S1 (1999) art. no. G3.40
94. N. Sakagami and M. Wada: Yogyo Kyokai Shi 82(8) (1974) 405 (in Japanese)
95. K. Maeda, M. Sato, I. Niikura, and T. Fukuda: Semicond. Sci. Technol. 20 (2005) S49
96. E. Ohshima, H. Ogino, I. Niikura, K. Maeda, M. Sato, M. Ito, and T. Fukuda: J. Cryst. Growth 260 (2004) 166
97. F. Masuoka, K. Ooba, H. Sasaki, H. Endo, S. Chiba, K. Maeda, H. Yoneyama, I. Niikura, and Y. Kashiwaba: Phys. Status Solidi C 3 (2006) 1238
98. H. Sasaki, H. Kato, F. Izumida, H. Endo, K. Maeda, M. Ikeda, I. Niikura, and Y. Kashiwaba: Phys. Status Solidi c(3), 4(2006)1034
99. A. Suzuki, M. Ikeda, H. Matsunami, and T. Tanaka: J. Electrochem. Soc. 122 (1975) 1741
100. K. Arai and S. Yoshida: Basics and Applications of SiC Devices, Ohmsha, Tokyo, 2003 (in Japanese)
101. H. Matsunami (ed.): Technology of Semiconductor SiC and Its Application, The Nikkan Kogyo Shimbun, Tokyo, 2003 (in Japanese)
102. R.S. Kern, K. Jarrendahl, S. Tanaka, and R.F. Davis: Phys. Status Solidi B: Basic Res. 202 (1997) 379
103. H. Matsunami and T. Kimoto: Mater. Sci. Eng.: Rep. R20 (1997) 125
104. N. Kuroda, K. Shibahara, W. Yoo, S. Nishino, and H. Matsunami: The 19th Conference on Solid State Devices and Materials, Tokyo, 25–27 August 1987, Extended Abstracts, p. 227
105. T. Kimoto and H. Matsunami: J. Appl. Phys. 75 (1994) 850
106. D.J. Larkin, P.G. Neudeck, J.A. Powell, and L.G. Matus: Appl. Phys. Lett. 65 (1994) 1659
107. T. Kimoto, S. Nakazawa, K. Hashimoto, and H. Matsunami: Appl. Phys. Lett. 79 (2001) 2761
108. T. Dalibor, G. Pensl, H. Matsunami, T. Kimoto, W.J. Choyke, A. Schoer, and N. Nordell: Phys. Status Solidi A: Appl. Res. 162 (1997) 199

109. K. Danno, T. Kimoto, and H. Matsunami: Appl. Phys. Lett. 86 (2005) 122104
110. P.G. Neudeck and J.A. Powell: IEEE Electr. Device Lett. 15 (1994) 63
111. I. Kamata, H. Tsuchida, T. Jikimoto, and K. Izumi: Jpn J. Appl. Phys. Part 1 39 (2000) 6496
112. I. Kamata, H. Tsuchida, T. Jikimoto, and K. Izumi: Jpn J. Appl. Phys. Part 2 40 (2001) L1012
113. S. Ha, M. Skowronski, and H. Lendenmann: J. Appl. Phys. 96 (2004) 393
114. J. Senzaki, K. Kojima, T. Kato, A. Shimozato, and K. Fukuda: Silicon Carbide and Related Materials 2004. Materials Science Forum Proceedings, Vol. 483–485, Trans Tech Publications, Zurich–Uetikon, 2005, pp. 661–664
115. S. Ha, P. Mieszkowski, M. Skowronski, and L.B. Rowland: J. Cryst. Growth 244 (2002) 257
116. S. Nishino, J.A. Powell, and H.A. Will: Appl. Phys. Lett. 42 (1983) 460
117. S. Nishino, H. Suhara, and H. Matsunami: The 15th Conference on Solid State Devices and Materials, Tokyo, August 30–September 27, 1983, Extended Abstracts, pp. 317–320
118. Y. Fujiwara, E. Sakuma, S. Misawa, K. Endo, and S. Yoshida: Appl. Phys. Lett. 49 (1986) 388
119. K. Shibahara, S. Nishino, and H. Matsunami: Appl. Phys. Lett. 50 (1987) 1888
120. Y. Ishida, T. Takahashi, H. Okumura, S. Yoshida, and T. Sekigawa: Jpn J. Appl. Phys. Part 1 36 (1997) 6633
121. F.Y. Huang, X.F. Wang, G.S. Sun, W.S. Zhao, Y.P. Zeng, and E.L. Bian: Thin Solid Films 484 (2005) 261
122. H. Nagasawa, T. Kawahara, and K. Yagi: Silicon Carbide and Related Materials 2001. Materials Science Forum Proceedings, Vol. 389–393, Trans Tech Publications, Zurich–Uetikon, 2002, pp. 319–322
123. H. Nagasawa, K. Yagi, and T. Kawahara: J. Cryst. Growth 237–239 (2002) 1244
124. A. Dadgar, J. Christen, T. Riemann, S. Richter, J. Blasing, A. Diez, A. Krost, A. Alam, and M. Heuken: Appl. Phys. Lett. 78 (2001) 2211
125. M. Akiyama, Y. Kawarada, and K. Kaminishi: Jpn J. Appl. Phys. Part 2 23 (1984) L843
126. H. Ishikawa, G.Y. Zhao, N. Nakada, T. Egawa, T. Jimbo, and M. Umeno: Jpn J. Appl. Phys. Part 2 38 (1999) L492
127. H. Ishikawa, B.J. Zhang, T. Egawa, and T. Jimbo: Jpn J. Appl. Phys. Part 1 42 (2003) 6413
128. T. Egawa, T. Moku, H. Ishikawa, K. Ohtsuka, and T. Jimbo: Jpn J. Appl. Phys. Part 2 41 (2002) L663
129. T. Egawa, B. Zhang, and H. Ishikawa: IEEE Electr. Device Lett. 26 (2005) 169
130. X.H. Wu, C.R. Elsass, A. Abare, M. Mack, S. Keller, P.M. Petroff, S.P. DenBaars, J.S. Speck, and S.J. Rosner: Appl. Phys. Lett. 72 (1998) 692
131. B.J. Zhang, T. Egawa, H. Ishikawa, Y. Liu, and T. Jimbo: Appl. Phys. Lett. 86 (2005) art. no. 071113
132. S. Arulkumaran, T. Egawa, S. Matsui, and H. Ishikawa: Appl. Phys. Lett. 86 (2005) art. no. 123503
133. S. Arulkumaran, T. Egawa, and H. Ishikawa: Solid State Electron. 49 (2005) 1632
134. M. Hikita, M. Yanagihara, K. Nakazawa, H. Ueno, Y. Hirose, T. Ueda, Y. Uemoto, T. Tanaka, D. Ueda, and T. Egawa: IEEE Trans. Electr. Devices 52 (2005) 1963

135. H. Okushi and S. Yamasaki: Oyo Butsuri 74 (2005) 1227
136. H. Watanabe, D. Takeuchi, S. Yamanaka, H. Okushi, K. Kajimura, and T. Sekiguchi: Diam. Relat. Mater. 8 (1999) 1272
137. H. Okushi: Diam. Relat. Mater. 10 (2001) 281
138. H. Okushi, H. Watanabe, S. Ri, S. Yamanaka, and D. Takeuchi: J. Cryst. Growth 237–239 (2002) 1269
139. H. Watanabe and H. Okushi: Jpn J. Appl. Phys. Part 2 39 (2000) L835
140. H. Shiomi, Y. Nishibayashi, and N. Fujimori: Jpn J. Appl. Phys. Part 1 30 (1991) 1363
141. S. Yamanaka, H. Watanabe, S. Masai, D. Takeuchi, H. Okushi, and K. Kajimura: Jpn J. Appl. Phys. Part 2 37 (1998) L1129
142. S. Koizumi, M. Kamo, Y. Sato, H. Ozaki, and T. Inuzuka: Appl. Phys. Lett. 71 (1997) 1065
143. M. Katagiri, J. Isoya, S. Koizumi, and H. Kanda: Appl. Phys. Lett. 85 (2004) 6365
144. H. Kato, S. Yamasaki, and H. Okushi: Appl. Phys. Lett. 86 (2005) 222111
145. T. Tachibana, Y. Yokota, K. Miyata, T. Onishi, K. Kobashi, M. Tarutani, Y. Takai, R. Shimizu, and Y. Shintani: Phys. Rev. B 56 (1997) 15967
146. W.P. Kang, J.L. Davidson, Y.M. Wong, and K. Holmes: Diam. Relat. Mater. 13 (2004) 975
147. H. Okushi: New Diam. 21 (2005) 2 (in Japanese)
148. H. Amano, N. Sawaki, I. Akasaki, and Y. Toyoda: Appl. Phys. Lett. 48 (1986) 353
149. I. Akasaki, H. Amano, Y. Koide, K. Hiramatsu, and N. Sawaki: J. Cryst. Growth 98 (1989) 209
150. K. Hiramatsu, S. Itoh, H. Amano, I. Akasaki, N. Kuwano, T. Shiraishi, and K. Oki: J. Cryst. Growth 115 (1991) 628
151. H. Amano and I. Akasaki: Oyo Butsuri 68 (1999) 768
152. I. Akasaki: J. Cryst. Growth 221 (2000) 231
153. I. Akasaki (ed.): Group III Nitride Semiconductors – Category 1: Electronic Materials, Physical Properties, Devices. Advanced Electronics Series, Vol. I-21, Baifukan, Tokyo, 1999 (in Japanese)
154. M. Iwaya, T. Takeuchi, S. Yamaguchi, C. Wetzel, H. Amano, and I. Akasaki: Jpn J. Appl. Phys. Part 2 37 (1998) L316
155. A. Usui, H. Sunakawa, A. Sakai, and A.A. Yamaguchi: Jpn J. Appl. Phys. Part 2 36 (1997) L899
156. A. Sakai and A. Usui: Oyo Butsuri 68 (1999) 774
157. K. Hiramatsu, K. Nishiyama, M. Onishi, H. Mizutani, M. Narukawa, A. Motogaito, H. Miyake, Y. Iyechika, and T. Maeda: J. Cryst. Growth 221 (2000) 316
158. K. Hiramatsu, K. Nishiyama, A. Motogaito, H. Miyake, Y. Iyechika, and T. Maeda: Phys. Status Solidi A: Appl. Res. 176 (1999) 535
159. K. Linthicum, T. Gehrke, D. Thomson, E. Carlson, P. Rajagopal, T. Smith, D. Batchelor, and R. Davis: Appl. Phys. Lett. 75 (1999) 196
160. R.F. Davis, T. Gehrke, K.J. Linthicum, P. Rajagopal, A.M. Roskowski, T. Zheleva, E.A. Preble, C.A. Zorman, M. Mehregany, U. Schwarz, J. Schuck, and R. Grober: MRS Internet J. Nitride Semicond. Res. 6 (2001) 1
161. M. Kuramoto, C. Sasaoka, Y. Hisanaga, A. Kimura, A.A. Yamaguchi, H. Sunakawa, N. Kuroda, M. Nido, A. Usui, and M. Mizuta: Jpn J. Appl. Phys. Part 2 38 (1999) L184

162. Y. Oshima, T. Eri, M. Shibata, H. Sunakawa, K. Kobayashi, T. Ichihashi, and A. Usui: Jpn J. Appl. Phys. Part 2 42 (2003) L1
163. S. Nakamura, M. Senoh, S. Nagahama, N. Iwasa, T. Yamada, T. Matsushita, H. Kiyoku, Y. Sugimoto, T. Kozaki, H. Umemoto, M. Sano, and K. Chocho: Jpn J. Appl. Phys. Part 2 36 (1997) L1568
164. S. Nakamura, M. Senoh, S. Nagahama, T. Matsushita, H. Kiyoku, Y. Sugimoto, T. Kozaki, H. Umemoto, M. Sano, and T. Mukai: Jpn J. Appl. Phys. Part 2 38 (1999) L226
165. A. Sakai, H. Sunakawa, and A. Usui: Appl. Phys. Lett. 71 (1997) 2259
166. J.F. Falth, M.N. Gurusinghe, X.Y. Liu, T.G. Andersson, I.G. Ivanov, B. Monemar, H.H. Yao, and S.C. Wang: J. Cryst. Growth 278 (2005) 406
167. X.Q. Shen, M. Shimizu, T. Yamamoto, Y. Honda, and H. Okumura: J. Cryst. Growth 278 (2005) 378
168. M. Sumiya and S. Fuke: MRS Internet J. Nitride Semicond. Res. 9 (2004) 1
169. E.S. Hellman: MRS Internet J. Nitride Semicond. Res. 3 (1998) 11
170. M. Sumiya and S. Fuke: Oyo Butsuri 70 (2001) 178
171. R. Di Felice and J.E. Northrup: Appl. Phys. Lett. 73 (1998) 936
172. B.W. Seo, Y. Ishitani, and A. Yoshikawa: Phys. Status Solidi C 0 (2003) 2570
173. S. Sonoda, S. Shimizu, Y. Suzuki, K. Balakrishnan, J. Shirakashi, and H. Okumura: Jpn J. Appl. Phys. Part 2 39 (2000) L73
174. R. Dimitrov, M. Murphy, J. Smart, W. Schaff, J.R. Shealy, L.F. Eastman, O. Ambacher, and M. Stutzmann: J. Appl. Phys. 87 (2000) 3375
175. K. Xu, N. Yano, A.W. Jia, A. Yoshikawa, and K. Takahashi: Phys. Status Solidi B 228 (2001) 523
176. Y.S. Park, H.S. Lee, J.H. Na, H.J. Kim, S.M. Si, H.M. Kim, T.W. Kang, and J.E. Oh: J. Appl. Phys. 94 (2003) 800
177. A.R. Smith, R.M. Feenstra, D.W. Greve, M.S. Shin, M. Skowronski, J. Neugebauer, and J.E. Northrup: Appl. Phys. Lett. 72 (1998) 2114
178. H.M. Ng: Appl. Phys. Lett. 80 (2002) 4369
179. C. Chen, V. Adivarahan, J. Yang, M. Shatalov, E. Kuokstis, and M.A. Khan: Jpn J. Appl. Phys. Part 2 42 (2003) L1039
180. A. Chitnis, C. Chen, V. Adivarahan, M. Shatalov, E. Kuokstis, V. Mandavilli, J. Yang, and M.A. Khan: Appl. Phys. Lett. 84 (2004) 3663
181. M.D. Craven, F. Wu, A. Chakraborty, B. Imer, U.K. Mishra, S.P. DenBaars, and J.S. Speck: Appl. Phys. Lett. 84 (2004) 1281
182. N.F. Gardner, J.C. Kim, J.J. Wierer, Y.C. Shen, and M.R. Krames: Appl. Phys. Lett. 86 (2005) 111101
183. T. Nagamoto, T. Kuboyama, H. Minamino, and O. Omoto: Jpn J. Appl. Phys. Part 2 28 (1989) L1334
184. S. Nakamura, M. Senoh, and T. Mukai: Appl. Phys. Lett. 62 (1993) 2390
185. K. Nakajima (ed.): Epitaxial Growth Mechanisms. Crystal Growth Dynamics Series, Vol. 3, Kyoritsu Shuppan, Tokyo, 2002 (in Japanese)
186. A. Koukitu and H. Seki: Nihon Kessho Seicho Gakkaishi 25 (1998) 81 (in Japanese)
187. I. Akasaki (ed.): Group III Nitride Semiconductors – Category 1: Electronic Materials, Physical Properties, Devices. Advanced Electronics Series, Vol. I-21, Baifukan, Tokyo, 1999 (in Japanese)
188. S. Keller, U.K. Mishra, S.P. Denbaars, and W. Seifert: Jpn J. Appl. Phys. Part 2 37 (1998) L431

189. T. Sugahara, M. Hao, T. Wang, D. Nakagawa, Y. Naoi, K. Nishino, and S. Sakai: Jpn J. Appl. Phys. Part 2 37 (1998) L1195
190. T. Takeuchi, H. Takeuchi, S. Sota, H. Sakai, H. Amano, and I. Akasaki: Jpn J. Appl. Phys. Part 2 36 (1997) L177
191. T. Takeuchi, S. Sota, M. Katsuragawa, M. Komori, H. Takeuchi, H. Amano, and I. Akasaki: Jpn J. Appl. Phys. Part 2 36 (1997) L382
192. A. Kaneta, T. Izumi, K. Okamoto, Y. Kawakami, S. Fujita, Y. Narita, T. Inoue, and T. Mukai: Jpn J. Appl. Phys. Part 1 40 (2001) 110
193. K. Okamoto, J. Choi, Y. Kawakami, M. Terazima, T. Mukai, and S. Fujita: Jpn J. Appl. Phys. Part 1 43 (2004) 839
194. A. Kaneta, K. Okamoto, Y. Kawakami, S. Fujita, G. Marutsuki, Y. Narukawa, and T. Mukai: Appl. Phys. Lett. 81 (2002) 4353
195. Nikkei Electronics 818, March 25 (2002) 26–28 (in Japanese)
196. Nikkei Electronics 844, March 31 (2002) 128–133 (in Japanese)
197. K. Tadatomo, H. Okagawa, Y. Ohuchi, T. Tsunekawa, H. Kudo, Y. Sudo, M. Kato, and T. Taguchi: J. Light Vis. Environ. 27 (2003) 140
198. K. Tadatomo, H. Okagawa, Y. Ohuchi, T. Tsunekawa, Y. Imada, M. Kato, and T. Taguchi: Jpn J. Appl. Phys. Part 2 40 (2001) L583
199. URL: http://www.cree.com, 1 August 2006, Cree, Inc., Durham, NC
200. M. Yamada, T. Mitani, Y. Narukawa, S. Shioji, I. Niki, S. Sonobe, K. Deguchi, M. Sano, and T. Mukai: Jpn J. Appl. Phys. Part 2 41 (2002) L1431
201. D. Morita, M. Sano, M. Yamamoto, T. Murayama, S. Nagahama, and T. Mukai: Jpn J. Appl. Phys. Part 2 41 (2002) L1434
202. S. Watanabe, N. Yamada, M. Nagashima, Y. Ueki, C. Sasaki, Y. Yamada, T. Taguchi, K. Tadatomo, H. Okagawa, and H. Kudo: Appl. Phys. Lett. 83 (2003) 4906
203. K. Tadatomo, Y. Ohuchi, H. Okagawa, H. Itoh, H. Miyake, and K. Hiramatsu: MRS Internet J. Nitride Semicond. Res. 4 (1999) art. no. G3.1
204. S.F. Chichibu, T. Sota, K. Wada, O. Brandt, K.H. Ploog, S.P. DenBaars, and S. Nakamura: Phys. Status Solidi A: Appl. Res. 183 (2001) 91
205. Y. Koide, N. Itoh, K. Itoh, N. Sawaki, and I. Akasaki: Jpn J. Appl. Phys. Part 1 27 (1988) 1156
206. K. Itoh, K. Hiramatsu, H. Amano, and I. Akasaki: J. Cryst. Growth 104 (1990) 533
207. H. Amano, M. Iwaya, N. Hayashi, T. Kashima, S. Nitta, C. Wetzel, and I. Akasaki: Phys. Status Solidi B 216 (1999) 683
208. T. Kashima, R. Nakamura, M. Iwaya, H. Katoh, S. Yamaguchi, H. Amano, and I. Akasaki: Jpn J. Appl. Phys. Part 2 38 (1999) L1515
209. Y. Ohba and H. Yoshida: Jpn J. Appl. Phys. Part 2 37 (1998) L905
210. T. Shibata, Y. Kida, T. Nagai, S. Sumiya, M. Tanaka, O. Oda, H. Miyake, and H. Hiramatsu: in Proceedings of MRS Symposium: GaN and Related Alloys (J.E. Northrup, J. Neugebauer, S.F. Chichibu, D.C. Look, and H. Riechert, eds.), Materials Research Society Proceedings, Vol. 693, MRS, 2002, Boston, 26–30 November 2001, p. 541
211. T. Shibata, K. Asai, S. Sumiya, M. Mouri, M. Tanaka, O. Oda, H. Katsukawa, H. Miyake, and K. Hiramatsu: Phys. Status Solidi C 0 (2003) 2023
212. H. Amano, N. Sawaki, I. Akasaki, and Y. Toyoda: Appl. Phys. Lett. 48 (1986) 353
213. S. Nakamura: Jpn J. Appl. Phys. Part 2 30 (1991) L1705

214. S. Yoshida, S. Misawa, and S. Gonda: Appl. Phys. Lett. 42 (1983) 427
215. T. Shibata: Ph.D. Thesis, Faculty of Engineering, Mie University, Tsu, Japan, 2004
216. K. Hiramatsu, S. Itoh, H. Amano, I. Akasaki, N. Kuwano, T. Shiraishi, and K. Oki: J. Cryst. Growth 115 (1991) 628
217. K. Hiramatsu, S. Itoh, I. Akasaki, N. Kuwano, T. Shiraishi, and K. Oki: Nihon Kessho Seicho Gakkaishi 20 (1993) 346
218. K. Hiramatsu and H. Miyak: in Proceedings of the ECS Meeting, 5th Nitride and Wide Bandgap Semiconductors for Sensors, Photonics, and Electronics Symposium (H. Ng and A.G. Baca, eds.), Electrochemical Society Proceedings, Vol. PV 2004-06, Electrochemical Society, Honolulu, HA, 3–8 October 2004, pp. 472–483
219. Y.-H. Liu, T. Tanabe, H. Miyake, K. Hiramatsu, T. Shibata, M. Tanaka, and Y. Masa: Jpn J. Appl. Phys. Part 2 44 (2005) L505
220. K. Osamura, K. Nakajima, Y. Murakami, P.H. Shingu, and A. Ohtsuki: Solid State Commun. 11 (1972) 617
221. T.L. Tansley and C.P. Foley: J. Appl. Phys. 59 (1986) 3241
222. T. Inushima, V.V. Mamutin, V.A. Vekshin, S.V. Ivanov, T. Sakon, M. Motokawa, and S. Ohoya: J. Cryst. Growth 227–228 (2001) 481
223. V.Y. Davydov, A.A. Klochikhin, R.P. Seisyan, V.V. Emtsev, S.V. Ivanov, F. Bechstedt, J. Furthmüller, H. Harima, A.V. Mudryi, J. Aderhold, O. Semchinova, and J. Graul: Phys. Status Solidi B 229 (2002) r1
224. J. Wu, W. Walukiewicz, K.M. Yu, J.W. Ager III, E.E. Haller, H. Lu, W.J. Schaff, Y. Saito, and Y. Nanishi: Appl. Phys. Lett. 80 (2002) 3967
225. T. Matsuoka, H. Okamoto, M. Nakao, H. Harima, and E. Kurimoto: Appl. Phys. Lett. 81 (2002) 1246
226. Y. Saito, H. Harima, E. Kurimoto, T. Yamaguchi, N. Teraguchi, A. Suzuki, T. Araki, and Y. Nanishi: Phys. Status Solidi B 234 (2002) 796
227. K. Xu, W. Terashima, T. Hata, N. Hashimoto, Y. Ishitani, and A. Yoshikawa: Phys. Status Solidi C 0 (2003) 377
228. Y. Ishitani, H. Masuyama, W. Terashima, M. Yoshitani, N. Hashimoto, S.B. Che, and A. Yoshikawa: Phys. Status Solidi C 2 (2005) 2276
229. A. Yoshikawa, Y. Ishitani, S.-B. Che, K. Xu, X. Wang, M. Yoshitani, W. Terashima, and N. Hashimoto: in Proceedings of Symposium E: GaN, AlN, InN and Their Alloys (C. Wetzel, B. Gil, M. Kuzuhara, and M. Manfra, eds.), Materials Research Society Proceedings, Vol. 831, Warrendale, PA, 2004, p. E4.1
230. O. Ambacher, M.S. Brandt, R. Dimitrov, T. Metzger, M. Stutzmann, R.A. Fischer, A. Miehr, A. Bergmaier, and G. Dollinger: J. Vac. Sci. Technol. B: Microelectr. Nanomet. Struct. 14 (1996) 3532
231. Y. Nanishi, Y. Saito, and T. Yamaguchi: Jpn J. Appl. Phys. Part 1 42 (2003) 2549
232. X. Wang and A. Yoshikawa: Prog. Cryst. Growth Charact. Mater. 48–49 (2004) 42
233. A.G. Bhuiyan, A. Hashimoto, and A. Yamamoto: J. Appl. Phys. 94 (2003) 2779
234. K. Xu and A. Yoshikawa: Appl. Phys. Lett. 83 (2003) 251
235. K. Xu, W. Terashima, T. Hata, N. Hashimoto, M. Yoshitani, B. Cao, Y. Ishitani, and A. Yoshikawa: Phys. Status Solidi C 0 (2003) 2814
236. C.H. Swartz, R.P. Tomkins, T.H. Myers, H. Lu, and W.J. Schaff: Phys. Status Solidi C 2 (2005) 2250

237. V.Y. Davydov, A.A. Klochikhin, V.V. Emtsev, D.A. Kurdyukov, S.V. Ivanov, V.A. Vekshin, F. Bechstedt, J. Furthmüller, J. Aderhold, J. Graul, A.V. Mudryi, H. Harima, A. Hashimoto, A. Yamamoto, and E.E. Haller: Phys. Status Solidi B 234 (2002) 787
238. J. Wu, W. Walukiewicz, K.M. Yu, J.W. Ager III, E.E. Haller, H. Lu, and W.J. Schaff: Appl. Phys. Lett. 80 (2002) 4741
239. W. Terashima, S.-B. Che, Y. Ishitani, and A. Yoshikawa: Jpn J. Appl. Phys. Part 2 45 (2006) L539
240. S.-B. Che, Y. Ishitani, and A. Yoshikawa: Phys. Status Solidi C 3 (2006) 1953
241. M. Kurouchi, H. Naoi, T. Araki, T. Miyajima, and Y. Nanishi: Jpn J. Appl. Phys. Part 2 44 (2005) L230
242. S.-B. Che, W. Terashima, Y. Ishitani, A. Yoshikawa, T. Matsuda, H. Ishii, and S. Yoshida: Appl. Phys. Lett. 86 (2005) 261903
243. A. Yoshikawa, N. Hashimoto, N. Kikukawa, S.B. Che, and Y. Ishitani: Appl. Phys. Lett. 86 (2005) 153115
244. X. Wang, S.-B. Che, Y. Ishitani, and A. Yoshikawa: Phys. Status Solidi C 3 (2006) 1561
245. Y. Kuga, T. Shirai, M. Haruyama, H. Kawanishi, and Y. Suematsu: Jpn J. Appl. Phys. Part 1 34 (1995) 4085
246. M. Haruyama, T. Shirai, H. Kawanishi, and Y. Suematsu: International Symposium on Blue Laser and Light Emitting Diodes (ISBLLED'96), Chiba, Japan, 5–7 March 1996, p. 106
247. T. Takano, M. Kurimoto, J. Yamamoto, and H. Kawanishi: J. Cryst. Growth 237–239 (2002) 972
248. T. Takano, H. Kawanishi, M. Kurimoto, Y. Ishihara, M. Horie, and J. Yamamoto: in Proceedings of MRS Symposium: GaN and Related Alloys (U. Mishra, M.S. Shur, C.M. Wetzel, B. Gil, and K. Kishino, eds.), Materials Research Society Proceedings, Vol. 639, MRS, 2001, Boston, November 27–December 1, 2000, p. G12.9.1 (837797).
249. T. Takano, M. Kurimoto, J. Yamamoto, Y. Ishihara, M. Horie, and H. Kawanishi: Proceedings of International Workshop on Nitride Semiconductors. IPAP Conference Series, Vol. 1, Institute of Pure and Applied Physics, Nagoya, 24–27 September 2000, pp. 147–149
250. S. Watanabe, T. Takano, K. Jinen, J. Yamamoto, and H. Kawanishi: Phys. Status Solidi C 0 (2003) 2691
251. T. Yao, M. Ogura, S. Matsuoka, and T. Morishita: Jpn J. Appl. Phys. Part 2 22 (1983) L144
252. A. Yoshikawa, S. Yamaga, K. Tanaka, and H. Kasai: J. Cryst. Growth 72 (1985) 13
253. K. Yoneda, Y. Hishida, T. Toda, H. Ishii, and T. Niina: Appl. Phys. Lett. 45 (1984) 1300
254. S. Myhajlenko, J.L. Batstone, H.J. Hutchinson, and J.W. Steeds: J. Phys. C: Solid State Phys. 17 (1984) 6477
255. J.M. DePuydt, H. Cheng, J.E. Potts, T.L. Smith, and S.K. Mohapatra: J. Appl. Phys. 62 (1987) 4756
256. K. Menda, I. Takayasu, T. Minato, and M. Kawashima: Jpn J. Appl. Phys. Part 2 26 (1987) L1326
257. T. Yao, Z.Q. Zhu, K. Uesugi, S. Kamiyama, and M. Fujimoto: J. Vac. Sci. Technol. A: Vac. Surf. Films 8 (1990) 997

258. T. Yao and T. Takeda: Appl. Phys. Lett. 48 (1986) 160
259. K. Mohammed, D.A. Cammack, R. Dalby, P. Newbury, B.L. Greenberg, J. Petruzzello, and R.N. Bhargava: Appl. Phys. Lett. 50 (1987) 37
260. H. Mitsuhashi, I. Mitsuishi, and H. Kukimoto: Jpn J. Appl. Phys. Part 2 24 (1985) L864
261. W.A. Harrison, E.A. Kraut, J.R. Waldrop, and R.W. Grant: Phys. Rev. B 18 (1978) 4402
262. Q.D. Qian, J. Qiu, M.R. Melloch, J.J.A. Cooper, L.A. Kolodziejski, M. Kobayashi, and R.L. Gunshor: Appl. Phys. Lett. 54 (1989) 1359
263. Q.D. Qian, J. Qiu, M. Kobayashi, R.L. Gunshor, M.R. Melloch, and J.J.A. Cooper: Proceedings of the 16th Annual Conference on the Physics and Chemistry of Semiconductor Interfaces. Journal of Vacuum Science and Technology Proceedings, Vol. 7, AVS, Bozeman, MT, 1989, pp. 793–798
264. J. Qiu, Q.D. Qian, R.L. Gunshor, M. Kobayashi, D.R. Menke, D. Li, and N. Otsuka: Appl. Phys. Lett. 56 (1990) 1272
265. D. Li, J.M. Gonsalves, N. Otsuka, J. Qiu, M. Kobayashi, and R.L. Gunshor: Appl. Phys. Lett. 57 (1990) 449
266. D.R. Menke, J. Qiu, R.L. Gunshor, M. Kobayashi, D. Li, Y. Nakamura, and N. Otsuka: J. Vac. Sci. Technol. B: Microelectr. Nanomet. Struct. 9 (1991) 2171
267. M.C. Tamargo, J.L. de Miguel, D.M. Hwang, and H.H. Farrell: J. Vac. Sci. Technol. B: Microelectr. Nanomet. Struct. 6 (1988) 784
268. J. Qiu, Q.D. Qian, M. Kobayashi, R.L. Gunshor, D.R. Menke, D. Li, and N. Otsuka: J. Vac. Sci. Technol. B: Microelectr. Nanomet. Struct. 8 (1990) 701
269. L.H. Kuo, L. Salamanca-Riba, B.J. Wu, G. Hofler, J.M. DePuydt, and H. Cheng: Appl. Phys. Lett. 67 (1995) 3298
270. D.J. Chadi and M.L. Cohen: Phys. Status Solidi B 68 (1975) 405
271. J.O. McCaldin, T.C. McGill, and C.A. Mead: Phys. Rev. Lett. 36 (1976) 56
272. W.A. Harrison: J. Vac. Sci. Technol. 14 (1977) 883
273. A.D. Katnani and G. Margaritondo: Phys. Rev. B 28 (1983) 1944
274. A.D. Katnani and G. Margaritondo: J. Appl. Phys. 54 (1983) 2522
275. C.K. Shih, A.K. Wahi, I. Lindau, and W.E. Spicer: J. Vac. Sci. Technol. A: Vac. Surf. Films 6 (1988) 2640
276. C.K. Shih and W.E. Spicer: Phys. Rev. Lett. 58 (1987) 2594
277. C.G. van de Walle: Phys. Rev. B 39 (1989) 1871
278. M. Cardona and N.E. Christensen: Phys. Rev. B 35 (1987) 6182
279. G. Landwehr, A. Waag, F. Fischer, H.J. Lugauer, and K. Schull: Physica E 3 (1998) 158
280. J.A. van Vechten and T.K. Bergstresser: Phys. Rev. B 1 (1970) 3351
281. J.E. Nicholls, J.J. Davies, N.R.J. Poolton, R. Mach, and G.O. Muller: J. Phys. C: Solid State Phys. 18 (1985) 455
282. Y. Tokumitsu, H. Kitayama, A. Kawabuchi, T. Imura, and Y. Osaka: Jpn J. Appl. Phys. Part 2 28 (1989) L349
283. R. Hill and D. Richardson: J. Phys. C: Solid State Phys. 6 (1973) L115
284. I.K. Sou, K.S. Wong, Z.Y. Yang, H. Wang, and G.K.L. Wong: Appl. Phys. Lett. 66 (1995) 1915
285. T. Yokogawa and T. Narusawa: J. Cryst. Growth 117 (1992) 480
286. A. Ebina, M. Yamamoto, and T. Takahashi: Phys. Rev. B 6 (1972) 3786
287. A.K. Ghosh, K.K. Som, S. Chatterjee, and B.K. Chaudhuri: Phys. Rev. B 51 (1995) 4842

288. R. Hill: J. Phys. C: Solid State Phys. 7 (1974) 521
289. I. Suemune, K. Yamada, H. Masato, Y. Kan, and M. Yamanishi: Appl. Phys. Lett. 54 (1989) 981
290. K. Mohammed, D.J. Olego, P. Newbury, D.A. Cammack, R. Dalby, and H. Cornelissen: Appl. Phys. Lett. 50 (1987) 1820
291. K. Shahzad, D.J. Olego, and C.G. Van de Walle: Phys. Rev. B 38 (1988) 1417
292. N. Teraguchi, Y. Takemura, R. Kimura, M. Konagai, and K. Takahashi: J. Cryst. Growth 93 (1988) 720
293. T. Nakayama: Physica B: Condens. Matter 191 (1993) 16
294. Y. Rajakarunanayake, R.H. Miles, G.Y. Wu, and T.C. McGill: Phys. Rev. B 37 (1988) 10212
295. Y.-H. Wu, S. Fujita, and S. Fujita: J. Appl. Phys. 67 (1990) 908
296. C. Priester, D. Bertho, and C. Jouanin: Physica B: Condens. Matter 191 (1993) 1
297. M. Ukita, F. Hiei, K. Nakano, and A. Ishibashi: Appl. Phys. Lett. 66 (1995) 209
298. H.J. Lozykowski and V.K. Shastri: J. Appl. Phys. 69 (1991) 3235
299. Y. Yamada, Y. Masumoto, J.T. Mullins, and T. Taguchi: Appl. Phys. Lett. 61 (1992) 2190
300. Y. Kawakami, S. Yamaguchi, Y.-H. Wu, K. Ichino, S. Fujita, and S. Fujita: Jpn J. Appl. Phys. Part 2 30 (1991) L605
301. T. Nakayama: Jpn J. Appl. Phys. Part 2 33 (1994) L211
302. Q. Fu, D. Lee, A. Mysyrowicz, A.V. Nurmikko, R.L. Gunshor, and L.A. Kolodziejski: Phys. Rev. B 37 (1988) 8791
303. Y. Hefetz, J. Nakahara, A.V. Nurmikko, L.A. Kolodziejski, R.L. Gunshor, and S. Datta: Appl. Phys. Lett. 47 (1985) 989
304. W.J. Walecki, A.V. Nurmikko, N. Samarth, H. Luo, J.K. Furdyna, and N. Otsuka: Appl. Phys. Lett. 57 (1990) 466
305. T. Miyajima, F.P. Logue, J.F. Donegan, J. Hegarty, H. Okuyama, A. Ishibashi, and Y. Mori: Appl. Phys. Lett. 66 (1995) 180
306. S. Fujita, Y. Kawakami, and S. Fujita: Physica B: Condens. Matter 191 (1993) 57
307. T. Yao: in II–VI Semiconductor Materials and Their Applications (M.C. Tamargo, ed.), Optoelectronic Properties of Semiconductors and Superlattices, Vol. 12, Taylor and Francis, Ann Arbor, MI, 2002, pp. 67–112
308. Y. Chen, H.-J. Ko, S.-K. Hong, and T. Yao: Appl. Phys. Lett. 76 (2000) 559
309. A. Setiawan, H.-J. Ko, S.-K. Hong, Y. Chen, and T. Yao: Thin Solid Films 445 (2003) 213
310. S.-K. Hong, T. Hanada, H.-J. Ko, Y. Chen, T. Yao, D. Imai, K. Araki, M. Shinohara, K. Saitoh, and M. Terauchi: Phys. Rev. B 65 (2002) 115331
311. D.C. Oh, J.J. Kim, H. Makino, T. Hanada, M.W. Cho, T. Yao, and H.J. Ko: Appl. Phys. Lett. 86 (2005) 042110
312. S.-K. Hong, T. Hanada, Y. Chen, H.-J. Ko, T. Yao, D. Imai, K. Araki, and M. Shinohara: Appl. Surf. Sci. 190 (2002) 491
313. T. Minegishi, J. Yoo, H. Suzuki, Z. Vashaei, K. Inaba, K. Shim, and T. Yao: J. Vac. Sci. Technol. B 23 (2005) 1286
314. S. Limpijumnong and W.R.L. Lambrecht: Phys. Rev. B 63 (2001) 104103
315. R.J. Guerrero-Moreno and N. Takeuchi: Phys. Rev. B 66 (2002) 205205
316. J.M. Recio, M.A. Blanco, V. Luaña, R. Pandey, L. Gerward, and J. Staun Olsen: Phys. Rev. B 58 (1998) 8949

317. A. Ohtomo, M. Kawasaki, T. Koida, K. Masubuchi, H. Koinuma, Y. Sakurai, Y. Yoshida, T. Yasuda, and Y. Segawa: Appl. Phys. Lett. 72 (1998) 2466
318. I. Takeuchi, W. Yang, K.S. Chang, M.A. Aronova, T. Venkatesan, R.D. Vispute, and L.A. Bendersky: J. Appl. Phys. 94 (2003) 7336
319. Z. Vashaei, T. Minegishi, H. Suzuki, T. Hanada, M.W. Cho, T. Yao, and A. Setiawan: J. Appl. Phys. 98 (2005) 054911
320. T. Makino, Y. Segawa, M. Kawasaki, A. Ohtomo, R. Shiroki, K. Tamura, T. Yasuda, and H. Koinuma: Appl. Phys. Lett. 78 (2001) 1237

Index

D

J

K

L

T

U

V

Printing: Krips bv, Meppel
Binding: Stürtz, Würzburg